Dung Beetle Ecology

Dung Beetle Ecology

Edited by
Ilkka Hanski and
Yves Cambefort

PRINCETON UNIVERSITY PRESS

PRINCETON, NEW JERSEY

Copyright © 1991 by Princeton University Press
Published by Princeton University Press, 41 William Street,
Princeton, New Jersey 08540
In the United Kingdom: Princeton University Press, Oxford

Library of Congress Cataloging-in-Publication Data

Dung beetle ecology / edited by Ilkka Hanski and Yves Cambefort.
p.   cm.
Includes bibliographical references and indexes.
1. Dung beetles—Ecology. I. Hanski, Ilkka.
II. Cambefort, Yves, 1945-.
QL596.S3D86   1991   595.76′49—dc20   90-44682

ISBN 0-691-08739-3 (cloth : acid-free paper)

This book has been composed in Linotron Times Roman

Princeton University Press books are printed on acid-free paper,
and meet the guidelines for permanence and durability of
the Committee on Production Guidelines for Book Longevity of the
Council on Library Resources

Printed in the United States of America by Princeton University Press,
Princeton, New Jersey

10   9   8   7   6   5   4   3   2   1

Contents

List of Contributors

Y. CAMBEFORT
Muséum National d'Histoire Naturelle
Entomologie (U.A. 42 C.N.R.S.)
45, Rue de Buffon, 75005 Paris, France

B. M. DOUBE
CSIRO, Division of Entomology
Black Mountain, Canberra, ACT, Australia

B. D. GILL
Department of Biology, Carleton University
Ottawa, Canada K1S 5BD

I. HANSKI
Department of Zoology, University of Helsinki
P. Rautatiekatu 13, SF-00100 Helsinki, Finland

A. A. KIRK
USDA/ARS Montpellier Station
ENSA.M, 2 place P. Viala, 34060 Montpellier, France

B. KOHLMANN
Instituto de Ecología
A.P. 18-845, 11800, México D.F., México

J. KRIKKEN
Rijksmuseum van Natuurlijke Historie, Coleoptera Section
P.O. Box 9517, Leiden, The Netherlands

J.-P. LUMARET
Laboratoire de Zoogéographie, Université Paul Valéry
34032 Montpellier, France

A. MACQUEEN
CSIRO, Division of Entomology, Private Bag No. 3
Indooroopilly, Qld. 4068, Australia

T. J. RIDSDILL-SMITH
CSIRO, Division of Entomology, Private Bag
P.O. Wembley, Perth W.A. 6014, Australia

C. Rougon and D. Rougon
Laboratoire d'Ecologie animale, Université d'Orléans
B.P. 6759, 45067 Orléans, Cedex 2, France

N. Stiernet
Département de Morphologie, Systématique et Ecologie animales
Institut Van Beneden, Université de Liège, 4020 Liège, Belgium

P. Walter
Laboratoire d'Endocrinologie des Insectes
Faculté des Sciences, Université de Nantes
44072 Nantes, Cedex 03, France

T. A. Weir
CSIRO, Division of Entomology
P.O. Box 1700, Canberra, ACT 2601, Australia

Preface

MANY BOOKS on insects could be justified by their sheer numbers of species and by their wonderful diversity of life-styles. This one cannot. There are only some 7,000 described species of dung beetles in relatively well-known taxa, a figure comparable to the global diversity of birds (8,600 species) and an insignificant 0.02% or so of all animal species. There is little fundamental variation in the life-styles of dung beetles, though it is wonderful that many of them, like birds, construct nests in which they care for their young. Heavy investment in parental care is associated with low fecundity, and some birds and dung beetles produce just one offspring per breeding season. Many birds and dung beetles have more density-dependent population dynamics than most other animals. Dung beetles are not birds, however, but they do provide a model example of populations exploiting ephemeral resources in a patchy environment. Dung beetles also provide many examples of very competitive communities and pose challenging questions about coexistence of many similar competitors.

This volume summarizes our current knowledge about dung beetle ecology. It complements the monographs by Halffter and Matthews (1966) on the natural history of dung beetles and by Halffter and Edmonds (1982) on their breeding biology. Apart from consolidating the existing knowledge for the specialist, we hope that the book will also make interesting reading for population and community ecologists not yet enthusiastic about dung beetles. We also hope that the gaps in our knowledge that will be disclosed, and the new and often obvious research opportunities that will be outlined, will stimulate a new wave of research on dung beetle ecology.

The kind of ecology presented in this volume is largely observational, based on the trapping of beetles with baited pitfalls. Experimental studies on dung beetles are yet in short supply. Dung beetle ecology is currently stronger in questions about communities than about populations, partly because dung beetles always occur in multispecies assemblages. There is no need for us to review systematically the extensive body of information on dung beetle breeding biology, as this area is covered admirably in the comprehensive work by Halffter and Edmonds (1982); but we do make an attempt to relate foraging behavior and breeding biology in dung beetles to the ecology of their populations and communities.

This book is divided into three parts. The first part consists of a general introduction to the ecology of insects using ephemeral resources in patchy habitats, outlines the evolution of the dung-feeding habit in beetles, and describes briefly their breeding biology and somewhat more thoroughly their biogeography. Some of the material is new, although the introductory chapters have been written with the nonspecialist in mind. The second part consists of eleven chapters on regional dung beetle assemblages, aiming at a comprehensive, worldwide coverage. We used a regional basis for these chapters because we realize that different types of dung beetles, and different types of ecological inquiries, are most fascinating in different biomes, from deserts to tropical forests and from north temperate pastures to African savannas. The third part attempts to synthesize much of the material in the regional chapters into simplified patterns and working hypotheses about processes. Such a framework has been lacking for dung beetle ecology, and we hope that our efforts are of wider interest to ecologists who work with insects living in ephemeral and patchy habitats.

The book can be read in two ways. Ecologists might want to read Parts 1 and 3 first and consult the chapters in Part 2 as needed. Entomologists, especially those working with dung beetles, may want to start with Part 2, though we would be disappointed if they find nothing of interest in Parts 1 and 3.

Ilkka Hanski
Yves Cambefort
June 1990

Acknowledgments

THE IDEA FOR this book grew from the preparations for the 17th International Congress of Entomology in Hamburg in 1984, where Orrey P. Young organized a symposium on dung beetle biology. Unfortunately, Orrey Young was unable to pursue this project further. The two of us, reflecting on the current state of knowledge about dung beetle taxonomy, biogeography (YC), and ecology (IH), decided to combine our efforts in 1986 to edit the present volume.

This project would have been impossible to complete without the help of numerous colleagues. We wish to express our particular thanks to Orrey Young for organizing the Hamburg symposium. We also thank the following individuals for reading and commenting on parts of the manuscript: B. M. Doube (Chapters 16 to 18), T. Fincher (Chapter 5), R. D. Gordon (Chapter 5), G. Halffter (Chapter 7), P. M. Hammond (Chapter 10), H. F. Howden (Chapters 1 to 5), A. R. Ives (Chapter 1), O. Järvinen (Chapter 1), P. J. Kuijten (Chapter 5), J.-P. Lumaret (Chapters 2 and 3), T. J. Ridsdill-Smith (Chapter 17), M. Tyndale-Biscoe (Chapter 7), B. Wellington (Chapter 7), and M. Zunino (Chapter 4). We thank V. and G. Halffter for making available unpublished material for Chapter 7. The Indonesian Institute of Sciences (LIPI) made possible the fieldwork on which Chapter 10 is largely based, while P. M. Hammond and his colleagues from the British Museum and J. Huijbregts and J. de Vos from Leiden helped in collecting and processing this material. Figures 2.1 and 3.3 were provided by G. Halffter and J.-P. Lumaret, respectively.

For the photographs we thank J.-P. Lumaret (Plates 3.1, 3.2, 3.5), M. Imamori (Plates 3.6 to 3.11), and P. B. Edwards (Plate 3.3); Plate 3.4 was photographed by Y. Cambefort. Tuula Landén and Hannu Pietiäinen drew most of the figures using the facilities provided by Samuel Panelius and his staff. The cover picture and drawings of beetles were skillfully executed by G. Hodebert (Muséum National d'Histoire Naturelle, Paris). We are indebted to Judith May, Emily Wilkinson, and Alice Calaprice of Princeton University Press for their assistance and encouragement. IH thanks Eeva Furman for her support. Two field trips to the Ivory Coast by YC in 1985 and 1987 were funded by CNRS (ECOTROP). Two trips by IH to Paris and two trips by YC to Helsinki were funded by the National Research Council of Finland and by Muséum National d'Histoire Naturelle (Paris).

PART ONE

Introduction

THE FOUR CHAPTERS in Part 1 lay the foundation for this book by reviewing four perspectives of dung beetle ecology: the population ecology of insects that use ephemeral resources in patchy habitats; the evolution of coprophagy from saprophagy in beetles; the breeding biology of dung beetles; and the biogeography and evolution of these beetles.

We start with an overview of patchy and ephemeral microhabitats, of which animal droppings are a prime example. For dung beetles and dung-breeding flies, droppings are concentrations of high-quality resources, for which competition is often severe. In Chapter 1 we describe two contrasting models of competitive dynamics in species using ephemeral resources in patchy habitats: the lottery model and the variance-covariance model. This distinction remains one of the important concepts throughout the book.

Dung beetles generally occur in multispecies communities, and it is impossible not to consider their ecology from this viewpoint, whether or not other viewpoints are also considered. In Chapter 1 we outline the structure of the entire dung insect community, in which dung beetles, dung-breeding flies, and predatory beetles are most often the numerically or functionally important members. Dung beetles frequently interact most strongly among themselves, and are generally superior in competition with dung-breeding flies.

In Chapter 2 we describe the evolution of coprophagy—that is, the habit of feeding on dung—in beetles. It is evident that the evolution of coprophagy in dung beetles is closely associated with the evolution of mammals, by far the most significant dung producers of the past and present. The evolution of nesting behavior (Chapter 3), the most striking feature in dung beetle biology, is closely related to the evolution of coprophagy; but we cite a recent discovery of a nest of a primitive saprophagous "dung beetle," suggesting that nest building may have been present in some of the saprophagous forms that turned to coprophagy when mammalian dung became available in large quantities.

The resource patches are often fiercely contested, and it is not surprising that dung beetles have evolved means of outmaneuvering one another, including moving all or part of the resource to a safe location. It is convenient to make a distinction between three basic types of dung beetles, based on the mode of resource relocation: the rollers, the tunnelers, and the dwellers. The first two terms are widely used in the entomological literature, while the third one is our new suggestion.* The dwellers, which have the least evolved be-

---

* Entomologists tend to use fancy technical terms to confuse the uninitiated. Thus rollers are called telocoprids, tunnelers are paracoprids, and those dwellers which construct a nest inside a dropping are endocoprids. Gill (Chapter 12) elaborates on some further types of beetles which are largely restricted to forest habitats in South America.

havior, do not relocate food at all, but the beetles and their larvae either live freely in droppings (Aphodiinae), or females construct a nest for their offspring inside a dropping (*Oniticellus*). The tunnelers dig burrows and construct nest chambers below the food source and push chunks of dung down into the tunnels. The even more sophisticated rollers make balls of dung, which they roll away from the food source—and from the reach of other beetles—before concealing the ball in the soil. A more detailed account of these behaviors and their functional significance is given in Chapter 3.

Dung beetle biogeography adds another perspective to their ecology. In Chapter 4 we summarize the geographical distribution of the different tribes of Scarabaeidae, the most highly evolved dung beetles. It is hypothesized that their distributions can be related to the evolution and biogeography of the major mammalian taxa. As with mammals, we observe that advanced groups of dung beetles have largely replaced the older ones in Africa; but in South America, Madagascar, southern Africa, and Australia the more primitive tribes still reign.

CHAPTER 1

The Dung Insect Community

Ilkka Hanski

ANIMAL DROPPINGS are one of the best examples of what Charles Elton (1949) called minor habitats, "centres of action in which interspersion between populations tends to be complete and ecological dynamic relations at their strongest." In short, dung beetle ecology is about competitive exploitation of nutritionally rich resources in one such "minor habitat" by species with an elaborate breeding behavior. I do not wish to suggest that all dung beetles live in equally competitive communities, nor do all dung beetles exhibit equally intricate nesting behavior. The chapters in this volume will focus on a range of dung beetle populations and communities with respect to these and other ecological attributes. Nonetheless, my own interest in the ecology of dung beetles does stem from these elements: patchy and ephemeral habitat, severe competition, and complex behavior in many similar species living together. Though dung beetles may not comprise a model system for a large number of other communities, they do comprise one of the animal populations most worthy of the notion of community.

Dung beetles are often exceedingly abundant. Thousands of individuals and dozens of species may be attracted to single droppings in both temperate and tropical localities. Anderson and Coe (1974) counted 16,000 dung beetles arriving at a 1.5 kg heap of elephant dung in East Africa, and these beetles ate, buried, and rolled this "minor habitat" away in two hours! No wonder that much of the ecological interest in dung beetles has been generated by questions about coexistence of competitors using the same resource, in an apparent contradiction to the "principle of competitive exclusion," also known as Gause's law (Gause 1934; Hardin 1960; McIntosh 1985). The "competition paradigm" was established by the early work of Lotka, Volterra, Kostitzin, and Kolmogoroff in the 1920s and 1930s (Scudo and Ziegler 1978), and it was much elaborated in the 1960s and 1970s (MacArthur 1972; May 1973; Roughgarden 1979), when the principle of competitive exclusion was transformed into the theory of limiting similarity, that is, that competitors which are too similar cannot coexist. The competition-centered community ecology met stern criticism in the late 1970s and early 1980s (for example, see many chapters in Strong et al. 1984 and references therein), primarily due to a lack of rigorous demonstration of interspecific competition in natural species assemblages. It has subsequently been shown that competition occurs regularly in a wide variety of taxa (Schoener 1983; Connell 1983). These reviews do not report on competition in dung beetles. But as we have indicated above and shall abundantly demonstrate later in this volume, the reason is not that competition does not regularly occur in many dung beetle assemblages; the reason is the virtual lack of experimental studies on competition in dung beetles.

Species diversity of dung beetles is not all that great in comparison with many other groups of insects. The true dung beetles include the family Scarabaeidae, with some 5,000 species; the subfamily Aphodiinae of Aphodiidae, with about 1,850 species, most of which belong to a single genus, *Aphodius*; and the subfamily Geotrupinae of Geotrupidae, with about 150 species. Some species of Hybosoridae, Chironidae, Trogidae, and Cetoniidae also use dung at the adult or larval stage, and there are numerous small species of Hydrophilidae and Staphylinidae that use some components of the decomposing material or the microorganisms in dung pats (these latter species are not usually referred to as dung beetles). The relative scarcity of dung beetle species is in striking contrast with the often great abundance of individuals. One is tempted to speculate that competition limits the number of extant dung beetle species worldwide (Chapter 19).

This chapter begins with an overview of patchy and ephemeral microhabitats, that is, Elton's minor habitats, in order to place animal droppings in a wider ecological framework and to present two contrasting population dynamic models that define the range of possible competitive interactions in insects living in patchy habitats. Next comes a discussion of the physical characteristics and the chemical properties of droppings and the nutritive value of the resources available to insects in this microhabitat. Finally, we must keep in mind that dung beetles are seldom alone in droppings: they belong to a complex community comprising other beetles and flies as well as a multitude of other organisms, for example, mites and nematodes. I will outline the taxonomic composition and the structure of the entire dung insect community for the benefit of readers not familiar with these insects.

1.1. PATCHY AND EPHEMERAL MICROHABITATS: AN OVERVIEW

The microhabitats we are concerned with here share several attributes: relatively small size, scattered spatial occurrence, and short existence or durational stability, generally not more than one insect generation. Many, though not all, of such microhabitats stand out in the matrix of the surrounding environment as "islands" of high-quality resources, which to a large extent explains the diverse assemblages of insects and other invertebrates that colonize the particular microhabitat (Chapter 2).

Density of patchy and ephemeral microhabitats in the surrounding environment varies enormously. At one extreme are fallen leaves that may carpet an entire forest floor and merge, for all practical purposes, into one contiguous habitat. The same thing almost happened to cattle pats in Australia before the introduction of foreign dung beetles (Chapter 15). Generally, droppings in pastures are so numerous that between-patch movement is not a great problem for dung beetles. The other extreme is exemplified by the fruiting bodies of rare macrofungi. Locating one of them would be time consuming, which surely contributes to the high degree of polyphagy observed in fungivorous insects (reviewed in Hanski 1989c). The droppings of larger mammals are a scarce microhabitat in many ecosystems, and we would not expect widespread specialization by many insects to any particular kind. The specialization that does exist occurs primarily among the major types of dung, in particular between omnivore and large herbivore dung (Chapters 9 and 18). Unlike perhaps some mushrooms and other living microhabitats, droppings do not "defend" themselves, though interactions may occur between the microbial populations and insects in droppings (Hanski 1987a and references therein).

Table 1.1 puts forward a hypothesis about a continuum of competitive interactions that I perceive in patchy and ephemeral microhabitats. This contin-

TABLE 1.1
A continuum of ephemeral and patchy microhabitats with varying degrees of dominance by some species due to preemptive resource competition.

Resource use not dominated by a few species	Resource use dominated by a few species
Patterns and processes:	
Mixed community	Relatively few species
High diversity	Low diversity
Predation and/or competition	Resource competition
Variance-covariance dynamics	Lottery dynamics
Examples from different types of microhabitats:	
Cattle pats where large Scarabaeidae absent	Cattle pats in southern Africa: *Kheper*
Large carcasses where vertebrate scavengers unimportant	Small carcasses: *Nicrophorus*
Polyporaceae conks	*Leccinum* sporophores: *Pegomya*
Individual plants and animals as host individuals for many macro-parasites	Individual insects as hosts for many parasitoids

Sources: Price (1980); Hanski (1987a, 1989a,c); Edwards (1988a).

uum is to a large extent correlated with the relative size of individual habitat (resource) patches. The key point is whether a small number of individuals is able to dominate a habitat patch to the exclusion of other individuals. In the ideal situation, of which there are examples, each habitat patch is dominated by a single individual or by a breeding pair of individuals. Such resource dominance is easier in the case of small than large microhabitats, though "smallness" and "largeness" naturally depend on the size of the consumers as much as on the size of the habitat patches themselves. Table 1.1 lists some examples. Cattle pats may be dominated by the largest tropical dung beetles but not by the mostly small northern temperate beetles. Small carcasses may be buried by *Nicrophorus* beetles (Silphidae), but large carcasses cannot be so dominated by any one insect species. Though some mushrooms are rapidly consumed by fungivorous fly larvae, perennial conks typically have a diverse community of insects that only slowly destroy the microhabitat. Considering host individuals as microhabitats for parasites (Price 1980), ectoparasites of plants and animals are not generally able to exclude others from host individuals, but parasitoids (insect parasites of other insects) provide good examples of dominance of habitat patches (Hassell 1978).

The mechanisms of resource dominance are varied (Hanski 1987a). In the most obvious case, exemplified by the burying beetles *Nicrophorus* and by the large *Heliocopris* dung beetles, a pair of beetles removes the entire habitat patch (a small carcass or a dung pat) to an underground nest, away from the reach of competitors, predators, and adverse climatic conditions. In other cases, habitat patches may be dominated by territoriality, as described for the carrion-breeding fly *Dryomyza anilis* by Otronen (1984a, 1984b), for the dung-breeding fly *Scatophaga stercoraria* by Borgia (1980, 1981) and for the fungivorous beetle *Bolitotherus cornutus* by Brown (1980). Some specialist species are able to exploit all the resources in a habitat patch extremely fast and thereby reduce the chances of reproduction by inferior competitors. A probable consequence of the race to resource preemption is the habit of depositing first-instar larvae instead of eggs by the carrion-breeding *Sarcophaga* (Denno and Cothran 1976) and Australian *Calliphora* flies (Levot et al. 1979; for a general discussion see Forsyth and Robertson 1975; Hanski 1987a). The dung beetles that construct a nest and provision it rapidly with food for their larvae achieve the same result even faster. In other species, rates of larval development appear to have been pushed to the physiological upper limit: *Lucilia illustris*, a carrion-breeding calliphorid fly (Hanski 1976), and *Hydrotaea irritans*, a dung-breeding muscid fly (Palmer et al. 1981), can complete their larval development in the amazingly short period of 50 hours under optimal conditions. The insect communities in which the use of resources in individual habitat patches is strongly dominated by a small number of individuals are here called "dominance communities" and are structured by lottery dynamics. The meaning of this concept will become clear in a moment.

The other extreme of the continuum in Table 1.1 is found in microhabitats where resource patches cannot be dominated by a small number of individuals, usually because the patches provide such a large amount of resource compared with the size of the insect consumers that resource dominance by burial or by other means is not possible. In this case a large assemblage of many species may accumulate in the same patches. The cattle pat community in Europe is a case in point, with up to one hundred species of insects occasionally crowding in a single pat. The opposite extreme to the dominance community is a "mixed community" (of many species) structured by variance-covariance dynamics. We shall now turn to the two population-dynamic models corresponding to the end points of this continuum.

Lottery Dynamics

The first papers that explicitly dealt with lottery competitive dynamics were written by Sale (1977, 1979), who worked with coral reef fishes, which defend interspecific territories and hence compete for space. Deaths of adult fishes create empty lots available for colonization by young individuals from the planktonic pool. The lottery mechanism operates at this stage: all individuals of all species are assumed to have an equal chance of obtaining a vacant territory. Sale (1977) assumed that there is no density dependence in recruitment—in other words, that both locally rare and common species have equal probabilities of securing a territory. This assumption is not essential for lottery dynamics, and it is difficult to accept unless the recruits arrive from a larger area than the "community" under study, in which case the apparent density independence is an artefact of scale.

Sale (1977, 1979) originally conjectured that the lottery mechanism would allow coexistence of a large number of equal competitors—the coral reef fishes in his example. But this is not correct (Chesson and Warner 1981; Chesson 1985; Hanski 1987b). Assuming that the competitors are identical in every respect, no one species will deterministically replace the others, but nothing prevents the relative frequencies of the species from entering a process of random walk: by chance, one species does better than the others in a particular generation, increasing its frequency in the community. These changes are often reversed, as every species has the same chance of increasing; but given sufficient time, one species after another is lost from the community. The greater the spatial scale and the number of habitat patches, the slower the process of species elimination, but the point is that no mechanism exists in the lottery dynamics itself that would tend to maintain species richness.

Chesson and Warner (1981) showed that in iteroparous species, which may reproduce several times in their lifetime, environmental stochasticity can promote regional coexistence in lottery-competitive systems. For such species, even grand failures in some lotteries (breeding seasons) are not catastrophic, because the lost opportunities to increase in frequency can be made up by successful reproduction in subsequent breeding seasons; the great longevity of the species buffers their populations against severe declines (Chesson's "storage effect"). For this mechanism to work, different species must not have complete temporal correlation in reproductive success (Chesson and Huntly 1988). (There is no storage effect or only a slight one in dung beetles [see Chapter 17], hence the lottery dynamics is not expected to facilitate coexistence in dung beetles even with environmental stochasticity.)

To go back to insects colonizing patchy and ephemeral microhabitats, lottery dynamics in the purest form occurs when the first individual (female), or a pair of breeding individuals, that locates a habitat patch is able, in one way or another, to exclude any later-arriving individuals. Locating the habitat patches is the lottery, at which some species may be better than others but which nonetheless involves an essential component of randomness. The resource owners reproduce, and the newly emerged individuals take part in the next round of the lottery.

Dynamics in the burying beetles *Nicrophorus* and in some large dung beetles (for example, *Heliocopris*) closely approximate the lottery model, as one pair of beetles may completely dominate a resource patch, a carcass (*Nicrophorus*), or a dung pat (*Heliocopris*). In other cases, droppings are so large compared to the individual beetles and their resource requirements that one dropping has resources for more than one individual or a pair of beetles. However, due to their nesting behavior (Chapter 3), Scarabaeidae dung beetles are able to rapidly remove a portion of the resources in a dropping for their exclusive use, essentially on a first arrived–first served basis, until the whole resource is used up. The lottery dynamics applies to this situation, but now the contested resource units correspond to the chunks of resource required by individual beetles, or by pairs of beetles, rather than to the entire resource patches themselves.

Variance-Covariance Dynamics

Not all dung beetle communities have dynamics that even approximately fit the lottery model. We need another model for communities in which species do not rapidly preempt resources in habitat patches but stay and possibly compete for a prolonged period of time. An essential feature of most such species assemblages is extensive variation in the numbers of individuals between habitat patches. Atkinson and Shorrocks (1981) and Hanski (1981) originally developed the idea of incorporating the spatial variances and covariances of populations in competition models (see also Lloyd and White 1980). Further work on these and related models can be found in Ives and May (1985), Ives (1988a, 1988b, 1990), Shorrocks and Rosewell (1986, 1987), and Hanski (1987b).

To contrast with the communities characterized by lottery dynamics, we will now focus on assemblages in which a small number of individuals cannot monopolize habitat patches. On the contrary, large numbers of individuals may accumulate in single patches; but for a number of reasons, as discussed by Atkinson and Shorrocks (1984), Green (1986), Hanski (1987b; Chapter 5), Hanski and Cambefort (Chapter 16), and Ives (1988b), species composition tends to vary greatly across individual patches. The spatial patterns are suc- cinctly summarized by species' spatial variances and pairwise covariances cal- culated across patches, hence the term "variance-covariance dynamics." The key point here is that as the density-dependent interactions within and among species are localized into individual habitat patches, the varying densities lead to varying strengths of interactions in different patches. Hence the spatial vari- ances and covariances must affect the regional (community-level) outcome of these interactions.

Competition within patches may or may not be strong in communities char- acterized by variance-covariance dynamics. If competition occurs, it will be most severe in the patches that have the greatest number of colonists or off- spring of colonists. Increasing intraspecific aggregation (variance) across patches amplifies intraspecific competition relative to interspecific competi- tion. As the coexistence condition always hinges on the relative magnitudes of intraspecific and interspecific competition, increasing the relative strength of intraspecific aggregation tends to facilitate coexistence. Increasing covari- ance works in the opposite direction, as it tends to return the relative magni- tudes of intraspecific and interspecific competition to the level they would have without aggregation. Ives (1988a), however, has shown that coexistence is still greatly facilitated by intraspecific aggregation even if spatial covariance between two species is large (but not complete).

Competition is not always strong, and it is even less often the only impor- tant factor structuring dung insect communities. Variance-covariance dynam- ics has important implications also for communities in which generalist pred- ators or parasitoids play a significant role, as they do in many dung insect communities (Fig. 1.3 gives one example). Assuming that the prey species have more or less independently aggregated distributions and that the natural enemies tend to inflict the highest rate of mortality in the patches with the highest pooled density of prey, we can show that, at the regional level, the predators cause heavier per capita mortality in the more abundant prey species, even if they are entirely unselective while preying within patches (Hanski 1981; Comins and Hassell 1987). In this case it is the nonrandom spatial search of aggregated prey that leads to frequency-dependent predation, which generally tends to maintain prey species diversity (Murdoch and Oaten 1975; May 1977).

Generalist predators may facilitate the coexistence of their prey in variance- covariance communities also for an entirely different reason. The relationship between the spatial variance and the mean abundance (calculated across habitat patches) is typically such that the level of intraspecific aggregation decreases with increasing mean abundance (aggregation measured as explained in Ives 1988b; also see Chapter 16). Assuming that the prey species are independently aggregated, decreasing mean abundance due to predation increases intraspecific aggregation and intraspecific competition relative to interspecific competition, and again tends to facilitate prey coexistence (Hanski 1990). This mechanism is especially likely to work in cases where some species are competitively superior to others; the increase in their spatial aggregation indirectly due to predation makes invasion by inferior competitors easier.

The Two Kinds of Dynamics Compared

At the level of individual habitat patches, there is a clear difference between the two kinds of dynamics: ideally only one species occurs per habitat patch in the lottery dynamics, while the species number may be dozens per patch in the variance-covariance dynamics. What we are ultimately interested in is community structure at the level of many habitat patches during many generations. The purpose of this section is to demonstrate, with simple numerical models, that the within-patch difference in species number leads to a parallel difference in regional community structure. In the following models we assume that the insect populations have discrete, nonoverlapping generations.

   Lottery dynamics. Let there be one hundred habitat patches that can be colonized in each generation. We assume that the first individual to arrive at a patch is able to exclude all later-arriving individuals, and it produces ten offspring. Let us assume that there are initially ten species, all equally good at locating habitat patches; hence each dispersing individual, regardless of species, has the same probability of colonizing a vacant patch. Fig. 1.1 shows that with these assumptions, species are quickly eliminated from the community until a single species remains. Increasing the number of habitat patches would slow down the rate of species elimination, while differences among the species would make it faster.

   Variance-covariance dynamics. In this case the habitat patches are larger and allow successful reproduction by many individuals; we assume that there are ten such patches available to beetles in each generation. As with lottery dynamics, let us assume that there are initially ten species in the model community. Let $N_i(t)$ be the number of individuals of species i colonizing habitat patches in generation t. In the case of *uniform* colonization (no spatial variance), each patch receives $N_i(t)/10$ individuals of species i. In the case of *aggregated* colonization, let us assume for simplicity that five randomly chosen patches receive one individual each, while the remaining five patches receive $N_i(t)/5 - 1$ individuals each. There is thus large intraspecific aggregation (variance) in the numbers of each species, unless the species is rare, but there is no

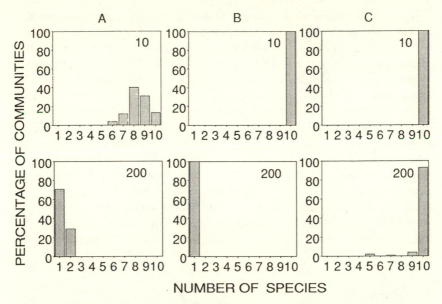

Fig. 1.1. Simulation results of the lottery and variance-covariance models described in the text. Each panel shows the results for one hundred replicate simulations after ten (upper panel) and two hundred (lower panel) generations. (A) The lottery dynamics in identical species lead to an eventual elimination of all but one species. The rate of species elimination is increased by differences between the species and decreased by the increasing number of resource patches. (B) The model of Eq. (1.1), with uniform colonization of patches. Competition coefficients were selected from uniform distribution between 0 and 2; $r = 1.5$ and $K = 100$. All but the best competitor are quickly excluded. (C) As in (B) but with aggregated colonization: the rate of exclusion is much slowed down and many species may coexist at equilibrium or for long periods of time.

interspecific spatial correlation (covariance). Following colonization, individuals compete within patches. The following model gives the number of surviving offspring of species i in patch k in generation t, given that n_{ik} individuals of species i had colonized the patch (other competition models could be used instead):

$$\text{Number of offspring produced by species } i \text{ in patch } k = n_{ik} \exp\left(r_i(1 - \Sigma\alpha_{ji}n_{jk}/K_i)\right), \tag{1.1}$$

where α_{ji} is the competition coefficient between species j and i, r_i is the population growth parameter, and K_i is the environmental carrying capacity for species i (r_i and K_i are assumed to be equal for all species).

Without spatial aggregation (uniform colonization), the competitive situation is essentially the same as in a single large habitat patch, and, as could be expected from the classical competition theory, the best competitor quickly excludes the others (Fig. 1.1). With spatial aggregation, things are different. Even though competition remains the regulating factor, species are excluded much more slowly, and depending on the strength of within-patch competitive interactions and the variances and covariances, many species may coexist at equilibrium (Fig. 1.1). In summary, variance-covariance dynamics tends to facilitate coexistence, and several species may coexist even if competition between them is strong.

It would be misleading to suggest that natural dung beetle assemblages can always be easily assigned to one of the two models, but I do suggest that this classification captures an essential element of the dynamics of competitive populations inhabiting patchy and ephemeral microhabitats. In actual dung beetle communities there are other factors, such as various forms of resource partitioning, which need to be taken into account. The ultimate challenge is to integrate the effects of different types of processes into the same model of community structure (Chesson and Huntly 1988).

Resource Partitioning

The scheme in Table 1.1 assumes that all habitat patches are the same and consist of one resource type. In reality, habitat patches may be heterogeneous, creating the possibility for within-patch resource partitioning, or there may be several kinds of habitat patches, or there may be differences among the species that affect their resource use in one way or another. All these aspects of resource partitioning in dung beetles will be dealt with in many of the chapters in Part 2 and are summarized in Chapter 18. It suffices here to indicate briefly their possible importance in dung beetle communities, and their relationship to the continuum from lottery to variance-covariance dynamics.

Within-patch resource partitioning would be an extension of the continuum from lottery to variance-covariance dynamics, tacked on to its variance-covariance end, but within-patch resource partitioning is relatively unimportant in dung beetles, because dung pats do not consist of distinct resource types recognized as such by dung beetles. One exception is the large droppings in situations where exploitative competition is weak or nonexistent, and the droppings remain available for colonization for weeks or months. In this case, different sorts of insects colonize the droppings at different times (Koskela and Hanski 1977; Hanski 1987a), presumably because different species use different resources that become successively available during the dropping's decay, or they use the same resource in a different way.

Resource partitioning based on different types of habitat patches exists in dung beetles, but primarily among the few main types of dung. The dynamic models in Table 1.1 apply separately for each resource type. There are also differences among species in macrohabitat selection (forest versus grassland), in occurrence on different soil types (especially in rollers and tunnelers), in diel activity (diurnal versus nocturnal species), in seasonality, and in size and its various correlates (Chapter 18). All such differences may directly or indirectly facilitate regional coexistence of competitors, and hence increase regional species diversity. To take an example from *Nicrophorus* with putative lottery dynamics, there are typically three to five species that coexist in one locality in northern Europe and North America (Anderson 1982), but the species show substantial interspecific differences in seasonality and macrohabitat selection (Pirone 1974; Shubeck 1976; Anderson 1982; Wilson et al. 1984). Two species of *Nicrophorus* are abundant in the nonseasonal tropical forests of Sulawesi (Indonesia), but they have nonoverlapping elevational ranges, one species occurring below 1,400 m and one above (Chapter 10).

1.2. THE DUNG MICROHABITAT

Table 1.2 summarizes the key characteristics of the dung microhabitat from the beetles' perspective. The size of a dropping is correlated with its durational stability; small droppings dry out faster than large ones and quickly become unsuitable for dung-inhabiting insects. Small droppings can be secured by a pair of beetles, and the community dynamics tend toward the lottery type. Large droppings cannot often be so dominated by dung beetles or by other coprophagous insects, and many species usually occur together there. Large droppings, if not consumed rapidly by dung beetles, change in quality with time, and give rise to a characteristic succession of fungi (Harper and Webster 1964), nematodes (Sudhaus 1981), flies (Mohr 1943), and beetles (Koskela

TABLE 1.2
General characteristics of the dung microhabitat.

Patch size	is generally small but varies by many orders of magnitude from small mammal pellets (0.0001 kg or less) to elephant droppings (10 kg)
Patch distribution	is haphazard to very aggregated
Patch density	varies from very low to very high (e.g., in many pastures)
Durational stability	Droppings generally last for one insect generation only; physical and chemical conditions in droppings change rapidly, and many species show distinct preferences for certain stages in the decomposition of droppings
Nutritional quality	is generally high, though there is a major distinction between herbivore and omnivore dung; herbivore dung is a mixture of relatively low-quality undigested plant remains and a high-quality microbial suspension (Chapter 2)

and Hanski 1977). The occurrence of multiple species is facilitated by the activities of other species that colonize droppings earlier. For example, tunnels made by dung beetles improve the oxygen supply to coprophagous flies but also provide runways so predatory staphylinids can get at the flies (Valiela 1974).

From the beetles' nutritional standpoint, the main types of dung to be consumed are large herbivore and omnivore dung, with carnivore and small mammal dung used by a relatively small number of species. The large herbivore dung is the quantitatively most important resource for dung beetles in most regions; if a single type must be singled out in this category, it would be cattle dung, now available in large quantities all over the world. Herbivore dung consists of two components, the relatively low-quality undigested plant remains and a high-quality component consisting of the mammalian gut fauna and flora and their products. Use of the two kinds of resources requires different feeding morphologies and physiologies; the undigested plant material is comparable to other decomposing plant material, while the microorganisms occur largely in a liquid suspension. Dung beetles typically use both components but with a developmental difference, larvae using the solid, low-quality component and adults the liquid, high-quality component (Chapter 2).

Herbivore dung varies in nutrient concentrations, moisture content, consistency, and so on (Hanski 1987a). Some such differences are reflected in differences in dung beetle assemblages; for example, cattle and deer dung have quite distinct faunas in North America (Gordon 1983). The quality of herbivore dung varies seasonally, depending on the kind of forage available to the herbivore, and such changes are reflected in seasonally varying breeding success in many dung-breeding insects (Chapter 15). Other similar examples are described in many of the chapters in Part 2.

1.3. The Dung Insect Community

Figure 1.2 and Table 1.3 present an overview of the food web and the taxonomic composition of the insect community that colonizes large herbivore (cattle) dung in Europe. Note the diversity of beetles and flies inhabiting this microhabitat. Many dung-frequenting insect families and genera, and even some species, are nearly cosmopolitan, though their relative numbers vary from one biome to another (Part 2). Lack of important geographical variation in resource characteristics probably explains the relative taxonomic uniformity of dung insects worldwide. Though we have not considered organisms smaller than insects, they nonetheless significantly add to the total species number and complexity of interactions in the community. For example, with a mere eighteen samples Sudhaus (1981) recorded thirty-four species (13,440 individuals) of bacterial-feeding, fungivorous, mixed-diet, and predatory nematodes in cattle pats in Europe, with apparent successional replacements of species from

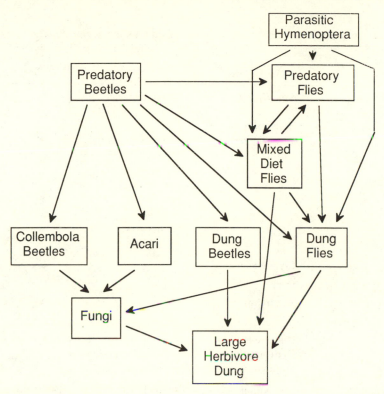

Fig. 1.2. A simplified food web of the insect community inhabiting cattle dung in Europe (modified from Skidmore 1985).

fresh to older pats. Mites representing a number of families utilize dung beetles as a means of transport from deteriorating to fresh dung pats. The most numerous and diverse taxon is the Macrochelidae, which consists of predatory mites feeding mostly on insect eggs in droppings. Some species are specific to their beetle carriers (Krantz et al., in prep.)

Three kinds of flies exist in the dung community (Fig. 1.2). Most species have entirely coprophagous larvae, mostly feeding on the microorganisms present in the dung, but the family Muscidae includes both facultative predators (usually predators in the final larval instar, e.g., *Muscina*) and entirely predatory larvae (e.g., *Mydaea*). As if this were not complex enough, one of the most prominent dung insects in the northern hemisphere, the yellow dung fly, *Scatophaga stercoraria*, has coprophagous larvae but the adults are voracious predators, inflicting heavy mortality on flies and beetles in pastures.

Faced with this great diversity of insects and their interactions, one may ask what is the point in isolating only a small subset—the dung beetles—for spe-

TABLE 1.3
Insect families with taxa feeding and/or breeding in the dung microhabitat.

Food-Web Position	Coleoptera	Diptera	Hymenoptera
Coprophages	**Aphodiidae** **Geotrupidae** **Hydrophilidae** **Scarabaeidae** **Staphylinidae**	Anthomyiidae Ceratopogonidae Chironomidae **Muscidae** Psychodidae **Scatophagidae** Scatopsidae Sciaridae **Sepsidae** **Sphaeroceridae** Stratiomyidae	
Mycophages	Cryptophagidae Ptiliidae		
Saprophages	**Staphylinidae**		
Predators	Carabidae **Histeridae** **Hydrophilidae** **Staphylinidae**	Muscidae	
Parasitoids	Staphylinidae	Bombyliidae	**Braconidae** **Eucoilidae** **Ichneumonidae** **Pteromalidae**

Note: Taxa that are most often numerically or functionally dominant are in boldface.

cial consideration. Shouldn't we instead make an attempt to study comprehensively the entire dung insect community?

There are two explanations for our more modest approach. One could indeed gain a better understanding of the workings of the dung insect community if one were able to consider simultaneously all the species occurring in this microhabitat. Work done especially in Australia demonstrates clearly how dung beetles and dung flies have strong symmetric or, more often, asymmetric competitive interactions (Bornemissza 1970b; Ridsdill-Smith 1981; Ridsdill-Smith and Matthiessen 1984; Ridsdill-Smith 1990). Unfortunately, studies on competition and other forms of interaction among the different taxa of dung-inhabiting insects are yet few and generally limited to only two or a few key species. The Australian cattle dung communities are exceptionally amenable to experimental work because of the impoverished fauna: the introduced resource (cattle dung) is used primarily by two species of abundant flies (of which one was accidentally introduced into Australia in 1825) and some twenty purposely introduced dung beetles (Chapter 15). To take one example, Ridsdill-Smith and Matthiessen (1988) were able to show that the introduction of two new species of dung beetles, *Oniticellus pallipes* and *Onthophagus binodis*, into an area in southwestern Australia reduced the numbers of the fly *Musca vetustissima* to about 10% of its previous level during the time of year when the introduced dung beetles were most active. The more typical dung insect communities have dozens or even hundreds of species with complex interactions. As one example, Fig. 1.3 shows the food web relationships between coprophagous flies and their parasitoids in northern North America. Remember that here we are looking at a species-level magnification of only a small part of the food web of the entire dung insect community (Fig. 1.2).

Another important point is that not all the species in the dung insect community are always equally numerous, nor do all the species enter into strong interactions with one another. Dung beetles in particular interact primarily

PARASITOIDS

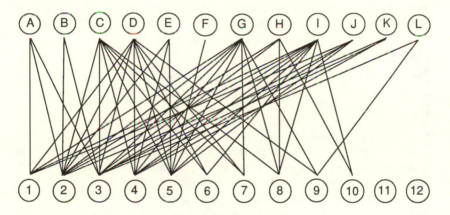

HOST SPECIES

Fig. 1.3. The food web of dung-breeding flies and their parasitoids in northern North America. The host flies are: 1, *Paragle cinerella*; 2, *Orthellia caesarion*; 3, *Ravinia* spp.; 4, *Sepsis biflexuosa*; 5, *Saltella sphondylii*; 6, *Musca autumnalis*; 7, *Ravinia lherminieri*; 8, *R. querula*; 9, *Gymnodia arcuata*; 10, *Musca domestica*; 11, *Myospila meditabunda*; and 12, *Sargus cuprarius*. The parasitoids are: A, *Aleochara* sp. (Staphylinidae); B, *Alysia ridibunda*; C, *Aphaereta pallipes*; D, *Eucoila* sp.; E, *Kleidotoma* sp.; F, *Cothonaspis* sp.; G, *Trichopria* sp.; H, *Trichomalopsis dubia*; I, *Muscidifurax* sp.; J, *Spalangia haematobiae*; K, *S. nigra*; and L, *S. nigroaenea*. The results are based on the rearing of 96,372 fly puparia (data from Figg et al. 1983).

with three other taxa (Fig. 1.2). First and most important, they compete with various types of flies. This interaction is generally asymmetric, because dung beetles, unlike flies, are able to remove dung into underground nests (Ridsdill-Smith 1990). Second, predatory beetles prey upon dung beetles, but this interaction is often rather weak, with most predatory beetles using fly eggs and larvae instead of dung beetles. The field experiments of Doube, Macqueen, and Fay (1988) conducted in southern Africa and in Australia demonstrated that dung beetles and insect predators and parasitoids caused up to 98% mortality of the dung-breeding flies *Haematobia thirouxi* (Africa) and *H. irritans* (Australia). In many regions of the world, vertebrates are more important predators of dung beetles than are insects, for example in the African savannas, where omnivorous mammals feed on the often large concentrations of large dung beetles (Chapter 9). Third, hundreds of mites (Halffter and Matthews 1971) and nematodes (Sudhaus 1981) use beetle hosts to move from one habitat patch to another. Although these phoretic relationships pose interesting evolutionary questions (Costa 1964, 1969; Springett 1968; Wilson 1980, 1982), the phoretic mites and nematodes are probably of minor consequence to beetles.

The chapters in Part 2 describe in greater detail the relationships between dung beetles and the other taxa in some geographical regions. In anticipation of these results we conclude that dung beetles often form a distinct guild of their own, and the primary interactions with other insects are asymmetric: where dung beetles are particularly dominant, they suppress the other members of the dung insect community to a secondary role.

1.4. SUMMARY

In this chapter I have outlined two contrasting models that define a range of possible competitive interactions in insects using nonrenewable resources in patchy habitats. In lottery dynamics, a few individuals, in some cases only a breeding pair of beetles, are able to monopolize an entire resource patch. Lottery dynamics does not facilitate coexistence of many similar competitors, hence regional species richness is expected to remain low in assemblages obeying lottery dynamics. In variance-covariance dynamics, rapid resource preemption does not occur, but large numbers of individuals accumulate in habitat patches. Typically, the species composition varies from one patch to another, hence the strengths of intraspecific and interspecific competitive interactions also vary across patches. If species are more or less independently aggregated in resource patches, the relative strength of intraspecific competition increases and coexistence of competitors becomes more likely. In assemblages obeying variance-covariance dynamics, regional species richness may be high in spite of competition. I have given examples from insect communities covering the continuum from lottery dynamics to variance-covariance dynamics.

The physical characteristics and the chemical properties of animal droppings have been briefly reviewed. Dung beetles seldom occur alone in droppings, but they belong to a complex community comprising other beetles and flies as well as a multitude of other organisms, including mites and nematodes. The taxonomic composition and the structure of the entire dung insect community has been described, using insects that colonize cattle pats in Europe as an example. Nonetheless, dung beetles often comprise a distinct guild of their own and exhibit the strongest interspecific interactions among themselves. Interactions with other insects are typically asymmetric: where dung beetles are particularly dominant, they suppress the other members of the dung insect community to a secondary role.

CHAPTER 2

From Saprophagy to Coprophagy

Yves Cambefort

DUNG BEETLES comprise a small number of families in the superfamily Scar-abaeoidea, section Laparosticti: Scarabaeidae, Geotrupidae, Aphodiidae, and, to a lesser degree, Chironidae, Hybosoridae, and Trogidae (see the appendix at the end of this chapter for the taxonomy used in this book). Many of the families clustered in this section are poorly known, but it seems likely and has been assumed by many authors (Janssens 1954; Balthasar 1963; Halffter and Matthews 1966, 1971) that the majority are saprophages, both in the larval and adult stages. Saprophagy and coprophagy are not fundamentally different feeding habits: the pioneering experiments by Fabre (1910) demonstrated that some saprophagous beetles can be reared on dung, while some coprophagous species can be successfully reared on humus.

In his portrayal of the ancestral beetle, Crowson (1981) assumed that the archetypal coleopteran structure—a somewhat flattened form, with relatively short legs and nonprojecting coxae and short antennae inserted low down on the sides of the head—is related to the insect's life-style, with the species habitually taking refuge under loose bark on dead trees (Hamilton 1978). Scar-abaeoidea Laparosticti appeared much later than the first Coleoptera, but they may have retained the ancestral structure and habits, feeding with their prim-itive grinding (biting) mandibles on rotten vegetable matter under the bark of fallen trees.

Today, Crowson's (1981) four ancestral characteristics are probably never preserved all together in the same species. Although saprophagy and the use of microbial organisms still occur in many laparostict families, important vari-ations have evolved from the ancestral type. The first change was probably from hard, biting mandibles to soft, filtering ones, with a correlated change in adult diet. Second, some groups have evolved a more or less sophisticated nesting behavior. And third, we envision an ecological radiation from ances-tral saprophagy to xylophagy, mycophagy, predation, and, most importantly in terms of the number of extant species, to coprophagy.

2.1. THE FIRST EVOLUTIONARY TREND: FROM HARD TO SOFT DIET

While concluding a taxonomic paper with some ecological reflections, Steb-nicka (1985) made the pertinent distinction between "hard" and "soft" saprophagy. She gives these definitions: hard saprophagy, the use of "hard organic substances, for example, dead wood, leaf litter, mushrooms, spores"; and soft saprophagy, the use of "liquid organic contents, for example, vegetal juice, dissolved albuminous substances and/or bacterial albumens in decaying humus." Saprophagous Lamellicornia (and Pectinicornia) can indeed be divided into two groups according to their diet and to the form of adult mouthparts, especially the mandibles, which can be of different shapes. The primitive type is the most undifferentiated: it has short, strong, and biting mandibles, today found in the Chironidae, Orphnidae, and others. One evolutionary line appears to have led to the monstrous, hypertelic mandibles of male Lucanidae, while Passalidae and female Lucanidae have retained strong but not disproportionally large mandibles used for feeding on dead wood.

Another evolutionary line has led to the soft, filtering mandibles of adult Aphodiidae and Scarabaeidae (Fig. 2.1), some taxa of which are still saprophagous in the adult stage. The morphology of the soft mandibles, characterized by two parts, has been described by Hata and Edmonds (1983). The upper part, or incisor lobe, is a modified filtering apparatus, which strains food of particles that are too large for ingestion. The retained particles are small (2–200 μm), which has suggested the term microphagous beetles, but even these tiny objects must be broken down before they enter the digestive tract. The lower part of the mandible, or molar lobe, is used for this purpose. Miller (1961) was the first to point out that the ability of microphagous dung beetles to triturate their food resides in the peculiar structure of the molar lobes of their mandibles. A system of submicroscopic scrapers forms an abrasive surface that breaks up all but the very smallest (<1 μm) particles passing between them. Besides being suited for trituration of food particles, the mandibles are also highly modified for working together with the other mouthparts to gather soft, semiliquid materials.

We shall refer to the laparostict dung beetles whose adults possess hard, biting mandibles as "hard-diet consumers," and the ones whose adults possess soft, filtering mandibles as "soft-diet consumers." Some species use fine-grained, soft humus in the larval stage, though the adults have retained biting mandibles (e.g., many North American and Australian Bolboceratinae; Howden 1955a): we therefore consider them hard consumers. Table 2.1 gives a list of all the families and subfamilies of Scarabaeoidea Laparosticti, arranged alphabetically into these two groups. One family, Aulonocnemidae, cannot be placed in either of the two groups (Cambefort 1987a). In this taxon, adult mandibles are of the intermediate type: the incisor lobe possesses one or two apical teeth, similar to the apex of the hard consumer type, followed by the filtering structure of the soft consumers. This particular shape may give us a model of evolution from the ancestral taxa with presumably the most primitive structures (as we can see today, for example, in the Chironidae) to the advanced type of soft consumers.

But why change from the biting, hard consumption to the filtering, soft

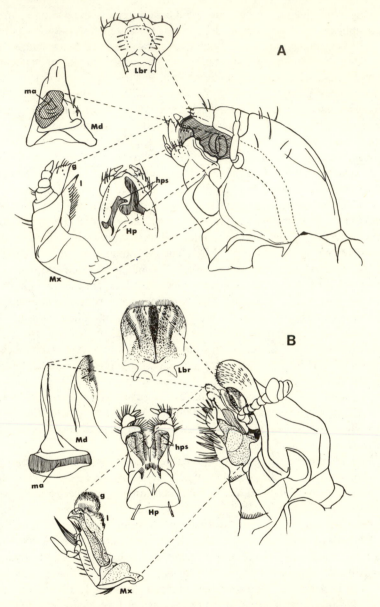

Fig. 2.1. (A) Lateral view of the head of *Phanaeus* larva showing the different mouthparts and their positions in the middle and on the right (left mouthparts removed). The middle part is also shown separately. (B) The same for adult *Phanaeus*. Abbreviations: g, galea; Hp, hypopharynx; hps, hypopharyngeal sclerites; l, lacinia; Lbr, labrum-epipharynx; ma, molar area of mandible; Md, mandible; Mx, maxilla (from Halffter and Matthews 1966).

TABLE 2.1

List of families and subfamilies of Scarabaeoidea Laparosticti, arranged in three nutritional groups: hard-diet consumers (larvae and adults feeding on hard, coarse-grained diet), intermediate consumers, and soft-diet consumers (all adults and some larvae feeding on soft, fine-grained diet).

Family Subfamily	Diet
Hard Consumers	
Aegialiidae	
Aegialiinae	Roots, humus
Eremazinae	?
Allidiostomidae	?
Belohinidae	?
Ceratocanthidae	Humus
Chironidae	Humus, dung
Diphyllostomidae	?
Geotrupidae	
Athyreinae	?
Bolboceratinae	Humus (larvae); fungi (adults)
Geotrupinae	
(incl. Taurocerastinae)	Humus, dung (larvae); dung, fungi (adults)
Lethrinae	Leaves
Glaresidae	?
Hybosoridae	
Dynamopinae	?
Hybosorinae	Humus, dung
Orphninae	Humus, dung
Ochodaeidae	
Chaetocanthinae	?
Ochodaeinae	?
Pleocomidae	Conifer mycorhizae (larvae)
Trogidae	Dry carrion, carnivora dung; locust eggs (larvae)
Intermediate Consumers	
Aulonocnemidae	Humus under bark or on the ground
Soft Consumers	
Aphodiidae	
Aphodiinae	Dung, humus (larvae); dung (adults)
Corythoderinae	Litter in termite nests
Eupariinae	Humus, dung
Odochilinae	Humus
Psammodiinae	Humus
Rhyparinae	Humus, dung
Termitoderinae	Litter in termite nests
Thinorycterinae	Roots, humus
Scarabaeidae	
Coprinae	Dung, carrion, humus
Scarabaeinae	Dung, carrion, humus
Termitotrogidae	Litter in termite nests

consumption? Plant litter or humus is a heterogeneous matter of decomposing organic, mostly vegetable substances. In the beginning of the decomposition process, humus is composed of relatively large units, such as dead leaves, pieces of bark, wood, and so forth. As decomposition proceeds, the pieces become smaller and the decomposing agents more numerous (Dickinson and Pugh 1974). When the habitat is saturated with water, the decomposing humus has a pasty consistency, with fine grains and a cementing liquid containing both decomposing agents and other microorganisms preying on them. This liquid is very rich in carbohydrates and especially in proteins. For saprophagous organisms, the most nutritient-rich humus is the most decomposed one, and, within it, the richest part is the liquid. One is tempted to postulate a selection pressure in saprophagous Laparosticti to selectively use the most nutritious humus, and especially its richest, liquid component. Today the rich humus is used by the intermediate (Aulonocnemidae) and some soft consumers (Aphodiidae: Eupariinae, and a few Scarabaeidae genera, e.g., *Bdelyrus*, *Bdelyropsis*, and *Paraphytus*), with highly evolved mouthparts and digestive tracts adapted to exploit this rich but little concentrated diet (Edmonds 1974).

Why should such nutrition not be more widespread? We think that the nutritionally richest humus occurs in the tropical regions, and in very specific and infrequent microhabitats, for example, under the loose bark of fallen trees and the leaf bases of various plants, especially palms and bromeliads. In addition to being patchy, these microhabitats are also ephemeral. Even under the most favorable circumstances, presently found in the evergreen rain forest, this particular resource cannot sustain large numbers of consumers. The same situation may have prevailed in the past, and only a small number of mostly small species has ever been able to occupy this niche. Most of them still belong to the families of Laparosticti Scarabaeoidea listed in Table 2.1.

2.2. THE SECOND TREND: EVOLUTION OF NESTING BEHAVIOR

The laparostict larval mouthparts are always of the biting type, lacking the microtriturating molar lobe (Fig. 2.1). Laparostict larvae, like all Scarabaeoidea larvae, are able to process almost any type of food. From a behavioral viewpoint, one may divide laparostict larvae into two groups. In the first group, larvae live freely in the microhabitat, while in the second one they live in "nests" separated from the main microhabitat, very often in a burrow dug in the soil. In this second group, one can make a further distinction between the taxa in which the burrow is dug by the larva itself and the ones in which it is prepared by the female or a pair of beetles before oviposition. Taxa with free-living larvae are common in Scarabaeoidea. This is the rule when the larva lives and feeds in a habitat where the resources are scattered, for example, Pleurosticti of the family Melolonthidae (true "white grubs": see, e.g.,

Ritcher 1966). In contrast, when the larva feeds on a patchy resource, free-living larvae have to be small in comparison with the food source. This is the case for most Aphodiinae with larvae living freely in large dung pats. Some *Aphodius* larvae make their own burrows (Bernon 1981; Rojewski 1983; Chapter 14), and the same is true about some Trogidae (Baker 1968; Lumaret 1984b).

In most cases where burrowing occurs, the burrow is made in advance of larval development by the female or by a bisexual pair of beetles. Such nesting behavior is well known in other beetles, for example in *Nicrophorus* (Silphidae). In this necrophagous genus, a pair of beetles secures an entire small mammal carcass by burying it into the ground (Halffter, Andugua, and Huerta, 1982, 1984). Burying by adults is nearly universal in Laparosticti, which dig into the ground using their front legs and head. In the soft-diet consumers, membranaceous and fragile mouthparts are covered by a cephalic anterior shield called the episome, which has a protective function but is also used for digging. Nesting behavior is the rule in Scarabaeidae, whose larvae are modified to feed and live within a relatively small and close space (Chapter 3).

It is not obvious whether nesting behavior evolved before, after, or in parallel with evolution of coprophagy from saprophagy. We suggest the working hypothesis that nesting behavior evolved first, which begs the question about the advantage of nest building in saprophagous species.

Let us first consider two saprophagous, nest-building species. *Geotrupes stercorosus* (Geotrupidae) provides its nest with moldy forest litter, rolled up tightly and buried at a depth of 20–30 cm in a previously dug burrow (Rembialkowska 1982). In this hard consumer, both adults and larvae are able to process almost any type of plant litter, but packing the litter in the nest facilitates the growth of decomposing microorganisms, which may significantly contribute to larval nutrition. The nest is also likely to have a protective role for the large, defenseless larvae.

The second example, *Paraphytus*, is one of only two saprophagous Scarabaeidae whose nest has been described so far (Cambefort and Walter 1985). The adults live freely, together with Eupariinae and other insects, in the rather thin layer of humus between the loose bark and the wood of fallen trees in tropical forests in Africa and Asia. Such humus belongs to the richest type referred to above. The female makes her nest without digging a burrow but by simply forming a brood ball on the spot and laying an egg in it. Laying only one egg at a time indicates low fecundity and that the few offspring are well protected.

The important role of the brood ball inside the nest has been demonstrated only for coprophagous species, but we may assume that the same function applies to saprophagous species as well. Malan (in Goidanich and Malan 1962, 1964), who was the first to study the microbiology of dung beetle nutrition, demonstrated that the microflora in the food and intestine of the beetle larva share the same components: bacteria (Schizomycetes), which in turn include anaerobic clostridia and aerobic bacilli; yeasts (Blastomycetes); and fungi (Hyphomycetes). The bacterial content of the food material is high when the food is fresh but drops to about half in the uneaten part of the brood ball. However, once the food is ingested by the larva, the bacteria build up from two or three up to seventeen times the basic level due to cellulose-degrading activity. After the material has passed out as larval feces, the bacterial content grows back to about the same level as in fresh food. The fecal material with high bacterial content is re-eaten several times by the larva. Recent biochemical measurements have shown that the amino acid composition of the inner part of the brood ball is greatly improved through larval ingestion (Rougon 1987). By re-eating its own feces, the larva takes advantage of this improvement. The repeated ingestion of the food material thus enables efficient resource utilization, in a manner analogous to caecotrophy in some mammals, especially rabbits and lemurs (Charles Dominique and Hladik 1971).

In brief, larval development in a nest prepared by the adults offer protection for the offspring and facilitate efficient utilization of the secured food resources. Construction of the nest involves great investment of time and energy from the parents and is clearly related to reduced fecundity and low mortality in the offspring (Chapter 3).

2.3. THE THIRD EVOLUTIONARY TREND:
FROM SAPROPHAGY TO COPROPHAGY

The ecology, biology, and diet of most of the laparostict families are poorly known (Table 2.1). Many taxa have retained the assumed primitive saprophagy both in larval and adult diets, but most of the species in the largest family, Scarabaeidae, have coprophagous larvae and adults.

We have suggested that the availability of a pasty, fine-grained humus rich in proteins (decomposing agents and microorganisms preying on them) has selected for highly specialized mandibles in some small saprophages, allowing them to use efficiently the humus with the highest quality. Another trend has been the evolution of nesting behavior that both protects the offspring and further increases the efficiency of food utilization. We have also noted that despite the high nutritive value of the rich humus, only a few small species have specialized to use it, possibly because of the relative scarcity and patchy occurrence of this resource.

The explosive evolution of mammals (Chapter 4) changed the situation dramatically by bringing along a new resource, the mammalian dung, in vast quantities. Mammalian dung is not very different, in its composition, from rich humus. The pasty texture is almost the same, but fresh dung especially is even richer in proteins than humus. We suggest that not only the quality but also and perhaps chiefly the great quantity of mammalian dung promoted the evolution of coprophagy from saprophagy. As dung occurs in very patchy and ephemeral units, its exploitation has probably been possible only for species that were adapted to the manipulation and utilization of similar substances. Scarabaeidae "dung beetles," which had already mastered the processing of rich humus, were in the best position to take advantage of the now widely available mammalian dung. Other families have attempted to occupy a part of this niche. The few dung-feeding Geotrupidae (mostly subfamily Geotrupinae) have mandibles equipped not only with a biting apex but also with a microgrinding molar lobe. With such mouthparts, they were able to process any sort of humus and humuslike material. The Aphodiidae were soft diet consumers, equipped to use the highest quality humus. Adults and larvae are small and live freely in the microhabitat. They were preadapted to live in large dung pats, but only some taxa turned to coprophagy, especially the genus *Aphodius* in the subfamily Aphodiinae.

Today, the large coprophagous Geotrupinae, with some 150 species, occur predominantly in the cooler regions. They are probably not good enough competitors to share the coprophagous niche with the nearly 5,000 species of Scarabaeidae in the tropical regions, where competition is most severe. In contrast, the 1,850 species of Aphodiinae, which are much smaller, have become successfully established also in the tropics, where they coexist with Scarabaeidae, though the main distribution of Aphodiinae lies in the cooler regions. A few species of Aphodiinae have reached medium or large size and some species construct primitive nests (Bernon 1981; Rojewski 1983; Lumaret 1984b).

So far we have not discussed the most striking behavioral characteristic of dung beetles, the ability to make and roll balls. The reason is that none of the present roller species are saprophages. However, there is one remarkable genus, the Australian *Cephalodesmius*, which belongs to rollers from the systematic point of view. In *Cephalodesmius*, bisexual pairs of beetles occupy fixed, subterranean nests on the floor of the rain forest (Monteith and Storey 1981). Males forage outside the nest for leaves and other plant material. These food stuffs are processed by females into a malleable, pasty material, which is used to construct balls in which the eggs are laid. Such behavior resembles nesting in Lethrinae (Geotrupidae), whose adults are equipped with very large, chisel-shaped mandibles with which they cut young leaves (hence their name "bud-cutters" in Hungary, where they were considered vineyard pests). Cut leaves are chewed by the female and packed in burrows in the shape of sausages, similar to those of Geotrupinae (Popovici-Baznosanu 1932; Nikolajev 1966). *Cephalodesmius* does not have biting mandibles, and the beetles have been forced to develop alternative mechanisms of leave harvesting and mastication, using tibiae and clypeal prongs. *Cephalodesmius* represents an evolutionary trend that is parallel to the evolution of coprophagy. *Cephalodesmius* have solved the problem of low availability of high-quality humus by synthesizing it from the abundant low-quality plant litter. *Aphodius tasmaniae* apparently does the same thing: the adults make individual burrows, into which they take grass blades cut at the soil surface. This beetle is known as a pest in Australian pastures (McQuillan and Ireson 1983).

Although classified in a roller group, *Cephalodesmius* do not actually make nor roll balls outside the nest. Matthews (1974) makes a clear distinction between the ability to make a ball from a larger food mass and the ability to roll naturally subspherical objects. It is true that rolling more or less spherical pieces of dung is a behavior not exclusive to rollers. For instance, rolling has been found in some Scarabaeidae tunnelers, for example, *Dichotomius* (Halffter and Matthews 1966), *Phanaeus* (Halffter et al. 1974) and *Phalops* (Cambefort unpublished observations in the Ivory Coast). Some dung beetles do not roll the pieces of excrement but they butt them with the head, or push them with the head and forelegs, a behavior similar to that observed in the ''true'' roller *Pachylomera* (Walter 1978; Bernon 1981). Rolling has also been found in dung beetles other than Scarabaeidae, for example in Geotrupinae (Zunino and Palestrini 1986). Nonetheless, ball rolling is most likely to have evolved in beetles that, for some reason, started to transport pieces of food. It could have evolved in some saprophagous species such as *Cephalodesmius*, or in nest-building coprophagous species transporting pieces of dung down to their burrows, or in species exploiting natural dung balls—for example, goat or rodent pellets—which are today used in South America by the ''true'' roller Eucraniini (Howden, pers. comm.) and by the tunneler *Canthidium* (Gill, Chapter 12).

Table 2.2 summarizes the two evolutionary shifts from low-quality to high-quality food and from hard (coarse-grained) to soft (fine-grained) diets. The low-quality diet, that is, vegetable litter in the beginning of the decomposition process, is always coarse-grained and is used by larvae and adults with grinding mandibles. The high-quality diet, either decomposed humus or dung, is more fine-grained but may still contain large particles. It may be used as such, including both large and small particles, by larvae and adults with grinding mandibles; but some adults have evolved specialized, filtering mandibles, enabling them to use only the soft, fine-grained part of the highest nutritive quality. A few dung-feeding larvae, though equipped with the usual grinding mandibles, seem to specialize on the fine-grained component. We shall now turn to this ultimate nutritional adaptation in dung beetles.

2.4. FROM CELLULOSOBIONTIC TO COPROBIONTIC DIET

Dung beetle larvae repeatedly eat the provisions they have in the brood balls. Malan (in Goidanich and Malan 1962), while describing the microflora present in dung and in the intestine of dung beetle larvae, referred to the cellulose-

TABLE 2.2
The shift from low-quality to high-quality diet in Scarabaeoidea Laparosticti.

Stage	Low-Quality Diet	High-Quality Diet	Examples
1	Larva H Adult H		*Chiron, Ochodaeus, Orphnus*
2	Larva H	Adult H	Some Hybosoridae
3		Larva H Adult H	*Geotrupes*
4		Larva H Adult S	Almost all Aphodiidae and Scarabaeidae
5		Larva S Adult S	*Scarabaeus, Euoniticellus*

Note: The 5 stages are assumed to represent an evolutionary trend toward the use of higher-quality resources. H and S refer to hard (coarse-grained) and soft (fine-grained) parts of the food resource.

degrading clostridia as "cellulosobionts," and to the remaining flora, representing ubiquitous saprophytes found in dung, as "coprobionts." Although this latter term does not reflect the fact that these microorganisms are probably also present in high-quality humus, we shall use it here for convenience and because Malan's observations were made on true dung beetles.

Halffter and Matthews (1971 and references therein; see also Rougon 1987 and references therein) have given the following summary of larval and adult diet in dung beetles. The larvae that have a "fermentation chamber" in their intestine depend primarily on a dense culture of cellulose-digesting bacteria for their nutrition (although they can be reared under sterile conditions in a rich medium: Charpentier 1968). This is a form of symbiosis similar to that in other cellulose-digesting organisms, and Halffter and Matthews (1971) suggest that it is a primitive character in the Scarabaeoidea, shared with many other saprophagous and xylophagous Lamellicornia. Larvae with this intestinal structure use primarily various kinds of herbivore excrement. On the other hand, Halffter and Matthews (1971) and Rougon (1987) point out that some Scarabaeidae larvae and, by anatomical analogy, all the adults whose alimentary tract has been studied so far (Edmonds 1974) base their diet on other microorganisms ingested with the food ("coprobionts" of Malan), and they do not establish their own enteric cultures to the same extent. This is the diet postulated by Halffter and Matthews (1966) for adult dung beetles, and it appears to be a derived form of feeding, more specialized than the cellulose-culture method, because it restricts the diet to those foodstuffs that can support large microbial populations aerobically. Examples are nonherbivore excrement, decaying fungi, carrion, and nonruminant herbivore dung, although, of course, the noncellulose portion of the ruminant dung can also be utilized.

Adopting the terms coined by Malan, Halffter and Matthews (1971) referred to two types of feeding in Scarabaeidae as the "primitive cellulosobiontic" type and the "derived coprobiontic" type. Judging by the intestinal structure, the former is represented by most larvae and the latter by all adult beetles. Many intermediate stages between the two types can be expected to occur, manifested by a progressive reduction in the intestinal diameter. The primitive digestion of cellulose by most Scarabaeidae larvae is associated with the retention of generalized lamellicorn morphological characteristics, such as the structure of the mouthparts (Fig. 2.1) and of the intestine, which have disappeared in the adult. But the primitive structures of the intestine have also disappeared in some specialized larvae which, like the adults, feed directly on the high-quality component of their food material.

Not all dung beetles feed on herbivore dung, which nonetheless is the main or only source of food for most species. Some species preferentially feed on nonherbivore dung, which is rich in nitrogen, and the search for proteins probably explains the switch in some species to partial or complete necrophagy. The adults of some necrophagous species have strong clypeal prongs and front legs enabling them to process carrion (Edmonds 1972; Cambefort 1984) and even vertebrate bones (Young 1980b). From necrophagy, a few species have evolved to predators (Halffter and Matthews 1966). At the same time, some coprophagous species have returned to saprophagy, which should not be confused with primitive saprophagy. In these derived plant eaters the nest is still generally provisioned with dung, as, for example, in some North American *Onthophagus*, the adults of which are often collected in rotten fruits (Halffter and Matthews 1966). Another peculiar case is myrmecophilous species that feed on the debris in ant nests, "mixtures of the consistency of wet sawdust containing primarily stems, bits of dry leaves and ant corpses" (Edmonds and Halffter 1972). Finally, a few dung beetles live on the hair of mammals, close to the anus, probably feeding on dung particles that stain the fur (on monkeys: Halffter and Matthews 1966; on wallabies: Matthews 1972; on sloths: Ratcliffe 1980). *Zonocopris*, a South American monotypic genus, lives on large snails and probably feeds on foam and organic particles glued to it.

Although dung beetle larvae are efficient in utilizing their food, the fact that the amount of food is strictly limited in the brood ball sets a severe constraint on what can be achieved during larval development. In particular, there is apparently little scope for building up body reserves for the adult life during larval development (Hanski 1987a). This has led to an often prolonged "maturation feeding," especially characteristic for dung beetles (Halffter and Matthews 1966). The coprobiontic feeding type in necrophagous and in some coprophagous larvae is, however, an exception. These larvae use, like the adult beetle, the richest possible diet, the microorganisms present in the food. Two examples are *Scarabaeus* (Halffter and Matthews 1971) and *Euoniticellus* (Halffter and Edmonds 1982). In the latter genus females can lay eggs soon after eclosion (five days: Halffter and Edmonds 1982). Thus the highly evolved larval biology allows a fast larval development and a high population growth rate.

In summary, the evolution of coprophagy from saprophagy entails a shift from the use of generally abundant but scattered and low-quality resource to the use of often scarcer but patchy (concentrated) and high-quality resource. It is not surprising that the nutritional shift has been more widespread and complete in the adult beetles than in their larvae: the great mobility of the adults enables them to locate the ephemeral food patches and to use the highest quality food in these patches. The larvae cannot generally be so selective, because the highest-quality food is also the most ephemeral component in dung pats and would probably be too unreliable a food source for the relatively immobile larvae. A second-best solution to this nutritional problem is larval development in the protected nest, where low-quality food may be slowly but efficiently utilized with the aid of microorganisms, almost without any threat from competitors and natural enemies. (The evolution of the nesting behavior in dung beetles will be placed in a broader context in the next chapter.)

2.5. SUMMARY

The true dung beetles comprise three families in the superfamily Scarabae-oidea, section Laparosticti: Scarabaeidae, Geotrupidae, and Aphodiidae. The extant primitive relatives of true dung beetles feed on humus. Both larvae and adults possess primitive, biting mandibles suited to process coarse-grained material, and the larvae live freely in their habitat. Larval mortality and adult fecundity are probably relatively high in these species. In contrast, the true dung beetles in the family Scarabaeidae feed on dung, and the adults possess highly evolved mandibles, allowing them to retain only the most nutritious part of the food, which their digestive tract is specialized to use. The larvae do not possess similar mouthparts but they spend all their life in separate, mostly individual units of the resource, the brood balls, located in a nest constructed by the female or by a pair of beetles. The nest provides protection to the offspring and facilitates efficient use of the food resources. The investments of time and energy by the parents are high and fecundity is low to extremely low. Intermediate cases exist between primitive saprophages and evolved coprophages, one example being taxa of true dung beetles feeding on fine-grained humus resembling dung. This chapter has examined in turn the probable evolutionary shifts from the diet and behavior of primitive saprophages to the highly evolved biology of coprophages: change from the use of coarse humus to the use of fine-grained humus; evolution of nesting behavior; and change from saprophagy to coprophagy, where the ultimate stage is represented by species with larvae as well as adults using the richest possible diet, the microorganisms present in the food.

Appendix: Dung Beetle Taxonomy

We have chosen to use in this book a classification of Scarabaeoidea Laparosticti based on Balthasar (1963) with minor modifications. The relevant families and subfamilies are listed in Table 2.1. Compared with the classification of Janssens (1949), used by Halffter and Edmonds (1982), Balthasar's classification has the advantage that the main groups of true dung beetles have their own families, namely, Aphodiidae, Geotrupidae, and Scarabaeidae, which recognize the specificity of each of them. Other authors who have used this classification include Ferreira (1972; summary of afrotropical Scarabaeidae in Balthasar's sense), Baraud (1977, 1985; West European and North African faunas), and Dellacasa (1988; world catalog of Aphodiidae).

As far as Scarabaeidae (in Balthasar's sense) is concerned, this classification has two advantages. First, the family is divided into two equivalent taxa, which correspond to the biological groups of rollers (subfamily Scarabaeinae) and tunnelers (subfamily Coprinae; Table 2.3). This is logical in the phylogenetic frame: if we consider that the two biological groups of true dung beetles are also two sister groups in the Hennigian sense, they must be given an equivalent rank. Therefore, the category of subfamily instead of tribe is used for the rollers as well as for the tunnelers. It would not be satisfactory to cluster all the rollers within the "tribe" "Scarabaeini," as the "subtribes" "Can-

TABLE 2.3
The division of the family Scarabaeidae s.str. into subfamilies and tribes (in alphabetical order).

Subfamily	Tribe
Coprinae	Coprini
	Dichotomiini
	Oniticellini
	Onitini
	Onthophagini
	Phanaeini*
Scarabaeinae	Canthonini
	Eucraniini
	Eurysternini
	Gymnopleurini
	Scarabaeini
	Sisyphini

* Considered a subtribe of Onitini by Zunino (1985).

thonina,'' ''Gymnopleurina,'' ''Scarabaeina,''and ''Sisyphina'' are at least as different from one another as are, for example, the Oniticellini and Onthophagini. It seems more appropriate to give all of them an equal tribal rank: Canthonini, Gymnopleurini, Scarabaeini, Sisyphini, and so on.

Finally, the classification adopted here allows the use of the subtribal rank within respective tribes. For example, Matthews (1974), in his revision of Australian rollers, treated the taxon as ''subtribe Canthonina.'' But he recognized two groups in this subtribe, groups to which he could give neither a taxonomic rank nor a formal name. Within Balthasar's system, his subtribe is raised to tribal rank (Canthonini), and it is now possible to divide it formally into two subtribes, Canthonina and Mentophilina.

CHAPTER 3

Dung Beetle Population Biology

Yves Cambefort and Ilkka Hanski

THE SINGLE most particular biological feature of dung beetles is their breeding behavior, often involving the formation of bisexual pairs of beetles for shorter or longer periods, construction of one or more nests in the soil, and time-consuming if not otherwise costly parental care. Although not all dung beetles make nests, and although some of the above characteristics can be found in other insects, for example in Dermaptera (Caussanel 1984) and in other beetles, such as Silphidae (Halffter, Anduaga, and Huerta 1984) and Passalidae (Reyes-Castillo and Halffter 1984), the sheer range of variation in and the degree of perfection of nesting is greatest in the dung beetles.

The nesting behavior of one dung beetle, *Scarabaeus sacer*, greatly impressed the ancient Egyptians, perhaps because this beetle united three elements of significance to the Egyptians (Cambefort 1987b): the sun, the soil, and the cattle. By a remarkable coincidence, this large beetle, which was the most obvious insect attracted to cattle dung, reminds one of the sun in its morphology (the head is similar to the rising sun; Fig. 3.1) and in its behavior (a dung ball is shaped early in the morning, and it is rolled across the plain, like the sun travels across the sky). The beetle's development in the underground nest chamber, culminating in the appearance of the new scarab from the nymph, also resembles the daily resurrection of the sun, a promise of new life to the Egyptians (Fig. 3.1). Cambefort (1987b) suggests that the mummy and the mummylike god Osiris were inspired by the beetle's pupa (the nymph). And what else were the pyramids but idealized dung pats!

From the beetles' perspective, the crucial advantage of nesting is protection: against competitors, against predators, and against unfavorable climatic conditions. Before we go into any further detail, we want to remind the reader about our classification of dung beetles into three behavioral or functional groups: the dwellers, the tunnelers, and the rollers. Aphodiinae, the relatively small and typically temperate dung beetles, comprise the bulk of dwellers: they eat their way through the dung, and most species deposit their eggs in dung pats without constructing any kind of nest or chamber. Geotrupinae and many tribes of Scarabaeidae are tunnelers: they dig a more or less vertical tunnel below the dung pat and transport dung into the bottom of the burrow; this resource may be used either for adult feeding or for breeding. Finally, many Scarabaeidae have evolved the ultimate dung beetle skill of making a

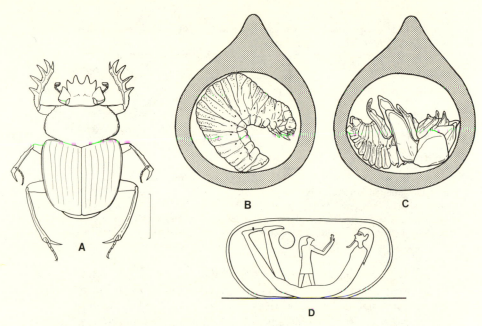

Fig. 3.1. The sacred scarab, *Scarabaeus sacer*. Adult (A; scale: 1 cm); the larva (B) and the pupa (C) inside the brood ball; and a drawing from the period of Ramesses VI, depicting Osiris, Horus, and the sun (D). Cambefort's (1987b) interpretation: Osiris (the god of the earth and of the dead) is the scarab larva (note the body form), from which his son Horus (the god of the glorious sun, but also the new scarab) emerges, with arms bent to the shape of the scarab's forelegs. Osiris and Horus are inside the ''brood ball.''

ball of dung, a transportable resource unit, and rolling it for a shorter or longer distance before burying it into a suitable spot. These are the rollers.

Table 3.1 summarizes the potential for competition in dwellers, tunnelers, and rollers. Competition may occur and often does occur for both food and space (in the dung pat) in the adults and larvae of dwellers. In the species that dig an underground burrow, larval competition is largely removed, in the sense that the food that the adult beetles provide for each larva remains uncontested by other larvae, with the exception of parasitic dung beetles (below). Competition is here restricted to the adults. In the tunnelers, competition occurs both for the food and for the space in the soil beneath the dropping (Fig. 3.2). Competition for food cannot be avoided if there are many beetles; but competition for space has been resolved by the rollers, which are able to locate an uncrowded place for their balls, which can be used either for feeding or breeding. Some adult tunnelers and some rollers feed directly in dung pats but many others feed only on their relocated dung reserves.

TABLE 3.1

Potential for competition for food and space in the adults and larvae of dung beetles belonging to the three main functional groups.

Functional Group	Adults		Larvae	
	Food	Space	Food	Space
Dwellers*	Yes	Yes	Yes	Yes
Tunnelers	Yes	Yes	—	—
Rollers	Yes	—	—	—

* The largest group of dwellers consists of dung-breeding *Aphodius*, which is the numerically dominant group in temperate dung beetle assemblages. In two other groups of dwellers competition among larvae is much reduced or is nonexistent for two reasons: (1) some *Aphodius* have saprophagous larvae that feed on decaying plant material in the soil, where space and food are less limiting than in dung pats; (2) *Oniticellus* (Scarabaeidae) construct proper nests in dung pats, which much reduces the chances of larval competition (as in tunnelers).

Finally, the kleptoparasites, "parasitic" dung beetles, do not relocate dung by themselves but they use a part of the reserves buried by the tunnelers and the rollers. They pose a competitive threat to the host species, but their significance is believed to be generally small. One might assume from what was said above and summarized in Table 3.1 that competition is generally severest in the dwellers and weakest in the rollers, but in fact exactly the opposite is

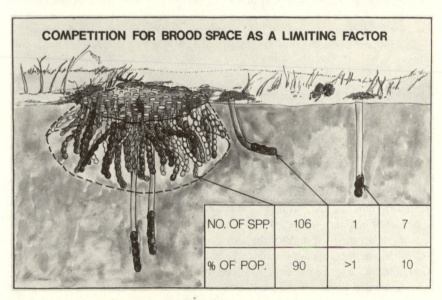

COMPETITION FOR BROOD SPACE AS A LIMITING FACTOR

NO. OF SPP.	106	1	7
% OF POP.	90	>1	10

Fig. 3.2. Competition for space in the soil (from Bernon 1981).

true. We shall discuss and review competition in dung beetles in detail in Chapter 17.

In the next sections we present more details about the biology of the four groups of beetles—the dwellers, the tunnelers, the rollers, and the kleptoparasites—with an emphasis on their breeding biology. Plates 3.1 to 3.11 give some representative illustrations.

3.1. THE DWELLERS: APHODIUS

The vast majority of the coprophagous species in the family Aphodiidae belong to the subfamily Aphodiinae and to the single genus *Aphodius* (nearly 1,650 described species). Most of these species are relatively small, less than 10 mm in length. They are the characteristic dung beetles of north temperate regions (Chapters 4 and 5), though they are present in substantial numbers in subtropical and tropical regions as well.

The entire egg, larval, and pupal development of *Aphodius* typically takes place in dung pats. Some large species in the subgenus *Colobopterus* (and related subgenera) are exceptional and construct primitive but true nests in the soil (Bernon 1981; Rojewski 1983). Although most species are small, they prefer to use large droppings, especially bovine dung, where they interact with a large range of insects apart from conspecifics and other *Aphodius* (Chapter 1). The adults of most species occur in fresh to relatively fresh droppings, while the slowly developing larvae are easiest to find in droppings that are a couple of weeks old, when most insects have already left the dropping.

Aphodius females have two ovaries with a few ovarioles each (Fig. 3.3), and their lifetime fecundity is relatively high, typically of the order of one hundred eggs or more, though some large species have lower fecundity (Table 3.2). Most species are believed to lay their eggs singly, though they may lay several eggs in the same pat, but some species lay small clutches of four to six eggs, for example, *Aphodius rufipes* (Holter 1979). Most temperate species have one generation per year. Hibernation may occur at the egg, larval, pupal, or adult stages, generating an orderly seasonal succession of spring, early summer, late summer, and autumn species (Chapter 5). As a measure of phenological flexibility, *Aphodius constans* occurs only in winter in the Mediterranean region, taking advantage of the season when the competitively superior Scarabaeidae dung beetles are inactive (Lumaret 1975; Chapter 6).

One small genus of tunnelers, *Oniticellus*, with six species, is functionally really a dweller, in the sense that it constructs its nests in the dung pat and therefore may suffer some competition at the larval stage (Table 3.1). It will become apparent in several chapters in this book that *Oniticellus*, like *Aphodius*, are really common only in situations when the competitively superior tunnelers and rollers are, for some reason, uncommon.

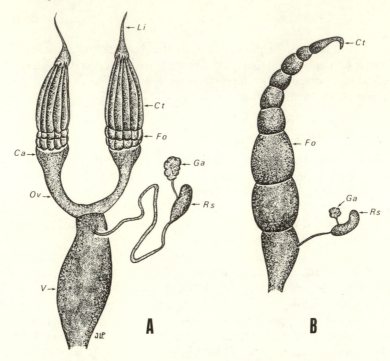

Fig. 3.3. Ovaries of *Aphodius* (A; Aphodiidae: Aphodiinae) and *Onthophagus* (B; Scarabaeidae: Coprinae). Abbreviations: Ca, calyx; Ct, terminal chamber; Fo, follicle; Ga, accessory gland; Li, ligament; Ov, oviduct; Rs, receptaculum seminis; and V, vagina (from Lumaret 1980b).

3.2. THE NESTERS: SCARABAEIDAE AND GEOTRUPINAE

The life cycle of the nesters (i.e., the tunnelers and the rollers) commences with the adult beetle emerging from the subterranean nest, usually in the rainy season. On emergence, the beetles are very frail and light, and they often must feed for several weeks, or even for some months. This period has been called the "maturation feeding period," from the German *Reifungsfrassperiode* (Halffter and Matthews 1966). Maturation feeding may be viewed as a cost of nesting: each brood ball in the protected nest contains only a limited amount of food for the larva, sufficient to complete the larval development but perhaps not enough to build up reserves for adult life (Hanski 1987a).

The food is located primarily by its odor, though even visual and even auditive cues may play a role in some species using fresh monkey droppings (Howden and Young 1981). Many dung beetles do not generally travel long distances to reach their food, and in the savannas, for example, they may wait

Plate 3.1. *Gymnopleurus*.

Plate 3.2. Pair of *Scarabaeus typhon* rolling a ball (Corsica).

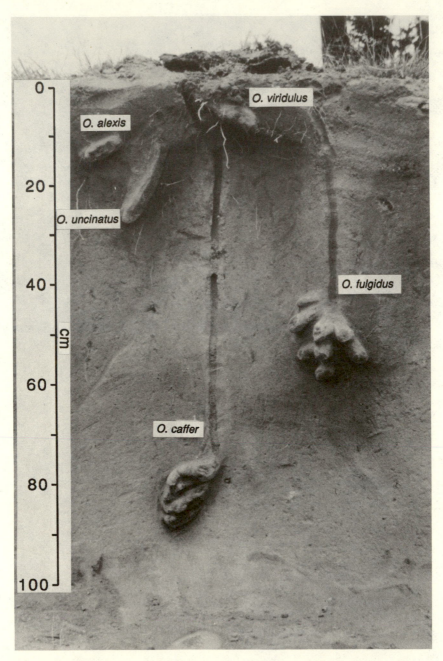

Plate 3.3. Vertical section under a dung pat to which pairs of *Onitis viridulus*, *O. folgidus*, *O. alexis*, *O. caffer*, and *O. uncinatus* had been added two weeks earlier.

Plate 3.4. Tunnels of *Digitonthophagus gazella* (West Africa).

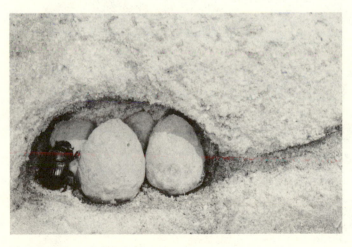

Plate 3.5. Nest of *Copris hispanus*.

Plate 3.6. Six completed brood balls in the nest of
Copris acutidens (Japan).

Plate 3.7. An egg in the brood ball of *C. acutidens*.

Plate 3.8. A final instar larva of *C. acutidens*.

Plate 3.9. A male pupa of *C. acutidens*.

Plate 3.10. *Kheper platynotus* and a ball of elephant dung (East Africa).

Plate 3.11. *Kheper aegyptiorum* rolling a ball of eland dung (East Africa).

TABLE 3.2
Examples illustrating the range of variation in some life-history parameters in dung beetles.

Species	Lifetime Fecundity	Eggs per Nest	Adult Longevity (months)	Reference
Dwellers				
Aphodius elegans	139	No nest	1	Yasuda (1987)
Aphodius haroldianus	12	No nest	2	Yasuda (1987)
Tunnelers				
Copris diversus	30?	10		Tyndale-Biscoe (1983)
Copris interioris	18?	6	12+	Cambefort (1984)
Copris lunaris	10	5	12+	Klemperer (1982a)
Digitonthophagus gazella	90+	44		Blume & Aga (1975)
Euoniticellus intermedius	120	8	1.5	Tyndale-Biscoe (1978)
Heliocopris dilloni	5+	2–7	6	Kingston & Coe (1977)
Liatongus phanaeoides	19		4	Yasuda (1986)
Oniticellus egregius	65	7	4–5	Davis (1977, 1989)
Oniticellus formosus	60–70	20	5–9	Davis (1977, 1989)
Oniticellus planatus	50–150	10	7–13	Davis (1977, 1989)
Onitis alexis	130	24	2	Tyndale-Biscoe (1985)
Onitis belial		20		Klemperer (1982c)
Onitis caffer		12+	3	Edwards (1986b)
Onthophagus binodis		35	2–5	Ridsdill-Smith et al. (1982)
Onthophagus lenzii	8–16		4	Yasuda (1986)
Rollers				
Circellium bacchus	2+	1		Tribe (1976)
Kheper lamarckii	3+	1–3		Edwards (1988a)
Kheper nigroaeneus	3+	1	24	Edwards (1988a)
Kheper platynotus	6+	1–4	24	Sato & Imamori (1987)
Neosisyphus calcaratus	41	1	4–8	Paschalidis (1974)
Neosisyphus fortuitus	55	1	5–9	Paschalidis (1974)
Neosisyphus infuscatus	56	1	4–6	Paschalidis (1974)
Neosisyphus mirabilis	47	1	5–9	Paschalidis (1974)
Neosisyphus rubrus	36	1	4–6	Paschalidis (1974)
Neosisyphus spinipes	44	1	4–7	Paschalidis (1974)
Sisyphus seminulum	26	1	4–5	Paschalidis (1974)
Sisyphus sordidus	10–20	1	12–24	Paschalidis (1974)

while shallowly buried in the soil. In tropical forests, some species sit on leaves not far above the ground, apparently waiting for the odor trails to reach them (Howden and Nealis 1978; Young 1984; Chapters 10 to 12). Small species perch at a lower level than large ones, and the largest species (>10 mm) tend to fly rather than perch (Howden and Nealis 1978). But it is in the grasslands that the largest species, especially, tend to search for dung by cruise flight. Once located, the food source is generally approached by flight. Not surprisingly, flightless dung beetles are rare and mostly confined to special areas, such as mountain tops (Hanski 1983; Chapter 12).

After reaching the food source, some beetles may commence to feed on dung directly (e.g., *Gymnopleurus*), but in most cases a part of the dropping is first buried on the spot (tunnelers) or some distance away from the food source (rollers). During the maturation feeding, high nitrogen content of dung is probably more important than its carbohydrate content, which explains why immature adults are often attracted to the nitrogen-rich omnivore excrements even though they use the larger, carbohydrate-rich droppings of herbivorous mammals for breeding (Cambefort 1984). Cambefort (1984) found that the total nitrogen content decreased from 0.8% (dry weight) in uneaten, buried cattle dung to 0.3% in the feces of beetles. The necrophagous feeding habit of many tropical forest species (Chapters 10 to 12) may have the same explanation. Both males and females require nitrogen to build up their muscles, but the females need, in addition, nitrogen to maturate the large eggs, up to 5% of female body weight each. In some species the larval diet is enriched by microorganisms, and the larvae may contribute to the adult body reserves, shortening the maturation feeding period (Halffter and Edmonds 1982; Chapter 2). At the end of the feeding period, the individual body weight has substantially increased, for example by a factor of 1.7 in the small roller *Sisyphus* (Cambefort 1984). At the same time, the female ovary reaches maturity.

Nesting

The key behavioral trait of dung beetles is their ability to make nests, in which the offspring find food and shelter. A trend toward nesting is barely discernible in Aphodiinae, while nesting is universal in Geotrupinae and reaches the highest level of sophistication in Scarabaeidae. Most of our knowledge about nesting in Scarabaeidae has been reviewed and synthesized by Halffter (1977) and Halffter and Edmonds (1981, 1982). We will here give a short summary and add a few new facts.

Scarabaeidae dung beetles establish a pair bond for a short or long period. The two sexes usually meet at the dung pat or in its close vicinity. It is not well established whether long-distance sex pheromones are used or are important in dung beetles generally (Tribe 1975, 1976; Burger et al. 1983), but at close distances chemical communication seems important. Species and sex recognition probably does not necessitate a direct contact and is achieved with volatile short-range pheromones. Individual recognition, on the other hand, is probably based on physical contact; for example, in *Gymnopleurus*, members of a pair have been observed to touch each other's mouthparts repeatedly (Cambefort 1984). Recent studies have demonstrated interspecific and intersexual differences in the numbers and locations of tegumentary glands, which are likely to secrete pheromones. Some glands are situated in the mouthparts, for example, in the mandibles (Pluot-Sigwalt 1984, 1986, 1988; see also Houston 1986).

One nesting sequence consists of locating the food source, establishing the pair bond, constructing the nest, provisioning the nest with dung, forming the brood ball (of dung), ovipositing, and often, though not always, continuously protecting the brood ball or balls in the nest during larval development. The nesting sequence may be repeated as often as every three days in, for example, *Sisyphus*, in which up to five generations may develop per year (Paschalidis 1974). In *Sisyphus*, females copulate before each nesting sequence (Paschalidis 1974); but in the large tunnelers such as *Copris* and *Phanaeus*, females apparently mate only once in their lifetime (Halffter and Lopez 1977; Klemperer 1982a, 1983b). These species, as well as some large rollers (e.g., *Circellium* and *Kheper*), usually have only one nesting sequence per year, the nest containing as little as one offspring (Tribe 1976; Huerta et al. 1981; Klemperer 1982a; Edwards 1984, 1988a; Sato and Imamori 1986a,b, 1987, 1988). Not surprisingly, these large species may live up to three years (Klemperer 1982a, 1982b). All Scarabaeidae females have only one ovary with a single ovariole (Pluot 1979; Fig. 3.3), and even in the smaller species lifetime fecundity tends to be low (Halffter and Edmonds 1982; Kirk and Feehan 1984; Yasuda 1986). Fecundity is moderately low and the gonads are partially reduced also in Geotrupinae (Halffter et al. 1985).

Table 3.2 illustrates the range of variation in some life-history parameters in dung beetles (see also Table 8.7). A relationship between physiology and maternal care has been demonstrated: as long as the female cares for her offspring, the ovariole development is inhibited and the ovocytes that were present in the vitellarium are reabsorbed (*Copris*: Huerta et al. 1981; Anduaga et al. 1987; *Oniticellus*: Klemperer 1983c; *Kheper*: Sato and Imamori 1987). Such a relationship seems to exist also in the male when he is involved in parental care, for example in the carrion-feeding *Canthon cyanellus*: during the period of parental care, both ovarian and testicular follicles degenerate (Benitez and Martinez 1982; Halffter, Halffter, and Huerta 1984; Martinez and Caussanel 1984).

3.3. The Rollers

The subfamily Scarabaeinae comprises the rollers (see the Index for a list of the genera). In the species whose breeding biology is known, making of a brood ball is initiated by one individual, the "active partner" (Halffter and Matthews 1966). Either the brood ball that is being made or the completed ball may act as a sexual display for the other member of the pair, the "passive partner." The active partner is most often the male in the tribes Scarabaeini and Canthonini, but it is the female in the Gymnopleurini and Sisyphini (e.g., Paschalidis 1974). In some cases (e.g., *Gymnopleurus*, *Sisyphus*), the passive partner is immediately accepted by the active one, but the recognition of a possible mate is not self-evident. In *Canthon pilularius*, Matthews (1963) describes cases where the male first and repeatedly rebuffs the female before finally accepting her; in *Kheper*, the female sometimes rebuffs the male (Sato and Imamori 1987). In some species, the male offers a food ball (''nuptial ball'') to the female (Halffter and Matthews 1966; Halffter and Edmonds 1982; Sato and Imamori 1986a, 1986b). Mate choice does not seem to depend on size (Sato and Imamori 1987), though Cambefort (1984) reports an almost significant correlation between the sizes of the two partners. Making the ball is generally easier for small than for large rollers, and, among the latter, it is most difficult for the nocturnal species, which are often faced with severe competition from large nocturnal tunnelers (below). In some large nocturnal rollers, success in ball making is related to thoracic temperature (endothermy) and size (Bartholomew and Heinrich 1978; Heinrich and Bartholomew 1979a, b). Making a ball can take only a few minutes in some nocturnal rollers (Kingston 1977) but up to one hour in diurnal *Kheper* (Sato and Imamori 1986b). The weight of the ball can be up to fifty times or more the weight of the beetle (Chapter 17).

The brood ball is often rolled together by the two partners, but in some cases the female climbs onto the ball and is rolled off with it by the male. During the rolling process, other beetles often attempt to steal the ball. The attacker is most often a male. Fights and combat are common between members of the same species, but combat also occurs between different species, belonging even to different genera (fighting about balls is discussed in Chapter 17). After a combat, the separated members of a pair join together, if possible, touching their mouthparts to recognize each other, probably through chemical communication. Rollers may also protect their balls against other insects. The male of *Canthon cyanellus* secretes a chemical on the ball that repels calliphorid flies (Bellés and Favila 1984). In most species, sexual cooperation lasts until copulation in the burrow, after which the male leaves the female and the brood ball. Balls that are abandoned can be used by other species, for example *Canthon* balls are used by *Onthophagus* (Young 1969) or even by *Trox* (Ratcliffe 1983).

Several patterns of nesting have been described in the rollers (Halffter and Edmonds 1982; Edwards 1984, 1988a; Sato and Imamori 1986a,b).

Type 1. Copulation occurs at the food source, and the female alone makes the ball, rolls it away, and buries it. This is the only known type among rollers in which there is no sexual cooperation. This type has been described for *Megathoposoma*.

Type 2. The male assumes the more active role in ball rolling, though the female may be more active in ball shaping. The incipient brood ball is buried a few centimeters beneath the soil surface. Copulation occurs after burial. After copulation the sexes separate, and the female reshapes the ball into one or two "pears." Preparing more than one pear out of the rolled ball is considered to be an evolved trait by Klemperer (1983a), since it saves the time and energy needed to make more than one nest, but it also necessitates a larger amount of dung per ball, which may become difficult to move (Chapter 17). An egg is laid in the narrow upper end of each pear. After oviposition, the female abandons the nest to make another one. This is the most common type among rollers and has been described in Scarabaeini, Canthonini, Gymnopleurini, and Sisyphini. When the nesting sequence is repeated several times during the breeding season, this type may be associated with relatively high fecundity for rollers (Table 3.2). At the other extreme, only one nesting sequence occurs per year in the very large, flightless and relict canthonine *Circellium bacchus*, and the nest contains only one brood ball (Tribe 1976). Some *Neosisyphus* species do not dig a nest at all, but the brood ball is simply left at the soil surface or it is attached to grass stems or twigs (Arrow 1931; Paschalidis 1974). This may save energy but exposes the ball to desiccation and predation.

Type 3. The female makes one to four pyriform brood balls and remains in the nest, caring for them until her offspring emerge. In this remarkable case, described for the genus *Kheper* (Tribe 1976; Edwards 1984, 1988a; Sato and Imamori 1986a,b, 1987, 1988), fecundity can be as low as only one offspring per female per breeding season (often a year).

Type 4. After having rolled and buried one ball together with the female, the male stays with her and the two partners bring up to five more balls to the nest. The female, often accompanied by the male, remains in the nest, caring for the balls, until the emergence of progeny. Each female prepares several nests in her lifetime, but fecundity is nonetheless low. This nesting type has been described for a few, mostly necrophagous, species of *Canthon*.

Type 5. The rolling behavior has been lost. Two contrasting explanations for this derived character may be suggested: (1) high availability of dung and hence little competition for it (Eurysternini), and (2) low availability of dung suitable for ball making but presence of pellets that may be rolled without ball making (a species of *Canthon*; Halffter and Halffter 1989). In the Eurysternini, nesting is accompanied by complex sexual and maternal behavior (Halffter et al. 1980).

3.4. THE TUNNELERS

Apart from the rare cases of very primitive nests in Aphodiinae, this behavior is confined to Geotrupinae and Coprinae of Scarabaeidae (see the Index for a list of genera).

Before describing the nesting behaviors, we introduce a useful division of the species into two groups: small and large tunnelers. Field observations

(Cambefort 1984) have shown that the large tunnelers (more than 13 mm in length), represented especially by the tribes Coprini, Phanaeini, Onitini, and some Dichotomiini, are mainly nocturnal (many Phanaeini are diurnal). Small tunnelers are all (Oniticellini) or mostly (Dichotomiini, Onthophagini) diurnal. But the main difference is that small tunnelers have a less reduced fecundity than large tunnelers. The small species make shallow nests with numerous brood masses; they do not provide maternal care and they probably live for only one season. In contrast, large tunnelers make deep nests with fewer brood masses, cared for by the female in Coprini and in some Dichotomiini. Some Geotrupinae can also be included in this group. Large tunnelers usually live for more than one year. Fewer than the maximum possible number of eggs are laid in one season, which may increase the chances of survival to the following year. Survival of the adult female must be higher in the nest than outside, which may have contributed to the evolution of maternal care (Klemperer 1983a).

The role of the male is always secondary in tunnelers (with the exception of some Geotrupidae: Howden 1955a). The initiative in nesting is taken by the female, which is responsible for most of the work in excavating the nest and in forming the brood masses or balls. In addition, she may care for the brood balls until the emergence of the offspring. Like in the rollers, the two sexes encounter at the dung pat. Mating may occur rapidly, in which case the burrow is dug by the pair, or the female begins to dig and the burrow may act as a sexual display. In some North American *Geotrupes*, mating takes place in burrows dug by the male, with the female entering the burrow later (Howden 1955a). Each burrow is probably marked with a pheromone to allow its recognition by its owner and by other beetles. Marking has a repellent effect on other beetles, and the nests under a dung pat become well spaced out.

Various types of nests and their variants have been described for the tunnelers. The Geotrupinae nests are the most primitive, consisting of simple burrow filled with dung, the "sausages" containing usually one egg each (Howden 1955a; Klemperer 1978, 1979, 1984; Brussaard 1983, 1985a,b; Halffter et al. 1985; Klemperer and Lumaret 1985). In Coprinae the nests can be complicated (Masumoto 1973; Cambefort 1982a; Halffter and Edmonds 1982; Klemperer 1981, 1982a,b, 1983a, 1983b; Cambefort and Lumaret 1984). Detailed classifications have been produced for the many types of nests that are supposedly characteristic to particular taxa. However, it has also been demonstrated that a certain amount of variation can occur within species, for example, according to the texture and humidity of the soil (Walter 1980; Rougon and Rougon 1982b; Rougon 1987). Therefore, it is difficult to be completely certain about the assignment of a particular nest type to some particular taxon. We present here a tentative scheme of the nesting patterns in Coprinae, but we do not ascribe any phylogenetic significance to these "types" (Fig. 3.4).

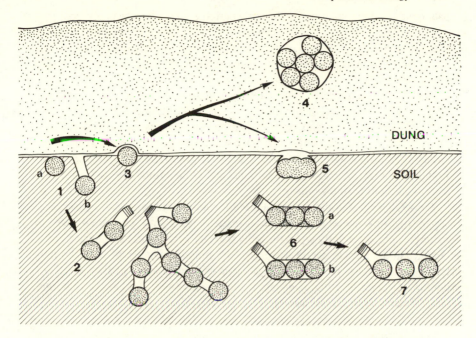

Fig. 3.4. Seven nest types (1–7) and their variations among the tunnelers.

Type 1. There is a single, more or less shallow brood mass (single nest). Females nest repeatedly and there is no bisexual cooperation or maternal care. Fecundity is relatively high. Examples are found in Dichotomiini, Onitini, Oniticellini, and Onthophagini.

Type 2. The nest contains several to many cylindrical or spherical brood masses (compound nest). Sexual cooperation may occur but there is no maternal care and fecundity is relatively high. Examples can be found in Dichotomiini, Onitini, Oniticellini, and Onthophagini. This is probably the most frequent type since it is represented by most Onthophagini, the tribe with most species.

Type 3. The shallow brood mass is surrounded by soil in its lower half and by dung in its upper half. Fecundity is moderately high. One example in Oniticellini (Davis 1977).

Type 4. Nest is constructed in the dung pat ("endocoprid" nesting). The preceding and the following types involve preliminary digging of a burrow in which the secured dung is at least partially buried. In the present type, the tunnel and the brood chamber are dug into the pat. The typical example is *Oniticellus* (Bornemissza 1969; Davis 1977; Cambefort 1982a; Rougon and Rougon 1982b; Cambefort and Lumaret 1984). Nesting takes place in large pats (cattle and elephant dung) during a period (dry season) or in a place (forest) where competition is not severe (Chapter 17). Male cooperation has not been observed. Fecundity is moderate to low. There is moderate to extensive maternal care with correlated physiological changes in the beetle (Davis 1977; Klemperer 1983c).

Type 5. A few more or less coalescent, shallow brood masses. Maternal care and low fecundity. One example in Oniticellini (Cambefort 1981).

Type 6. The nest contains several spherical brood masses arranged in a single or branched tunnel, without or with separation between them (Fig. 3.4). Generally no maternal care and fecundity is high to moderate. Examples are found in Coprini, Dichotomiini, Onitini, and Oniticellini. The two subtypes may occur in the same genus (e.g., *Onitis*).

Type 7. The nest contains only a few (down to one) brood balls, physically separated from one another. There is sexual cooperation, and often there is maternal care, especially when the balls are few, with correlated physiological changes in the female beetle. Examples are found in Phanaeini (with no maternal care, even if the female or the pair of beetles stays in the nest), Dichotomiini (with or without maternal care), and Coprini (extensive maternal care). Fecundity is usually low.

3.5. KLEPTOPARASITES

Some species of dung beetles are not dwellers, nor do they make any effort to dig and provision a nest: they simply use a part of the food resources secured by other beetles. The term kleptoparasite was coined by Paulian (1943) for these species. Examples are found both in Aphodiinae and Scarabaeidae (Rougon 1987 and references therein). Kleptoparasitic Aphodiinae parasitize Geotrupinae (Howden 1955b; Brussaard 1987) but also Scarabaeidae, especially in dry environments, for example in the Sahel (D. Rougon and C. Rougon 1980; Rougon 1987; Chapter 13). All kleptoparasitic Scarabaeidae belong to the tunnelers, Coprinae, but they may be parasites of either rollers or tunnelers. Most kleptoparasites are specialized taxa of generic rank, for example, *Pedaria* (Cambefort 1982c), *Cleptocaccobius* (Cambefort 1984), *Onthophagus* subgen. *Hyalonthophagus* (Palestrini, pers. comm.), and another, still undescribed subgenus of *Onthophagus* (Hammond 1976; Zunino 1981b). The smaller species (e.g., *Cleptocaccobius*) are parasites of rollers. They are attracted to the balls being rolled, which they follow, flying very close to the ground in a zigzag manner. Eventually they reach the ball, into which they immediately plunge. Cambefort (1984) observed up to 37 individuals of six species in one ball of a large *Neosisyphus*. The larger kleptoparasites use the reserves of large tunnelers. Cambefort (1984) counted 109 Scarabaeidae of five species and 15 Aphodiinae of two species in one *Heliocopris* nest (Chapter 9).

3.6. LARVAL DEVELOPMENT

Scarabaeidae larvae develop in the brood balls prepared by the female or by a pair of beetles. It has been observed that the female defecates on the brood ball during its preparation, probably to inoculate the microbial strains necessary for larval digestion. We have described in Chapter 2 how the larva repeatedly eats its own feces, thus enriching its intestinal microflora.

In some species with very low fecundity, the female or the pair stays with the brood and takes care of it. In these species, stridulation occurs not only in the adult beetles (Halffter and Edmonds 1982), but also, though very feebly, in larvae (*Cephalodesmius*: Paulian et al. 1984; *Copris*: Palestrini, pers. comm.; Geotrupinae: Palestrini and Zunino 1987). It is possible that the female and her offspring exchange auditive signals through the upper pole of the ball, which is only loosely packed. Stridulation in larvae sharing the same dung reserves could also be a territorial signal (Palestrini and Zunino 1987). Chemical communication is also possible (both maternal-larval and territorial), but we do not have any detailed information about this (Rougon 1987). Apart from communication among beetles, the tunnelers use stridulation when their nest is excavated, perhaps thereby disturbing insectivorous mammals (Masters 1980), for example, the ratel *Mellivora capensis* (Kingston and Coe 1977). During its development, the larva consumes its ball from inside. The feces are voided and partly re-eaten, but a part of the feces is accumulated in the terminal part of the intestine and can be used as a "stercoral cement" to repair any damages suffered by the ball (Halffter and Edmonds 1982; for Geotrupinae, see Klemperer 1978, 1984). The largest larvae may display active defensive behavior: *Heliocopris* larvae eagerly try to bite the intruder with their powerful mandibles (Cambefort, unpublished observation in the Ivory Coast).

At the end of its development, the larva usually constructs a pupation chamber using its excrements and thus voiding its intestine before pupation. The new adult will wait in this chamber until emergence, which sometimes cannot take place before the ball and the pupation chamber have been softened by increased soil moisture during the following rainy season.

3.7. SUMMARY

The nutritional shift from saprophagy to coprophagy in dung beetles has been associated with the evolution of highly sophisticated breeding behavior. An important driving force has been the often severe competition for dung, when the problem is to secure a part of the resource for the offspring and to create conditions that make efficient use of the resource. One part of the solution is larval development in individual, in most cases completely isolated, units, the brood balls, made of dung by the female. Brood balls are usually located in underground nests, providing further protection. Two main techniques are used to make and relocate brood balls. A nest may be dug in advance, then filled with dung, out of which the brood balls are made; or alternatively a ball is prepared first, then translocated by rolling and buried some distance away from the food source. These two techniques are characteristic of the two principal groups of true dung beetles that construct nests and prepare brood balls: the tunnelers (subfamily Coprinae) and the rollers (subfamily Scarabaeinae). The parental investment of time and energy in digging nests and making brood balls necessarily reduces potential fecundity. In some species minimally low fecundity is associated with maximal parental care during the entire period of larval development. In this chapter we have contrasted the rollers and the tunnelers and described their reproductive biology from an ecological point of view. Some primitive or deviant dung beetles do not make nests. These are the dwellers, which simply stay and breed in dung pats, and the kleptoparasites, which live as parasites in the nests prepared by other dung beetles.

CHAPTER 4

Biogeography and Evolution

Yves Cambefort

FOSSIL BEETLES are known from deposits formed some 250 million years BP. The earliest Scarabaeoidea Laparosticti, which includes the dung beetles, is *Aphodiites* from lower Lias (lower Jurassic) from Switzerland, 180 million years BP (Crowson 1981). *Aphodiites* was an "undifferentiated Laparosticti" with short grinding mandibles similar to the ancestral type described for beetles in general by Crowson (1981). The ancestral Scarabaeoidea were probably close to the Dascilloidea, a small group of mostly tropical floricolous beetles.

Fossil dung beetles are scarce (Paulian 1943; Balthasar 1963; Crowson 1981). The oldest fossils belong to the primitive families Geotrupidae (Jurassic) and Hybosoridae (lower Cretaceous). Eocene (50 million years BP) fossils include an *Aegialia* from North America, a *Saprosites* (Aphodiidae: Eupariinae) from London clay, and a *Bolboceras* from Central France. True dung beetles are not known before the Lower Oligocene (40 million years BP), with an *Aphodius* in Baltic amber and *Scarabaeus*, *Sisyphus*, and *Onthophagus* in French and Swiss deposits. The first fossil dung balls in a nest are also from this period, and they are similar to those of the extant genus *Megathopa*, from the Patagonian Deseadan (Frenguelli 1938; Halffter 1972). During the Miocene, many present genera are represented in fossil material.

Jeannel (1942) suggested that dinosaur troops were probably followed by dung beetles, and Halffter (1972) believes that dung beetles were well established and made nests during the dinosaur era, the Jurassic and Cretaceous periods. Unfortunately, the few fossils known from that time, almost all of which are Geotrupidae and Hybosoridae, do not throw any light on these suggestions. Only very few of the extant dung beetles search for reptile dung (Halffter and Matthews 1966; Young 1981; Chapter 12), the huge majority of species showing a clear preference for mammalian dung. During the early Cretaceous period, small mammals began to expand on a large scale. It is tempting to speculate that the dramatic increase in mammalian dung led to the shift from saprophagy toward coprophagy in the laparosticts (Chapter 2). True dung beetles (Aphodiinae, Scarabaeidae) may have thus appeared at the end of the Mesozoic. The oldest Scarabaeidae groups, Canthonini and Dichotomiini, probably diverged at that time, perhaps concurrently with the shift in their nutritional ecology (Chapter 2).

Climatic changes taking place in the Cenozoic may have played a role in dung beetle evolution. Forests regressed and were partly replaced by grass-lands; the first records of pollen grains of grasses date back from the Middle Eocene (van der Hammen 1983). Dung beetles, and especially the Scarabaei-dae, are today more numerous in open habitats than in forests (Halffter and Matthews 1966; Chapter 11). These climatic changes also coincide with the movements of the earth's crust, which led to the separation of South America from Africa 105 million years BP, isolation of Australia by the end of the Cretaceous period, isolation of South America from North America in the early Cenozoic (64 million years BP), repeated connections and disconnec-tions between Africa-Eurasia and Eurasia-North America, and finally to the reestablishment of the land connection between North America and South America at some time between 10 and 5 million years BP (Darlington 1957; Keast 1972).

Pangaea existed as a single continent up to the Triassic period, when it split into two blocks, Laurasia in the north and Gondwana in the south. Some taxa may date back from before the splitting. This is probably the case with Geo-trupidae, one of the oldest laparostict taxa (Crowson 1981), with two major subfamilies, Geotrupinae (including Taurocerastinae; Howden and Peck 1987; though see Zunino 1984a,c) and Bolboceratinae. The present nutritional ecol-ogies of these subfamilies are different: adult Geotrupinae are basically co-prophagous, with a few mycetophagous taxa, while Bolboceratinae are basi-cally mycetophagous. The geographical distributions of the subfamilies match the "microtypal" and "macrotypal" classes of Coleoptera (Murray 1870, cited in Crowson 1981). Murray suggested that smaller beetles are especially abundant in the cooler regions, while larger ones would dominate in the warmer regions. Crowson (1981) showed that the relationship between body size and climatic requirements is not so simple, but it remains true that tem-perate taxa tend to be microtypal species, while both microtypal and macro-typal species occur in warmer regions. Microtypal species are presumed to have broad climatic tolerances but are poor competitors, excluded by other species in the warmer zones. In Geotrupidae, the coprophagous Geotrupinae cannot match the competitive ability of subtropical and tropical Scarabaeidae, which may explain why this subfamily is restricted to temperate zones. The mycetophagous Bolboceratinae have been able to survive in the warmer bi-omes, but they may have been decimated by the glaciations from the Northern Hemisphere.

In Aphodiidae, Eupariinae is a basically southern and saprophagous group, paralleling in this respect Bolboceratinae in Geotrupidae. In contrast, Apho-diinae are mostly coprophagous and are most conspicuous in temperate bi-omes, paralleling the coprophagous Geotrupinae. Aphodiinae are firmly estab-lished also in tropical regions; but possibly because of fierce competition with the highly evolved coprophagous Scarabaeidae, Aphodiinae have not been able to diversify in the tropics to the extent they have done in the temperate zones of the northern hemisphere. We shall now turn to Scarabaeidae, the dung beetles proper.

4.1. OLD AND NEW FAUNAS: AN OVERVIEW

The family Scarabaeidae has a bipolar geographical distribution, with two "southern" tribes: Canthonini and Dichotomiini, and several "northern" ones, for example, Coprini and Onthophagini. Halffter (1972) assumes a Jurassic age for Scarabaeidae. As there do not seem to exist any Laurasian taxa, we suggest that Scarabaeidae have probably differentiated in Gondwana after the Pangaea split. The family may have then diverged into two sister groups, Canthonini and Dichotomiini, of which only the first one has retained a typical Gondwanian distribution until the present. These two taxa may have originally had saprophagous feeding habits in tropical forests (Chapter 2), where they possibly turned to the dung of primitive small mammals that were forest dwellers. Some members of these most primitive dung beetle tribes still use small mammal dung—for example, the small canthonine *Panelus* reported by Paulian (1945) from rat feces in Vietnam.

In the middle of the Cretaceous period marsupials appeared in North America and placentals in Eurasia, two sister groups probably separated by geographical isolation (Hoffstetter 1976). Both groups diversified rapidly and began to spread southward. It is possibly in the course of these migrations that they came in contact with Scarabaeidae, some of which became coprophages. At that point, dung beetles themselves began to expand and spread until the end of the Mesozoic.

At the beginning of the Tertiary, a dramatic ecological radiation took place, Africa being the main theater of radiation of placental mammals (Darlington 1957). Herbivorous mammals differentiated in step with vegetation differentiation. Forest and bush gradually and partly changed into grasslands from the middle Eocene onward (van der Hammen 1983). Together with the expanding grasslands, grazing mammals began to gain importance over the older browsers, and the older dung beetles, which presumably specialized in the dung of forest browsers, gave rise to species specializing in the dung of open habitat grazers. The older, southern dung beetles largely retained their old Gondwanian distribution, while the more modern taxa acquired a "Holarctic," basically northern, distribution.

During their evolution, mammal and dung beetle taxa have experienced successive cycles of expansion and contraction. In the expansion phases, mammals and dung beetles have migrated across the available regions; in the contraction phases, they have been restricted as "relicts" to small refugia, from which they expanded their ranges again in the next expansion phase (Jeannel 1942; Darlington 1957; Howden 1985). The distributions of living organisms are the results of all the geographical and evolutionary processes that have occurred since the differentiation of the taxa, but in many cases the distributions probably reflect the more recent events. In other words, although dung beetles may date back to the late Mesozoic and were well differentiated during the Cenozoic, their present status and biogeography mostly reflect events that took place during Miocene, Pliocene, and Pleistocene (Haffer 1979; van der Hammen 1983; Sinclair 1983). During these more recent periods, the climate has fluctuated, and especially the average temperature has changed several times from 2°C above to perhaps 5°C below the present-day values. It is not certain whether lowering of the average temperature decreases or increases aridity, but it is generally assumed that there have been several advances and retreats of forests and dry zones, with extensions and regressions of grasslands in between (Livingstone 1975). The process of adaptive radiation resulting from these fluctuations in vegetation types has been likened to an "evolutionary pump" (Morton 1972; Sinclair 1983): every time the habitat expanded and contracted, animals that were adapted to it experienced alternating phases of relaxing and intensifying selection pressures. New types of animals could evolve to fill the newly expanded habitats. During contraction, the populations would become fragmented and isolated. As habitats then expanded again, the various isolated populations would have met and, perhaps through reinforcement, diverged further into true species if they were already sufficiently differentiated. Studies on some beetle groups have shown that an isolation period of 5,000 years may be long enough for the development of genetically isolated species (Nagel 1986).

The end result of all these events is that, in a certain group of species within a given geographical area, taxa of different ages and origins coexist. I shall present in this chapter an overview of the distribution of dung beetles on different continents. Although the distinction between microtypal and macrotypal taxa is not practical, I find it useful to divide the taxa into "small" (less than 13 mm in body length) and "large" ones. Small species can be either diurnal or nocturnal, but most large species are nocturnal (Chapter 3). Small species have relatively high fecundity, but large species have relatively low to extremely low fecundity (Chapter 3). The smaller taxa seem to have been and still are better dispersers, with the result that they are much more widespread than the large taxa. A few large dung beetles, however, are conspicuously dominant in the regions where they occur—for example, *Dichotomius carolinus* in Central America and *Catharsius molossus* in tropical Asia.

Another important distinction is based on dung beetle foraging and nesting behavior, that is, the distinction between the rollers and the tunnelers (Chapter 3). The two classifications can be combined to give four groups of dung beetles: the small and large rollers, and the small and large tunnelers. Plates 4.1 to 4.8 summarize the geographical distributions in these four groups of species, taking further into account the relative ages of the taxa: the old (Gondwanian) taxa, intermediate taxa, and modern (Holarctic) taxa (which will be

Dung beetles of the world. The plates give the genera of Scarabaeidae in the six zoo-geographical regions, with the number of known species and their average size in mm. Endemic genera are marked with an asterisk (*). The genera have been divided into old, intermediate, and modern taxa, as explained in the text. The four groups of small (<13 mm) and large rollers and tunnelers have been treated separately. (Note that a genus has been placed in ''small'' or ''large'' species depending on the average size of the species; some species could have been placed in the other group. There is also some variation between the regions; for example, *Copris* is a small tunneler in the Nearctic region but a large tunneler in the Neotropical, Palearctic, Afrotropical, and Oriental regions.)

Plate 4.1. Small rollers.

NEARCTIC REGION

Old fauna
*Melanocanthon	4	8.0
Pseudocanthon	1	4.3
2	5	

Intermediate fauna
0

Modern fauna
0

AUSTRALIAN REGION

Old fauna
*Amphistomus	21	7.0
*Anonthobium	6	4.2
*Aptenocanthon	3	3.8
*Baloghonthobium	1	6.8
*Boletoscapter	2	6.5
*Caeconthobium	1	2.0
*Diorygopyx	8	7.3
*Cephalodesmius	3	12.5
*Falsignambia	1	8.0
*Ignambia	1	4.3
*Labroma	3	10.0
*Lepanus	26	4.1
*Mentophilus	2	8.0
*Monoplistes	6	5.3
*Oficanthon	1	5.0
*Onthobium	13	6.5
*Paronthobium	1	3.8
*Penalus	1	3.8
*Pseudignambia	2	2.7
*Pseudonthobium	2	5.3
*Pseudophacosoma	1	3.2
*Saphobiamorpha	1	12.5
*Saphobius	13	4.3
*Sauvagesinella	3	4.0
*Temnoplectron	15	8.5
*Tesserodon	13	6.5
26	150	

Intermediate fauna
0

Modern fauna
0

NEOTROPICAL REGION

Old fauna
*Agamopus	4	5.0
*Anisocanthon	4	7.8
*Canthochilum	13	5.2
Canthon	129	12.5
*Canthonella	11	3.4
*Canthonidia	2	7.8
*Canthotrypes	1	6.3
*Cryptocanthon	12	3.5
*Hansreia	1	9.0
*Holocanthon	1	5.6
*Paracanthon	4	5.3
Pseudocanthon	8	5.0
*Scybalocanthon	16	9.0
Scybalophagus	5	10.7
*Sinapisoma	1	2.9
*Sylvicanthon	5	7.8
*Tetraechma	2	6.9
*Vulcanocanthon	1	4.6
*Xenocanthon	1	5.6
*Zonocopris	1	4.3
20	222	

Intermediate fauna
*Ennearabdus	1	11.5

Modern fauna
Sisyphus	2	7.2

Plate 4.2. Small rollers.

PALEARCTIC REGION

Old fauna
Panelus	3	2.5

Intermediate fauna
Gymnopleurus	6	11.5

Modern fauna
Sisyphus	1	9.3

AFROTROPICAL REGION

Old fauna
Aleiantus	20	3.4
Aphengoecus	2	5.0
Apotolamprus	8	6.5
Arachnodes	69	9.0
Bohepilissus	2	2.4
Byrrhidium	3	10.8
Cambefortantus	1	2.0
Endroedyolus	1	3.4
Epactoides	2	4.3
Epirinus	23	8.3
Hammondantus	1	5.5
Haroldius	3	1.9
Janssensantus	2	4.3
Madaphacosoma	3	2.9
Namakwanus	1	8.0
Nanos	10	3.0
Nesovinsonia	1	4.5
Odontoloma	20	2.9
Outenikwanus	1	1.9
Panelus	1	2.5
Peckolus	2	2.1
Peyrierasantus	1	4.3
Phacosomoides	1	5.0
Pseudarachnodes	3	8.3
Pycnopanelus	2	3.8
Scybalophagus	1	12.5
Sikorantus	1	2.6
Sphaerocanthon	23	7.0
Tanzanolus	1	2.5
29	209	

Intermediate fauna
Gymnopleurus	36	10.5

Modern fauna
Neosisyphus	22	10.0
Nesosisyphus	4	3.5
Sisyphus	18	7.8
3	44	

ORIENTAL REGION

Old fauna
Cassolus	7	3.6
Haroldius	11	2.8
Larhodius	1	2.5
Panelus	18	2.3
Phacosoma	10	5.3
Phaedotrogus	1	2.9
Ponerotrogus	1	3.2
Pycnopanelus	1	2.7
8	50	

Intermediate fauna
Gymnopleurus	10	11.5

Modern fauna
Neosisyphus	2	10.0
Sisyphus	9	7.0
2	11	

Plate 4.3. Large rollers.

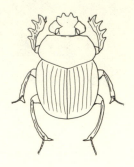

NEARCTIC REGION

Old fauna

Canthon	17	16.0
Deltochilum	2	22.5
2	19	

Intermediate fauna
0

Modern fauna
0

AUSTRALIAN REGION

Old fauna

*Aulacopris	3	19.0
*Canthonosoma	3	14.5
*Coproecus	1	15.0
3	7	

Intermediate fauna
0

Modern fauna
0

NEOTROPICAL REGION

Old fauna

*Deltepilissus	2	13.5
Deltochilum	79	23.0
*Eudinopus	1	32.5
*Malagoniella	9	21.8
*Megathopa	2	20.5
*Megathoposoma	1	23.0
*Streblopus	2	15.8
7	96	

Intermediate fauna

*Anomiopsoides	8	22.5
*Eucranium	7	28.0
*Eurysternus	22	13.5
*Glyphoderus	3	22.5
4	40	

Modern fauna
0

Plate 4.4. Large rollers.

PALEARCTIC REGION
Old fauna
  0
Intermediate fauna
  *Mnematidium   1   27.0
   Mnematium   3   20.5
   Scarabaeus   20   26.5
   3   24
Modern fauna
  0

AFROTROPICAL REGION
Old fauna
  *Anachalcos   8   25.0
  *Circellium   1   31.0
  *Epilissus   1   13.0
  *Gyronotus   5   14.5
   4   15
Intermediate fauna
   Allogymnopleurus   17   15.5
  *Drepanopodus   2   22.0
   Garreta   11   17.5
   Kheper   19   32.5
  *Madateuchus   1   21.5
   Mnematium   1   48.5
  *Neateuchus   1   29.5
  *Neomnematium   1   22.5
  *Pachylomera   2   38.5
  *Pachysoma   11   28.0
   Scarabaeus   74   21.6
  *Sceliages   6   17.5
   12   146
Modern fauna
  0

ORIENTAL REGION
Old fauna
  0
Intermediate fauna
   Allogymnopleurus   2   13.0
   Garreta   9   17.5
   Kheper   3   22.0
  *Paragymnopleurus   13   17.0
   Scarabaeus   5   26.5
   5   32
Modern fauna
  0

Plate 4.5. Small tunnelers.

Plate 4.6. Small tunnelers.

PALEARCTIC REGION
Old fauna
0
Intermediate fauna
0
Modern fauna

Caccobius	22	6.0
Euoniticellus	3	8.8
Euonthophagus	17	9.0
Liatongus	10	9.0
Onthophagus	220	9.0
5	272	

AFROTROPICAL REGION
Old fauna

Delopleurus	3	5.4
*Macroderes	13	10.0
Paraphytus	6	4.3
*Pedaria	34	7.7
*Sarophorus	4	7.5
*Xinidium	3	10.8
6	63	

Intermediate fauna
0
Modern fauna

*Alloscelus	4	6.9
*Amietina	4	3.5
Caccobius	44	6.6
*Cambefortius	9	4.6
Cleptocaccobius	18	2.7
*Cyptochirus	4	10.5
Digitonthophagus	1	10.0
*Dorbignyolus	1	5.5
Drepanocerus	18	4.5
*Drepanoplatynus	1	9.0
Euoniticellus	14	8.5
Euonthophagus	16	8.5
*Eusaproecius	7	8.5
*Falcidius	1	6.0
*Helictopleurus	53	9.0
*Heteroclitopus	8	7.5
*Heterosyphus	1	5.5
*Hyalonthophagus	9	7.8
*Krikkenius	1	10.5
Liatongus	15	10.8
*Megaponerophilus	1	5.3
Metacatharsius	61	9.5
*Milichus	8	7.0
*Mimonthophagus	2	8.5
*Neosaproecius	2	4.5
Oniticellus	5	11.5
Onthophagus	790	8.0
Phalops	33	10.0
*Pinacopodius	1	10.0
*Pinacotarsus	3	9.0
Proagoderus	96	12.0
*Pseudocopris	2	10.0
*Pseudosaproecius	12	6.0
*Scaptocnemis	1	10.5
*Stiptocnemis	2	8.0
*Stiptopodius	9	6.5
*Stiptotarsus	2	6.3
Strandius	4	11.5
Tiniocellus	4	8.5
*Tomogonus	2	5.5
*Walterantus	1	3.7
41	1,270	

ORIENTAL REGION
Old fauna

Delopleurus	3	5.0
*Parachorius	4	7.0
Paraphytus	7	6.3
3	14	

Intermediate fauna
0
Modern fauna

*Anoctus	5	5.0
Caccobius	21	5.0
Cleptocaccobius	6	3.2
*Cyobius	2	5.0
Digitonthophagus	17	10.0
*Disphysema	1	8.0
Drepanocerus	8	5.0
Euoniticellus	2	8.0
Liatongus	13	11.0
Metacatharsius	1	12.5
Oniticellus	4	8.0
Onthophagus	345	9.0
Phalops	4	11.0
*Sinodrepanus	6	9.5
Strandius	13	9.0
15	448	

Plate 4.7. Large tunnelers.

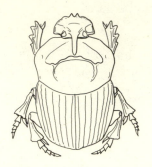

NEARCTIC REGION

Old fauna
  Dichotomius   2   25.0
Intermediate fauna
  Phanaeus   9   16.0
Modern fauna
  0

AUSTRALIAN REGION

Old fauna
  0
Intermediate fauna
  0
Modern fauna
  Coptodactyla   13   13.5

NEOTROPICAL REGION

Old fauna
  Chalcocopris   1   14.5
  Dichotomius   148   23.0
  Holocephalus   3   32.0
  Isocopris   3   27.0
  Ontherus   36   13.5
  5   191
Intermediate fauna
  Bolbites   1   19.0
  Coprophanaeus   28   35.0
  Diabroctis   3   33.0
  Gromphas   4   17.0
  Homalotarsus   1   14.0
  Oruscatus   2   20.0
  Oxysternon   15   23.0
  Phanaeus   51   19.0
  Sulcophanaeus   14   27.0
  Tetramereia   1   14.0
  10   120
Modern fauna
  Copris   17   20.0
  Liatongus   1   16.0
  2   18

Plate 4.8. Large tunnelers.

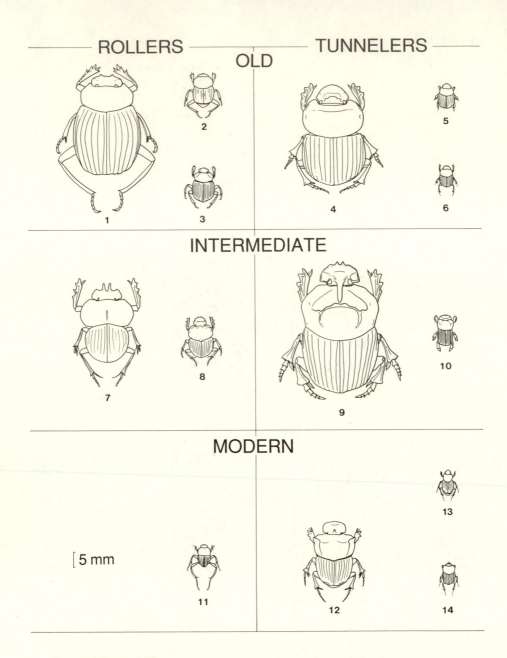

ROLLERS — TUNNELERS

OLD

INTERMEDIATE

MODERN

⌈5 mm

1 2 3 4 5 6 7 8 9 10 11 12 13 14

PLATES 4.9 AND 4.10.

Comparison between dung beetle faunas in America (Plate 4.9) and in Africa-Eurasia (Plate 4.10), giving examples of the tribes present in the two regions. The beetles have been arranged in small (<13 mm) and large rollers and tunnelers, and in three presumed evolutionary categories: old, intermediate, and modern tribes. All figures have been drawn to scale. Species in Plate 4.9: 1, *Deltochilum dentipes* (Canthonini); 2, *Canthon indigaceus* (Canthonini); 3, *Hansreia affinis* (Canthonini); 4, *Holocephalus eridanus* (Dichotomiini); 5, *Ateuchus ampliatum* (Dichotomiini); 6, *Canthidium batesi* (Dichotomiini); 7, *Eucranium arachnoides* (Eucraniini); 8, *Ennearabdus lobocephalus* (Eucraniini); 9, *Coprophanaeus (Megaphanaeus) ensifer* (Phanaeini); 10, *Dendropaemon*

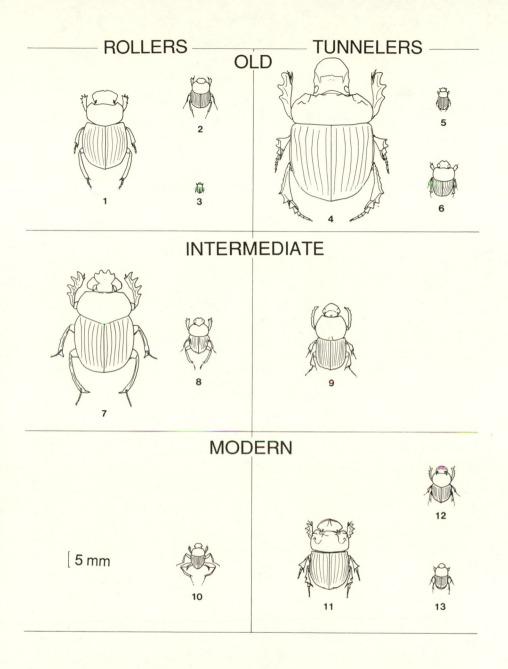

ROLLERS — TUNNELERS

OLD

INTERMEDIATE

MODERN

[5 mm

planum (Phanaeini); 11, *Sisyphus mexicanus* (Sisyphini); 12, *"Liatongus" monstrosus* (Oniticellini); 13, *Anoplodrepanus reconditus* (Oniticellini); 14, *Onthophagus sharpi* (Onthophagini). Species in Plate 4.10: 1, *Anachalcos convexus* (Canthonini); 2, *Epirinus flagellatus* (Canthonini); 3, *Odontoloma endroedyi* (Canthonini); 4, *Heliocopris antenor* (Dichotomiini); 5, *Pedaria nigra* (Dichotomiini); 6, *Coptorhina subaenea* (Dichotomiini); 7, *Scarabaeus goryi* (Scarabaeini); 8, *Gymnopleurus puncticollis* (Gymnopleurini); 9, *Onitis violaceus* (Onitini); 10, *Neosisyphus armatus* (Sisyphini); 11, *Copris interioris* (Coprini); 12, *Digitonthophagus gazella* (Onthophagini); 13, *Oniticellus planatus* (Oniticellini).

discussed in detail below). Table 4.1 and Fig. 4.1 summarize the numbers of species and genera of extant dung beetles on different continents. As expected, the Australian fauna is most distinct, the South American and North American faunas are relatively similar, and the rest of the regions form the third major cluster (Fig. 4.1).

4.2. AMERICA

This is the most complicated story because a land connection between North America and South America has been absent for a long period of time during the history of mammals and dung beetles, from the early Cenozoic up to the middle or late Pliocene. North America has repeatedly exchanged species with Eurasia through the Bering bridge: so-called Holarctic species have arrived from the west and have often undergone some evolution in North America. In contrast, South America, after its separation from Africa in the late Mesozoic, has not had a land connection except with North America. A particular evo-

TABLE 4.1

Numbers of genera and species (in parentheses) in different tribes of dung beetles in the main zoogeographical regions (major division is between rollers and tunnelers, within which species have been divided into old, intermediate, and modern faunas).

		Africa and Eurasia	Madagascar and Mauritius	America	Australia
Rollers					
Old	Canthonini	25 (133)	14 (143)	28 (341)	29 (157)
Int	Scarabaeini	9 (143)	3 (3)	—	—
	Gymnopleurini	4 (104)	—	—	—
	Eucraniini	—	—	4 (19)	—
	Eurysternini	—	—	1 (26)	—
Mod	Sisyphini	2 (52)	1 (4)	1 (2)	—
Tunnelers					
Old	Dichotomiini	10 (146)	—	20 (566)	1 (14)
Int	Onitini	18 (195)	—	—	—
	Phanaeini	—	—	12 (148)	—
Mod	Coprini	8 (331)	—	1 (26)	2 (17)
	Oniticellini	11 (109)	2 (54)	3 (6)	—
	Onthophagini	33 (1,790)	2 (7)	1 (133)	1 (280)

Note: There are differences between numbers of genera and species here and in the Index of Genera because some taxa occur in more than one region.

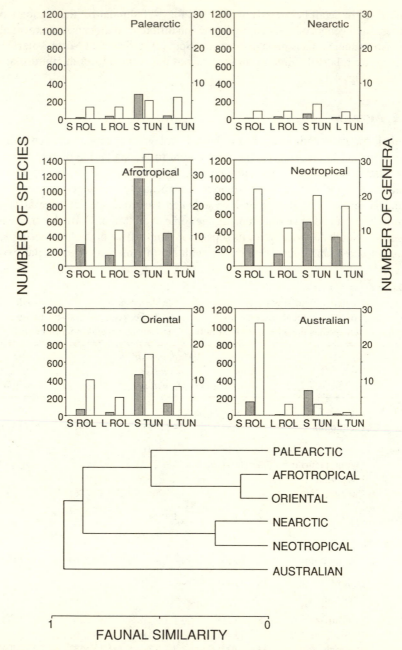

Fig. 4.1. Numbers of species and genera of dung beetles (Scarabaeidae) in small (S, <13 mm) and large (L) rollers (ROL) and tunnelers (TUN) in six zoogeographical regions. The dendrogram indicates the generic similarity of the six regions.

lution of both mammals and dung beetles took place in South America during the 50 to 60 million years of complete isolation. When the Panamanian bridge was reopened, many taxa came in contact with others that were ecologically similar, and against which they had to defend their niches. All the faunal migrations must have passed through the "Mexican transition zone," which has been studied in detail by Halffter (1962, 1964, 1972, 1976, 1977, 1978, etc.). The following information is based largely on Halffter's results.

Three faunas of mammals and true dung beetles (Scarabaeidae) coexist today in North and South America: the old, Gondwanian fauna (mammals: marsupials); the American endemic fauna (mammals: "Henotherida" of Hoffstetter 1972, Edentates, etc., and "Neotherida": primates, caviomorph rodents, etc., possibly of African origin); and the modern, Holarctic fauna (mammals: modern carnivores and ungulates, etc.). Each of the three respective dung beetle faunas can be divided into rollers and tunnelers (Plate 4.9).

The Old Dung Beetle Fauna

The original two tribes of dung beetles, Canthonini and Dichotomiini, are very well represented in South America. Canthonini have a limited distribution everywhere except in tropical America, Australia, and Madagascar (Table 4.1). In the Neotropical region, Canthonini are the only rollers, except for the Argentinian Eucraniini and two *Sisyphus*, and Canthonini have had the practical monopoly of this niche for a very long period of time, at least since the rolling of the *Megathopa*-like balls in the Patagonian Oligocene (Frenguelli 1938). Canthonini are widespread throughout South America, both in forests and grasslands, on lowlands and mountains, reaching the 50th parallel of southern latitude. Some Canthonini may have migrated northward before the breakage of the Panamanian bridge in the early Cenozoic; in any case, the tribe has diversified to endemic taxa in North America, including the genus *Melanocanthon* (Halffter 1961). Canthonini are mostly small (Plates 4.1, 4.3), coprophagous dung beetles, with some necrophagous or even predatory species (Halffter and Matthews 1966). Some coprophagous species have specialized in particular host species. For instance, some *Canthon* follow monkey troops that they are able to find either by visual or auditory cues (Howden and Young 1981; Chapter 12). A few species are "arboreal," making and rolling dung balls on branches and leaves (Howden and Young 1981; Chapter 12). Some *Glaphyrocanthon* seem to be ectoparasites on monkeys or tapirs (Halffter and Matthews 1966). The monotypic genus *Zonocopris* lives on large snails.

Northern and more modern rollers have attempted to invade America from Asia through the North Pacific (Bering) bridge, as some mammals have done (Darlington 1957). We do not know whether there were one or more groups of northern rollers attempting expansion, but they did not succeed. The only remains of these invasions are the two "relict" *Sisyphus* discussed below. We do not know whether *Sisyphus* ever reached South America. Canthonini may have been too firmly established in the Americas to become excluded by other taxa, notwithstanding the "modern" status of the invaders.

The second old tribe, Dichotomiini, consists of basically forest-dwelling tunnelers, which may have coexisted with the rollers from the very beginning of dung beetle evolution. It is only in South America where Dichotomiini does not have a relictual distribution. On the contrary, Dichotomiini is very successful in South America, where it includes both small and large tunnelers, the former being, as usual, much more numerous in species (Plates 4.5, 4.7). Most species are coprophagous, but a few have retained the ancestral saprophagy, notably the genera *Bdelyrus* and *Bdelyropsis*, the systematic position of which has been in dispute (Halffter and Martinez 1977; Huijbregts 1984). A few species of *Uroxys* and *Trichillum* ride in the fur of sloths (Halffter and Matthews 1966; Ratcliffe 1980; Chapter 12). Like Canthonini, Dichotomiini have migrated to North America, where they became established probably in the Pliocene after the reopening of the Panamanian bridge. But the position of Dichotomiini in North America is not as assured as that of Canthonini. Dichotomiini have to share the large tunneler niche not only with Phanaeini but also with *Copris*, and they have to share the small tunneler niche with *Onthophagus* (below). Nevertheless, the large *Dichotomius* and *Ontherus* and the smaller *Ateuchus* and *Canthidium* are successful genera. *Dichotomius carolinus*, a very common species in all warmer grasslands of North America and Central America, is a good example of a general trend in the family from forests to open grassland habitats. More than in Canthonini, the situation with Dichotomiini in the Americas is similar to that of the older mammals, which, after having evolved in secluded South America for 50–60 million years, suddenly had to face the northern invaders (Herschkovitz 1972; Patterson and Pascual 1972). In some cases they succumbed, in others they gained ground.

American Endemic Dung Beetle Fauna

Three dung beetle tribes do not occur outside the Americas, and two of them, Eucraniini and Eurysternini, are restricted to tropical regions. According to Halffter (1972), both of them are Gondwanian and relict, but they could as well be endemic, younger groups, analogous to edentates and South American ungulates in mammals. This faunal element, whatever its origin, evolved in almost complete isolation from the rest of the world.

Eucraniini (Plate 4.9), which are rollers, are now restricted to marginal arid environments in southern South America, where they search especially for lagomorph pellets. The group may have separated from Canthonini (although see Zunino 1985, who relates it to Onitini-Phanaeini) to specialize in lagomorph dung. Eurysternini (Plate 4.10) are more successful (Halffter et al. 1980). They are widespread in tropical South American forests and occur as far north as southern Mexico. They probably crossed the Panamanian bridge in the Pliocene-Pleistocene, and it is not possible to describe them as "relicts." They belong to a roller group but have lost the rolling behavior. Instead, they have developed a great ability to make balls in large numbers. A most distinctive feature is the habit of making what Halffter et al. (1980) call "experimental nests," which are constructed before true nesting begins. Apparently, Eurysternini have evolved in association with large mammals that used to be common in South America.

With the third endemic tribe, the Phanaeini, which are tunnelers, there is a question about its origin. In mammals, caviomorph rodents, some other mammals, and especially primates build up what Hoffstetter (1972) calls "Neotherida." The origin of these mammals is in dispute. Although South America has been isolated since the beginning of the Cenozoic, these animals do not appear in the fossil deposits before the Oligocene (Patterson and Pascual 1972). Darlington (1957) assumed that they arrived on rafts from Africa, an idea that has been further advocated by Hoffstetter (1972, 1974, 1976), who claims that this explanation, as unlikely as it seems at first, is the only one that takes into account the shared uniqueness of both African and tropical American primates. In the Eocene, Africa and South America were separated by a shorter distance than they are today, and the equatorial currents were east-west and naturally brought rafts from Africa toward South America.

Zunino (1985) has suggested a close relationship between the Phanaeini and an Old World dung beetle tribe, the Onitini (below). He assumes that a migration occurred by the middle Cretaceous from America to Africa, through one of the last land connections between the two continents. It is probably more likely that Phanaeini ancestors came from Africa by the same means as primates, in the late Eocene. Phanaeini may have successfully invaded South America in spite of the presence of another large group of tunnelers, the Dichotomiini (Plate 4.9), which is basically a forest taxon. If immigrant Phanaeini were grassland dwellers, as most Onitini still are in Africa and Eurasia, they may have been better competitors in this habitat. Once established, some species invaded forests by switching from coprophagy to necrophagy, which is especially characteristic for the (sub)genus with the largest species, *Megaphanaeus*. Similar to Canthonini and Dichotomiini, Phanaeini took part in migrations northward through the Panamanian bridge and established itself in North America. Halffter (1964:77) has suggested that this migration took place in the early Cenozoic, which does not match my own hypothesis of the tribe's late arrival in South America. If I am right, the northward migration did not take place before the Pliocene and the reopening of the Panamanian bridge. In this case, Phanaeini may have migrated together with Dichotomiini. In some Mexican species groups, Edmonds (1972) has described a "siblingness unapproached in any other phanaeine taxon." This may be an indication of recent dispersal to this area. One species has reached Jamaica.

The Holarctic Dung Beetle Fauna

After reopening of the Panamanian bridge, from the middle to late Pliocene, the genera that were present in all of Eurasia and North America began to migrate southward. These taxa are probably still migrating and struggling against the older, southern groups of beetles.

At present, only two Holarctic tribes have successfully crossed the Panamanian bridge and reached South America: Coprini and Onthophagini. Another two tribes have not reached farther south than the Mexican transition zone or the West Indies: Sisyphini and Oniticellini. Halffter (1964, etc.) assumes that *Sisyphus* fits his "paleoamerican" distribution pattern, namely, that they entered America from Eurasia through the Bering bridge between the end of the Cretaceous and Cenozoic (65 million years BP). He describes the American distribution of the genus as "relict." This suggestion must be compared with two facts. First, one of the two American *Sisyphus* has been regarded up to a recent date (e.g., Haaf 1955) as a mere subspecies of a taxon widespread in Afrotropical and Oriental regions. In any case, the two American species (Howden 1965) are very closely related to the Old World "*costatus* group" (Paschalidis 1974), one of the most abundant and vagile of the extant dung beetles, widespread in African savannas and in India. The second fact is that we know from Mauritius, in the Indian Ocean, a very characteristic genus closely allied to *Sisyphus*: *Nesosisyphus* (Vinson 1946), with four endemic species. This genus is restricted to a few mountains of the island and has a distribution that may well qualify as "relict" (Vinson 1951). Mauritius is only about 8 million years old (McDougall and Chamalaun 1969), and *Nesosisyphus* could not have settled there until a vegetation cover had developed and at least some vertebrates had colonized the island (probably birds, since there are no native mammals in Mauritius). In any case, the genus cannot date back from earlier than the Pliocene. Coming back to the "relict" American *Sisyphus*, which have not differentiated into a new genus and hardly into new species, we must conclude that they are probably not older than Pliocene, or even Pleistocene. They indeed have a relict distribution, but the relict status does not necessarily imply great antiquity. African *Sisyphus* of the "*costatus* group" are attracted to bovine dung, notably to that of the smaller species. Darlington (1957:355) concludes that "Bovidae was one of the last families of mammals to become dominant and to disperse." They have never reached South America. It is reasonable to suppose that *Sisyphus* has followed some small Bovidae, now extinct, and are presently declining.

The same story may well hold for Oniticellini, with its relict status in California, Mexico, and the West Indies. In Africa this group is attracted to bovine dung even more than *Sisyphus*. However, there is an endemic genus in Jamaica (*Anoplodrepanus*), and another species—probably of another endemic, undescribed genus—is extremely peculiar, both in its morphology and ecology, with myrmecophilous habits (Edmonds and Halffter 1972). This tribe may have migrated from Asia to America at an earlier date than Sisyphini.

Finally, we are left with the two "modern" tribes that have reached South America. Coprini belong to the group of large tunnelers. Only one genus, *Copris*, is found in America (Matthews 1976; though see Halffter 1974, and Halffter and Edmonds 1982). The genus is well diversified in North America, especially in the Mexican transition zone, and it reaches Ecuador in the south (Matthews 1962). Halffter (1964 and elsewhere) has suggested that *Copris* was distributed throughout the Americas according to his "paleoamerican pattern." It is more likely that the genus is a newcomer, having reached North America rather recently and crossed the Panamanian bridge during the Pliocene or Pleistocene. It could still be migrating. A fossil species has been described from Pleistocene asphalt deposits in California (S. E. Miller et al. 1981; Matthews and Halffter 1968), where it was associated with mammoths.

The second northern invader, Onthophagini, is also represented by just one genus, *Onthophagus*, representing small tunnelers. It has been more successful than *Copris* in spreading to South America, and is now present down to the 40th parallel. The success of *Onthophagus* supports the view that smaller dung beetles are better dispersers than large ones. The Panamanian isthmus has nonetheless played the role of a "filter bridge" in *Onthophagus*: while thirty-eight species are known from the United States (Howden and Cartwright 1963) and many more from Mexico (Zunino, pers. comm.), only about twenty-five species are reported from Brazil (Boucomont 1932) and three from Argentina (Martinez 1959).

4.3. AUSTRALIA

The dung beetle species of Australia have been recently revised by Matthews (1972, 1974, 1976). As in South America, the Australian dung beetles belong to different faunas, but only two are present in the area, as already pointed out by Howden (1981): the older, Gondwanian fauna, with tribes Canthonini and Dichotomiini; and the more modern, Holarctic fauna, with Coprini and Onthophagini. There are no endemic dung beetle tribes in Australia.

The Gondwanian Dung Beetle Fauna

Matthews (1974) assumes that the Canthonini "must have been derived from ancestors which invaded a long time ago, perhaps together with the early marsupials." He insists that the group has a close relationship with American taxa rather than with African or Madagascan ones. It is hence probable that Canthonini have followed marsupials along their dispersal route down to Australia (Hoffstetter 1976). The tribe is still plentiful in Australia (Table 4.1, Fig. 4.1). Some primitive species can roll pieces of dung but are not able to make balls.

Most of them are strongly attracted to marsupial feces, but the genus *Cephalodesmius* makes balls out of dead, "masticated" leaves and could be a remnant of the primitive saprophagous stock (Chapter 2). Canthonini are also widespread in the Oceanian region, outside mainland Australia; there are endemic genera in New Guinea, New Caledonia, and New Zealand. It is noteworthy that no native mammals (except for a few small bats) occur in the latter two archipelagos. New Caledonian dung beetles have been well studied (Paulian 1984, 1986, 1987a, 1987b), and they amount to eight genera with about twenty-five species. The New Zealand dung beetles are less well known and consist of two genera and at least fifteen species, all of which are saprophages (Paulian 1987b) or use bird droppings, though some of them have recently shifted to use mammalian dung as well.

The other old tribe, Dichotomiini, is represented in Australia by one genus, *Demarziella*. The tribe is also of Gondwanian origin, though its present distribution is more relictlike than that of Canthonini, possibly because of ecological interaction with Onthophagini. The smaller Dichotomiini and *Onthophagus* are both small tunnelers, and it is possible that the dichotomiines are losing in competition. Matthews (1974) assumed that Dichotomiini have reached Australia through a route different from the one used by Canthonini, but this needs confirmation. Dichotomiini do not seem to exist on the islands off Australia.

The Holarctic Dung Beetle Fauna

Contrary to Canthonini and probably also to Dichotomiini, which came from the south, the tribe Coprini arrived through the northern route. The two Australian genera are closely allied to Holarctic *Copris*. The following evolutionary drama has been reconstructed by Matthews (1976). An old Coprini stock came from the north and managed to survive in the Australian dead end. It is now represented by the genus *Thyregis*, whose four species have a clear relict status. One of them is probably ancestral to *Coptodactyla*, the other Australian genus, and another one is probably closely related to the ancestor of *Copris*. The occurrence of Coprini in Australia underscores the preserving nature of dead ends. Away from the tremendous evolutionary pressures that took place in the Holarctic and other regions, the distributional dead ends, where competition was not so severe, have conserved remnants of the ancestral taxa. *Thyregis* is still in existence, though out of its four species, one has not been collected in this century and may be extinct and two of the remaining three species are extremely rare. The genus is now suffering from competition with other small tunnelers, the more modern and more aggressive *Onthophagus*.

The tribe Onthophagini, probably the most recent element in the indigenous Australian dung beetle fauna, is represented by some 190 species in Australia and another 90 species in New Guinea, all belonging to *Onthophagus*. Matthews (1972; see also Palestrini 1985) suggested that the genus has reached Australia from the north, through thirty-four "original invasions" (since Matthews has divided the Australian species into thirty-four supposedly monophyletic groups). The species are grouped along eleven distribution patterns and are widespread on the continent, including the central desert area. The Australian species are mostly coprophagous (86%), and many of them are attracted to marsupial dung. Some species are mycetophagous (8%) while less than 1% are necrophages. A remarkable specialization has occurred in the *Macropocopris* species group (Matthews has made this name synonymous with *Onthophagus*, but Palestrini, pers. comm., will restore it). The six known species have been reported to occur as ectoparasites on marsupials, especially on wallabies (*Macropus*).

4.4. AFRICA AND EURASIA

During the course of mammalian history, land connections have been discontinuously present between Africa and Eurasia and between Eurasia and North America (Darlington 1957). All of this huge area has shared the same mammal as well as dung beetle pool. As stressed above, although dung beetles are an old group, their present distribution patterns reflect largely the most recent geological events. The result of the "evolutionary pump" is that in Africa there are at present as many ungulate species in the family Bovidae as there are species of mice (Muridae), 78 of each. In addition to the Bovidae, notable dung producers include the Perissodactyles (horses, rhinoceroses, elephants). The pump has played its role in the differentiation of dung beetles as well, and Africa is today the continent richest in large mammals as well as in dung beetles. Speciation has been especially wild in the small tunnelers, with 47 genera and nearly 1,350 species in Africa, compared with 21 genera and about 500 species in tropical America.

In Asia, the same processes did not occur at such a large scale as in Africa, apparently due to the patchy nature of the region, dissected as it is by mountain ranges and seas. Some Asian taxa have invaded Africa, where conditions have always been more favorable. The same is true about the Palaearctic region, where the fauna was largely decimated by Pleistocene glaciations. In the Palaearctic region, two other groups of dung beetles are present, the subfamily Geotrupinae of Geotrupidae and the subfamily Aphodiinae of Aphodiidae. They may have been present before the Scarabaeidae and, being probably more cool-resistant than the latter, have exploited the opportunity created by the large-scale extinction in this generally dominant family of dung beetles. Geotrupinae and Aphodiinae now occupy a significant part of the coprophage niche in northern temperate regions.

On all of the vast African-Eurasian continent, as in America, we recognize three dung beetle faunas: the old, Gondwanian fauna; the African-Eurasian endemic, intermediate fauna; and the modern, Holarctic fauna. It is interesting to compare the respective faunas in South America and Africa-Eurasia (Plates 4.9 and 4.10). In some cases, the same niche is occupied by the same genus (e.g., *Copris* and *Onthophagus*), while in other cases different genera of apparently different ages play the same role.

The Gondwanian Dung Beetle Fauna

Like South America and Australia, Africa and India were parts of Gondwana and presumably possessed an initial fauna of Canthonini and Dichotomiini. These two tribes are still represented by many genera in the region (Plates 4.2, 4.4, 4.6, and 4.8), though they are most numerous in southernmost Africa.

On mainland Africa, the Canthonini have, except for one genus, a typical relictual distribution, with the taxa largely restricted to South Africa (Howden and Scholtz 1987; Scholtz and Howden 1987a, 1987b). About 35 species are known in the Cape Province, 15 in Natal-Transvaal, 6 in Zaire, 5 in the Sudan and Ivory Coast, 4 in Somalia, and only 2 in Senegal. In Asia the tribe is relictual everywhere. The richest place is Madagascar, which has a particularly diverse fauna of 14 genera and 143 species, compared with the 20 genera and 80 species on the African mainland. In the rest of the region, competition with the modern rollers (Scarabaeini, Gymnopleurini and Sisyphini) has undoubtedly caused the regression of Canthonini. Outside Madagascar, the only widespread genus is the large *Anachalcos*, with 4 of the 8 species having a wide distribution all over the continent. These are opportunistic species, more or less associated with man.

The Dichotomiini has a similar distribution, centered on southern Africa with three endemic genera. There is a secondary center of species richness in Asia, with two endemic genera. At present, all dichotomiines are savanna dwellers, except for *Paraphytus*, an Afro-Asian taxon with 13 species, all retaining the ancestral saprophagy and living in dead trunks in the rain forest (Chapter 11). The genus *Pedaria* is entirely Afrotropical, and the 34 species are widespread, not showing the South African-centered distribution. The success of *Pedaria* is probably due to the fact that it lives as a kleptoparasite of large tunnelers, notably *Heliocopris* (Chapter 3). This association may be an old one, though it exists only in Africa. Other interesting genera are the mycetophagous *Delopleurus* and *Coptorhina*, both considered ancestral and very primitive by Zunino (1984b). The former has a relictual distribution in Africa and India, while the latter occurs only in Africa, where it has a south-centered distribution. The genus *Heliocopris* includes the largest existing dung beetles. Only four species live in tropical Asia, with the remaining 45 species being Afrotropical. *Heliocopris* exhibits a strong attraction to the dung of elephant and other Perissodactyle, with secondary specialization to buffalo and cattle dung in some species. *Heliocopris* is probably a remnant of large dung beetles associated with the rich and diverse mammalian megafauna that wandered in Africa, Eurasia, and North America in the Mio-Plio-Pleistocene, before glaciations and faunal extinctions (Owen-Smith 1987). At that time *Heliocopris* had a more extensive distribution than it has today, being known from the Japanese Miocene (Fugiyama 1968). Some extant species apparently cannot survive without elephant dung, and many must have disappeared. Others will unfortunately do so, as well, along with their host species.

The Afro-Eurasian Dung Beetle Fauna

Three tribes are not found outside the African-Eurasian region: the rollers Gymnopleurini and Scarabaeini and the tunneler Onitini (Plate 4.10).

The Gymnopleurini are often very abundant in the savannas for a short period at the beginning of the rainy season. As they are diurnal and often brightly colored, they are very conspicuous. Some species of other groups, especially Onthophagini, seem to be involved in mimicry complexes with the Gymnopleurini (Chapter 9). In Asia, *Paragymnopleurus* is the only large roller on the large islands of Indonesia. One species has reached Sulawesi, east of Wallace's line, but there are no Gymnopleurini in New Guinea.

Scarabaeini are, on average, larger than Gymnopleurini, and their taxonomy is more complex and less well known. The large genus *Scarabaeus* ranges all over Africa, the Mediterranean region, and continental Asia, but it is badly in need of revision. It consists of two different groups: one of day-flying, smaller species, and another of more or less nocturnal, larger species. The first group shows the relict pattern described for Canthonini and Dichotomiini: the Cape Province has about 18 species, Transvaal-Natal has 15 species, and the rest of Africa has 10, with 4 in Somalia and one in Senegal. All these species can apparently tolerate heavy drought and are among the best competitors in dry environments.

The group of larger *Scarabaeus* are not drought specialists. They make and roll balls at night and during early morning. They reach northward to the Atlantic coast of France and to Peking in the east. A group of 10–12 large species is widespread in the Mediterranean region, eastward to the Central Asian steppes. This group includes the famous sacred scarab of Egypt, *Scarabaeus sacer* (Fig. 3.1). A few flightless taxa close to *Scarabaeus* have interesting distribution patterns, being localized to dry, sandy areas in southwest or northeast Africa and Arabia. One of them is remarkable for its size, being the largest roller, and is probably strictly dependent on elephant dung in Angola (*"Mnematium" cancer*). The tribe has reached Madagascar, where two winged and one flightless species occur.

The tribe Onitini is probably the sister group of American Phanaeini (Zunino 1985; see also above). This taxon belongs to large tunnelers with a basically African savanna distribution, but the large genus *Onitis* is widespread and occurs in Africa, the Mediterranean region, Asia, and eastward to New Guinea. It has mostly savanna and some forest species. A few of the former are highly efficient in burying large amounts of large herbivore dung, for which reason they have been or are being introduced into Australia (Chapter 15). Some genera are exclusive forest dwellers (*Allonitis, Lophodonitis*), while others live exclusively in savannas—for example, *Cheironitis*, which occurs from South Africa to India, and *Pleuronitis*, which has a unique distribution pattern, being localized in the west African savannas.

The Holarctic Dung Beetle Fauna

This group includes four tribes: Sisyphini, Oniticellini, Coprini, and Onthophagini. All of them have reached North America and still survive there. We do not know whether other tribes also reached North America but subsequently went extinct.

Sisyphini are small rollers. They share their niche with Gymnopleurini in most of the region. There are not many species, but some have a large geographical distribution and are exceedingly abundant. Only Gymnopleurini can successfully compete with Sisyphini, apparently because the former are more drought tolerant. The *Sisyphus* are more specialized in small-mammal dung, whereas the *Neosisyphus* prefer herbivore dung. I have already referred to *Nesosisyphus*, which are endemic to Mauritius. Two of the 4 known species have extremely restricted distributions (Vinson 1951).

The Oniticellini are strongly associated with large herbivore dung. They are represented in Madagascar by the endemic subtribe Helictopleurina, with 2 genera and 54 species, which are almost the only tunnelers, small and large, on the island. The tribe has a second distribution center in the "Chinese transition zone" (Palestrini et al. 1987), where one finds one endemic genus and many species (Simonis 1985). Some taxa are most efficient in burying and dispersing large-herbivore and especially cattle dung, and they have been introduced into foreign countries—for example, *Euoniticellus intermedius* in Australia and the United States (Chapters 15 and 16).

The last two tribes, which we have already encountered many times, are the most important, vagile, and widespread of the Afro-Eurasian dung beetles. The two main genera of Coprini do not seem to have the same requirements. Both are large tunnelers, but *Copris* is less tolerant to drought while more tolerant to cold than *Catharsius*. Good tolerance of cold apparently allows *Copris* to be widespread in the Palearctic and Nearctic regions. A peculiar genus, *Metacatharsius*, seems to represent an attempt by the tribe to occupy the small tunnelers' niche. These species are found mostly in arid, tropical regions. An Asian genus, *Synapsis*, has a disjunct distribution in central Asia and from eastern China to Indonesia.

The Onthophagini is the most numerous tribe of all. In Africa the "evolutionary pump" has lead to an extreme diversification of this tribe, with some 29 genera and 1,100 species (9 genera and 415 species in Asia; 3 and 259 in the Palearctic region). Most of them are small tunnelers and coprophagous. A few species are saprophagous, mycetophagous, or frugivorous (Chapters 10 and 12). Some particular genera live with ants or termites, or their ecology is completely unknown, all the specimens having been taken with light-traps or by chance. The tribe covers all the available space, though most species have a rather restricted range. The widespread *Digitonthophagus gazella* has been imported to America and Australia (Chapter 15).

4.5. SUMMARY

The few known fossil dung beetles are not sufficient to allow a satisfactory reconstruction of the history of dung beetles. One may only speculate on the basis of the present distributions of the known species. In Chapters 2 and 3 we have suggested that the most important force in the evolution of dung beetles has been their association with mammalian dung. For this reason, it is tempting to relate dung beetle biogeography to the evolution and biogeography of mammals, and to supplement what we can't find out about the history of dung beetles with the known facts about the history of mammals. Another basic idea explored in this chapter is a distinction between primitive and derived taxa. It is universally agreed that two tribes of dung beetles, the rollers Canthonini and the tunnelers Dichotomiini, are primitive; they are here called "old." The old tribes have a southern, Gondwanian distribution, and I have suggested that the tribes with a northern, Holarctic distribution are the most derived ones, here called "modern." Most of the modern taxa have subsequently invaded many of the more southern regions, often probably replacing equivalent old taxa in competition. Apart from the old and modern tribes, there are taxa endemic to particular regions, and they are possibly of intermediate age. In this chapter I have described in some detail dung beetle biogeography in the three zoogeographical regions of the world: America, Australia, and Africa-Eurasia.

PART TWO

Regional Dung Beetle Assemblages

This table lists the sections and chapters of this book with the most substantial data, analysis, or discussion on a particular topic in dung beetle ecology.

Topic	Sections
Behavior	
Foraging behavior	5.4, 10.4, 12.4
Perching on vegetation	10.4, 11.3, 12.4
Strategies of dung use	6.4, 12.5
Breeding behavior	3, 5.4, 8.5, 13.4, 15.6
Population ecology	
Interference competition	7.5, 17.1
Resource competition	8.5, 10.4, 15.6, 17.1
Predation on dung beetles	12.4
Resource partitioning	
Habitat selection	5.3, 6.3, 8.4, 11.5, 18.4
Soil type selection	6.3, 8.4, 18.4
Food selection	5.5, 6.3, 8.4, 9.5, 10.4, 11.3, 12.3, 15.2, 18.1
Size differences	5.5, 7.4, 9.5, 18.2
Spatial processes and patterns	
Small-scale aggregation	5.5, 7.5, 9.6, 16.1
Large-scale distributions	9.6, 16.2
Dung beetle biogeography	4, 6.2, 7.2, 8.2, 10.3, 11.2, 12.2, 14.2, 15.2
Introduced dung beetles	5.3, 7.5, 15.7, 16.3
Elevational distributions	10.4, 14.2
Temporal patterns	
Diel activity	5.5, 8.4, 9.5, 12.5, 18.3
Seasonality	5.5, 6.4, 8.4, 9.5, 10.4, 12.4, 13.3, 14.4, 15.4, 18.3
Stability of populations	5.5, 6.2, 9.7
Community patterns	
Dung insect communities	1, 5.2
Species richness	11.4, 14.5, 19
Abundance distributions	5.5, 10.4, 12.5, 14.5, 15.4, 16.2

THE FOLLOWING eleven chapters describe the kind of dung beetle assemblages that one can expect to encounter in different parts of the world, from the north temperate regions to the tropics. Although such information exists in the specialist literature, it has never been presented in one volume, nor in a more general ecological context. We have compiled a detailed appendix of the ecology of particular beetle communities from different biomes, including information for about six hundred species of Scarabaeidae, or more than 10% of the known species. Apart from allowing many comparisons, some of which are presented in Part 3 of this volume, this information makes a contribution toward documenting what exists at present. Large-scale changes taking place in all world biomes inevitably mean that the kind of beetle assemblages we will have in the near future and later are going to be different from the ones described here, in some cases radically different, for example, in places where tropical forest is being destroyed.

The eleven chapters do not divide the world evenly. The emphasis is in subtropical and tropical biomes, where most dung beetles occur and where the taxonomic differentiation between the continents is greatest (Chapter 4). There is little information for temperate and subtropical Asia and tropical American grasslands, but otherwise the geographical coverage is good.

Chapter 5, on north temperate regions, covers Eurasia and North America, with most of the information coming from Europe, where most temperate studies have been conducted. The north temperate region is the realm of *Aphodius* dung beetles, to which Chapter 5 is largely devoted. Chapter 14, on montane dung beetles, is based on studies in the European mountain ranges and concerned with species assemblages not very different from the ones in north temperate regions. On the transition to south temperate regions, *Aphodius* give way to the generally larger and more highly evolved dung beetles, Scarabaeidae. Chapters 6 and 7 cover the Mediterranean region and the American south temperate and subtropical regions, respectively. These areas are exceptionally well studied from the biogeographical point of view by European entomologists and by G. Halffter's "school" of dung beetle entomologists in Mexico.

Chapters 8 and 9 describe dung beetle ecology in subtropical and tropical savannas, the biomes that owe their exceptionally rich life of dung beetles to the equally exceptional abundance of large mammals. It is here that dung beetle activity can be seen in its fiercest form, when thousands and thousands of beetles attack the dung pats that trail the herds of large herbivores. In the savanna ecosystems, in particular, the role of dung beetles is crucial in nutrient cycling. Cambefort (1984) has estimated that dung beetles bury one metric ton

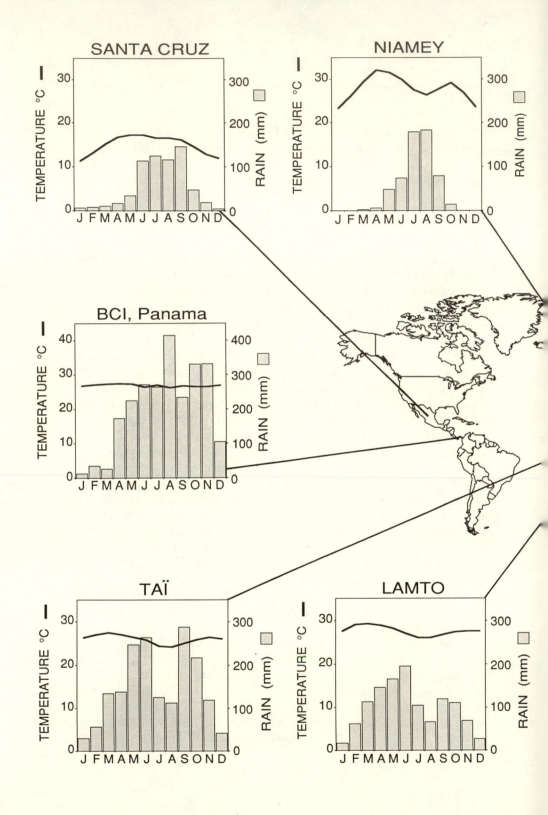

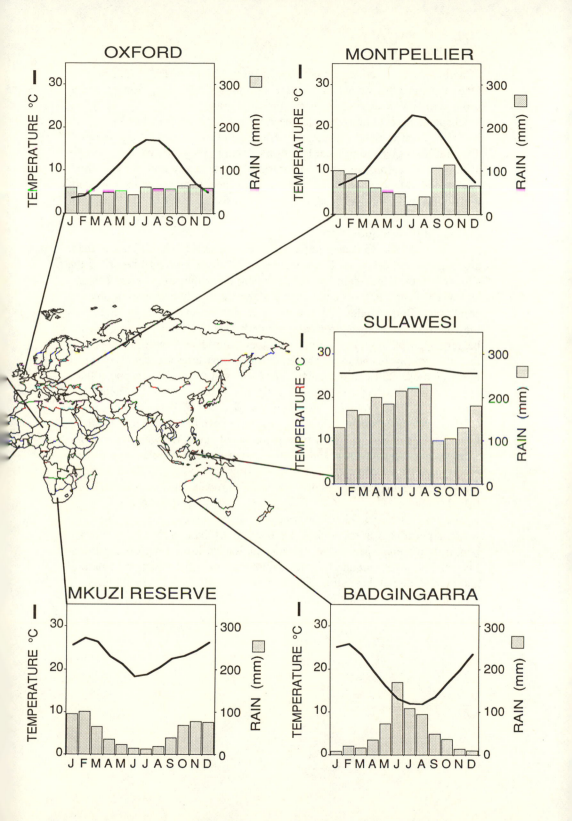

OXFORD

MONTPELLIER

SULAWESI

MKUZI RESERVE

BADGINGARRA

of herbivore dung per hectare per year in West African savannas (Chapter 9). One may argue that we would not have savannas without dung beetles, which provide the link between large herbivores and grasses.

Chapters 10 to 12 describe dung beetle ecology in the three blocks of tropical rain forest in South-East Asia, Africa, and Central and South America. The emphasis given to tropical forests may seem excessive. We defend it for a number of reasons: the tropical forests better than any other biome allow continental comparisons; tropical-forest dung beetles pose interesting problems in foraging behavior; and as tropical forests are being rapidly destroyed on all continents, there is an urgent need to add to our knowledge about the ecology of tropical-forest species.

Chapters 13 and 15 deal with particular ecosystems of significance to man: arid regions and the Australian pasture ecosystems. Chapter 13 describes the situation in the Sahel region in Africa, where desertification is a great threat, and where dung beetles need to cope with exceptionally harsh environmental conditions. Dung beetles play a key role in nutrient cycling in the Sahel, as they do in subtropical and tropical grasslands. The Australians do not need to be reminded of this. The European settlers introduced cattle to Australia, thereby creating an enormous problem of fouled pastures and clouds of dung-breeding muscid flies, as the native dung beetles in Australia were and still are poorly adapted to cope with cattle dung (Bornemissza 1979; Chapter 15). Since 1964 these problems have been tackled by a Dung Beetle Unit in CSIRO. Dozens of species of dung beetles have been introduced into Australia from Africa and Europe. Apart from its practical value, the work done by the Dung Beetle Unit represents an exciting experiment in population and community ecology. Chapter 15 is the most complete and comprehensive summary of the ecological aspects of dung beetle introductions into Australia published so far.

The map on pages 72–73 indicates the locations of the study sites, for which Appendix A at the back of the book gives data on mammals and Appendix B presents species-specific information on local dung beetle assemblages. The inserts around the map summarize seasonal variation in rainfall and temperature at these sites. The table on page 70 should help the reader in locating the most substantial analyses and discussions on particular topics in Chapters 5 to 15.

CHAPTER 5

North Temperate Dung Beetles

Ilkka Hanski

5.1. INTRODUCTION

The characteristic feature of dung insect communities in north temperate regions is the absence of Scarabaeidae—the dung beetles proper—and the dominance of the small *Aphodius* (Aphodiidae) species, which often coexist with a few larger *Geotrupes* (Geotrupidae) species (Balthasar 1963; Halffter and Matthews 1966). Unlike at lower latitudes, where Scarabaeidae may completely dominate the entire dung insect community, the north temperate communities are typically diverse mixtures of dung beetles and dung flies, preyed upon by hundreds of predatory and parasitic mites, beetles, flies, and wasps (Hanski 1987a; Table 5.1). Counting all the species of insects in the dung community, one arrives at the unusual result that the north temperate communities have as many species as the subtropical and tropical communities.

The difference in the breeding biology of Aphodiidae and Scarabaeidae (Chapter 3) has important population-dynamic consequences. Scarabaeidae are often engaged in severe competition in attempting to secure a portion of the food resources for their exclusive use in underground nests. Numbers of *Aphodius* in relation to resource availability are typically though not always lower, and interspecific interactions among them are less competitive, even if both adults and larvae mix freely in droppings; the influence of abiotic factors and perhaps also predation are more significant in Aphodiidae than in Scarabaeidae. As the north temperate region is the only region where *Aphodius* are the dominant dung beetles, this chapter is largely a review of the ecology of this genus, with an emphasis on community patterns and a bias toward the European species, with which I am most familiar.

The north temperate regions were drastically affected by the Pleistocene glaciations, with the vegetation zones repeatedly shifting, contracting, and expanding according to the changing climate. Dung beetles and beetles in general provide good fossil material, which shows that both in Europe (Lindroth 1948; Coope 1962, 1970, 1978) and in North America (Morgan 1972; Ashworth 1977) most insects "migrated" with changing climate and vegetation rather than became extinct. A dramatic example is *Aphodius holdereri*, a species now known only from the high plateau of Tibet (Chapter 14), but which was by far the most abundant large dung beetle in the British Isles during the last glaciation (Coope 1973).

TABLE 5.1
Species composition at the family level, rough numbers of species in the regional species pool, and very rough estimates of typical numbers of individuals feeding and breeding in cattle pats in northern Europe (the latter cannot even be guessed for many groups of insects, for example adult flies).

Trophic Level	Family	Order	Number of Species in N. Europe	Individuals	
				Adults	Larvae
Coprophages	Anisopodidae	D	1		10
	Anthomyiidae	D	10		
	Aphodiidae	C	40	100+	200+
	Ceratopogonidae	D	5		10
	Chironomidae	D	5		10
	Geotrupidae	C	10	+	
	Hydrophilidae	C	15	100	
	Muscidae	D	15		10
	Psychodidae	D	10		500+
	Scarabaeidae	C	10	+	
	Scatophagidae	D	5		50
	Scatopsidae	D	5		10
	Sciaridae	D	5		
	Sepsidae	D	15		500
	Sphaeroceridae	D	50	100+	200+
	Staphylinidae	C	10	100+	
	Stratiomyidae	D	10		10
	Trichoceridae	D			50
Fungal feeders	Cryptophagidae	C	10		
	Ptiliidae	C	20		
	Staphylinidae	C	5		
Predators	Carabidae	C	10	+	
	Histeridae	C	10	10	
	Hydrophilidae	C	5		
	Muscidae	D	40		10
	Staphylinidae	C	100+	100+	
Parasitoids	Braconidae	H	5		
	Eucoilidae	H	5		
	Ichneumonidae	H	5		
	Pteromalidae	H	5		
	Staphylinidae	C	5	10+	

Sources: Data from Laurence (1954), Papp (1976), Hanski & Koskela (1977), Mäkelä (1983), and Skidmore (1985).

Note: D, Diptera; C, Coleoptera; H, Hymenoptera.

Another environmental change that has affected dung beetles occurred when the largely forested land was turned into an agricultural landscape in most of Europe and in large parts of North America. The timing has been different on the two continents, however: extensive clearance of forests and the great expansion of cattle is much more recent in North America than in Europe. The *Aphodius* assemblages on the two continents show striking differences in their habitat selection, which can partly be ascribed to this historical factor (Section 5.3). Most recently, the great intensification and modernization of agriculture during this century has once again changed the spatial pattern of dung availability in the most populated parts of the north temperate region.

Section 5.2 describes the ecological setting and the general characteristics of the north temperate dung insect communities. Section 5.3 compares the biogeography of Europe and North America, including the ecology of dung beetles accidentally introduced from Europe to North America. From there I move on to population ecology of *Aphodius* (Section 5.4), which is not especially well known, partly because the species are embedded in complex communities, making single-species studies difficult. In Section 5.5 I will examine four kinds of community patterns: niche differences between coexisting species, spatial distributions at different scales and how these may affect population dynamics, abundance distributions, and temporal stability of populations. Finally, in Section 5.6 I will attempt to tie these results together, and I will make some suggestions for further research.

5.2. NORTH TEMPERATE DUNG INSECT COMMUNITIES

Most temperate ecosystems have been altered by man, usually to the detriment of wild animals and plants. Dung beetles are often an exception, as availability of large herbivore dung has increased in many areas with increasing numbers of cattle. Different types of grasslands often have different dung insect faunas (e.g., see Merritt and Anderson 1977; Chapter 8), but such differences are not so important for *Aphodius*, which breed in droppings and are not directly affected by factors such as the soil type and vegetation cover. The dung of other domestic mammals, especially horses and sheep, is also used by *Aphodius*, and in wooded habitats wild herbivores, such as deer, are often important dung producers. Small-mammal dung is little used by dung beetles in Europe, but in North America there are dozens of species specializing on this resource in open grasslands, deserts, and montane regions (Section 5.3).

Table 5.1 gives an overview of species composition and diversity, as well as rough estimates of the numbers of individuals, in cattle pats in northern Europe. There is hardly a need to emphasize the complexity of this community, with more than four hundred species in the regional species pool and often thousands of insects in single droppings. The full food web for the community has 10^5 possible interactions between pairs of species! In this perspective, it is clear that a discussion of dung beetles in isolation of the rest of the community is a simplification, even if dung beetles largely use different resources in droppings than coprophagous flies, and even if most predatory insects prey upon flies rather than beetles (Chapter 1). This simplification is unavoidable at this time because of the paucity of ecological information for coprophagous flies (but see, e.g., Papp 1976 and references therein).

5.3. BIOGEOGRAPHY: EUROPE AND NORTH AMERICA COMPARED

The European and North American *Aphodius* have contrasting habitat selection (Table 5.2). In northern and central Europe, the clear majority of species are generalist coprophages, and at least two thirds of the species breed in pastures (including many of the montane species listed in Table 5.2). Only a few species occur exclusively in forests, where they use deer dung (Landin 1961; Kronblad 1971; Krikken 1978). In contrast, in eastern North America, 40% of the species are restricted to deer dung in forests; 20% are saprophages or coprophagous microhabitat specialists; and fewer than 20% have been classified as generalist coprophages (Gordon 1983). The latter species are those that inhabit pastures, where their numbers are augmented by twelve accidentally introduced species (Brown 1940; Gordon 1983, pers. comm.), which generally dominate the local *Aphodius* assemblage (Table 5.3). In the western and midwestern United States, where climatic conditions are for the most part unlike those of Europe, most species are either microhabitat specialists, associated with burrows and nests of rodents and gopher tortoises, or they are sap-

TABLE 5.2
Comparison of the *Aphodius* faunas in northern and central Europe and eastern North America.

	Northern and Central Europe	Eastern North America
Approximate number of species	75	50
Pasture species	48	8 + 12*
Forest species	5	20
Specialists	15	10

Notes: Some pasture species prefer sheep to cattle dung. The specialists in Europe include 10 montane species, many of which are pasture species. The remaining specialists are either saprophages or coprophages with special macrohabitat selection, e.g., littoral species, or food and microhabitat specialists, e.g., *A. porcus* parasitizing *Geotrupes* brood balls in Europe (Chapman 1869), and *A. badiceps* living in tree squirrel nests in North America (Gordon 1983).

* Introduced pasture species, mostly from Europe.

TABLE 5.3
Aphodius species colonizing cattle pats in pastures in North America.

Status	Species	Locality							
		1	2	3	4	5	6	7	8
Introduced	*haemorrhoidalis*	1,801	+	**	+		**	13	1,309
	granarius	625	+	*		***	*		471
	fimetarius	282	+	***	+	***	**	3	707
	distinctus	68	+			***	***		451
	fossor	28	+		+	*			54
	prodromus	2			+				285
	lividus			*				3,356	
	erraticus				+	*			777
Native	*vittatus*	109	+	**			*	8	
	coloradensis	42							
	stercorosus	34			+	**	*		1
	congregatus		+						
	pectoralis		+						
	tenellus		+						
	pardalis			*					
	terminalis					*	*		
	rubeolus							31	
	tenuistriatus							455	
	femoralis						**		
	ruricola	49					*		5
	bicolor						*		
	lutulentus							3	
	kirni							1	
	lentus								1
	walshi								2

Notes: Note the numerical dominance of the introduced European species. Most studies do not report quantitative data. The numbers of asterisks (*) indicate relative abundances, while the pluses (+) indicate presence.

Localities and data from: 1, East-Central South Dakota (Kessler and Balsbaugh 1972); 2, British Columbia (Macqueen and Beirne 1974); 3, California (Merritt 1974); 4, New York (Valiela 1974); 5, New Jersey (Wilson 1932); 6, Illinois (Mohr 1943); 7, East-Central Texas, swine feces (Fincher et al. 1986); 8, Minnesota (Cervenka 1986).

rophages (Gordon 1983). Blume (1985) lists twenty-six native *Aphodius* species associated with bovine dung in pastures in America north of Mexico, but most of these species are generalists, not bovine dung specialists.

The twelve *Aphodius* in North America were introduced from Europe; the only certain exception is the Oriental *A. rectus*, which has a restricted distribution in Oregon and Washington (the cosmopolitan *A. lividus* and *A. gran-*

arius may also have been introduced from outside of Europe). The introduced species conspicuously do not include southern European *Aphodius*, a fact that has two plausible explanations: the best chances of introduction may have been from northern and central Europe; and while the northern and central European *Aphodius* are mostly generalist pasture species, the southern European species include many more specialized taxa. Most of the introduced species are ''core'' species (Hanski 1982) in Europe; in other words, they are widely distributed and locally abundant in pastures. Species not fitting this description are *A. lividus*, *A. scrofa*, *A. granarius*, and *A. subterraneus* (though *subterraneus* is common in southern France). It is noteworthy that while most of the introduced species are widespread in North America (Table 5.3), *A. scrofa* and *A. subterraneus* are restricted to eastern seaboard localities (Gordon 1983), where they were first introduced in the 1920s or earlier (Brown 1940). Thus the less widespread and abundant species in Europe tend to show the same characteristics in North America. *Aphodius granarius* is a special case as its larvae may be primarily saprophagous or feed even on living roots of grass (Jerath and Ritcher 1959). This species is a good colonizer, being the only known *Aphodius* from the Azores (Landin 1960), and it is frequently found in large numbers in various types of organic-rich deposits from ancient towns, especially in the samples from the Viking period (H. Kenward, pers. comm.). *Aphodius rufipes* occurs in North America in forested mountain regions from New York to Virginia (Gordon 1983). Although this species is common in European woodlands (Hanski 1979), it is better known as a pasture species, often very abundant in late summer (Holter 1979, 1982; Hanski 1980a). It is surprising that it has failed to become abundant in North American pastures.

The difference in the habitat selection of *Aphodius* in northern and central Europe and in eastern North America may be due to the diverse histories of these regions since the last glaciation and earlier. Human impact on the landscape, with the attendant cattle, horses, and other domesticated mammals, has been significant in Europe for thousands of years (Birks 1986 and references therein). In contrast, native Americans had no domesticated mammals except dogs (Delcourt 1987), and large-scale clearing of forests did not take place before the westward expansion of the American frontier between 1790 and 1880 (Delcourt and Delcourt 1987). The cumulative Indian impact on the landscape in North America over millennia increased the size of old fields and of early successional forests, which led to an increase in the populations of the white-tailed deer, a major food source for Indians (Delcourt 1987). Environmental conditions have long been favorable for deer-dung specializing *Aphodius* in forested North America. These species have been unable to colonize the recent pasture ecosystems, probably because of their general ecophysiological adaptations to forest habitats. Landin (1961) has shown that *Aphodius zenkeri*, one of the few European forest species restricted to deer dung, can tolerate only moderately high temperatures, both as a larva and as an adult, and he demonstrated that high temperatures on pastures are often lethal to this species.

When discussing the impoverished fauna of the native pasture species in North America, it is reasonable to ask what was the fauna of bison dung and what happened to it. Not much is known about this (S. A. Elias, pers. comm.). Asphalt-impregnated sediments at Rancho La Brea in California have provided late Pleistocene insect material from a locality with a diverse and abundant large-mammal fauna. Three dung beetles, *Copris pristinus*, *Onthophagus everestae* and a species of *Phanaeus*, appear to have gone extinct, and it is possible that the reduction in dung availability due to disappearance of the large-mammal fauna contributed to their demise (Miller et al. 1981; Miller 1983). Three extinct species are not many, but it is noteworthy that documented Pleistocene insect extinctions are practically nonexistent (Miller 1983; Coope 1978). We should also note, however, that the climatic conditions in the Great Plains are generally unfavorable for dung beetles that use bovine droppings: precipitation is minimal, humidity is low, and the dung pats consequently dry out very quickly (R. D. Gordon, pers. comm.). The specialist species of *Aphodius* living in rodent tunnels in the Great Plains may have survived because of their more protected habitat. Modern North American rodents and their burrowing habits evolved during the Miocene in response to increasing aridity (R. D. Gordon, pers. comm.).

5.4. POPULATION ECOLOGY

In this section I will focus on two sets of questions, after the presentation of a brief outline of the biology of *Aphodius* (see also Chapters 2 and 3). First, how do individual beetles track resources, how do they find mates, and how do females spread their eggs in dung pats? These are crucial questions for species living in a markedly patchy and ephemeral habitat. Second, what are the important ecological interactions in dung beetles in north temperate regions? At the end of the section, I shall briefly summarize the population ecology of coprophagous Geotrupidae, the other group of dung beetles that occurs in north temperate regions (for their breeding biology see Chapter 3).

Biology of Aphodius

Most coprophagous Aphodiidae belong to the cosmopolitan genus *Aphodius*, which has more than one thousand species. The genus is still rather poorly known taxonomically, especially in the tropics. Compared with Scarabaeidae and Geotrupidae, *Aphodius* and Aphodiidae in general represent a simple breeding biology without sexual cooperation and nest construction (Chapter 3). Most *Aphodius* are relatively small, less than 10 mm in length. They have high fecundity in comparison with the nest-building dung beetles, of the order of one hundred eggs per female (Holter 1979; Yasuda 1987), though some smaller values of lifetime egg production have been reported for large species (Yasuda 1987; Table 3.2). Typical development times are 3–5 days for eggs, 4–6 weeks for larvae, and 1–4 weeks for pupae (Landin 1961; Holter 1975; Stevenson and Dindal 1985). There is much variation in the stage of hibernation, which may be the egg, third instar larva, prepupa, or adult (White 1960; Landin 1961; Hanski 1980e; Yoshida and Katakura 1985). This variation is related to conspicuous segregation of species' breeding seasons (next section). If one wishes to apply the r-K species concept to dung beetles, *Aphodius* are clearly located at the r-end of the continuum in comparison with most Scarabaeidae.

Although most temperate *Aphodius* are coprophagous and breed in the dung of herbivorous mammals, several species have herbivorous or saprophagous larvae, feeding on the roots of grasses and on decomposing plant material. Examples include *A. pardalis* (Downes 1928; Ritcher and Morrison 1955) and *A. distinctus* in North America (Christensen and Dobson 1976), and *A. howitti*, destructive to pastures in Australia (Swan 1934; Carne 1950). Some abundant European species, such as *A. prodromus* and *A. contaminatus*, have generalist saprophagous larvae (White 1960), which may explain their frequent occurrence in various archaeological samples (H. Kenward, pers. comm.). *Aphodius* provides good examples of the evolutionary shift from saprophagy to coprophagy (Chapter 2). More than that, we may here discern probable intermediate stages in the evolution of larval feeding habits. Some species may develop both in vegetable matter and in herbivore dung, for example, the above-mentioned *A. prodromus* and *A. contaminatus*. Other species lay their eggs beneath droppings, and the larvae may move between the soil and the dropping. *Aphodius obscurus* (Chapter 14) and *A. luridus* (Chapter 6) are two temperate examples. One may imagine a species with initially saprophagous larvae ovipositing in the soil beneath the adult feeding station (dropping), the larvae thereby coming in contact with a new food resource—that is, the vegetable remains in the herbivore dung—enriched by microorganisms and their products.

Some *Aphodius* are kleptoparasitic, breeding in the brood balls of nest-building Scarabaeidae and Geotrupidae. Chapman's (1869) description of *A. porcus* parasitizing *Geotrupes stercorarius* nests is the classical example, which, remarkably, has not been followed up by later workers (Howden 1955b; the host species may have been *G. spiniger*: Main 1917). Chapman (1869) observed how an adult *A. porcus* entered the egg cavity of *Geotrupes* brood balls, destroyed the *Geotrupes* egg by eating or otherwise, and then laid her own eggs, using the food supply intended for *Geotrupes* larvae. Other instances of *Aphodius* developing in the brood balls of larger dung beetles are described by Fabre (1918), Howden (1955b), and Hammond (1976).

Movements between Droppings

General evolutionary considerations suggest that the beetles' between-dropping movement behavior has evolved to maximize each individual's fitness. Males should generally distribute themselves in droppings in such a way that the number of matings they achieve is maximized (see Parker 1970, 1974; Parker and Stuart 1976, for the yellow dung fly). Immature females feed and mature ones breed in droppings, and in both cases success can be expected to decrease with the density of beetles. In both sexes, movement decisions are expected to be density-dependent, though males and females, and feeding and breeding females, may use different criteria in making their movement decisions.

Simple experiments demonstrate that emigration from droppings is density-dependent at sufficiently high densities. Landin (1961) conducted laboratory experiments with several species of *Aphodius*. In one series of experiments, twenty *A. zenkeri* and twenty *A. ictericus* were placed in containers with 55, 35, or 15 cm^3 of sheep dung. Nearly all beetles stayed in the largest droppings, about one third moved away from the medium-sized droppings, while most beetles moved out from the smallest droppings. The differences were highly significant. Landin (1961) calculated that adult *Aphodius* show increased emigration rates if individuals have less than twenty-five to seventy times their volume of dung. The corresponding figure for the larvae is even higher, two to four hundred times their volume in the second instar larvae (Landin 1961). Densities reaching these limits are not found in all droppings, but even higher densities are not uncommon in nature (Landin 1961), which is sufficient for competition to occur and possibly affect the dynamics of populations. Yasuda (1987), working in Japan, found density-dependent effects on the rate of emigration; in some species intraspecific effects exceeded interspecific ones, but in other species the opposite was true.

Holter's (1979) work on *Aphodius rufipes* provides evidence for density-dependence in the reproductive behavior in the field. He found that the distribution of eggs in cattle pats was significantly less aggregated than the distribution of adult beetles. Laboratory experiments confirmed that the rate of oviposition decreases with the increasing density of beetles (Holter 1979). In *A. haroldianus* and *A. elegans*, the number of eggs laid per female decreased with density even at the lowest densities (1–10 beetles per pat; Yasuda 1987). Hanski (1980c) found that in four species of *Aphodius* mature females stayed longer in cattle pats than immature females (see also Hanski 1980d), a pattern also found in the common dung-breeding hydrophilids *Sphaeridium lunatum* and *S. scarabaeoides* (Otronen and Hanski 1983). Mature females may need time to select suitable oviposition sites, or they may simply spend time in droppings to mature another batch of eggs. *Aphodius* lay their eggs singly or in small clutches (White 1960; Landin 1961), but it is not known how many eggs females typically lay into one dropping.

In conclusion, between-dropping movements in *Aphodius* are density-dependent at densities that are commonly observed in the field. On the other hand, the optimal foraging model predictions (e.g., Stephens and Krebs 1986) about immigration to and emigration from droppings are not supported (Hanski 1980f, 1990). Beetles may not be able to optimize their stay-times in droppings because they do not have the necessary information to do that; and they cannot obtain such information because they cannot sample a sufficient number of droppings before factors such as weather have changed the qualities of droppings, regardless of what the beetles and other insects do.

Interactions

Unlike dung beetles in many subtropical and tropical environments, north temperate dung beetles are not often conspicuously abundant in relation to the amount of resource available to them (but see Table 5.6). It has been suggested that *Aphodius* are unlikely to compete because the cumulative larval consumption generally amounts to only a fraction of the matter present in droppings (Holter 1982). However, as *Aphodius* larvae move freely in droppings, part of which they can potentially consume, it is hardly reasonable to conclude that there is no competition if all the resource is not used. As Landin's (1961) experiments demonstrated, second instar *Aphodius* larvae require two to four hundred times their volume of dung for optimal development, and densities as high as this are not uncommon in the field. This argument however makes competition only a likely factor; it does not indicate how frequently it occurs in the field, and what the consequences are.

There is an even greater lack of knowledge about predation and parasitism. *Aphodius* larvae have long development times for dung-inhabiting insects (Hanski 1987a), and most predatory beetles have left droppings before the *Aphodius* larvae reach the later larval stages. Egg predation may be important, but nothing is known about it.

Geotrupinae

The subfamily Geotrupinae, which includes the coprophagous Geotrupidae, has a circumpolar distribution in the Northern Hemisphere (Chapter 4). Geotrupinae are relatively large beetles, typically from 20 to 30 mm in length, and they burrow deep shafts, down to 270 cm (Howden 1952), under or near the food source (Howden 1955a and Brussaard 1983 give the observed shaft depths in many species). There are several brood chambers that are provisioned with food material, usually collected by the male and processed underground by the female. The life cycle is completed in one (*Geotrupes*) to two or more years (e.g., *Typhaeus typhoeus* in Europe; Brussaard 1983).

Geotrupinae include a wide variety of species in terms of larval food source. Most European *Geotrupes* are coprophagous, but *Geotrupes stercorosus* often makes brood balls of moldy forest litter (Rembialkowska 1982). The North American species have significantly more varied food selection, and often the larval and adult food is different (Howden 1955a). For instance, the adults of *G. ulkei* and *G. splendidus* feed commonly on fungi while their larvae develop on decaying leaves, and *G. egeriei* and *G. hornii* have fungivorous adults but coprophagous larvae (Howden 1955a).

It is noteworthy that while local assemblages of *Aphodius* often have fifteen to twenty species, there are usually only one or a few coexisting coprophagous Geotrupinae in north temperate regions. A contributing factor may be the difference in the population dynamics in the two groups of beetles: *Aphodius* have typical variance-covariance dynamics, which tend to increase intraspecific competition in relation to interspecific competition and hence to facilitate coexistence, while Geotrupinae approach lottery dynamics with the possibility for strong preemptive competition (Chapter 1). Density of single species of Geotrupinae may be very high. *Typhaeus typhoeus*, a common species on dry heathlands in western Europe, may occur at densities up to one to two pairs per square meter (Brussaard 1983), and because a breeding pair of beetles typically clears an area of 0.6 square meters around the nest of rabbit pellets (Brussaard 1985a), their chief food source, some populations may occur at or near their environmental carrying capacity. A detailed mark-recapture study of *Geotrupes stercorosus* revealed densities up to 0.1 beetles per square meter in an oak-hazel forest in Belgium (Desière 1970).

5.5. COMMUNITY PATTERNS AND PROCESSES

This section will cover four kinds of community patterns: niche differences, spatial distributions, abundance distributions, and temporal stability, largely based on the results from one well-studied local community in southern England (Hanski 1979). Appendix B.5 gives data for this community.

Niche Differences

Food selection. Most species in northern Europe live in pastures and use the dung of domestic mammals—cattle, horses, and sheep—without much discrimination (e.g., Rainio 1966). It is not clear which characteristics of droppings are decisive in food selection, since most resource characteristics are correlated, but the moisture content and consistency are probably generally important. Ruminant dung has a particular microflora and microfauna, which may affect food selection by adult beetles with soft mandibles (Chapter 3).

Dung pats change rapidly in quality. They dry up quickly in dry weather, and the activities of insects and microorganisms affect droppings (Hanski 1987a). There are consistent differences among the species in the times of immigration to and emigration from droppings (Hanski 1980c,d, 1986; Holter 1982), which may reflect differences in the resources used by the species. There is more of a tendency for the larger species to occur in older droppings than for the smaller species (Holter 1982).

Habitat selection. Landin (1961) divided *Aphodius* into eurytopic, oligotopic, and stenotopic species with respect to their habitat selection. Most species occur widely in different habitat types, including open and shaded habitats. In Landin's (1961) study area in Sweden, only *A. zenkeri* was classified as a stenotopic species, restricted to forests by its limited tolerance to high temperatures. He found that during an exceptionally hot and dry spell of weather many species of *Aphodius* moved from open to more shaded habitats.

The five species—*A. prodromus*, *A. sphacelatus*, *A. sabulicola*, *A. contaminatus*, and *A. obliteratus*—comprise a group of similar species with poorly known ecology. The adults are often exceedingly abundant in droppings in autumn and/or spring, but the larvae are probably primarily saprophagous. Near Oxford in southern England, four of the five species coexisted, but with conspicuous differences in habitat selection (Table 5.4). These differences are not explicable by dung availability, and may indicate unknown differences in larval ecology. Hanski and Kuusela (1983) report a similarly striking difference in the occurrence of *A. prodromus* and *A. sabulicola* in southern Finland.

Diel activity. Differences in diel activity are not important, as most species colonize droppings during several days and are able to adjust their flight activity according to prevailing weather conditions (especially temperature; Landin 1968; Bernon 1981; Chapter 8). Though *Aphodius* prefer darkness to daylight (Landin 1968), most species in northern Europe are active during the warmer

TABLE 5.4
Numbers of four similar species of *Aphodius* in five pastures, in a nearby woodland, and in the center of a town (Oxford) in southern England.

Species	Pastures					Woodland	Town
	1	*2*	*3*	*4*	*5*		
A. prodromus	845	643	3,448	148	1,340	1,293	177
A. sphacelatus	21	37	479	67	34	4	15
A. contaminatus	212	776	855	573	220	1	1,666
A. obliteratus	—	1	1	2	636	2	22

Source: Data from Hanski (1979).

hours of day (Koskela 1979) and fly more or less at the same time (Landin 1968; Koskela 1979). *Aphodius rufescens* and *A. rufipes*, two abundant late-summer species, are nocturnal, and *A. rufipes* is most abundant in pats deposited late in the evening (Holter 1979). Coprophagous flies, which are diurnal, are likely to have their lowest numbers in these pats, which have a dry crust the following morning, preventing or making oviposition by flies more difficult (Campbell 1976). North America has many nocturnal *Aphodius* in arid regions, typically living in rodent burrows and nests (R. D. Gordon, pers. comm.).

Seasonality. In contrast to slight differences in diel activity, local assemblages of *Aphodius* show conspicuous phenological differences (White 1960; Holter 1982; Yasuda 1984; Yoshida and Katakura 1985; Hanski 1986). Most *Aphodius* are univoltine (Landin 1968; Holter 1982), but at any time from early spring to late autumn there are some species active, with species number typically peaking in early summer. The early summer species have overwintered as third instar larvae or adults (Hanski 1980e). About one half of the species in Sweden (Landin 1961) and in northern England (White 1960) overwinters at the adult stage, which has the advantage that the mobile adult may both locate a suitable overwintering site in the autumn and a good breeding site in the following summer (Hanski 1980e).

Phenological differences entirely eliminate interspecific interactions in coprophagous insects, but whether the observed differences in *Aphodius* are (Hanski 1980a) or are not (Holter 1982) due to (past) competition remains unresolved. To take an example of the kinds of observations that are available and support the competition hypothesis, *Aphodius borealis* and *A. equestris* are two species that occur commonly in woodlands. Hanski (1980a) compared the mean phenological occurrence of *Aphodius* in two communities in England and Finland. The only substantial difference was in *A. borealis*, which occurred 6 weeks earlier, on average, in Finland than in England. Hanski (1980a) suggested that this difference may be due to competition with *A. equestris*, which was absent in Finland but was very abundant in late summer in England, at the time when *A. borealis* flies in Finland.

Size. Another niche dimension along which coexisting *Aphodius* show marked differences is size—for example, from 1.6 (*A. merdarius*) to 40 mg (*A. fossor*) in the eighteen species studied by Hanski and Koskela (1977) in southern Finland. It is not clear exactly what difference size makes, but undoubtedly size affects many aspects of a species' ecology. Fig. 5.1 compares the pairwise size differences with combined season-habitat niche differences among the eighteen species in Finland. There are many similar-sized species with much niche overlap, but most of such species pairs involve one or two uncommon species; only *A. fimetarius* and *A. rufescens* are both abundant, similar in size, and have great niche overlap. But even this case is somewhat special, as *A. fimetarius* is bivoltine (Hanski 1979; Holter 1982). The comple-

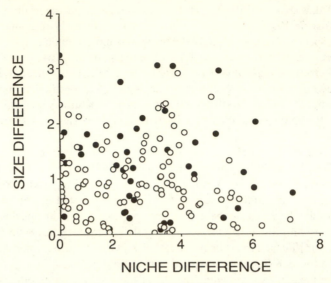

Fig. 5.1. Plot of pairwise size ratios against habitat-season niche differences in eighteen species of *Aphodius* in southern Finland (data from Hanski and Koskela 1979; logarithmic transformation of size). The values among the ten most abundant species have been denoted by black dots. There are five pairs of abundant species (out of 45) with values on both axes < 1.5, while there are thirty-one such pairs involving one or two uncommon species (out of 107). The difference is significant ($\chi^2 = 5.46$, P<0.02), indicating a complementary effect between size and niche differences.

mentary size and niche differences in Fig. 5.1 are largely due to two sets of common species with similar seasonality and habitat selection but with conspicuous size differences: an early summer group consisting of *A. fossor* (40 mg), *A. ater* (6 mg), and *A. merdarius* (1.6 mg); and a late summer group consisting of *A. rufipes* (33 mg), *A. rufescens* (7 mg), and *A. borealis* (1.9 mg). Hanski (1980a) and Holter (1982) have discussed the equivalent early and late summer species groups in England and Denmark, respectively.

In conclusion, phenological and size differences lead to a nonrandom structure in at least some assemblages of *Aphodius*: the most abundant species have more overdispersed niches than expected by chance.

Spatial Patterns

Perhaps the most interesting level of spatial variation in dung beetles occurs between droppings in small areas. At this level, inquiries directed at three questions are helpful: (1) the *behavior* of adult beetles that move between droppings; (2) the *distribution* of adult beetles in droppings, which is continuously changing because of movements, and the distribution of eggs and lar-

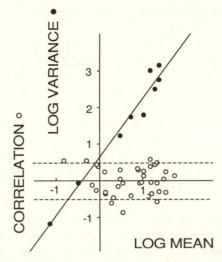

Fig. 5.2. The black dots give the log variance against the log mean abundance regression for nine species of *Aphodius* in fifteen cattle pats. The regression slope is 1.6. The circles give correlation coefficients for pairs of species (log-transformed abundances), plotted against their average log abundance. The broken lines indicate the 95% confidence interval for the correlation coefficients. Note that correlations are mostly nonsignificant and unrelated to the average abundance of the species (from Hanski 1986).

vae, which is determined by patterns of egg laying and by processes within droppings; and (3) the role of spatial distributions of adult beetles and larvae in population *dynamics*. Questions about adult movement behavior were discussed in the previous section, and I shall now turn to the spatial patterns themselves and their population-dynamic consequences.

Spatial distributions. Emigration from droppings is often density-dependent in *Aphodius* (previous section), which should decrease variance in the numbers of beetles between droppings. Nonetheless, between-dropping distributions of *Aphodius* are markedly aggregated, and spatial variance increases rapidly with increasing mean abundance (Fig. 5.2). This result is not unexpected, as most animals, regardless of their ecology and habitat, have more or less aggregated distributions at most spatial scales (Taylor et al. 1978). In the case of dung beetles, an obvious cause of aggregated distributions is the difference among various droppings. While it is clear that such differences will affect beetle distributions in the field, it is unlikely that this is the only or even the most important explanation of aggregated distributions, as the same aggregated patterns are found in experiments using artificial, homogenized dung pats (Hanski 1979; Holter 1982). Holter (1982) suggests that intraspecific aggregation occurs because it facilitates mate finding. Finding a mate may indeed be a problem in small populations of dung beetles (below), but there is no evidence for any mechanism (pheromone) enabling *Aphodius* to aggregate actively in the same droppings.

In one study, pairwise interspecific aggregation (covariance) increased from *Aphodius* to *Cercyon* to *Sphaeridium*, which is also the order of increasing ecological and morphological similarity of the species in the three genera (Fig. 5.3). In other words, the more similar the species are in their biology, the greater their spatial covariance is likely to be. This is not a surprising result, but note the corollary: any differences among species that directly or indirectly affect their movement behavior may affect their spatial covariance and hence their interactions. Other differences between species than the ones directly related to resource use can therefore be important in competition. Examples are slight differences in diel activity (Landin 1961; Koskela 1979), temperature tolerances (Landin 1968; Koskela 1979) and in the distribution of flight distances (see Otronen and Hanski 1983 for *Sphaeridium*).

Dynamics. As the interactions within and among species are localized in droppings, it is clear that the spatial distribution of individuals must affect population dynamics (the general argument is presented in Chapter 1). Increasing intraspecific aggregation increases the frequency of intraspecific interactions and hence amplifies intraspecific competition; increasing spatial correlation between two species increases interspecific interactions and com-

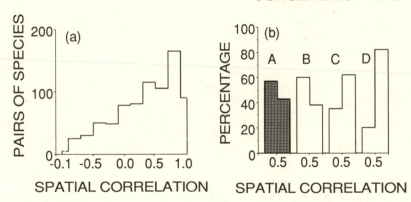

Fig. 5.3. (a) Pairwise correlation coefficients for twelve species of dung-breeding beetles trapped in four pastures with five pitfalls in each for one summer. Populations with less than ten individuals in the five traps have been omitted, leaving 795 pairs of species. (b) Comparison between noncongeneric (shaded histogram A) and congeneric pairs of species, with the values divided into two classes, correlation coefficient smaller or greater than 0.5. The genera and the results of χ^2 tests comparing histograms B to C with A are: B, *Aphodius* (6 species), $\chi^2 = 0.59$, NS; C, *Cercyon* (3), $\chi^2 = 7.96$, P<0.01; and D, *Sphaeridium* (2), $\chi^2 = 10.09$, P<0.001. For further explanation, see the text (from Hanski 1987b).

petition, but intraspecific aggregation facilitates coexistence even when there is substantial (but not complete) interspecific correlation (Ives 1988a). In the field, interspecific correlations tend to be low in *Aphodius* (Fig. 5.2; but see Holter 1982), and I conclude that aggregated spatial distributions are a factor that needs to be considered along with niche differences while analyzing community structure and dynamics in *Aphodius*.

Abundance Distributions

An assemblage of *Aphodius* sampled near Oxford, southern England, in 1977 had twenty-two species (Appendix B.5), exactly half of the British fauna. The distribution of species abundances in this sample is bimodal in the sense explained in Table 5.5. I suggest that the bimodality is due to the twenty-two species representing a mixture of local and nonlocal species. Presence of nonlocal species at a site is of interest because it indicates the potential for an establishment of new populations.

Three observations support the hypothesis that the six rarest species were not breeding in the study area in 1977. First, sex ratio in the pooled sample of the six rarest species was female-biased (71%), while the sex ratio was between 40% and 65% in the remaining sixteen species (the difference is significant; Hanski 1986). Because *Aphodius* females move longer distances and/or move long distances more frequently than males (Hanski 1980b), this result suggests that many individuals of the six rarest species had moved a long distance. Second, 32% of the individuals in the six rarest species were actually caught in the center of Oxford, far from the nearest pastures, while in the remaining sixteen species only 2% were caught in town (a significant difference; Hanski 1979). And third, simple calculations (Hanski 1979, 1986) suggest that if the rarest species have no special adaptations, such as long-distance

TABLE 5.5
Abundance distributions in local assemblages of *Aphodius*.

Abundance Classes															Species	Sample Size	Locality	Reference
1	2	3	4	5	6	7	8	9	0	1	2	3	4	5				
1	—	—	1	3	1	2	1	2	—	1	1	—	—	—	12 + 1	5,137	N. England	White (1960)
3	1	1	—	1	—	3	1	—	5	2	3	1	1	—	16 + 6	32,288	S. England	Hanski (1979)
1	—	—	—	1	—	1	2	3	2	—	4	1	—	—	13 + 1	20,659	Denmark	Holter (1982)
—	—	—	—	—	—	—	2	2	1	2	—	—	1	1	11 + 0	38,631	Japan	Yasuda (1984)

Notes: The number of species is given as the number of presumed local species + the number of presumed nonlocal species. Yasuda's (1984) study from Japan is from a south-temperate locality, where *Aphodius* coexist with many Scarabaeidae.

The numbers of species are given in logarithmic abundance classes: 1–2 individuals (class 1), 3–4 (2), 5–8 (3), etc.

sex pheromones to locate a mate (for which there is no evidence), their density in the study pastures was so low that they would quickly go extinct. Hanski (1979) estimated that such an "underpopulation" problem (Andrewartha and Birch 1954) would be serious at average densities below 0.5 beetles per dropping. It is encouraging that the density of the rarest presumed local species was around 0.5 beetles per dropping, and an order of magnitude lower in the most abundant of the six presumed nonlocal species. Table 5.5 gives the species abundance distribution for some other comprehensive samples of *Aphodius*. These results cannot be used to test the above prediction quantitatively, but they qualitatively support the conclusion about a bimodal distribution consisting of common local and rare nonlocal species.

Several species of *Aphodius* are regularly or occasionally so numerous in cattle pats that they disintegrate the pats in a matter of hours or days, cause mass mortality of fly larvae (Wolcott 1922; Hammer 1941; Merritt 1974), and make the pat entirely unsuitable for *Aphodius* larvae. Table 5.6 lists some of the best-known cases. Densities may reach several hundreds (Merritt 1974; Hanski, unpubl.) or even thousands per pat (Mohr 1943; Holter 1982). Some of the outbreak species are typical coprophages (e.g., *A. rufipes* and *A. fimetarius*) but others may have generalist or saprophagous larvae (e.g., *A. prodromus*, *A. contaminatus*, and *A. distinctus*), which may explain the exceptionally high adult densities in pats.

Temporal Stability

The climatic and environmental instabilities in north temperate regions during the last glaciation were associated with extensive movements of beetles with changing climate and vegetation zones: to the south when glaciers advanced, back to the north when they retreated (Coope 1975, 1977, 1978). One may ask how stable the insect communities remained in the midst of these fluctuations in species' ranges. Table 5.7 compares a Pleistocene fossil sample (43,000 BP) with a late Roman (1,600 BP) and two modern samples of *Apho-*

TABLE 5.6
Cases of mass occurrences of *Aphodius*.

Species	Locality	Season	Reference
A. lividus	Puerto Rico	Dry season	Wolcott (1922)
A. prodromus	N. Europe	Spring, autumn	Rainio (1966)
A. contaminatus	N. Europe	Autumn	Hanski (1979)
A. rufipes	N. Europe	Late summer	Holter (1982)
A. distinctus	Illinois, USA	Spring, autumn	Mohr (1943)
A. fimetarius	California, USA	Late spring	Merritt (1974)
A. fimetarius	Denmark	Spring	Hammer (1941)

TABLE 5.7

Four samples of coprophagous *Aphodius* from the British Isles, spanning from a temperate interlude in the middle of the last glaciation to the Late Roman period to the present.

Species	Isleworth, Middlesex, 43,000 BP	Barton Court, Oxfordshire, 1,600 BP	Oxford, Oxfordshire, 1977	Northern Pennines, 1955–1957
prodromus	4	2	19	4
rufipes	2	+	5	1
fimetarius	2	+	1	+
ater	8	—	7	14
constans	4	—	+	+
tenellus	38	1	—	7
fossor	2	—	3	—
erraticus	2	—	—	—
bouvoiloiri	1	—	—	—
costalis	1	—	—	—
contaminatus	—	19	12	1
sphacelatus	—	20	+	+
scybalarius	—	11	+	—
paykulli	—	10	+	—
distinctus	—	1	—	—
conspurcatus	—	+	—	1
pusillus	—	—	2	—
haemorrhoidalis	—	—	1	—
rufescens	—	—	3	—
equestris	—	—	8	—
obliteratus	—	—	3	—
lapponum	—	—	—	31
depressus	—	—	—	5
Number of species (±SD) in a sample of 64 individuals	10	7.9±1.1	11.9±1.2	8.4±1.3
Total number of species	10	11	22	14
Sample size	64	208	32,288	5,139

Notes: The modern samples are pooled annual samples. The numbers of individuals in the three largest samples are for one random draw of sixty-four individuals; + indicates that the species occurred in the original sample. Species diversity was calculated with rarefaction (according to Hanski 1986).

dius from the British Isles. The Pleistocene sample is from a short but warm interlude in the middle of glaciations, with climate similar to the present one (Coope and Angus 1975) and with abundance of large herbivores, for example, the reindeer and bison.

The four samples are remarkably similar (Table 5.7). Species diversity varies little, from eight to twelve species in a sample of sixty-four individuals (the size of the fossil sample). Seven and nine species are shared between the fossil and modern samples, and between the Roman and modern samples, respectively (though note that this comparison is affected by sample sizes). Two of the species in the fossil sample were "lost" from northern Europe during their last migration: *A. bouvouloiri* is now restricted to the Iberian peninsula (Chapter 14), while *A. costalis* occurs in an area north of the Caspian Sea (Balthasar 1963), but these are exceptions that confirm the rule: remarkable stability of species composition.

Can we conclude from these data that, when periods with similar climates are compared, species diversity of *Aphodius* has remained practically constant for tens of thousands of years? This conclusion is not warranted, because Table 5.7 does not tell us much about the rare species—the ones that are not likely to be found in a sample of sixty-four individuals—though they usually make a big contribution to the regional species pool. One example suggests that important changes in species diversity may occur at a much shorter time scale than thousands of years.

Johnson (1962) reports on the apparent changes in the dung beetles of Lancashire and Cheshire, in northwestern England, from the beginning of this century until the early 1960s. Of the thirty-three coprophagous *Aphodius* known from the area by 1912, about ten have disappeared. Only one new species has been discovered, *A. constans* (a species present in the fossil sample!). A telling point is that none of the ten species that Johnson (1962) claims have gone extinct between 1912 and 1960 are included in the fossil sample, while all of the species in the fossil sample excluding the above-mentioned *A. bouvoiloiri* and *A. costalis* were still present in Lancashire and Cheshire in 1962. This adds another aspect to the stability of the species in the fossil sample: these species have apparently retained their status as core species in the community for thousands of years, in spite of the "migrations" forced upon them by the fluctuating climate.

This result raises a paradox. The number and even the species composition of the abundant core species have remained relatively constant for 43,000 years, yet the size of the regional species pool in Johnson's (1962) study declined by 25% in 50 years. A probable solution is that the numbers of the common and rare species are not always regulated by the same factors. Although the number of the abundant core species (Hanski 1982) may be limited by, for example, competition, the extinction-prone rare species are more affected by even small changes in the environment, affecting the extinction-migration dynamics that one expects to be important in rare species. The causes of the extinctions in Johnson's (1962) example are not known, but they presumably involve, in one way or another, changes in cattle management.

Other examples of rapid abundance changes can be cited. In three studies conducted in southern Finland in 1932 (Pyhälahti 1934), in 1960–61 (Rainio 1966), and in 1966–67 (Koskela and Hanski 1977), the frequency of *A. fimetarius* declined from 83% to 30% to 14%. The management of cattle in Finland has changed during this period from largely forest and meadow grazing to entirely pasture grazing. Howden and Scholtz's (1986) observations about great abundance changes in a dung beetle community in Texas are also worth noting here, though this community is outside the north temperate region and is dominated by Scarabaeidae rather than Aphodiidae. Howden and Scholtz (1986) resampled a locality studied 10 years earlier by Nealis (1977). During this period the dominant species had changed dramatically: *Digitonthophagus gazella* (from 0% to 24% of all individuals collected) and *O. alluvius* (from 0% to 25%) had increased, while *O. pennsylvanicus* had decreased (from 55% to 12%). *Digitonthophagus gazella* is an introduced species in North America (Chapter 16) and had spread to the study area at a time between the two surveys, while *O. alluvius* may have benefited from a gradual change of vegetation (Howden and Scholtz 1986). The earlier dominant *O. pennsylvanicus*, as well as probably a few other species, may have declined because of competition with *D. gazella* (Howden and Scholtz 1986).

5.6. SUMMARY

The four kinds of community patterns reviewed in the previous section show two common elements. First, large samples of *Aphodius* from limited areas appear to be mixtures of common and very rare species; I suggested that the latter consist of species not breeding in the locality but present due to migration from elsewhere (Table 5.5). Second, ecological segregation of species is significantly better in common species than in rare ones (Fig. 5.1), and in northern Europe the common species have, for the most part, remained common for tens of thousands of years (Table 5.7) in spite of fluctuating climate and repeated range shifts in the species. In contrast, rare species may frequently go extinct and, presumably, establish new populations at new sites. It seems useful to propose, on this basis, a dichotomy between "core" and "satellite" species (Hanski 1982). The core species are present in all or nearly all suitable sites all the time; they interact strongly with other core species, and within-locality processes are crucial in their ecology. The satellite species play an extinction-immigration game at the regional level; they occur at densities where intraspecific competition is unimportant, and they may survive interspecific competition (when it occurs) because of aggregated distributions of the common species among droppings (Fig. 5.2). The satellite species are expected to be much affected by average distances between pastures, and they are likely to suffer from any large-scale changes in husbandry. In brief, different sorts of processes are important in the core and satellite species.

In the core species, further studies should be directed toward field experiments on between-dropping movements, better understanding of intraspecific aggregation, and demographic analyses. In the satellite species, studies should be conducted at the regional level. The changes that are presently taking place in the numbers and distribution of cattle in many areas of north temperate regions alter the regional pattern of resource availability and provide an excellent opportunity to study regional dynamics in rare species.

CHAPTER 6

South Temperate Dung Beetles

Jean-Pierre Lumaret and Alan A. Kirk

6.1. INTRODUCTION

Mediterranean dung beetles are well known taxonomically (Dellacasa 1983; Baraud 1985; Martin-Piera 1986) and the basic biology of most species has been worked out in great detail (Lumaret 1975, 1978, 1983a; Kirk 1983; Lumaret and Kirk 1987). Biogeographical and ecological studies have established the temperature, rainfall, and edaphic requirements of the species as well as their altitudinal and latitudinal distribution limits; many such studies have been conducted in Spain (Galante 1979; Martin-Piera 1982; Mesa 1985; Kirk and Ridsill-Smith 1986), in France (Lumaret 1978, 1978–79a, 1978–79b), and in Italy and Greece (Binaghi et al. 1969; Carpaneto 1974, 1981; Pierotti 1977; Pittino 1980). More recently, entomologists have focused on the structure of dung beetle communities, on the relationships among coexisting species and on interspecific competition (Lumaret 1980a, 1983a; Carpaneto 1986; Carpaneto and Piattella 1986; Lumaret and Kirk 1987). Special attention has been paid to quantifying the amount of dung buried by beetles (Lumaret 1980a, 1986) and to their ability to compete with dung-breeding flies. Some Mediterranean dung beetles have been introduced to Australia for the control of dung-breeding flies (Kirk 1983; Kirk and Feehan 1984; Ridsill-Smith and Kirk 1985; Kirk and Lumaret 1991; Chapter 15).

In this chapter, we describe and analyze the distributional ecology of Mediterranean dung beetles, including the influence of dung type, soil type, vegetation cover, and local climate on the occurrence of beetles. Section 6.4 examines dung beetle populations, emphasizing seasonality and the different strategies of dung use by different species. Most of the information is available from the Languedoc *sensu lato*, a Mediterranean province in southeastern France. This relatively small region consists of a mosaic of different types of localities into which the local assemblages of dung beetles are drawn from the regional species pool. This situation is particularly favorable for the study of factors affecting the assembly of local dung beetle communities.

6.2. BIOGEOGRAPHY OF THE LANGUEDOC

Languedoc, the meeting point of Mediterranean, Medio-European, and montane influences, comprises about 74,600 square kilometers between the Rhône

River and the Pyrenees Mountains (Fig. 6.1). Several concentrically arranged geographical entities may be distinguished from the coast to the Massif Central: the sandy shore, the coastal plains, the Garrigue region, the high plateaus of the Causse and the Cevennes, with Mont Aigoual (1,565 m), still less than 80 km from the sea. Most of these zones have a Mediterranean climate (Emberger 1954), characterized by low summer rainfall coinciding with maximum day length and maximum temperatures, a combination that creates a more or less pronounced drought. The seasonal drought period constitutes the most important ecological factor for Mediterranean organisms, including dung beetles.

The Garrigue region north of Montpellier, where most of the ecological studies on dung beetles have been conducted, has a cool, subhumid Mediterranean climate, with a strong north-south climatic gradient. The climate is characterized by extreme unpredictability, with the average daily minimum temperature of the coldest month ranging from $-4.0°C$ to $+3.3°C$ (in the years 1954–84). Frosts are common from November to April (40–70 days per year). Annual rainfall varies from 700 mm to 2,040 mm, with the average of 1,200 mm in 1954–84.

The extreme edaphic and climatic heterogeneity of the Languedoc province gives rise to a multiplicity of combinations of ecological factors, assigning a definite pool of species to different regions within Languedoc. Local assemblages of dung beetles are drawn from the regional species pools, depending on the edaphic conditions, microclimate, vegetation structure and availability of dung. The end result is a most interesting ecological situation, as two sites located only a few kilometers apart may support distinct communities drawn from the same species pool.

A detailed study of the dung beetle assemblages at 731 Languedoc sites led to a definition of thirteen main faunistic regions, each with its own pool of species (Lumaret 1978, 1978–79a,b; Fig. 6.1). Among the 103 species recorded in the Languedoc, Scarabaeidae represent 35%, Geotrupidae 9%, and Aphodiidae 56%. Species richness and the relative importance of the three families of dung beetles vary from one faunistic region to another. In the most typically Mediterranean communities, affected by summer drought, Scarabaeidae comprise the core of the species. The number of Geotrupidae species increases in the Atlantic communities at the expense of Scarabaeidae, whereas the montane communities are characterized by large numbers of Aphodiidae (see also Chapter 14).

On a larger spatial scale, there is a distinct gradient in the proportions of the three dung beetle families from northern Europe to Morocco (Table 6.1), with Aphodiidae (mostly dwellers) dominating in the north and Scarabaeidae (rollers and tunnelers) dominating in southern Europe. This gradient can be explained by the ecology of the beetles and by the drought in the south, which reduces the time available for dung utilization and thereby favors dung burying tactics (Section 6.4 and Lumaret 1987).

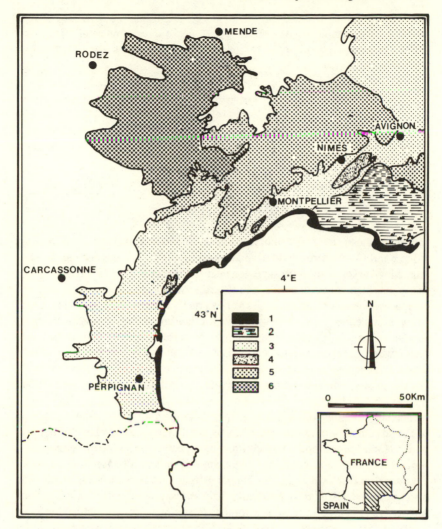

Fig. 6.1. The main faunistic regions in the Languedoc: 1, coastal dunes; 2, Camargue; 3, coastal plains; 4, coastal hills; 5, Garrigue; 6, Causse.

Languedoc Dung Beetles

The origin of Scarabaeidae (9 genera) present in the Mediterranean communities of Languedoc is diverse. Following La Greca's nomenclature (1964), the species belong to four main chorological patterns: (1) Euro-Centro Asiatic species (19% of species), distributed from Mediterranean Europe to central or eastern Asia; (2) Euro-Turanic species (35%), found from North Africa and Mediterranean Europe to Central Europe and Asia Minor; (3) Euro-Mediter-

TABLE 6.1

Percentage distribution of dung beetle species in the three families of Aphodiidae (Aph), Geotrupidae (Geo), and Scarabaeidae (Sca) from north to south in temperate regions.

Region	Aph	Geo	Sca	Reference
N. Europe	95	5	—	Chapter 5
Poland	73	7	20	Stebnicka (1976)
Belgium	71	9	20	Janssens (1960)
S. France	56	9	35	Lumaret (1978)
Spain	45	8	47	Kirk & Ridsdill-Smith (1986)
Morocco*	38	5	57	Maarouf (unpubl.)

* Casablanca region.

ranean species (16%), distributed from north Mediterranean Europe to Asia Minor; and (4) western Mediterranean species (30%). Languedoc can be considered as a southward passage for successive colonizers (Lumaret 1978), many of which are of Asiatic origin (Martin-Piera 1983).

Species richness of dung beetles varies greatly between the faunistic regions of the Languedoc. Four areas stand out for their great number of species: Camargue, Garrigue, Causse, and Albères (the extreme eastern end of the Pyrenees), where sheep and cattle rearing is an ancient practice. The coastal plains, where large areas are used for vine growing, have a low to intermediate number of species, while the shore dunes and coastal hills have only a few dung beetles.

The large number of dung beetle species in the Camargue is primarily due to the large numbers of horses and cattle and to the great diversity of habitats and edaphic conditions. The commonest species prefer deep, damp soils around lagoons, for example, *Geotrupes spiniger*, *Onthophagus taurus*, *O. vacca*, and *O. nuchicornis*. The Camargue has several Medio-European species, such as *Aphodius scrutator*, *A. erraticus*, and *Copris lunaris*, which are generally scarce in regions with severe summer drought, and there is an unusually large number of Aphodiidae (38 species). The adverse effects of low rainfall (less than 600 mm) and the 2-month drought are compensated for dung beetles by the damp soils (Lumaret 1978, 1978–79a,b).

The Garrigue supports the most Mediterranean dung beetle communities in the Languedoc. People have lived in this region for millennia, continuously changing the landscape, following the rises and recessions of the economy. The middle of the 19th century saw the highest population density, accompanied by extensive cutting of the remaining woodlands and large herds of sheep and goats. Tens of thousands of sheep and goats were driven each June from the Garrigue to Mont Aigoual to pass summer. Rural depopulation started at the end of the 19th century and was further intensified during the 20th century.

Although the Garrigue still has the highest dung beetle richness in the Languedoc, the number of species has diminished (presumably because of decreased availability of dung), and many species that were abundant in the beginning of the century have become very scarce.

A comparison of the collection records between the end of the last century and the present for Languedoc and Provence (Caillol 1908, 1954; Thérond 1975, 1980; Lumaret 1978; Lumaret and Kirk, unpubl.) reveals that there is an especially significant loss of the largest species, *Scarabaeus*, *Gymnopleurus*, *Onitis*, *Cheironitis*, and *Thorectes*. *Scarabaeus typhon*, common in many Provence and Languedoc localities at the turn of the century (Caillol 1908), has become rare and sporadic, with only a few individuals found annually in the Garrigue. *Scarabaeus pius*, once common in the Garrigue around Nimes and Avignon, seems to have completely disappeared since 1925 (Thérond 1980). *Gymnopleurus mopsus* and *G. flagellatus* were widely distributed from Alsace in the north to the western Atlantic coast (Charente-maritime) in the beginning of the century, and they were common and locally very abundant in the Garrigue around Avignon. Today they are very rare, as are the other *Gymnopleurus* species, *G. geoffroyi* and *G. sturmi*. *Thorectes intermedius* and *Cheironitis hungaricus*, collected regularly around Montpellier at the beginning of the century, have not been found for many years, while *Onitis belial* and *Scarabaeus sacer* are now restricted to a few localities in the Camargue.

Despite the general disappearance and decline of many species, some sites where sheep or cattle still graze regularly have preserved a rich fauna. Many western Mediterranean species occur in the Garrigue, such as *Scarabaeus laticollis*, *Copris hispanus* spp. *hispanus*, *Bubas bubalus*, *Onthophagus maki*, *O. emarginatus*, and *O. opacicollis*. Aphodiidae constitute half of the species, but while the distribution of most Scarabaeidae is relatively even across the Garrigue, Aphodiidae are more sporadic. One reason for this difference may be that, for example, *Onthophagus* have generally a wider food selection than Aphodiidae.

In spite of its more rigorous climate, the Causse (800–1,000 m altitude) support richer dung beetle communities than the Garrigue. Along the transition from the Garrigue to the Causse, many cases of apparent replacement of congeners are evident. For example, *Copris hispanus* is replaced by *Copris lunaris* or *C. umbilicatus*, and *Onthophagus opacicollis* is replaced by two sister species, *O. similis* and *O. fracticornis* (Lumaret 1978).

Kirk (unpubl.) compared three sites in the Languedoc area, St-Nazaire-de-Pezan in the Camargue, Pic-St-Loup in the Garrigue, and La Couvertoirade on the Causse du Larzac. Dung beetles were trapped simultaneously once a week over one year (1979) with cattle-dung baited traps (Table 6.2). The three sites, all of which are open and grazed by cattle and sheep, are characteristic of the main faunistic regions. Species requiring high temperatures are restricted to the Camargue and the Garrigue (*C. hispanus* and the genus *Bubas*).

TABLE 6.2
Comparison of three sites in the Languedoc: St-Nazaire-de-Pezan (Camargue), Pic-St-Loup (Garrigue), and La Couvertoirade (Causse du Larzac).

Species	Camargue	Garrigue	Causse	Length (mm)	Type of Species
Geotrupes niger	—	—	11	12–23	T
Geotrupes spiniger	5	—	3	18–26	T
Scarabaeus typhon	—	1	—	20–28	R
Scarabaeus laticollis	—	115	—	15–23	R
Sisyphus schaefferi	—	2	6	8–10	R
Copris lunaris	—	—	12	15–20	T
Copris hispanus	2	16	—	15–20	T
Bubas bison	46	—	—	13–18	T
Bubas bubalus	16	3	—	13–22	T
Euoniticellus fulvus	32	6	7	7–11	T
Caccobius schreberi	9	—	9	4–7	T
Euonthophagus gibbosus	—	—	24	7–12	T
Onthophagus taurus	28	2	2	5–11	T
Onthophagus emarginatus	—	99	—	5–7	T
Onthophagus furcatus	—	11	—	4–5	T
Onthophagus verticicornis	—	—	19	6–9	T
Onthophagus semicornis	1	—	—	5–6	T
Onthophagus grossepunctatus	—	—	168	4–5	T
Onthophagus ruficapillus	132	—	—	4–5	T
Onthophagus coenobita	1	—	—	6–10	T
Onthophagus fracticornis	—	—	29	7–10	T
Onthophagus lemur	1	293	14	5–8	T
Onthophagus maki	—	98	—	4–7	T
Onthophagus vacca	15	9	92	7–13	T
Aphodius erraticus	37	—	16	6–9	T,D
Aphodius scrutator	2	4	6	10–15	T,D
Aphodius fossor	—	—	20	10–11	T,D
Aphodius haemorrhoidalis	49	65	—	4–5	D
Aphodius luridus	2	24	1	6–9	T,D
Aphodius satellitius	45	—	—	6–8	D
Aphodius distinctus	—	2	143	4–6	D
Aphodius contaminatus	—	—	2	5–7	D
Aphodius prodromus	—	10	6	4–7	D
Aphodius sphacelatus	93	—	—	4–6	D
Aphodius consputus	557	—	32	4–5	D
Aphodius reyi	19	—	—	4–5	D
Aphodius merdarius	344	2	—	4–5	D
Aphodius paracoenosus	17	—	—	4–4	D
Aphodius fimetarius	—	6	347	5–8	D
Aphodius aestivalis	—	—	3	5–8	D

TABLE 6.2 (*cont.*)

Species	Camargue	Garrigue	Causse	Length (mm)	Type of Species
Aphodius scybalarius	6	—	—	5–8	D
Aphodius constans	2	10	—	5–6	D
Aphodius immundus	1	—	—	5–6	D
Aphodius varians	1	—	—	4–6	D
Aphodius granarius	49	20	2	3–5	D
Total number of species	27	21	24		
In a sample of	1,512	798	974		
Rarefied species number	23.7 ± 1.4	21	23.7 ± 0.5		

Notes: Species diversity was calculated with rarefaction as the number of species in a sample of 798 individuals (Simberloff 1979). R, T, and D denote rollers, tunnelers, and dwellers. Traps were examined once a week throughout 1979, and the results are given as the numbers of beetles per trap per year.

In contrast, *Onthophagus verticicornis* and *O. fracticornis*, absent in the Camargue and from the warmest Garrigue sites, are present on the Causse. The Aphodiidae are particularly numerous at St-Nazaire-de-Pezan and at La Couvertoirade (81% and 59%, respectively, of dung beetles trapped), benefiting from the edaphic and climatic conditions, which are more favorable than in the Garrigue (humid soil in the Camargue even in summer despite the drought; heavy precipitation in the Causse). Some species absent in the Garrigue (too dry) can be found at the same time in the Camargue and in the Causse (e.g., *A. erraticus*). Nevertheless, the most abundant species in the Camargue (*A. consputus*, *A. merdarius*) and in the Causse (*A. distinctus*, *A. fimetarius*) are different.

The coastal dunes, which occur as a narrow band between the sea and the littoral lagoons, constitute a fragile faunistic region in the Languedoc. This region, which is under constant pressure from the uncontrolled development of tourism, supports only eight species, including *Scarabaeus semipunctatus*, a stenotopic west Mediterranean species restricted to live in sand dunes (Lumaret 1978–79b). Behind the dune belt, in the rare spots where sand and silt are mixed together, *Scarabaeus sacer* may be found in horse droppings.

The coastal hills have very low species diversity due to their dryness and lack of cattle; hares and rabbits are the largest mammals in this region. The coastal hills represent the ancient emergent islands of the Miocene transgression, which flooded the lowlands of Languedoc. The current faunistic elements do not go back to this epoch, but the contrasting history of the dry and stony hills and the surrounding damp plains may partly explain their current faunistic differences. Nine species, well adapted to the prevailing edaphic and climatic constraints, occur on the hills, including *Onthophagus emarginatus* and *Aphodius bonnairei*, both colonizing the heaps of rabbit pellets that mark the territories of these small mammals (Lumaret 1978, 1978–79b).

6.3. ASSEMBLY OF LOCAL COMMUNITIES

Numerous ecological studies on dung beetles have been carried out in the Garrigue, at sites where sheep flocks and wild mammals, such as wild boars, badgers, and rabbits, still remain abundant. The landscape is a patchwork because of its diverse geology, a strong south-north climatic gradient, and varying physiognomy of vegetation, ranging from the grassland or *erme* landscape, where hemicryptophytes, chamaephytes, and annuals dominate, to entirely forested habitats. These three factors will be considered in turn below. The Garrigue has sixty-nine species of dung beetles, a large pool from which species assemble into local communities according to their ecological requirements.

Influence of Soil Type

Species composition and abundance of dung beetles was examined at two homologous sites differing only in the parent rock (Lumaret 1983a). The first site was a *Brachypodium ramosum* pasture on terra rossa and compact limestone, the second one was a *B. phoenicoides* pasture on a flinty limestone with marls. The vegetation structure was the same at the two sites, but both qualitative and quantitative differences in dung beetles were observed between the pastures. The Shannon-Weaver diversity and the rarefied species number were higher on marls than on compact limestone (Lumaret 1983a). However, the compact limestone site had 32% of the regional species pool, with rollers well represented (4 species, 16% of individuals), while rollers accounted for only 6% of the community on marls. Differences in sample size reflect differences in density: only 85 individuals were collected with two traps operated throughout one year on marls, compared with 1,766 individuals on the limestone.

This extremely rigorous sorting of dung beetles, most of which are burying species, is due to a difference in the mode of soil drainage in the two habitats. Whereas compact limestone sites dry quickly after a rainy period, those on marls or flinty limestone are poorly drained and become flooded periodically. This difference is fundamental because, for numerous species that nest in the soil, the capacity of the soil to retain water, the depth of the saturated horizons, and the period and duration of soak determine success or failure of reproduction. A soil that remains soaked with water for too long during the reproductive period is unfavorable for many species, and especially so for many tunnelers. The species that can cope with such conditions cannot build up large populations because of high larval mortality. One such species is *Bubas bison*, which is well adapted to heavy clay soils that are wet or flooded in winter

(Kirk 1983). The combined field mortality of the eggs and larvae of *B. bison* averaged 33% at a Camargue site (Kirk 1983).

The texture of the substrate can significantly influence the spatial distribution of dung beetles. Rollers provide good examples. Thus *Scarabaeus semipunctatus* is restricted to the coastal sand dunes; behind the dunes it is replaced by *S. sacer*, which prefers silty soils; and while *S. laticollis*, *S. typhon*, and *Sisyphus schaefferi* are more abundant on compact clay soils (Table 6.3). *Typhaeus typhoeus*, with tunnels occasionally reaching 1.4 m, prefers sandy soils that are well drained to a great depth. Brussaard (1985a) demonstrated by laboratory and field experiments that the length of the tunnel depends directly on the level of the water table and the upper level of the saturation zone, while the survival of the eggs and larvae was correlated with the soil's moisture content.

The influence of the substrate can be modified by climatic factors. For example, *Onthophagus emarginatus* is found on all types of substrate (sand, silt, marls, clay, etc.) in the areas of the Languedoc with summer drought. Elsewhere the influence of the substrate becomes more significant, and *O. emarginatus* becomes restricted to sandy soils that are dry and well drained (Lumaret 1978). The response of *Aphodius* to the substrate depends on the ecology of the species. Many dwellers require a humid substrate, which prolongs the time during which the dung pat remains suitable for the larvae. Some species oviposit at the dung-soil interface, and the larvae develop in tunnels in the soil (Section 6.4). In these cases substrate characteristics, especially the drainage and temperature, are important. Most of the sites where *A. biguttatus* and *A. quadriguttatus* have been collected are located on limestone (Lumaret 1978, 1980b).

TABLE 6.3
Presence of rollers on different soil types.

Species	Soil Type			Length (mm)
	Sandy	Silty	Clay	
Scarabaeus semipunctatus	+ + +	+	0	20–28
Scarabaeus sacer	+	+ + +	0	28–32
Scarabaeus typhon	0	+ +	+ + +	20–28
Scarabaeus pius	0	0	+ + +	20–28
Scarabaeus laticollis	0	+	+ + +	15–23
Gymnopleurus sturmi	0	0	+ + +	10–13
Gymnopleurus mopsus	0	0	+ + +	7–14
Gymnopleurus flagellatus	0	0	+ + +	8–11
Sisyphus schaefferi	0	+ +	+ + +	8–10

Note: Species' affinity with the soil type is indicated by 0 (no affinity), + (slight affinity), + + (moderate affinity), and + + + (high affinity).

The influence of the substrate on local and regional distribution of dung beetles is a general phenomenon, even if homologous soils do not have the same effect in different climates. Clay soils, because they retain moisture, support pastures for a longer period of time than other soils in the Languedoc. This means that cattle and sheep produce more dung for longer periods, a situation that is favorable for dung beetles. In contrast, Nealis (1977) found that in southern Texas clay soils were depauperate both in species number and biomass of beetles. In South Africa, comparisons between sites in bushveld on clay-loam and on sandy loam showed that the highest species number and the largest numbers of beetles were associated with the bushveld on sandy loam (Doube 1983; Chapter 8).

Influence of Dung Type

In the Garrigue, most dung beetles feed on all types of dung that are available, but a few species appear to be more specialized. Large species such as *Bubas bison*, *Aphodius fossor*, and *A. rufipes* prefer large droppings (mainly cattle dung), as do small *Aphodius* species that oviposit in droppings (Section 6.4; Lumaret 1987). In contrast, the small *A. bonnairei* is found only in heaps of rabbit pellets. Another specialist is *Onthophagus coenobita*, a medium-sized species, which is essentially restricted to human feces, though it has also been observed to nest and develop using rotten leaves of *Artemisia* (Compositae) (Moretto, pers. comm.). *Geotrupes niger* exploits mainly human feces in the Languedoc, but in Corsica it is found in many types of droppings (cattle, horse, human). Most species normally feed in fresh dung, but the adults of *Aphodius elevatus* are attracted by old, dry droppings, though the female still oviposits in fresh droppings and in the dung-soil mixture at animal resting places (Section 6.4).

Influence of Vegetation Cover

Dung beetles are generally more abundant in some habitats than in others, with vegetation cover influencing these preferences (Howden and Nealis 1975; Hanski and Koskela 1977; Nealis 1977; Doube 1983; Lumaret 1983a; Carpaneto 1986; Fincher, Blume, et al. 1986).

Lumaret (1984a) compared six sites in southern Languedoc on compact limestone belonging to the same vegetation succession (a mixed evergreen-deciduous oak formation). The six sites form a physiognomic gradient from the most open site to well-developed forest (Table 6.4). Dung was abundant at all sites, as most of the sites (sites 1 to 5) are regularly grazed by sheep and goats, and all sites are visited by wild boars, badgers, and people. More than ten thousand dung beetles belonging to thirty-four species were trapped during one year, the highest species number occurring at the most open sites. As soon

TABLE 6.4
Numbers of dung beetles trapped along a vegetation gradient from the most open site (1) to closed forest (6).

Species	Habitat Type						Length (mm)	Type of Species
	1	2	3	4	5	6		
Percentage Cover in Vegetation Classes <2 m:	80	70	85	100	85	75		
2–6 m:	0	10	15	80	85	55		
>6 m:	0	0	0	0	5	65		
Scarabaeus typhon	2	—	1	1	—	—	20–28	R
Scarabaeus laticollis	92	98	57	—	—	—	15–25	R
Gymnopleurus flagellatus	1	—	—	—	—	—	8–11	R
Sisyphus schaefferi	190	207	191	37	13	—	8–10	R
Copris hispanus	19	—	9	—	—	—	15–20	T
Euoniticellus fulvus	—	—	5	—	—	—	7–11	T
Caccobius schreberi	—	1	27	—	—	—	4–7	T
Euonthophagus amyntas	19	—	1	—	—	—	7–12	T
Onthophagus emarginatus	103	42	37	1	—	—	4–7	T
Onthophagus furcatus	3	6	7	—	—	—	3–4	T
Onthophagus verticicornis	33	1,016	15	190	91	—	6–9	T
Onthophagus grossepunctatus	—	57	—	3	1	—	4–5	T
Onthophagus joannae	6	810	330	709	38	—	4–6	T
Onthophagus coenobita	2	158	146	95	36	105	6–10	T
Onthophagus lemur	746	480	2,191	162	122	3	5–8	T
Onthophagus maki	98	71	331	1	—	—	4–7	T
Onthophagus vacca	429	—	25	1	—	—	7–13	T
Aphodius erraticus	—	2	—	—	—	—	6–9	T,D
Aphodius subterraneus	—	—	1	—	—	—	5–7	D
Aphodius haemorrhoidalis	—	1	58	—	—	—	4–5	D
Aphodius luridus	13	19	31	—	—	—	6–9	T,D
Aphodius paracoenosus	1	71	16	8	1	—	4–5	D
Aphodius pusillus	—	1	—	—	—	—	3–4	D
Aphodius distinctus	—	—	—	1	—	—	3–5	D
Aphodius prodromus	—	1	—	—	—	—	4–7	D
Aphodius quadrimaculatus	—	1	—	—	—	—	3–4	D
Aphodius biguttatus	—	31	18	7	—	—	2–3	D
Aphodius fimetarius	1	3	5	—	—	—	5–8	D
Aphodius constans	—	5	210	5	—	—	5–6	D
Aphodius granarius	4	—	3	1	1	—	3–5	D
Oxyomus silvestris	1	—	—	—	—	—	2–4	D
Typhaeus typhoeus	1	—	—	—	—	—	10–20	T
Geotrupes niger	2	13	12	2	2	—	12–23	T
Geotrupes spiniger	—	—	—	1	—	—	18–26	T
Sample size	1,766	3,094	3,727	1,225	305	108		
Number of species	21	22	24	17	9	2		
Rarefied species number	10.2	11.7	12.3	7.4	6.6	2		
Shannon-Weaver index	2.58	2.76	2.37	2.04	2.11	0.26		
Equitability	0.59	0.62	0.52	0.50	0.67	0.26		

Notes: R, T, and D denote rollers, tunnelers, and dwellers. The upper part of the table gives the vertical structure of vegetation, by percentage cover at each of three levels (<2 m, 2-6 m, >6 m; maximum is 100%). The rarefied species number is for a sample of 108 individuals.

as the shaded area projecting from the canopy exceeded 50% (sites 4 to 6), both species number and the pooled abundance of beetles decreased (Table 6.4). Scarabaeidae comprised the dominant group along the entire gradient, with Aphodiidae and Geotrupidae contributing only a few species and individuals. No specific forest species were found, since most species caught in the closed habitats belong to species with the highest numbers at open sites. Eight species were eurytopic, colonizing both open and forested habitats: *Geotrupes niger*, *Aphodius paracoenosus*, *Sisyphus schaefferi*, *Onthophagus lemur*, *O. joannae*, *O. grossepunctatus*, *O. verticicornis*, and *O. coenobita*. These species represented 80% of all the beetles at the six sites. At open sites the rollers, Scarabaeini and Sisyphini, accounted for one third of beetle biomass.

Lumaret (1980a) obtained similar results for dung beetles in the maquis formation in Corsica, with abundance and species number decreasing with increasing vegetation cover. However, in the maquis a small increase in species number was observed for the most forested sites, due to a progressive opening of the lowest vegetation stratum. The same phenomenon has been noted for dung beetles in Colombia and Texas (Howden and Nealis 1975; Nealis 1977). In Italy, Carpaneto (1986) observed nearly twice the number of species at a pasture than in an adjacent forested area, and very few species were exclusive to the forest. Species of steppic origin were relatively more abundant in the open area than in the forest.

These results clearly demonstrate that, to a large extent, the vegetation structure determines the species number of dung beetles. However, in contrast to what has been observed in tropical habitats (Howden and Nealis 1975; Walter 1978; Peck and Forsyth 1982), there are no two distinct groups of species in the Mediterranean region, that is, one from open areas and the other from forests. Here the vegetation cover acts like a filter that allows the penetration of several ubiquitous but fundamentally open habitat species into forests, at the same time drastically limiting their numbers (Lumaret 1983a).

Influence of Local Climate

The Garrigue zone of Languedoc is subjected to a strong south-north climatic gradient, which leads to local replacements of some typical Mediterranean species by other, more cold-tolerant species. Change in local climate may occur at the same altitude over very short distances, deeply modifying the structure of communities. Lumaret and Kirk (1987) compared two sites only 5 km apart, with a 0.3°C difference in the annual average temperature. They observed that the spring emergence of most species was much earlier at the warmer site, where their autumnal activity also continued longer than at the colder site. The most thermophilous species were absent from the colder site— for example, *Scarabaeus typhon*, *Copris hispanus*, and *Euonthophagus amyntas*—and the numbers of *Onthophagus lemur*, the most frequent species in the

Garrigue, decreased from 59% to 16%. In contrast, the frequency of two more montane species increased, from 9% to 26% in *O. joannae* and from 0.4% to 33% in *O. verticicornis*. These results show on a small scale the changes observed between the Garrigue and the Causse at the regional scale (Section 6.2 and Table 6.2).

6.4. Dung Beetle Populations

Seasonality

The activity of dung beetles at a site depends on the temperature and precipitation cycles and on the openness of the habitat. Mediterranean dung beetles possess phenological adaptations to cope with the summer drought, and are active mostly before and after the drought. Two peaks of activity are thus observed: a major one at the end of spring (May to June) and another, smaller one in autumn. The first peak corresponds to the oviposition period of many species, while the second peak is due to the emergence of immature beetles and takes place after the first heavy autumnal rains. This autumnal emergence, distinct at the most open sites, tends to be less marked or even disappears at the more shaded sites with generally smaller numbers of dung beetles (Fig. 6.2).

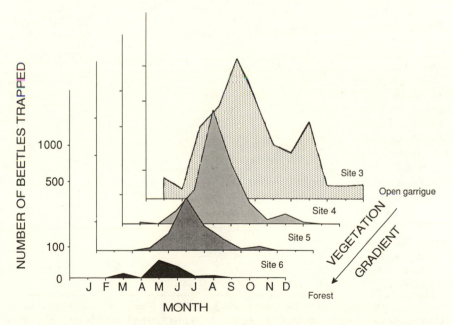

Fig. 6.2. Monthly variation in the numbers of beetles trapped at four compact limestone sites of Garrigue along a vegetation gradient from an open site to closed forest.

Most Aphodiidae oviposit in dung pats, leaving their eggs without any protection. In order to benefit from the periods when the desiccation rate of droppings is low, some *Aphodius* have completely reversed their seasonality, becoming active during the coolest and wettest months of the year. *Aphodius constans*, which has such "reversed" seasonality, is found mainly in cattle dung in the Garrigue throughout the winter. Females oviposit from December to March, and even during the rainy spring in April (Lumaret 1975).

At the three sites studied by Kirk (Section 6.2), the new generation of most species emerged in summer and early autumn. The exceptions were *Bubas bubalus* (Camargue) and *Onthophagus lemur* (Causse), which emerged in spring (Table 6.5). Most species overwinter as adults and delay egg laying

TABLE 6.5
Emergence and oviposition of dung beetles at sites varying in climate.

Species	Site	J	F	M	A	M	J	J	A	S	O	N	D	OS	DT
Group 1															
Bubas bison	1	-	-	-	-	-	-		*	-	-	-		Act A	7–11
Bubas bubalus	1		*	-	-	-	-							L,P,A	7–11
Geotrupes spiniger	1	--							*	-	-	-	-	L,P	7–11
Geotrupes spiniger	3								*	-	-	-		L,P	9–11
Geotrupes niger	3								*	-	-			L,P	10–11
Onthophagus lemur	3			*	-	-								L,P	10–11
Group 2															
Onthophagus lemur	2	O	-	-	-	-				*				A	3–7
Onthophagus maki	2			O	-	*								A	3
Onthophagus vacca	1			O	-	-	-	*						A	4
Onthophagus vacca	2			O	-	*								A	2
Onthophagus vacca	3				O	-	-	*						A	3
Onthophagus taurus	1			O	-	*								A	3
Onthophagus taurus	3					O	*							A	1
O. ruficapillus	1			O	-	-	-	*						A	5
O. emarginatus	2			O	-	*								A	3
Onthophagus furcatus	2						O	-	*					A	2
O. grossepunctatus	3				O	*								A	2
Euoniticellus fulvus	1				O	*								A	2
Euoniticellus fulvus	2						O	*						A	2
Euoniticellus fulvus	3				O	*								A	2
Scarabaeus laticollis	2				O	-	-	*						A	5
Copris lunaris	3						O	*						A	2

Notes: ------, oviposition period; *, emergence of new generation; O, emergence after overwintering. Group 1, species that oviposit after emergence. Group 2, species that overwinter without first ovipositing (see text). Act, active; A, Adult; L, Larva; P, Pupa. Site 1, St-Nazaire-de-Pezan (Camargue); site 2, Pic-St-Loup (Garrigue); site 3, La Couvertoirade (Causse). J to D, months January to December; OS, overwintering stage; DT, development time in months.

until the following spring or summer. However, *Geotrupes spiniger* (Camargue and Causse) and *G. niger* (Causse) oviposit throughout autumn and overwinter as larvae, pupae, and possibly as adults. *Bubas bison* oviposits from autumn to spring whenever temperatures are sufficiently high (Kirk 1983), while *Bubas bubalus* (Camargue) and *O. lemur* (Causse) oviposit only in spring (Table 6.5).

Species may modify their phenology depending on temperature and drought. The mean annual temperatures are 9.1°C at 800 m in the Causse, 13°C at 300 m in the Garrigue, and 13.5°C in the Camargue. *Onthophagus vacca* occurs at all three sites but with clear differences in the timing of oviposition and adult emergence. At the warmest site (Camargue), females begin to oviposit in March, in the Garrigue in April, and in the Causse in May. Rapid development in the Garrigue enables *O. vacca* to pass the dry, hot summer as an adult, capable of locating the scarce dung pats (Fig. 6.3). The pattern of emergence of *O. lemur* also reflects differences in climate. In the Causse, *O. lemur* is active in May and June only, but in the warmer Garrigue it is active from March to July and, after the drought, from September to February (Kirk, unpubl.). The new generation of *O. lemur* emerges in spring in the Causse and larvae or pupae overwinter, but in the Garrigue the new generation emerges after the summer drought and adults overwinter.

Strategies of Dung Use

Dung represents both a microhabitat with characteristics changing gradually in time and a resource which, although it may be abundant locally, is ephemeral in nature, only remaining suitable for dung beetles for a short period of

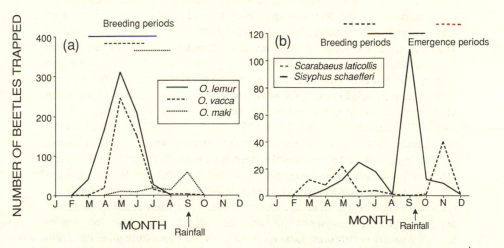

Fig. 6.3. Monthly activity of three species of tunnelers (a) and two species of rollers (b) at a Garrigue site.

time. The drying out of dung prevents any further utilization by most coprophagous insects.

Many *Aphodius* species using sheep droppings have abandoned the dweller habit typical for the genus (Chapter 3), apparently because sheep droppings tend to dry out very quickly under the Mediterranean conditions. Females deposit their eggs in the soil-excrement interface, and the larvae, as soon as they have hatched, dig a rudimentary, individual tunnel, which is enlarged as the larva grows and is filled with dung collected from the surface. The burrow is filled intermittently after a shower or dew has moistened the soil and the dropping. If the soil becomes too dry, the larvae become inactive and are able to fast for several weeks. Such behavior has been observed for *Aphodius luridus* in the Garrigue (Lumaret 1983b, 1987) and for several alpine species that exploit small droppings (e.g., *A. obscurus*; Chapter 14).

Several *Aphodius* species have adapted a primitive nesting behavior apparently to protect their eggs and larvae against the drought. Rojewski (1983) described the nesting habits of *Aphodius erraticus*. The female digs either vertical or oblique tunnels, 3–5 cm deep on average, directly under a dung pat. The tunnels are filled with dung after an egg has been deposited in the bottom of the burrow. At some sites, species have no lack of food as they exploit permanent accumulations of manure in sheep resting places. This natural mulch protects larvae from summer drought, and large populations may develop from spring until the end of the summer. *Aphodius elevatus* and *A. granarius* are common in such situations and may attain densities of up to five hundred adults (*A. elevatus*) per square meter in autumn (Lumaret and Bertrand 1985). In the Garrigue, sheep droppings deposited in early spring disappear rapidly; 50% of the dry matter is used by the dung fauna during the first month, and after eight months nothing remains at the soil surface. Droppings deposited in summer disappear more slowly, due to the summer drought, and 30% of the dry matter remains after eight months of exposure. Thirty percent of the dung deposited in winter remained for one year (Ricou 1984; Lumaret, unpubl.).

The Onitini *Bubas bison* and *B. bubalus* and the Onthophagini *Onthophagus vacca* are the numerically most important dung beetles in the pastures in the Camargue. These species bury dung only when females oviposit. The tunnels of *O. vacca* are 50–100 mm in length and are filled with 10–15 mm of dung, each mass weighing 0.9 g (dry weight) on average. During the spring months from April to June, when the three species are active at the same time, cattle pats are completely dispersed in 15–30 days. The same rate of dung burial was recorded in October, but the rate slowed down in November, even if some dung dispersal occurred every month from September to July (Lumaret and Kirk 1987). *Bubas bison* oviposits at the rate of 0.5 eggs per day in September, October, March, and April, when the soil temperature at 10 cm depth exceeds 10°C, but only 0.03 eggs per day in January, low temperatures (<5°C) slowing down oviposition and dung burial (Fig. 6.4). *Bubas bison* and *B. bubalus* bury dung in tunnels that are 150–230 mm in length; each dung mass weighs 10.5 g (dry weight) on average and is 100 mm long. The species bury 67% of the cattle dung available at the height of their activity in autumn and spring, and competition between the two species is reduced by a difference in their oviposition periods (Lumaret and Kirk 1987). *Bubas bison* emerges in September and continues to oviposit until May (Fig. 6.4), whenever soil temperatures are above 5°C (Kirk 1983). *Bubas bubalus* emerges in March and continues to oviposit until July; this species overwinters in the third larval and pupal stages (Fig. 6.4). Because of these differences, *B. bison* begins to oviposit when temperatures fall in autumn and *B. bubalus* when temperatures rise in spring. Dung dispersal is highest when the two species overlap in spring, when dung quality is high, and when temperature is rising.

At St-Nazaire-de-Pezan (Camargue), fresh dung pats were marked every week for 12 months. Those marked in May and October were destroyed within 10–14 days, mostly by *Bubas bubalus* and *B. bison*; the ones marked in March, April, and September disappeared within 21–35 days (*Onthophagus* species and the two *Bubas* species); and the ones marked in June, July, and August were destroyed in 75–130 days by several *Aphodius* and *Onthophagus* species. In winter, from the end of November to March, dung pats took more than 130 days to disappear (*B. bison* and *Aphodius* species; Kirk, unpubl.).

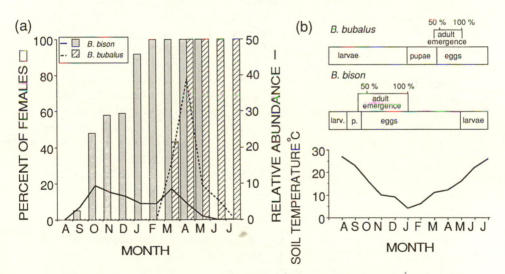

Fig. 6.4. Life cycles of *Bubas bison* and *B. bubalus* at St-Nazaire-de-Pezan (Camargue). (a) Relative abundances and percentage of females ovipositing, and (b) the presence of eggs, larvae, pupae, and adults in the populations. The lower part of (b) gives soil temperature.

6.5. Dung Beetle Communities

In an open *Brachypodium ramosum* pasture with twenty-four species in 1979, the most abundant species were four *Onthophagus* species (*O. lemur*, *O. maki*, *O. vacca*, and *O. emarginatus*), representing 76% of all dung beetles, and two rollers (*Scarabaeus laticollis* and *Sisyphus schaefferi*), accounting for 17% of the total (Lumaret 1983a). Analyses of their feeding habits, burrowing behavior, reproductive periods, and diel activity revealed differences that must significantly diminish interspecific competition. The coexistence of *O. lemur* and *O. vacca* is facilitated by a difference in their reproductive periods and by the use of distinct soil layers under the dung pat; the burrows of *O. lemur* are shallower than those of *O. vacca*. *Onthophagus lemur* and *O. maki* are ecologically very similar, except for their distinct reproductive periods (Fig. 6.3). *Onthophagus emarginatus* is active during most of the year, but it arrives at droppings a few days later than the other species, and it digs very shallow brood tunnels (2–5 cm depth) under droppings already containing deeper burrows of the earlier-arriving species. Competition between the rollers *Scarabaeus laticollis* and *Sisyphus schaefferi* is decreased by differences in their emergence and breeding periods (Fig. 6.3).

Appendix B.6 gives data on seasonal variation in the structure of a dung beetle community in the Garrigue (Viols-le-Fort site, altitude 240 m). Twenty-four species were collected in one year using two dung-baited pitfall traps of standard design (Lumaret 1979). Most of the beetles were collected in spring (61% of total, 20 species), and the lowest numbers were caught in winter (3%, 4 species). Three guilds of beetles compete for dung at this site: three rollers (Scarabaeini), thirteen tunnelers (12 Coprini, 1 Geotrupidae), and eight dwellers (Aphodiidae). Coprini constitute the dominant group both in numbers of individuals (84%) and in biomass (71%), with a maximum dominance in spring and summer. In winter the Coprini are replaced by Aphodiidae, which then represent 95% of all dung beetles, largely due to the activity of *Aphodius constans*, which oviposits during this period. The rollers are not very numerous (7%), but their large size (24% of beetle biomass) makes them a significant guild in the Garrigue community. The rollers reach a maximum in autumn, when the new generation emerges (Fig. 6.3). The role of Geotrupidae is negligible in the Garrigue, in contrast to what has been observed in the coastal plains or in the Camargue, where deep damp soils are common.

Each seasonal species assemblage is characterized by a few species that form a functional core; the remaining species often represent less than 20% of total numbers (Appendix B.6). In spring, *Onthophagus lemur* is the single core species (74%). Its frequency decreases in summer with the increase in the numbers of *O. maki* and *O. joannae*. The latter two species with *Sisyphus schaefferi* comprise the functional core in autumn, before being replaced by *Aphodius constans* in winter.

6.6. SUMMARY

The western Mediterranean region represents a transitional zone where dung beetles from diverse origins coexist. The cooler and wetter areas are characterized by relatively few Scarabaeidae, but they become increasingly dominant toward the south. Edaphic and climatic factors are important in assigning beetles to regional species pools, while local factors, such as vegetation cover and interspecific interactions, permit only a few of them to coexist in a local community. The climate imposes a marked seasonality in insect communities, with a decrease in activity during the summer drought at low altitudes, and the reverse on mountains and in the more mesic situations. Seasonality leads to significant differences in the rate of dung utilization by beetles throughout the year. Good understanding of the role played by dung beetles is essential for the proper management of Mediterranean pastures.

Changes in land use since the 1950s have dramatically reduced the numbers of cattle, sheep, and goats and hence the availability of dung to beetles. Many species have entirely disappeared while others have become rare. Similar changes are presently taking place in parts of northern Italy and Spain. Fortunately, in restricted areas of the Mediterranean basin the traditional pastoral practices are still followed, and dung beetle populations have been maintained at high levels of diversity and numbers. These reservoirs of beetles are an important potential resource from which a number of species have been selected and transferred to other parts of the world where beetles are required to facilitate dung dispersion in pastures without efficient native species (Chapter 15).

CHAPTER 7

Dung Beetles in Subtropical
North America

Bert Kohlmann

7.1. INTRODUCTION

Subtropical North America is characterized by great biogeographic complexity, and by the presence of plant and animal communities that are mixtures of old, established species and more recently arrived ones. It is impractical to define strict geographic boundaries to subtropical North America, but for the purposes of this chapter its territory is considered to comprise Mexico and the southeastern United States, from Texas to Florida and southern Georgia.

The unique biogeography of dung beetles in subtropical North America has been intensively studied by Halffter (1962, 1964, 1972, 1974, 1976, 1978, 1987). There are not many ecological studies of entire dung beetle communities from this region (Nealis 1977; Kohlmann and Sánchez-Colón 1984), but studies on the biology and natural history of particular species are numerous. The two important general works by Halffter and Matthews (1966) and Halffter and Edmonds (1982) on the natural history and nesting behavior of dung beetles, respectively, include much material from subtropical North America. Other more specific studies have been conducted by Lindquist (1933), Davis (1966), Stewart (1967), Fincher (1972, 1973a), Fincher et al. (1970, 1971), Gordon and Cartwright (1974), Blume et al. (1974), and Bellés and Favila (1984). There are numerous faunistic surveys (Poorbaugh et al. 1968; Blume 1970, 1972; Woodruff 1973; Fincher 1975a; Fincher and Woodruff 1979; Morón 1979; Morón and Zaragoza 1976; Morón and López-Méndez 1985; Harris and Blume 1986), and two further sets of papers focus on dung beetles as intermediate hosts and control agents of gastrointestinal parasites (Porter 1939; Miller 1954; Stewart and Kent 1963; Fincher 1973b, 1975b; Fincher et al. 1969) and on the introduction of dung beetles to Texas for the control of dung-breeding flies (Blume et al. 1970, 1973, 1974; Blume and Aga 1975, 1978; Lancaster et al. 1976; Fincher 1981, 1986; Fincher et al. 1983).

This chapter commences with a brief outline of the vegetation and climate of subtropical North America, followed by a description of the biogeographic origin of its dung beetle fauna and man-induced changes in it. In Section 7.3 we will examine the structure of dung and carrion insect communities in three kinds of habitats, while Section 7.4 gives an analysis of geographical variation in dung beetle communities. Aspects of population ecology are covered in Section 7.5, and Section 7.6 contains a discussion of an impoverished dung beetle community in a disturbed habitat.

7.2. BIOGEOGRAPHIC SETTING

Vegetation and Climate

It is most difficult to give a satisfactory description of the vegetation of subtropical North America because of its extreme variability, due to equally striking variation in the key environmental factors. Vegetation in this region includes a wide spectrum from tropical evergreen forests to xerophytic and alpine vegetation, but the different vegetation types are not well correlated with climatic factors (Rzedowski 1978). Floristically, subtropical North America is one of the richest regions in the world (Wulff 1937), comparable to Malaysia, Central America, and parts of South America (Rzedowski 1978).

During the early Tertiary, most of what is now subtropical North America was covered by a tropical flora (Axelrod 1979). During this period, the Laramide revolution started to lift the mountain systems of eastern North and Central America, allowing the penetration of holarctic vegetation elements toward the south. During the Eocene, North America became progressively more arid, and semideserts with xerophytic vegetation dominated during the Pleistocene interglacials (Axelrod 1979). During the glacial periods, temperate vegetation attained its maximum extension. Remnants of this vegetation type are still present as "islands" of relict communities in the "sea" of tropical forest (Toledo 1976). The distribution of tropical evergreen vegetation has greatly diminished in many areas during the Holocene—for example, on the the Pacific Coast of Mexico, where tropical forests have contracted by at least 700 km during the past three thousand years to their present boundary in Chiapas (González-Quintero 1980).

Several factors contribute to the great climatic variability in subtropical North America: the great range of altitudes and complex topography, the oceanic influence, and the geographical and relative position of the region to the Tropic of Cancer (Rzedowski 1978). With the exception of some very dry and wet areas, a wet and a dry season occur each year. In the region as a whole rainfall is very variable, the mean annual precipitation ranging from 50 to 5,500 mm; the wettest areas are located around the coast of the Gulf of Mexico, while the driest areas are inland. The amount of solar radiation is high throughout the year, and more than half of the region has an annual radiation value of 60%. The highest and lowest mean annual temperatures recorded are 30°C and −6°C, respectively, while common values vary between 10°C and

28°C. Altitude, not latitude, is the most important factor affecting temperature.

Origin of the Dung Beetle Fauna

Halffter (1962, 1964, 1972, 1976, 1978, 1987) has described five distribution patterns for the insect fauna of the Mexican Transition Zone that can be invoked to explain the origin of dung beetles in subtropical North America. The five distribution patterns are the Nearctic, Paleo-American, Neotropical, Mexican Highland, and Montane Meso-American (for a somewhat different perspective, see Chapter 4).

The Nearctic pattern is characterized by Holarctic and Nearctic genera of comparatively recent (Plio-Pleistocene) penetration along the Mexican and northern Central American mountains; these species typically occur in temperate conifer and oak-conifer forests as well as in high-altitude grasslands. This category includes Geotrupinae in the family Geotrupidae.

The Paleo-American distribution pattern is the result of an early immigration from the Old World to the American continent (Mesozoic, early Cenozoic). The taxa involved are taxonomically and ecologically more diverse in the Old World, though they are also widely distributed in America. Dung beetles belonging to this pattern include the genera *Copris* and *Onthophagus*, most probably many of the North American Aphodiidae, and the relictual genera *Liatongus* and *Sisyphus*.

The Neotropical pattern comprises elements of South American origin and of recent penetration (Plio-Pleistocene) into tropical and temperate humid lowlands. The following genera belong to this pattern: *Uroxys, Ateuchus, Canthidium, Cryptocanthon, Dichotomius, Ontherus, Scatimus, Eurysternus, Pseudocanthon*, and *Deltochilum*, as well as some species in the genera *Phanaeus* and *Canthon*. The subfamily Hybosorinae also fits this distribution pattern.

The Mexican Highland pattern is typical for species originating from South American elements that migrated during the Oligo-Miocene period, deeply invaded Mexico and the United States and adapted to Miocene and Plio-Pleistocene biomes, such as deserts, grasslands, pine-oak forests, and deciduous temperate forests. This category includes the genus *Melanocanthon*, the subgenus *Boreocanthon*, and temperate and subtropical species in the genera *Phanaeus* and *Canthon*.

The Meso-American Montane pattern is represented mainly by elements of South American origin (Oligo-Miocene), which evolved in the Central American Nucleus associated with temperate, tropical, and humid mountain forests and later expanded northward and southward during humid Pleistocene periods. The *Onthophagus clypeatus* lineage (Zunino 1981a) is an example of this pattern.

There is some segregation of species belonging to the different distribution patterns in particular biomes, which exemplifies a historical-biogeographical constraint on present-day species assemblages. For example, lowland tropical vegetation is occupied mostly by elements of the Neotropical pattern, arid areas by elements of the Mexican Highland pattern, and alpine and temperate conifer vegetation by elements of the Nearctic pattern. Species representing the Paleo-American pattern mix with other species to varying degrees, but it is mostly in the tropical and subtropical mountains where many of the distribution patterns break down and much mixing can occur. This is undoubtedly due to the highly fragmented distribution of different climatic conditions and vegetation types on the mountains.

Man-induced Changes

Human activity has had a major influence on the distribution of dung beetles at the regional scale, especially since the introduction of cattle by the Spaniards. Many of these changes are associated with the destruction of temperate and tropical forests and their replacement by grasslands with dung beetle communities that are poor in species number in subtropical North America. According to collections made in 1977 around the area of Bonampak in Chiapas, Mexico, the replacement of forest dung beetle species by tropical grassland species in forest clearings depends on the size of the clearing, its isolation from a grassland species source, and the age of the clearing (Table 7.1). These results are consistent with the equilibrium theory of island biogeography (MacArthur and Wilson 1967). Halffter and Halffter (pers. comm.) have col-

TABLE 7.1

Dung beetle assemblages in forest clearings and their edges around (a) the area of Bonampak, Chiapas (Kohlmann, unpubl.), and (b) the area of Palenque, Chiapas, Mexico (Halffter and Halffter, pers. comm.).

Origin of Clearing	Clearing		Distance to Other Clearings	N_{for}	N_{gra}
	Size	Age			
(a) Hurricane	40 m²	1 yr	17 km	10	0
Agricultural	1 ha	1 yr	15 km	3	0
Agricultural	2 ha	1 yr	100 m	2	3
Agricultural	2 ha	3 yrs	20 m	0	5
(b) Forest edge		many yrs	0 m	10	2
Archaeological	20 ha	many yrs	5 km	3	2

Notes: The original tropical forest community consists of 27 Scarabaeidae species in (a) and 23 Scarabaeidae species in (b). Distance to other clearings indicates the nearest continuous source area for dung beetle fauna adapted to forest-cleared vegetation, e.g., introduced tropical grasslands, agricultural terrains, etc. N_{for} and N_{gra} are the numbers of forest and open grassland species, respectively.

lected beetles in a nearby tropical forest region in Palenque, Chiapas, in 1965, with similar results. They found a steady decline in species number toward the edge of the forest, and a further decline from the edge to the clearing (Table 7.1).

7.3. DUNG INSECT COMMUNITIES

Grasslands

Dung is an example of a patchily distributed and ephemeral resource with a high nutritive value to insects but rapidly changing physical, chemical, and biological attributes, which select for fast resource use by insects. Great numbers of insects, mainly beetles and flies, frequent the dung microhabitat. The most characteristic beetle families are Scarabaeidae, Hydrophilidae, Staphylinidae, and Histeridae, while the most common fly families are Muscidae, Sarcophagidae, Sepsidae, and Sphaeroceridae (Hanski 1987a and Chapter 1). This generalized taxonomic composition of dung insects is broadly reflected in Blume's studies (1970, 1972) on cattle-dung associated insects in grasslands in Texas (Fig. 7.1). He reported twenty-three species of Scarabaeidae,

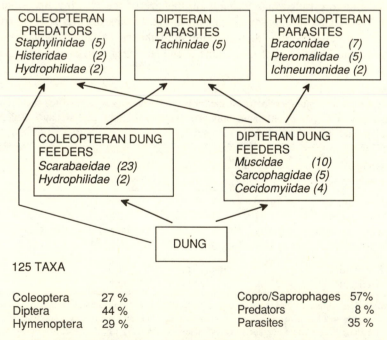

Fig. 7.1. Simplified food web for a subtropical grassland dung insect community in Texas. The percentages are calculated from the total number of species (based on Blume 1970, 1972).

but only five species of Staphylinidae, two species of Histeridae, and two species of Hydrophilidae. The last three families have fewer species in Texas than expected for a locality in temperate regions, which would have approximately seventy, five, and ten species, respectively (Hanski 1987a). The fly families Muscidae (ten species), Psychodidae (1), Scatopsidae (1), Sepsidae (3), and Sphaeroceridae (2) also have low species numbers in Texas compared with temperate localities, with the approximate numbers of twenty, five, five, ten, and thirty species, respectively (Hanski 1987a).

The subtropical dung insect communities have more species of large dung beetles than temperate communities. Large species such as *Dichotomius carolinus*, *Geotrupes opacus*, *Phanaeus triangularis*, *Canthon pilularius*, and *Canthon imitator* remove large quantities of dung, thereby reducing resource availability to dung-breeding flies. This may contribute to the relatively low species diversity of flies and their predators (staphylinid, histerid, and hydrophilid beetles) in Blume's results from Texas. Another possible explanation for the low diversity of dung-breeding flies is the high desiccation rate of dung pats in open fields in the subtropics. Predators may preferentially prey on parasitized fly larvae, hence a less diverse predator fauna could lead to an increase in parasite species richness. The parasitic Hymenoptera are represented by thirty-five species, whereas only ten species are expected to occur in a temperate locality (Hanski 1987a).

Deserts

In his study of the dung insect community in the Chihuahuan desert, Schoenly (1983) found that Coleoptera, Diptera, and Hymenoptera were the numerically dominant orders (Fig. 7.2), accounting for 93% of the total numbers of insects collected. In contrast to the previously mentioned grassland and temperate and tropical communities, where flies and beetles are the dominant taxa (Hanski 1987a), the desert community was dominated by ants, represented by five species and 74% of the individuals collected. Ants, spiders, and histerids were the most numerous predators, taking the place of staphylinid beetles in temperate communities (Hanski 1987a). The characteristic parasitoids, Braconidae and Ichneumonidae, were absent. During the later stages of decomposition, when the dung has dried out, termites are important coprophages in deserts (Johnson and Whitford 1975).

In general, the desert community is relatively poor in species and is dominated by omnivores. The very small number of dung specialists may be due to several factors: general scarcity of vertebrate dung; the type of dung available, which is dry and pelletlike, often being buried in the dung chambers of the desert-dwelling vertebrates; and the high rate of desiccation in deserts, which makes dung unusable to a large number of coprophagous insects. Some desert dung beetles possess adaptations to exploit dung in vertebrate burrows—for example, *Canthon (Boreocanthon) praticola*, found in prairie dog burrows in

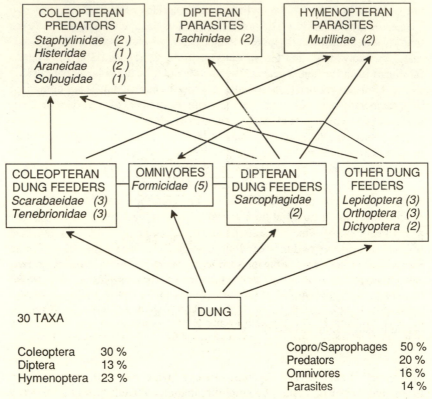

Fig. 7.2. Simplified food web for a desert dung-insect community in Texas. The percentages are calculated from the total number of species (based on Schoenly 1983).

Colorado and North Dakota—feeding on prairie dog dung pellets (Gordon and Cartwright 1974).

Subtropical Forests

There have been no studies of subtropical forest dung insect communities in North America, but several faunistic surveys of carrion insects have been published (Morón and Terrón 1984; Morón and López-Méndez 1985; Morón et al. 1986; Deloya et al. 1987). Many "dung beetles" are also found in carrion, and in fact a large fraction of the species in tropical forests in America are copro-necrophagous (Chapter 12). The large number of carrion-feeding Scarabaeidae has been attributed to the absence of large herds of herbivorous mammals in America, where there is a predominance of forests with a relatively low density of mammals and necrophagous insects like the silphids (Halffter and Matthews 1966; see also Chapter 18).

The high-level taxonomic composition of carrion insect communities in four subtropical localities in Mexico is similar to that of dung insect communities elsewhere, with Coleoptera, Diptera, and Hymenoptera being the numerically dominant orders (Morón and Terrón 1984; Morón and López-Méndez 1985; Morón et al. 1986; Deloya et al. 1987). The dominant family in biomass is Scarabaeidae, though Leptodiridae and Nitidulidae are also well represented. The most numerous predators are Staphylinidae and Histeridae, while Sarcophagidae, Phoridae, Sphaeroceridae, and Calliphoridae are the most obvious fly families. In contrast to the results of Cornaby (1974) from Costa Rican tropical forests, ants do not play any significant role in the carrion insect communities in Mexican subtropical forests.

7.4. DISTRIBUTIONAL ECOLOGY

Geographical Variation in Community Structure

Two interesting biogeographical questions are whether different biomes may have the same community structure, and whether examples of the same biome may have different community structures. To answer these questions, I analyzed a data set of 102 species from ten selected biomes in North America using the canonical variate analysis (I will provide the species lists on request). Five ecological variables, found to be useful in summarizing the structure of dung beetle communities in previous studies (Kohlmann and Sánchez-Colón 1984), were employed here: diel activity, food resource used by the adult beetle, the beetle's body length and weight, and the foraging mode. Aphodiidae were not included because of their poorly known taxonomy in this region.

The analysis revealed three main kinds of community structure (Fig. 7.3). One group (I, D, and J), with the largest number of species, is associated with the tropical forest biome, where the majority of species are nocturnal and co-pro-necrophagous. The mesophilous mountain forest (J, cloud forest) belongs to this group, rollers and small tunnelers being the dominant members of the community. Note that, in this analysis, two different biomes, the temperate magnolia forest (D, in Florida) and the tropical evergreen forest (I), have a similar assemblage of dung beetles. The phylogenetic origin of their faunas is different, as many of the species in Florida (D) represent elements that evolved in North America, whereas the evergreen tropical forest (I) and the mesophilous mountain forest (J) are dominated by species attributable to Plio-Pleistocene invasion from South America.

The case of the mesophilous mountain forest (J) is particularly interesting. This forest consists of a mixture of South American, North American, and Asiatic plant elements. There is no autochtonous dung beetle fauna in this forest type, but it is invaded by tropical and temperate species to varying degrees. A comparison of three mesophilous mountain forest communities of

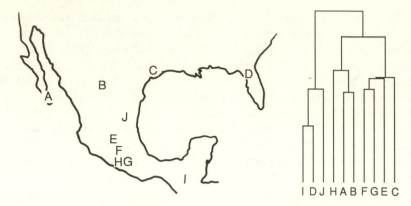

Fig. 7.3. A dendrogram for the similarity of dung beetle communities in ten different biomes in North America. A, Sierra de la Laguna, Baja California Peninsula, relictual tropical deciduous forest (data from Halffter and Halffter, pers. comm.); B, Mapimí, Durango, desert (data from Halffter and Halffter, pers. comm.); C, Welder Wildlife Refuge, Texas, chaparral (data from Nealis 1977); D, Gainesville, Florida, mixed magnolia forest (data from Woodruff 1973); E, Tuxpan, Michoacán, pine-oak forest (data from Kohlmann, unpubl.); F, Villa de Allende, Edo. Mex., temperate man-induced grassland (data from Morón and Zaragoza 1976); G, Teloloapan, Guerrero, tropical caducifolious forest (data from Kohlmann and Sánchez-Colón 1984); H, Teloloapan, Guerrero, tropical savannalike vegetation (data from Kohlmann and Sánchez-Colón 1984); I, Boca del Chajul, Chiapas, tropical evergreen forest (data from Morón et al. 1985); J, El Cielo, Tamaulipas, mesophilous mountain forest (data from Kohlmann, unpubl.). The insert gives the locations of the sites.

dung beetles with a tropical and a temperate community (Fig. 7.4) reveals that the mesophilous mountain forest community can vary from a tropical (El Cielo and Laguna Atezca localities) to a temperate (El Triunfo locality) structure. This demonstrates that the same biome can be occupied by two different community structures and that, in this case, interactions between the invading species or simply the proximity of the locality to different species pools can be more important in determining community structure than any local environmental parameters.

The group consisting of H, A, and B represents arid-environment dung beetle communities. These communities have generally only few species, and they are dominated by diurnal, coprophagous beetles. The Sierra de la Laguna locality (A) is a tropical deciduous forest that became isolated from the mainland and drifted northward with the Baja California Peninsula. It is now a relict surrounded by desert. The respective dung beetle community appears to have converged to a desertlike structure, where diurnal rollers are the dominant element. Note that the desert and the relictual tropical deciduous forest communities are closely related in structure to the dung beetle community in

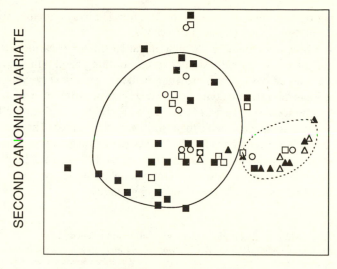

FIRST CANONICAL VARIATE

Fig. 7.4. Canonical correlation analysis of five dung beetle communities in Mexico. Each point represents one species. Localities: black square, Boca del Chajul, Chiapas, tropical evergreen forest (data from Morón et al. 1985); open circle, El Cielo, Tamaulipas, mesophilous mountain forest (data from Kohlmann, unpubl.); open square, Laguna Atezca, Hidalgo, mesophilous mountain forest (data from Kohlmann, unpubl.); open triangle, El Triunfo, Chiapas, mesophilous mountain forest (data from Kohlmann, unpubl.); black triangle, San Cristóbal de las Casas, Chiapas, pine-oak forest (data from Kohlmann, unpubl.).

tropical sabanoid vegetation (H). According to Axelrod (1979), the desert vegetation in North America has its origin in dry-adapted tropical plant lineages, probably similar to the ones found in the dry savannalike vegetation present at locality H.

The remaining sites F, G, E, and C in Fig. 7.4 include a mixture of temperate and tropical caducifolious forests and temperate grasslands. All have intermediate species richness and mostly nocturnal and coprophagous beetles.

Size Differences in Sympatric Species

Much of the recent debate in community ecology has focused on the significance of size differences in sympatric species and whether or not such differences constitute a proof of (past) competition (Strong et al. 1984). Simberloff (1983) concludes "that sizes of coexisting species in communities in general are not primarily a coevolutionary product of interspecific competition." Several authors have tested statistically whether the size distribution of coexisting species differs from a random distribution (Simberloff and Boecklen 1981), but it is not obvious that the random distribution is always a reasonable and "neutral" null hypothesis (Simberloff 1983).

The dung beetle communities of subtropical North America include many pairs of sympatric, functionally similar species of dung beetle (the same food preferences, diel activity, and burrowing method). This is very conspicuous in the tribe Phanaeini, where commonly two species of different sizes are present in the same locality, often in large numbers. Table 7.2 gives the known pairs of coexisting species, with their sizes and size ratios. The size ratios fluctuate around the average value of 1.38. Testing this size ratio against a null hypothesis of randomly distributed sizes (Schoener 1984) would be difficult, since the sample size is small and several combinations of species would be invalid, as not all the species have overlapping ranges.

TABLE 7.2

Average sizes, size ratios, sample sizes, and relative abundances in pairs of co-occurring Phanaeini species (P, *Phanaeus*; C, *Coprophanaeus*; and S, *Sulcophanaeus*).

Species Pair	Size (mm)	Size Ratio	Sample Size	Relative Abundance	Reference
P. igneus/P. vindex	20/19	1.05	2,785	0.99/0.01	4
			8,893	0.99/0.01	5
			1,370	0.13/0.87	2
			8,335	0.03/0.97	3
P. sallei/S. chryseicollis	19/16	1.19	26	0.04/0.96	1
P. quadridens/P. palliatus	20/16	1.25	33	0.97/0.03	6
			146	0.08/0.92	7
P. demon/P. daphnis	20/15	1.33			
P. pilatei/P. tridens	20/15	1.33			
P. amithaon/P. furiosus	22/16	1.37	174	0.32/0.68	11
P. quadridens/P. adonis	20/14	1.42			
C. telamon/P. endymion	23/16	1.44	2,074	0.67/0.33	8
			51	0.78/0.22	9
			321	0.30/0.70	12
P. mexicanus/P. daphnis	22/15	1.46	56	0.29/0.71	10

Notes: *Coprophanaeus telamon* and *P. endymion* use mainly carrion, the other species are entirely coprophagous. The species pairs *quadridens/palliatus* and *quadridens/adonis* are parapatric, the other pairs are sympatric.

Localities and data from: 1, Los Tuxtlas, Veracruz, Mexico (Morón 1979); 2, Tifton, Georgia, USA (Fincher et al. 1969); 3, Tifton, Georgia, USA (Fincher et al. 1971); 4, Blackbeard Island, Georgia, USA (Fincher 1975a); 5, Cumberland Island, Georgia, USA (Fincher and Woodruff 1979); 6, Villa de Allende, State of Mexico (Morón and Zaragoza, 1976); 7, Santa Cruz Acatlán, State of Mexico (Kohlmann, unpubl.); 8, Boca del Chajul, Chiapas, Mexico (Morón et al. 1985); 9, Los Tuxtlas, Veracruz, Mexico (Morón 1979); 10, Teloloapan, Guerrero, Mexico (Kohlmann and Sánchez-Colón 1984); 11, Ajijic, Jalisco, Mexico (Kohlmann, unpubl.); and 12, Cacaohatán, Chiapas, Mexico (Morón 1987).

A more useful approach may be to relate the relative abundances of the coexisting species to their body sizes. An ANOVA of the six size ratios with information on relative abundances (Table 7.2) shows significant differences in relative abundances among the groups (F[5,6] = 12.42, P<0.01). There are two groups with different relative abundances. One group is formed by pairs of species with size ratios from 1.05 to 1.25; in these pairs the average relative abundance of the rarer species is 0.05. The other group consists of pairs of species with the size ratio from 1.37 to 1.46, and the average relative abundance of the rarer species is 0.30. Thus if two similar-sized species occur in the same community, one of them tends to be very rare, a result that is consistent with the hypothesis that competition is strongest among species with similar size (Hanski 1982).

Small-sized *Phanaeus* species belong to lineages whose other members are also small, while large-sized *Phanaeus* species belong to lineages with large species (Edmonds 1972). This suggests that each species group has followed a different phylogenetic size path, and competition partly determines which species may coexist. This result supports Simberloff's (1983) above-mentioned conclusion that size ratios of coexisting species are not primarily a product of interspecific competition, but the relative sizes of species may affect their chances of coexistence.

7.5. POPULATION ECOLOGY

Dung Burial Rates

Lindquist (1933) undertook a pioneering investigation of the amount of dung buried by dung beetles in Manhattan, Kansas, in 1929 and 1930 (not a subtropical locality, but a locality that shares many species with the subtropics). Lindquist studied six species, measuring the dry weight of dung stored in their burrows. He found that *Dichotomius carolinus* (cited as *Pinotus carolinus*), with an average length of 26 mm, buried between 6 to 130 g of dung per individual (average 49 g). The amount of soil brought to the surface varied from 71 to 1,000 g (average 287 g). A census of the area established the average density of 490 burrows per hectare, which corresponds to the burial of 25 kg of air-dried dung, and the excavation of 140 kg of soil by *D. carolinus* per hectare and per year. *Copris fricator* (average length 41 mm, cited as *C. tullius*) buried 8 g of dung on average and excavated 37 g of soil per individual. In the genus *Phanaeus*, represented by two species, *P. difformis* and *P. vindex* (mean length 17 mm, the latter cited as *P. carnifex*), the amounts of dung buried and soil excavated per individual were 10 g and 94 g, respectively.

Two beetles burying dung can accomplish more work than a single beetle, the average ratios being 1:1.6 for *C. fricator* and 1:1.5 for *Phanaeus*. Lindquist's (1933) study covered the entire season from May to October, but since beetles can construct more than one burrow in their lifetime and population densities vary throughout the season, the exact amount of dung buried per year is difficult to estimate.

Diel Activity

Dung beetle communities typically have both diurnal and nocturnal species. Occasionally, for unknown reasons, diel activity is reversed, as in the case of the tropical *Megathoposoma candezei*, which has been recorded as a nocturnal species in Catemaco, Veracruz (Halffter and Matthews 1966), but I have seen it flying only in daytime in Bonampak, Chiapas (11 times).

There are few studies of diel activity of subtropical dung beetles in North America. Fincher et al. (1971) found that, in August in southern Georgia, *Phanaeus igneus* is active from 7 A.M. until noon, and again from 7 P.M. to 10 P.M. Another related species, *P. vindex*, flew uninterruptedly from 8 A.M. until 8 P.M., being most abundant between 10 A.M. to 6 P.M. The diel activity patterns were dramatically different in October, when the day length had decreased. Now *P. igneus* started to fly at 9 A.M. and continued until 8 P.M., while the flight period of *P. vindex* extended from 9 A.M. to 7 P.M. Unfortunately, it cannot be determined from this study whether the reduced daylight hours, cooler midday temperatures, or the reduced number of *P. vindex* individuals modified the diel activity of *P. igneus*.

Interference Competition

The following observations of interference competition between *Copris lugubris* and *Phanaeus tridens* were recorded in Palma Sola, Veracruz (Kohlmann, unpubl.). A male of *P. tridens* was seen entering a cavity made by a male of *C. lugubris* inside a cattle pat. Upon entry of the *Phanaeus*, the *Copris* male left its burrow and started hitting and ramming the intruder with his head until the latter retreated and flew away. Many similar incidents (14 cases) were observed between males of *P. tridens*, and in every instance the expelled individual was the intruder. After being repelled, the intruder went to another unoccupied part of the pat and started to dig its own feeding burrow unhindered. Such behavior tends to result in a relatively even distribution of individuals of the same species in a dung pat.

A somewhat different situation occurs with females of the same species. Their usual behavior is to cut off a portion of dung and to push it with their head some distance before burying it shallowly in a feeding burrow. Sometimes another female interrupts this process and attempts to steal the piece of dung; when the intruding female is bigger she generally succeeds. Several times (11 cases) female intruders tried to evict females who had already dug and provisioned a feeding burrow, but the owner always succeeded in remaining inside the burrow by blocking the entrance with her body.

When females are pushing fragments of dung away from the dung pat, flying males land near them to mate. When there are two males present, fierce competition between them develops, in which each tries to roll the opponent onto his back by inserting the head horn under the body of the opponent and quickly jerking the head upward. If several males land near a female at once, the ensuing competitive interactions result in a real melée.

Competition may also occur between beetles and flies for the common resource, at times taking take the form of chemical interference. *Canthon cyanellus*, a common dung beetle in the region, has been shown to secrete a chemical substance from the seventh abdominal sternite to the surface of its food balls, which are then discriminated against by *Calliphora* flies (Bellés and Favila 1984). The chemical mechanism of this process is not known.

Aggregation

Single dung pats are often occupied by several species of dung beetle with different resource-use strategies. Sometimes very large numbers of one species are found in one pat, and other beetles with a similar resource-use strategy, common in the same locality, are apparently excluded from the pat. Such exclusion has been observed in the genus *Ateuchus*, which is often very abundant in forest-grassland ecotones and disturbed situations (Kohlmann 1984; Kohlmann and Halffter 1988). It is not known whether the apparent exclusion is due to a pheromone, interference competition, lack of space due to large numbers of individuals, or differences in diel activities.

One series of observations was made over a period of three days in Teloloapan, Guerrero, Mexico (Kohlmann, unpubl.), where cattle pats similar in size, dryness, and age were found to have large numbers of *Ateuchus carolinae*, a small (7 mm) dichotomine species (Fig. 7.5). Cattle pats with twenty-seven to forty-eight individuals of *A. carolinae* had no species of *Onthophagus, Copris, Phanaeus* and *Dichotomius*, even if they were common in other pats in the same locality, including pats with fewer than seven individuals of *Ateuchus* (Fig. 7.5). The density of *A. carolinae* (3 density classes: 0, 1–7 and 27–48 individuals per pat) had a highly significant effect on the number of other species (Fig. 7.5; $F[2,20] = 8.72$, $P<0.005$).

Introduced Species

Several species of dung beetles have been introduced into North America, mostly from Europe (see also Chapter 5). Some introduced species have subsequently invaded subtropical localities with great success—for example, the hydrophilids *Sphaeridium scarabaeoides* and *Cercyon nigriceps*, and the dung

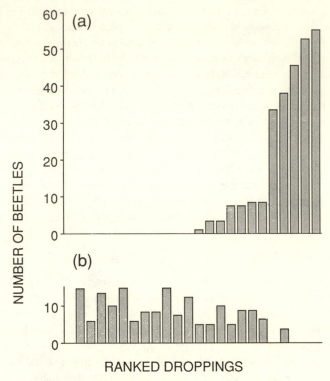

Fig. 7.5. (a) Number of individuals of *Ateuchus carolinae* in twenty-seven similar-looking cattle pats in a forest-grassland ecotone in Teloloapan, Guerrero, Mexico, in 1983 (Kohlmann, unpubl.); (b) the number of individuals of species other than *A. carolinae* in the same pats.

beetles *Aphodius lividus* (which has penetrated deeply into Mexico), *A. fimetarius*, *A. haemorrhoidalis*, and *A. granarius*.

Woodruff (1973) has reported on an African species, *Onthophagus depressus*, from Florida and Georgia. It was first recorded by Robinson (1948) in Florida, where it remained curiously localized in an inland rather than in a coastal locality for many years. Fincher and Woodruff (1975) recorded a common European dung beetle, *Onthophagus taurus*, from several localities in Florida and Georgia in 1974. Fincher and Woodruff (1975) believe that this species was first introduced into the Florida panhandle near Pensacola, from where it radiated in all directions, reaching the North Carolina–Virginia border (900 km to the northeast) and the Mississippi-Louisiana northern border (700 km to the west) by 1981 (Fincher et al. 1983). In 1974–75, the rate of spread was about 300 km per 8–9 months of adult activity (Fincher et al. 1983).

Fincher et al. (1983) also report on the spread of *Digitonthophagus gazella*, an Afro-Asian beetle first released in Texas in 1972 in a dung-control program. It has subsequently become established in California (Anderson and Loomis 1978) and in Georgia (Fincher 1981). From the Kleberg County releases in Texas in 1973, the beetle spread 16 km during the first year and an additional 32 km by the second year (Blume and Aga 1978). By 1981, *D. gazella* had colonized much of Texas and the coastal part of Louisiana, overlapping with *O. taurus* at the East Baton Rouge Parish. In August 1984, 12 years after the release, *D. gazella* was recorded in the Mapimí Biosphere Reserve, Durango, Mexico (Halffter and Halffter, pers. comm.), 700 km to the southeast from the original site of release.

Sphaeridium scarabaeoides was first recorded from Cuajimalpa, near Mexico City, in 1974 (Hendrichs 1975). In 1976 it was recorded from Tepotzotlán, State of Mexico (Kohlmann, unpubl.), 60 km to the northeast of the first locality; and in 1979 from Altotonga, Veracruz (Kohlmann, unpubl.), 230 km to the east of the first recording. This species appears to have been introduced into Mexico accidentally with dairy cattle from the United States (Hendrichs 1975).

7.6. An Impoverished Dung Beetle Community

In this section I examine an impoverished dung beetle community in an environment strongly modified by man's action. The locality is 8 km to the north-northwest of Mexico City, near the town of Santa Cruz Acatlán, in the spur of the Sierra de las Cruces (2,250 m altitude). The area was formerly occupied by an oak forest, which now remains only in the more inaccessible parts of the sierra. This type of forest is heavily used in Mexico for local purposes, and it was formerly used for charcoal production (Rzedowski 1978). The area used to be an agricultural and cattle-raising community until the late 1970s, when construction of suburban dwellings for the ever-expanding Mexico City began.

The agricultural fields around Santa Cruz Acatlán were sampled for one year with human-dung baited traps in 1975; carrion-baited traps produced no beetles at all. The results are given in Appendix B.7. Some species, including *Onthophagus mexicanus* and *Aphodius lividus*, started to emerge in the beginning of the year, albeit in low numbers, but the majority of species appeared in June. This is a common pattern over most of subtropical North America, where the adult emergence of dung beetles is strongly associated with the beginning of the rainy season. Large numbers of beetles are active throughout the rainy season until October, though the above-ground activity of nest builders such as *Phanaeus palliatus* is greatly diminished by September (Appendix B.7), in the beginning of their breeding season.

The oak forest remnants were sampled in the same year, and the following species were found: *Onthophagus lecontei*, *Phanaeus palliatus*, *P. adonis*, and *Ceratotrupes fronticornis*. A community with a similar if not richer species composition should have existed originally at the Santa Cruz Acatlán locality. The low numbers of *O. lecontei*, *P. adonis*, and *C. fronticornis* still collected suggest that they were either represented by migrants from the forest remnants or that the local populations were going extinct. In the same year, the following species were found at a perturbed grassland site (Kohlmann 1979) around the nearby town of Tlalnepantla: *Onthophagus mexicanus*, *Phanaeus quadridens*, *Aphodius lividus*, *Dichotomius carolinus*, and *Canthon humectus*. *Onthophagus mexicanus* and *P. quadridens*, two grassland species, had managed to invade the Santa Cruz Acatlán locality, but *Canthon humectus*, a roller, was still absent. *Aphodius lividus* and *D. carolinus* occur mostly in disturbed areas, and their presence indicates that the original habitat has been modified. In 1975 the Santa Cruz Acatlán locality was in a state of change, from the former forest community to a perturbed grassland dung beetle community.

In 1983 the same locality in Santa Cruz Acatlán was resampled. By then the agricultural and cattle-raising activities had stopped and the area had been converted into a park (the Naucalli Park). Collecting in the park produced only one species, *Aphodius lividus*. This species has been collected in dog dung, which has now become the main food source for dung beetles in this locality.

7.7. SUMMARY

The North American subtropics form an area of extreme environmental variability with considerable ecological and biogeographical mixing of communities and species. Intensive and profound changes have occurred since the Paleocene, resulting in several distribution patterns of dung beetles, related to past and present ecological conditions and paleoevents. Man-induced changes in the environment have lately played an important role in changing the distribution of species, mainly by increasing grassland and open-field communities at the expense of forest communities.

Different biomes such as tropical evergreen forest and mixed magnolia forest may have similar dung beetle communities, while localities representing the same forest type can have different community structures, as exemplified by the mesophilous mountain forest, with fauna strongly affected by invading species from the neighboring beetle communities. Relative abundances of coexisting pairs of *Phanaeus* species are related to the difference in their body size: if two species are similar in size, one of them is very rare, probably due to interspecific competition.

CHAPTER 8

Dung Beetles of Southern Africa

Bernard M. Doube

8.1. INTRODUCTION

In Africa, south of the Sahara, there are more than 2,000 species of dung beetles in the family Scarabaeidae. Southern Africa alone has 780 species, of which some 150 species may be found in one locality (Scholtz and de Villiers 1983; Doube 1987). Additionally, southern Africa has approximately 60 species of dung-dwelling Aphodiidae (Bernon 1981; A.L.V. Davis, pers. comm.), some dung-frequenting Hybosoridae and Trogidae (C. H. Scholtz, pers. comm.), but no coprophagous Geotrupidae. There are several hundred species of dung-frequenting staphylinid, histerid, and hydrophilid beetles, most of which are predators (Bernon 1981; Doube 1986; Davis et al. 1988). African dung beetles show considerable specialization with respect to soil type and vegetation cover, diel and seasonal activity, the age and the type of dung used, and in their foraging and reproductive behaviors (Halffter and Matthews 1966; Davis 1977, 1987, 1989; Endrody-Younga 1982; Halffter and Edmonds 1982; Doube 1983, 1986, 1987; Edwards 1986a, 1986b; Fay 1986; Mostert and Scholtz 1986; Edwards and Aschenborn 1987, 1988, 1989; Doube, Macqueen, and Fay 1988).

In this chapter we are primarily concerned with the ecology of dung beetles in the summer rainfall regions of the Transvaal highveld and the Natal lowveld in South Africa. Three main types of data will be used: (1) observational data on geographical distributions and habitat associations of beetles and similar data on their seasonal and successional occurrences; (2) field and laboratory data on the life history and ecology of individual species; and (3) experimental results on the role of competition in structuring dung beetle communities. A conceptual framework for this chapter is provided by a functional classification of the southern African dung beetles into seven guilds (Section 8.3).

8.2. DUNG BEETLE BIOGEOGRAPHY IN SOUTHERN AFRICA

The dung beetle fauna of southern Africa can be divided into three main components. The first one is a desert-adapted fauna, consisting of a group of highly mobile, wide-ranging species, and of a group of specialized, wingless species in the genus *Pachysoma*, largely restricted to the Namib Desert

(Scholtz, unpubl.). Many *Pachysoma* species can use vegetable matter as well as dry, pelleted dung, which they transport to tunnels excavated below the moisture line in the soil, where the organic matter rehydrates and can be used for feeding and breeding (Scholtz, unpubl.). These species are highly opportunistic and become active following rain.

The second faunal element is adapted to the Mediterranean climate of the Cape Province. This fauna comprises one group of species restricted to the winter rainfall regions and another group with distribution extending to the summer rainfall regions (Davis 1987). The species in the former group are active mostly during the moist spring and autumn but also in winter in the warmer regions; their local distribution is restricted largely to uncleared areas of indigenous vegetation. The latter species are active during the warmer months of the year, and they are most abundant in the pastures that have replaced the indigenous bush vegetation in many districts. The third faunal element in southern Africa is a highly diverse one, primarily associated with the summer rainfall regions of the eastern part of southern Africa, though the distribution of many species extends to the tropics.

The ecology and biology of the desert and winter rainfall species are poorly known, but the species from the summer rainfall regions have been intensively studied by the staff of CSIRO since 1970 (Bornemissza 1970b, 1976, 1979; Paschalidis 1974; Tribe 1976; Davis 1977, 1989; Bernon 1981; Fay and Doube 1983, 1987; Doube et al. 1986; Doube, Macqueen, and Fay 1988; Doube, Giller, and Moola 1988; Fay 1986; Edwards and Aschenborn 1987, 1988; Edwards 1988a; Giller and Doube 1989). Although many of these studies have focused on the fauna associated with cattle dung in pastoral ecosystems, the beetles of the game reserve environment have also received substantial attention; the latter studies have helped to provide a general conceptual framework within which to consider the results on the cattle dung fauna.

The large number of dung beetle species in southern Africa is primarily due to the great diversity of species in the summer rainfall regions, especially in the lowveld and highveld regions, where rainfall ranges from 300 to 750 mm (G. F. Bornemissza, pers. comm.). For example, about 140 species have been recorded from the Mkuzi Game Reserve alone (A.L.V. Davis, pers. comm.), and 147 species are known from the Hluhluwe region (Table 8.1). In contrast, species diversity is relatively low in the high montane regions of South Africa (e.g., the Drakensberg Mountains) and in the Cape Province. Twenty-five species were found in indigenous shrubland at Langabaan, 11 species in the pastures at Paarl, near Cape Town (Davis 1987), and 22 species at George on the southeast coast of the Cape Province (Breytenbach and Breytenbach 1986).

The striking diversity of dung beetles in Africa may be explained by episodic geographical isolation of populations during the Pleistocene, by the presence of a wide variety of habitats, and by the survival of a diverse fauna

TABLE 8.1
Numbers of species of dung-frequenting insects in climatically similar regions of Australia (Rockhampton, coastal central Queensland) and southern Africa (Hluhluwe, coastal Natal).

Order	Family	Subfamily/Tribe	Africa	Australia
Coleoptera	Scarabaeidae	Canthonini	2	—
		Coprini	16	—
		Dichotomiini	12	—
		Drepanocerina	7	—
		Gymnopleurini	5	—
		Oniticellina	7	2*
		Onitini	11	1*
		Onthophagini	62	16 + 2*
		Scarabaeini	11	—
		Sisyphini	13	2*
	Aphodiidae	Aphodiinae	43	3
	Staphylinidae	Aleocharinae	34	3
		Oxytelinae	19	6
		Paederinae	6	1
		Staphylininae	26	5
		Tachyporinae	4	?
		Xantholinae	3	1
	Hydrophilidae		13	4
	Histeridae		27	3
Hymenoptera			>3	12
Diptera	Putative predators		10	3
	Coprophagous species		>10	>11
Acarina	Macrochelidae		60	3
	Others		>40	>6

Source: Doube (1986).
* Introduced species.

of mammals up to the present time. During the past 1.7 million years there have been at least seventeen glacial and interglacial periods in southern Africa (Brain 1981), the present being a relatively brief warm interlude in what has been a predominantly cold Pleistocene period. Under the influence of repeated cold episodes during the past 3 million years, many African habitats became more open, with vegetation zones moving to and fro repeatedly in response to climatic changes (Brain 1981). Accordingly, the ranges of plants and animals, including dung beetles, must have expanded and contracted; such pulses served to break once continuous ranges into disjunct patches, centers for allopatric speciation. The chain of high mountains running from north to south along the eastern seaboard of Africa has exacerbated geographical isolation. Rapid evolution is known to have occurred in animals adapted to open country. For example, among plains-dwelling alcelaphine antelopes, the tribe to which hartebeest and blesbok belong, twenty of the twenty-eight or so species came to existence during the last 3 million years (Brain 1981).

Stone Age smelting sites in Natal indicate the presence of humans since 50,000 BP (Hall 1979a,b), while Iron Age agricultural man has been in the area for at least 1,600 years (Feeley 1980). Clearing, burning, and farming have extensively altered the pristine vegetation. During the time of early European settlement 100–200 years ago, the landscape consisted of a mosaic of agricultural and grazing land, with tracts of land set aside as private hunting parks for the Zulu kings (Hall 1979a; Feeley 1980). These parks have since become the Hluhluwe-Umfolozi Game Reserve complex, the dung beetle fauna of which will be discussed in this chapter. In recent times many mammalian species, especially the large pachyderms (elephants and rhinoceroses) and small mammals dependent on bushveld, have become extinct over wide areas, and in southern Africa they are now restricted to the game reserves. The same may be true about the species of dung beetles that depend on the dung of these threatened mammals.

The pristine condition of the Natal landscape is not known, but in historical times it has been dominated by savanna, interspersed with riparian gallery forests and patches of woodland. The determinants of the savanna physiognomy include climatic, edaphic and geomorphic factors, as well as fire, herbivory, and human activity, which restrain the succession from grassveld to bushveld (Bourlière and Hadley 1983), but there is as yet no general agreement as to the hierarchical relationships among these factors. In the Hluhluwe-Umfolozi complex, fire and herbivory have slowed down, but have not prevented, the succession of open savanna to bushveld, which covered 64–88% of the area in 1975, about double the area in 1937 (Watson and Macdonald 1983). Until very recently, elephants, which encourage open grassveld by knocking down trees (Barnes 1983; Pellew 1983), have been absent from the Natal game parks. In agricultural regions, habitat boundaries have become sharpened due to clearing and the use of fences, whereas in game parks the transition from grass cover to bush cover is usually gradual. In the studies reported here, *grassveld* refers to areas in which grasses are a major component of the vegetation, while the woody component is limited to scattered trees and small patches of bush; *bushveld* refers to areas with sparse grass cover and substantial cover of canopy. Forest habitats with no grass and closed canopy have received little attention.

For dung beetles, there are many similarities between the game reserve environment and the agricultural landscape around villages. Both systems comprise a mosaic of grassveld and bushveld, although bushveld dominates many game reserves whereas grassveld dominates village surroundings. Many types of dung are available in both environments, including large and moist dung pats (buffalo, cattle); pelleted dung (buck, goats); and nonruminant herbivore (zebra, horse), omnivore (baboon, man), and carnivore (lion, dogs) dung. Many of the environmental similarities between these two systems disappear in large-scale agriculture, and current agricultural practices have a major impact on the dung beetles of southern Africa, through modification of the habitat and the supply of dung.

8.3. THE DUNG INSECT COMMUNITY

During the moist summer season, more than 100 species of beetles can occur together in a single cattle pat in Transvaal and Natal. In Natal, Davis et al. (1988) recorded 100 staphylinid, 21 histerid and 13 hydrophilid species in pats exposed for only twelve hours. In Transvaal, Bernon (1981) found 161 species of beetles inhabiting fresh cattle pats, along with phoretic mites, dung-breeding flies, and a variety of other arthropods. Bernon (1981) found that the mean number of beetles colonizing fresh pats in twenty-four hours during summer ranged from 742 to 1,585, including Aphodiidae (35%), Scarabaeidae (42%), Histeridae (5%), Hydrophilidae (9%), and Staphylinidae (9%). Such a diversity of species involves a multitude of possible interspecific interactions, some of which have been studied in detail to identify species for biological control of the dung-breeding buffalo fly in Australia, where species diversity in the dung insect community is low (Table 8.1; Paschalidis 1974; Bornemissza 1976; Tribe 1976; Davis 1977; Bernon 1981; Fay and Doube 1983, 1987; Doube et al. 1986; Fay 1986; Doube and Huxham 1987; Doube and Moola 1987, 1988; Edwards and Aschenborn 1987; Wright and Mueller 1989; Wright et al. 1989).

Despite their relatively simple trophic structure, dung insect communities are complex systems, with numerous small species of coprophages (flies, mites, and staphylinid beetles), an array of predators (flies, mites, and beetles), parasites (beetles and wasps), numerous phoretic mites and some phoretic flies, and an extensive suite of dung beetles, all ultimately depending on the same food supply. There are not many records of parasitoids of dung beetles, but Davis (1977) reared a bombyliid fly and a pteromalid wasp from the broods of *Oniticellus formosus*. Predators of dung beetles are more common. A variety of vertebrates (baboons, monitor lizards, guinea fowl, owls) are known to eat dung beetles, and there are robber flies, wasps, and reduviid bugs that catch beetles and flies as they approach dung pats (Paschalidis 1974; Bernon 1981).

Functional Groups of Dung Beetles

The southern African dung beetles (Scarabaeidae) can be divided into seven functional groups (FG I to FG VII), based on the way the species use dung and thereby affect dung pats. The functional groups of dung beetles are comparable to guilds (Root 1967; Southwood et al. 1982; Stork 1987), and their main characteristics are illustrated in Fig. 8.1 and summarized in Table 8.2. This classification provides a conceptual framework within which to analyze the structure of dung beetle communities, the role of competition in structuring communities, and the interactions between beetles and other members of the community. The distributions of species, individuals and biomass among the functional groups (Fig. 8.2) contributes complementary insights to the structure of dung beetle communities.

There exists a clear hierarchy among the functional groups in the ability of the species to compete for dung. The competitively dominant groups are the large rollers (FG I) and the fast-burying tunnelers (FG III), which are mostly large, aggressive beetles that may rapidly remove dung from the pat. Small rollers (FG II) are also strong competitors, because they too can remove dung soon after arrival at the pat. Subordinate to these groups are the tunnelers (FG

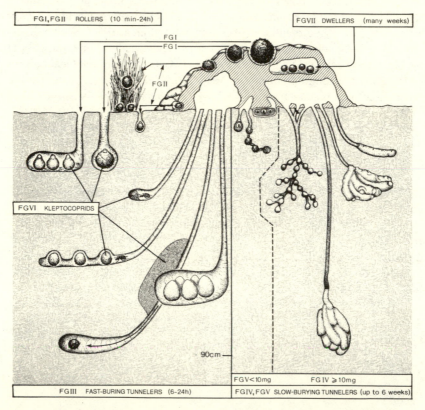

Fig. 8.1. Schematic illustration of the nesting habits of dung beetles belonging to the seven functional groups described in Table 8.2.

TABLE 8.2

Characteristics of the seven functional groups (FG) of dung beetles based on results on southern African Scarabaeidae and Aphodiidae.

FG	Description	Stay-Time in a Pat	Ratio of Beetle Weight to Amount of Dung Removed	Location of Maturation Feeding	Type and Location of Nest and Brood	Effect on Dung Pats When Abundant	Diel Flight Activity
I	Large rollers, >400 mg dry wt	2 to 3 h	1:5 to 1:20	Buried dung	Simple and compound subterranean nests with brood balls	Complete and rapid dung removal	Diurnal (most spp.)
II	Small rollers, <400 mg dry wt	2 to 3 h, up to 24 h in some spp.	1:5 to 1:20	Buried dung (most spp.)	Simple subterranean nests with brood balls	Complete dung removal over several days	Diurnal (most spp.)
III	Fast-burying tunnelers	6 to 24 h	up to 1:500	Buried dung	Compound subterranean nests with brood balls	Complete dung removal within 24 h	Crepuscular/ nocturnal
IV	Large slow-burying tunnelers, >10 mg dry wt	Up to 6 weeks	up to 1:1000	Surface and buried dung	Compound subterranean nests with brood masses	Complete burial (some spp.); complete shredding (other spp.)	Diurnal or crepuscular/ nocturnal
V	Small slow-burying tunnelers, <10 mg dry wt	Up to several weeks	unknown	Surface and buried dung	Simple or compound nests in the pat-soil interface or in shallow tunnels with brood balls	Shredding of pat with little or no breeding	Diurnal or crepuscular/ nocturnal
VI	Kleptoparasites	Highly variable	n.a.	Surface and buried dung	Not known	Shredding of dung and inhibition of breeding	Unknown
VII	Dwellers	Many weeks	n.a.	Surface dung	Compound nests in dung with brood balls	Unknown	Diurnal

Note: n.a., not applicable.

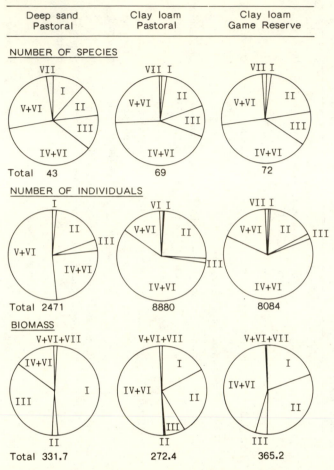

Fig. 8.2. An example of distributions of species, individuals, and biomass of dung beetles among the functional groups described in Table 8.2. Dung beetles were caught in pitfall traps baited with one liter of fresh cattle dung for 24 hours. Data are presented for three localities: a pastoral grassveld habitat on deep sand in Transvaal (pooled data for eight trapping occasions, three traps per occasion, in 1981–82; Doube, Macqueen, and Fay 1988); and grassveld habitats in pastoral and game reserve regions in the Hluhluwe district of Natal (pooled data for twenty-four trapping occasions, seven traps per occasion, in summer 1982–83; Doube, unpubl.).

IV and FG V), which remain in the pat or bury dung slowly over many days. Species that build shallow nests (FG V) and the ones that nest within the dung pat (the dwellers, FG VII) are especially likely to have their breeding activities disrupted by members of the other functional groups. Kleptoparasites (FG VI) use the dung buried by other species. Rearing beetles from dung buried by other species is the only way to determine whether a species is a kleptoparasite. Because little is known of the breeding biology of many of the smaller species, kleptoparasites have not been separated from the FG V and FG VII species in the analyses presented in this chapter.

The observed relative abundance of a species depends on the sampling technique employed, of which two have been used most commonly. One is the baited pitfall trap, and the other one consists of extracting beetles from dung pats. Doube and Giller (unpubl.) compared the two methods in a study of overnight colonization of dung pats by thirteen crepuscular/nocturnal tunnelers. In only three species was the result significantly different, and the rank order of species was nearly the same in the results obtained by the two methods (Spearman's rank correlation, 0.91). We may thus conclude that pitfall trapping, the most commonly used method, yields census data that closely reflect natural colonization of dung pats.

Most studies of dung beetle populations have been based on data from traps set in one locality, with the assumption that such data are representative of the same habitat type throughout a larger region. The degree to which one site is representative of other sites in the same habitat was examined in two extensive pastoral regions in the Natal lowveld in 1985–86 (Doube, unpubl.). There were six sites within 200 square kilometers of clay-loam soil and four sites within 100 square kilometers of deep sandy soil. Within soil types, the between-site differences in species composition and functional group composition were small when data were pooled for all trapping periods, though differences among the sites were often noticeable on individual trapping occasions.

8.4. RESOURCE PARTITIONING IN SPACE AND TIME

Tropical dung beetle communities are characterized by a wide range of species with little opportunity to partition resources within dung pats, but with a variety of nesting behaviors and modes of resource use (Halffter and Edmonds 1982; Doube, Macqueen, and Fay 1988; Giller and Doube 1989). There is evidence for below-ground partitioning of space (Edwards and Aschenborn 1987) and for substantial specialization with respect to dung type and age, seasonal activity, and habitat selection (Doube 1987). Nevertheless, many species make use of the same, limited resource, and one might expect competition to be a major determinant of community organization, as it is, for example, in carrion insects (Denno and Cothran 1976; Doube 1987; Hanski 1987c; Chapter 1). Before examining the role of competition in structuring southern African dung beetle communities, I shall review our knowledge of resource partitioning among these species.

Vegetation Cover

It is widely recognized that vegetation cover influences the local distribution of dung beetles (Nealis 1977; Doube 1983, 1987; Janzen 1983b). The effect of vegetation is well illustrated by changes in species abundances across man-made clearings in woodland, where there is no important variation in habitat parameters other than vegetation cover (Fig. 8.3). Doube (1983) demonstrated substantial changes in the abundances of thirty-four species across one such clearing in the Hluhluwe-Umfolozi Game Reserve in Natal. During overcast weather the distribution of bushveld species extended into the lightly wooded transitional zone between bushveld and grassveld, suggesting that the observed habitat preferences were determined, in part at least, by beetles' response to light intensity (Doube 1983). The vegetation preferences of beetles have been examined further in the Mkuzi Game Reserve (Appendix B.8), where many though not all species show a preference for grassveld over bushveld or vice versa.

The degree of habitat specificity varies with the structural characteristics of vegetation. For example, in the Hluhluwe-Umfolozi Game Reserve the tiny bushveld species *Sisyphus seminulum* showed a twentyfold decrease in abundance over 20 m at a sharp transition from bushveld to grassveld, and was virtually absent from grassveld which had been grazed bare (Fig. 8.3), but it was common (100–200 per trap) in another open region with a deep sward of grass. In the Mkuzi Game Reserve, *Allogymnopleurus thalassinus* showed an intense preference for grassveld over bushveld (Appendix B.8) but was scarce in nearby, small, and isolated patches of grassveld within bushveld.

While it is clear that the relative abundances of species change across vegetational boundaries (Fig. 8.3, Appendix B.8), the relative abundances of the different functional groups frequently remain more constant (Table 8.3). The Mkuzi study on which Appendix B.8 is based compared the functional group composition of beetles in eight contrasting habitats (bushveld and grassveld on four soil types). Analysis of variance for the number of species, the number of individuals, and beetle biomass showed no consistently significant effect of vegetation type, although the effect of soil type was highly significant (see next section). In contrast, data from a long-term study in the Hluhluwe district indicate that some aspects of community structure change with vegetation type (Table 8.3). For example, the numerical abundance and biomass of the FG III and FG VII beetles were consistently higher in grassveld than in bushveld in paired habitats (P<0.01, Table 8.3), indicating that the fast-burying tunnelers (mostly *Copris* and *Catharsius*) and dwellers are more abundant in the grassveld than in the bushveld.

Soil Type

The association of southern African dung beetles with different soil types has been examined at a number of localities (Endrody-Younga 1982; Doube 1983; Fay 1986; Davis 1987; Osberg 1988). Doube et al. (Appendix B.8) found significant differences between soil types in the average abundance of

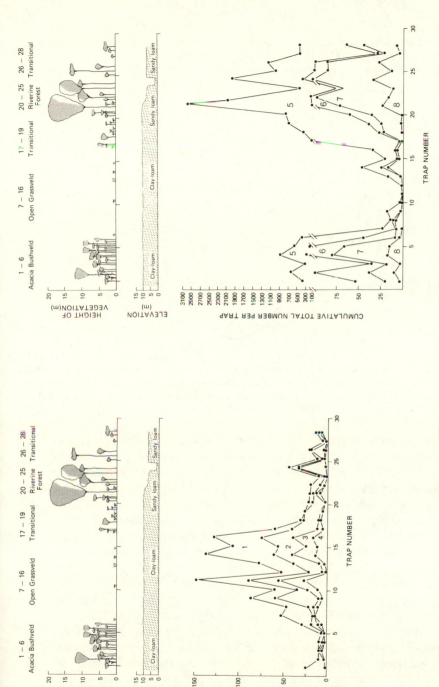

Fig. 8.3. Numbers of individuals in four grassveld and four bushveld species trapped along a transect through a man-made clearing in the Hluhluwe Game Reserve in January 1979. Species: 1, *Sisyphus costatus*; 2, *Onthophagus (Digitonthophagus) gazella*; 3, *Liatongus militaris*; 4, *Onthophagus beiranus*; 5, *Sisyphus seminulum*; 6, *Sisyphus seminulum*; 7, *Drepanocerus laticollis*; 8, *Neosisyphus mirabilis* (adapted from Doube 1983).

TABLE 8.3
The effect of soil type and vegetation cover on the biomass of dung beetles (g) in the different functional groups (Table 8.2) in the Hluhluwe region of Natal, South Africa.

Habitat	Functional Group						Total Biomass	N	S	S*
	I	II	III	IV + VI	V + VI	VII				
Deep sand: pastoral										
Grassveld	197	52	194	89	67	1	599	14	97	93
Bushveld	293	11	84	154	94	0	635	21	113	95
Deep sand: game reserve										
Grassveld	960	176	178	356	313	1	1,984	99	115	86
Bushveld	795	17	73	105	103	0	1,093	37	981	76
Clay-loam: pastoral										
Grassveld	73	137	27	318	18	3	575	19	98	90
Bushveld	59	58	15	218	22	0	372	17	102	98
Clay-loam: game reserve										
Grassveld	64	191	37	385	110	1	788	50	112	86
Bushveld	39	295	6	253	86	0	1,668	42	101	79
Total biomass	2,480	936	614	1,877	811	7	6,726			
N	2	37	2	59	198	0.2	8	299		

Notes: Pitfall traps (3 per locality) baited with cattle dung were set in grassveld and bushveld on deep sand and clay-loam regions in pastoral and game-reserve environments. Traps were set for 24 h once per week from April 1983 to June 1984. N is the number of individuals collected (in thousands), S is the species number, and $S*$ is the rarefied species number in a sample of 10,000 individuals.

most of the common species in the Mkuzi Game Reserve. For example, *Pachylomera femoralis*, *Allogymnopleurus thalassinus*, and *Sisyphus sordidus* were more than five hundred times more abundant in regions of deep sand than in regions with clay or loam soils (Appendix B.8). Other species strongly preferred other soil types; for example, 79% of *Sarophorus costatus* were trapped on skeletal loam and 99% of *Allogymnopleurus consocius* were trapped on clay soil (see also Osberg 1988). Some species, for example, *Kheper nigroaeneus*, *Sisyphus seminulum*, and *Euonthophagus carbonarius*, are widely distributed throughout the reserve and show only minor differences in abundance from one soil type to another (Appendix B.8).

A general feature of regions with deep sandy soil is the numerical and biomass dominance of the FG I and FG III beetles (Fig. 8.2, Table 8.3). Numerous unpublished records from the Kalahari region of Botswana, the Zambezi Valley in Zimbabwe, and from game parks and pastoral regions of South Africa support this generalization. Deep sandy soils are particularly favorable for a number of tunnelers, including several species of *Catharsius*, and for large rollers, including *P. femoralis*, *Scarabaeus goryi*, and *Kheper lamarcki*. Other large rollers—for example, *Kheper clericus*—are more abundant in regions of clay-loam soil, but are never abundant in general.

The structure of the habitat may affect the success of large rollers (FG I). For example, the biomass of large rollers is greater in regions of grassveld within a bush-grass mosaic than in areas of extensive grassveld, which may be related to the dung and microhabitat requirements of breeding beetles. The roller *K. nigroaeneus*, which readily uses many types of dung (Edwards and Aschenborn 1988), buries its balls in soft soil under a tree or bush, or against a grass tussock, but rarely in an unshaded area (Edwards and Aschenborn 1989). This species is abundant in the Mkuzi Game Reserve but scarce in the surrounding pastoral regions, irrespective of soil type.

The distribution of beetle biomass among beetle size classes differs among soil types. In the Hluhluwe region, large beetles (>1,024 mg dry wt) made up 65% of beetle biomass at sites on sandy soil, but only 18% at sites on clay-loam soil (Fig. 8.4). Medium-sized beetles (16–1,024 mg) were the dominant element on clay-loam soils (66%), while small beetles (<16 mg) made up 13–17% of the biomass, regardless of soil type.

The larger beetles (FG I and FG III) have the capacity to remove dung rapidly to their tunnels (Heinrich and Bartholomew 1979b; Doube, Macqueen, and Fay 1988; Edwards 1988a), hence the soil-related differences in community structure are reflected in the impact of beetles on dung pats. For example, in Natal in October–November 1984, a high proportion of dung pats on sandy soil was removed during the first 2 days, and the unburied remains were dry and fragmented, whereas on clay-loam soil only 20% of the dung was buried

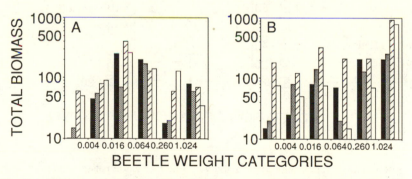

Fig. 8.4. Frequency distribution of beetle biomass according to beetle size at four clay-loam sites (A) and for four deep sand sites (B) in the Hluhluwe region of Natal (the four sites are distinguished by different shading of bars). Pitfall traps were baited with one liter of cattle dung for 24 hours at weekly intervals from May 1983 to June 1984. The sample size is nearly 300,000 beetles. Each set of four sites represents (from left to right) pastoral grassveld, pastoral bushveld, game reserve grassveld, and game reserve bushveld.

over 4 days, and many pats remained relatively intact with moist cores, suitable for breeding of coprophagous flies and dweller dung beetles (Fig. 8.5).

Osberg (1988) has analyzed in detail the distributions of two diurnal *Allogymnopleurus* species in the Mkuzi Game Reserve, where both species are restricted to specific habitats (Appendix B.8). *Allogymnopleurus thalassinus* is largely restricted to regions of deep sand, while *A. consocius* is abundant in only one region of self-mulching clay; both species are scarce on a sandy-clay area separating the two other soil types. Osberg (1988) showed that neither species could bury a dung ball in the sandy clay, except on the rare occasions when the soil was moist and soft; but the reasons for the absence of *A. consocius* from regions of deep sand and the scarcity of *A. thalassinus* in regions of self-mulching clay were not obvious. In a series of field experiments, Osberg (1988) found no difference among the species in their respective abilities to breed in wet and dry soil of both types (Fig. 8.6). She then examined the levels of immature survival in the two soil types at a range of moisture levels. Again no obvious differences were evident, except that *A. consocius* survived water-logged treatments better than *A. thalassinus* in both soil types. Periodic flooding of the self-mulching clay region may keep *A. thalassinus* rare. Osberg (1988) also found that while *A. thalassinus* could remove a dung ball within minutes of arrival, *A. consocius* took hours to form a ball, and it often remained in the pat overnight before rolling a ball away. Inferior competitive

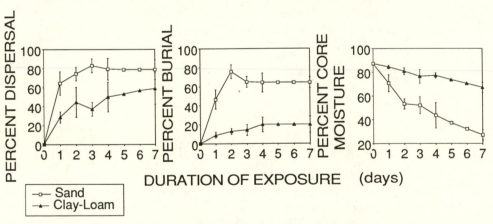

Fig. 8.5. The effect of dung beetle activity on one-liter cattle pats in regions of deep sand (near Bushlands Halt, Natal) and clay-loam soil (near Hluhluwe Village, Natal). Data originate from five experiments conducted during October and November 1984. Each experiment comprised twenty-five pats at each locality. Pats were collected 1, 2, 3, 4, and 7 days after deposition, and the percent burial was estimated volumetrically by comparison with control pats. The percent dispersal gives the amount of dung buried plus shredded. The percent moisture of the remaining dung was estimated using a sample of dung from the core of the pat.

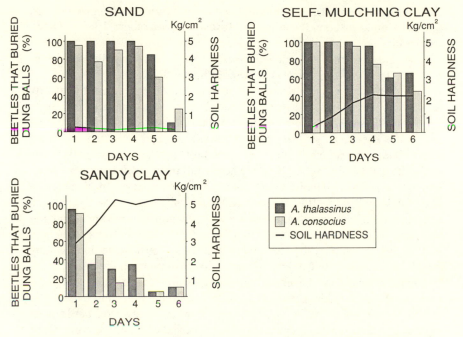

Fig. 8.6. The effect of soil hardness and soil type on the burial of dung balls by two *Allogymnopleurus* species during seven warm, dry days following 50 mm of rain (Mkuzi Game Reserve; adapted from Osberg 1988).

ability may account for the scarcity of *A. consocius* in communities where competition for dung is intense.

Dung Type

Many species of dung beetles are attracted to a wide variety of dung types, but there is little published information for the south African species. S. Endrody-Younga has kindly allowed me to quote his unpublished study conducted at Nylsvlei, on the Transvaal highveld. Some fifty species were caught in traps baited with cattle dung, human dung, carrion, or rotting fruit. Many species showed a high degree of bait specificity. Thus 99% of *Euoniticellus interme-dius* and *Onthophagus (Digitonthophagus) gazella* were caught in traps baited with cattle dung, while over 95% of *Anachalcos convexus* and *Gymnopleurus humanus* were taken in traps baited with carrion or human dung. Some species showed a preference for the fruit bait, while yet others (e.g., *P. femoralis*) were caught in moderate numbers with all types of bait. One of the top com-petitors, the roller *K. nigroaeneus*, uses a wide variety of dung types, includ-ing coarse nonruminant dung (zebra, horse, rhinoceros), nonpelleted ruminant dung (cow, wildebeest), pelleted ruminant dung (impala, kudu, giraffe), and omnivore dung (baboon, bush pig; Edwards and Aschenborn 1988). Numerous species are attracted to the carcasses of large herbivores, which provide carrion, gut contents, and dung. Braack (1984) found forty-five species of dung beetles colonizing impala carcasses in Kruger Park.

The reproductive performance of some African dung beetles is distinctly affected by the quality and type of dung. To take an example of the latter, Davis (1989) showed that the number of broods produced per chamber by *Oniticellus egregius* is several times greater when using horse dung than cattle dung. *Onthophagus binodis* produces many progeny when using good quality cattle dung in spring but far fewer when using poor quality summer dung. Other species, for example, *Onitis alexis*, are little affected by variation in dung quality (Tyndale-Biscoe 1985; Ridsdill-Smith 1986). The species that are sensitive to dung quality may have evolved in association with the dung of browsing herbivores or omnivores, with a relatively high nitrogen content in their dung. In contrast, the species that are not so sensitive to dung quality may have evolved with the dung of large grazing ruminant herbivores.

Elephants were hunted to extinction in Natal before the establishment of the game reserves, and they have only recently been reintroduced to the area. In the Natal reserves, large *Heliocopris* species are relatively rare, but in other regions in which elephant populations have persisted to the present time (e.g., Kruger Park and the Zambezi Valley in Zimbabwe) there are large populations of *Heliocopris* associated with elephant dung in sandy-loam regions. Cambefort (1982b) has attributed the virtual disappearance of some large *Heliocopris* species in the Ivory Coast to the severe decline in the numbers of elephants.

Temporal Patterns

Successional processes. The rate of colonization of dung pats varies with the age of the pat, but not all species are similarly affected. Some species preferentially colonize fresh pats, while other species prefer older pats and for some species the age of the pat appears to make little difference.

Diel activity. All species of dung beetle show some variation in their activity throughout the 24 hours. Some diurnal species (e.g., *P. femoralis*) fly from dawn to dusk, and some nocturnal species fly throughout the night (e.g., *Catharsius* species), but others have relatively restricted flight periods lasting only for a few hours (e.g., *Sisyphus* species; Paschalidis 1974; Tribe 1975). All but one of the twenty summer-active Aphodiidae studied by Bernon (1981) flew at dusk, whereas all the eight winter-active species flew during the afternoon. Most members of the functional groups FG II, FG V, and FG VI are diurnal, while most FG III species are crepuscular/nocturnal (Table 8.4). The remaining groups include both diurnal and crepuscular/nocturnal species.

The diel flight activity of dung beetles has two components, the initiation

TABLE 8.4
Diel activity of dung beetles in the Mkuzi Game Reserve.

FG	Number of Species		Number of Individuals		Biomass (g)	
	D	N	D	N	D	N
I	5	4	4,107	1,286	5,704	812
II	19	1	14,098	1	689	*
III	—	17	—	933	—	138
IV + VI	15	29	14,478	16,149	662	432
V + VI	20	8	48,508	40,430	156	221
VII	2	—	8	—	*	—
Total	61	59	81,199	58,799	7,211	1,603

Notes: Beetles were trapped over two 24-hour periods in December 1981. Pitfall traps were rebaited with fresh baits within 2 hours after dawn and 2 hours before dark (data from Doube, Macqueen, and Davis, unpubl.). D, diurnal; N, nocturnal.
* < 1.

of the flight and its duration. The tunneler *Onitis alexis* has a well-synchronized flight at dusk (Houston and McIntyre 1985) and again at dawn (A. Macqueen, pers. comm.). About 20–30 min before the flight at dusk, the beetles take up positions just below the soil surface, but they do not start flying before the light intensity has dropped below a threshold value. The evening flight of the crepuscular *Onitis* species is usually completed within 15–30 min of the appearance of the first beetle (Bernon 1981). Whether most beetles find dung during this short period of time is not known.

Seasonality. In the summer rainfall regions of southern Africa, most species emerge in abundance after the first spring rains in September–October. They are most active during the wet summer months, become scarce again in late autumn, and are rare during winter and the early, dry part of the spring. Dry periods during the wet season cause a temporary reduction in the numbers of active beetles. However, despite this general pattern of wet season activity, there are very few summer-active species with a unimodal pattern of activity. Multivoltine species show variable levels of abundance throughout the wet season, depending on their breeding cycle and weather. Some species have a bimodal pattern of seasonal activity; for example, the univoltine *K. nigroaeneus* usually has two peaks of abundance per year, often reaching fifty to hundred beetles per trap in the Mkuzi Game Reserve (Edwards 1988a; Section 8.5). Other species are most abundant in spring and early summer (Stickler 1979; Bernon 1981), and still others (e.g., *Onitis caffer*) are primarily active during autumn (Edwards 1986a). The dwellers (both Scarabaeidae and Aphodiidae) breed in the largely intact dung pats during winter (Davis 1977, 1989; Bernon 1981).

The net result of these complementary and overlapping activity patterns is the unimodal seasonality of beetle activity. This pattern contrasts sharply with the bimodal pattern of beetle activity in localities situated closer to the equator. Kingston (1977), working in the Tsavo National Park in East Africa, found that the numbers of beetles colonizing elephant dung were closely correlated with the annual bimodal rainfall pattern, and that beetles were extremely scarce during the dry seasons (Fig. 8.7).

Beetle activity is also influenced by short-term changes in temperature and rainfall. In Natal, South Africa, *K. nigroaeneus* passes the cool, dry winter in the soil as an adult in reproductive diapause. Once the diapause has been completed, the flight activity of the beetle is prompted by rainfall, and as little as 17 mm may result in a major emergence of beetles. The number of beetles flying on any one day is strongly related to ambient temperature. In the Hluhluwe-Umfolozi Game Reserve, the mean numbers of beetles per trap increased two to threefold on the day immediately following 10 mm of rain (Doube 1983).

In a 5-year study in the Hluhluwe-Umfolozi Game Reserve, there was substantial year-to-year variation in the relative abundance of individual species on clay-loam soil (Table 12.2 in Doube 1987). The 20 most abundant species in an assemblage of 114 species made up between 76% and 94% of individuals trapped, depending on the year, and the rank order of species varied widely between the years. The structure of the dung beetle community also varied; for example, the small rollers comprised from 9% to 53% of beetles in differ-

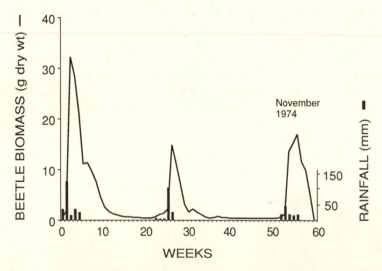

Fig. 8.7. Seasonal changes in dung beetle abundance in relation to rainfall. Data are for dung beetles attracted to elephant dung pats in the Tsavo National Park, Kenya, in 1973–74 (adapted from Kingston 1977).

ent years (Table 12.3 in Doube 1987). It is not known whether equally dramatic variation in the species composition occurs on sandy soils, where competition plays a major role in structuring the beetle community.

8.5. DETERMINANTS OF RELATIVE ABUNDANCE AND COMMUNITY STRUCTURE

Life-History Strategies

The r-K-A templet (Greenslade 1983; Southwood 1987) provides a useful scheme within which to view the startling diversity of life-history strategies shown by the southern African dung beetles. Some aspects of this diversity have been discussed earlier in the section on dung burial behavior, but there is also great variation in reproductive behavior. Edwards (1988b) has suggested that the summer rainfall regions of southern Africa provide a predictably favorable habitat for some long-lived species with low fecundity, but a less predictable habitat for short-lived, r-selected species.

The large roller *K. nigroaeneus* exhibits the highest level of parental investment possible for an insect (Edwards 1988a,b; Edwards and Aschenborn 1988, 1989). Adults emerge from the soil following the first rain in spring, usually in October, and after 1–3 weeks of feeding they begin to breed. Males release a pheromone to attract females (Tribe 1975; Burger et al. 1983; Edwards and Aschenborn 1988). On each nesting occasion the female *K. nigroaeneus* rears a single offspring in an underground nest, usually located in the shade at a depth of about 20 cm. Survival of the offspring is high, from 58% to 84%, and the time of development from egg to adult is about 12 weeks. Most nests are initiated between October and early December. Beetles may construct a second nest if the first offsping dies within 3 weeks of oviposition, or if the first offspring has completed its development by February. Adults live for 2 years or more, and the entire population spends the winter in the soil as adults in reproductive diapause.

Such an extreme level of parental investment has not been recorded previously for dung beetles, but there are many species of large rollers (e.g., *K. subaeneus*, *K. lamarcki*, *K. platynotus*: Sato and Imamori 1987; Edwards 1988a) and large tunnelers (e.g., *Copris* and *Heliocopris* species: Klemperer and Boulton 1976; Kingston and Coe 1977; Klemperer 1983a; Tyndale-Biscoe 1983, 1984), which produce a clutch of only a few young per breeding season. Many of these species allow the buried dung to ferment before they manufacture brood balls (e.g., Sato and Imamori 1987), which may kill insect competitors in the dung, such as fly larvae. These species are predictably abundant in their preferred habitat. A good example is the huge tunneler, *Heliocopris dilloni*, a specialist in elephant dung (Kingston 1977; Kingston and Coe 1977). In Kenya, the strongly bimodal rainfall (Fig. 8.7) controls the life-history events in this species. Some 50–100 mm of rain is sufficient to trigger adult emergence from the soil; beetles start breeding almost immediately, and the immature development is complete by the time of the next rainy season.

The predictability of a particular habitat is largely determined by a species' generation length (Southwood 1977). Since one good fall of rain is sufficient to drive population events in both *K. nigroaeneus* and *H. dilloni* for the next three months, the Tsavo and Mkuzi environments are more predictable for these species than they are for multivoltine species with higher fecundity and shorter generation time (Edwards 1988a). As a consequence, while populations of r-species (high fecundity, opportunistic breeding; e.g., *O. gazella*, *E. intermedius*, *Liatongus militaris*, *Sisyphus (Neosisyphus) spinipes*) have shown marked year-to-year variation in their relative and absolute abundances in the Hluhluwe region of Natal and at Rockhampton in Australia (Chapter 15), populations of K-species (low fecundity, seasonally restricted breeding) have varied little from one year to another.

Other species show a mixture of r and K life-history characteristics (Table 8.5). For example, a number of univoltine *Sisyphus* species have moderate fecundity and breed only during a restricted period in summer (Paschalidis 1974). Similarly, *Onitis caffer* has moderate fecundity and its autumnal emergence in the highveld is precisely determined by temperature and larval diapause (Edwards 1986a,b). This species is a poor competitor for dung, but it breeds in the Transvaal highveld at a time of year when there are few other dung beetles (Stickler 1979), and has shown similar patterns of seasonal activity in each of the 5 years for which data are available (Stickler 1979; Bernon 1981; Edwards 1986a).

A-selected species occupy predictably unfavorable habitats (Greenslade 1983). The dweller species show a mixture of A-selected and r-selected characteristics (Table 8.5). They colonize old dung, which is unattractive to most other species. Some species nest for long periods in the same dung pat, in which a female may rear up to twenty offspring, the number depending on dung characteristics (Davis 1977, 1989; Bernon 1981). In this niche there is little interspecific competition during winter, but during the summer, especially in regions on sandy soils, dung pats are buried so rapidly by other species that the dwellers have very limited breeding opportunities.

Intraspecific and Interspecific Competition

It is clear from data on the abundances of dung beetles at many localities in southern Africa, and from the knowledge of their dung burial behavior (Edwards 1986a; Doube 1987; Doube, Macqueen, and Fay 1988; Edwards and Aschenborn, 1988, 1989), that there are frequently times when competition for dung is intense. This conclusion is supported by much anecdotal evidence from field observations (Heinrich and Bartholomew 1979b), but until recently experimental evidence has been lacking. Competition for dung can take a

TABLE 8.5
Life-history parameters in a suite of dung beetles from southern and central Africa.

Species	TB	LE	BN	ID	LF	OS	GY	LO	
Kheper nigroaeneus	R	28	1	150	3+	NBA	1	700	(1100)
Kheper platynotus	R	30	1–4	?	6+	A	1	700	
Sisyphus sordidus	R	6	1	66	10–20	NBA	2	350	(700)
Sisyphus seminulum	R	4	1	47	26	BA	4	125	(159)
Neosisyphus mirabilis	R	10	1	77	47	BA	3	144	(265)
Neosisyphus fortuitis	R	10	1	73	55	BA	2	154	(268)
Neosisyphus spinipes*	R	9	1	52	44	BA	4	104	(202)
Neosisyphus rubrus*	R	8	1	64	36	BA	4	117	(183)
Heliocopris dilloni	T	50	2–7	160	5+	L	2	180	
Digitonthophagus gazella*	T	11	<45	30	90	A,L	many	60	
Onitis alexis*	T	15	24	56+	130	DL	1–2+	56	
Oniticellus egregius	D	8	7	?	65	BA	?	98	(165)
Oniticellus planatus	D	9	10	22	52–144	BA	?	223	(389)
Oniticellus formosus	D	8	20	<36	59–70	BA	?	142	(262)
Euoniticellus intermedius*	D	8	many	28	120	A,L	many	45	

Sources: Data from Paschalidis (1974); Blume and Aga (1975); Davis (1977, 1989); Kingston and Coe (1977); Tyndale-Biscoe (1985); Sato and Imamori (1987); Edwards (1988a).

Notes: The duration of immature development given here is the minimum recorded in the literature; this minimum duration is extended by changes in many factors, including larval diapause, temperature, and dung characteristics. The number of broods per nesting occasion is an average value, usually obtained for single pairs of beetles in 1–2 liters of cattle dung. The number of broods produced is influenced by numerous factors, including dung type and quality, competition for dung, and soil type. Species marked with an asterisk (*) have been introduced into Australia (Chapter 15).

The abbreviations are as follows: TB, type of beetle (R, roller; T, tunneler; D, dweller); LE, length in mm; BN, broods (offspring) per nesting occasion; ID, duration of immature development in days; LF, life-time fecundity; OS, overwintering stage (A, adult; BA, breeding adult; NBA, nonbreeding adult; L, larva; DL, diapausing larva); GY, number of generations per year; and LO, longevity in days. Values in parentheses are maximum values recorded.

number of forms ranging from direct combat, in which beetles fight over the possession of dung, through resource preemption, in which priority of access determines the winner, to scramble competition, in which the beetles' activity at high densities prevents most individuals acquiring sufficient resources for breeding.

Doube, Macqueen, and Fay (1988) and Giller and Doube (1989) have examined intraspecific and interspecific competition among twelve crepuscular/nocturnal tunnelers. The twelve species were selected from a diverse assemblage, the members of which coexist in the sandy soil regions of Natal, South Africa. Two distinct patterns of dung burial were recognized. In eight species of Coprini (FG III), dung burial was complete within 24–48 hours of the arrival of beetles at the pat, while in four species of Onitini (FG IV), dung was buried gradually over a period of 12 days. At high densities, Catharsius tri-

cornutus, *Copris elphenor* (FG III), and *O. alexis* (FG IV) removed 70–80% of 1.5-liter pats, and strong intraspecific competition for dung occurred at densities greater than two to four pairs per pat. Competition was asymmetric between the coprine and onitine species, the former being largely unaffected by the presence of the latter, while the amount of dung buried by *O. alexis* was markedly reduced in the presence of two or three pairs of the coprine beetles. Differences in the age of colonized dung create other asymmetries. For example, many species colonize fresh dung and begin burial soon after arrival, while other species (e.g., *O. viridulus*) colonize older pats and wait in the pat for several days before beginning to bury dung (Edwards and Aschenborn 1987; Doube, Macqueen, and Fay 1988). The latter species will be at an obvious disadvantage when dung is in short supply.

Species must occupy the same pat in order to compete with one another. Some species breeding in ephemeral resource patches (e.g., *Drosophila* in rotting fruit: Shorrocks and Rosewell 1987; blowflies in carrion: Hanski 1987c) show independently aggregated spatial distributions among habitat patches, which reduces the probability that potential competitors will occur together (the variance-covariance model in Chapter 1). Doube and Giller (unpubl.) examined the small-scale spatial distributions of eleven crepuscular or nocturnal dung beetle species in a local community of about eighty species by dispersing nine blocks of nine cattle pats over a tract of 400 ha (pats within a block were 5 m apart in a 3 x 3 grid). The pooled numbers of individuals of all eleven species extracted per pat ranged from ten to one hundred per night. There was marked variation between the blocks in the numbers of beetles caught, and the rank order of blocks remained relatively constant within 3-day sampling periods, though not during longer periods (3 weeks). Similar patterns were found for individual species, some blocks yielding up to twenty times more individuals than others. However, within blocks the distributions were random, and there was no evidence for independent spatial aggregations. Doube and Giller (unpubl.) conclude that spatial aggregation of beetles did not facilitate their coexistence in this case.

In regions with clay-loam soils, inferior competitors (FG IV, FG V, and FG VI) are well represented, suggesting that competition with the competitively dominant groups (FG I, FG II, and FG III) is not an important determinant of community structure in these regions. Furthermore, the density of beetles that bury dung at a moderate depth (FG I, FG II, FG III and FG IV) is low in comparison with adjacent regions on sandy soils. Why these species are not abundant in the African clay-loam soils is not clear, especially because in homoclimatic regions in Australia with similar soil, several FG II and FG IV species have become hugely abundant (see Table 15.3). It is possible that predation on buried dung by termites provides an explanation. Termites are scarce or absent in the regions of deep sandy soil but are abundant in the clay-loam soils in southern Africa. *Odontotermes* have been observed to remove broods and dung buried by *O. alexis*, *O. caffer*, and *C. tricornutus* in the field.

8.6. CONCLUSIONS

The following conclusions about the structure of southern African dung beetle communities apply primarily to assemblages associated with bovine dung, though it is likely that similar conclusions will apply to other assemblages as well:

1. Large rollers (FG I) are frequently a dominant group on sandy soils, especially in regions of bush-grass mosaic, where they may account for up to 80–90% of the biomass of beetles attracted to cattle dung. However, there are FG I species (mostly *Kheper* and *Anachalcos*) that are widespread or prefer clay-loam soils, but these are not abundant.
2. The fast-burying tunnelers (FG III) occur in all habitats, but they appear to be more abundant in the grassveld than in the bushveld, and in sandy regions than in clay-loam regions. They form a major component of the dung beetle assemblage in open pastoral grassveld on sandy soils.
3. The small rollers (FG II) and the remaining tunnelers (FG IV and FG V) are relatively evenly distributed across habitats, though individual species may show habitat specificity (Appendix B.8).
4. The dwellers (FG VII) occur primarily in the grassveld on clay-loam soils and are relatively scarce in situations where competition for dung is intense.

The structure of the dung beetle communities in southern Africa is thus determined primarily by soil type and to a lesser extent by vegetation cover and dung type. Sandy soil regions have a greater biomass of dung beetles than regions with heavier soils. The dung beetle communities on sandy soils are dominated in biomass by low-fecundity FG I and FG III beetles, and competition for dung is an important determinant of community structure. In regions with clay-loam soils and extensive grassveld, the smaller, high-fecundity beetles that nest in shallow tunnels (FG II, FG IV, and FG V) are a dominant group of beetles. In these systems there is little resource competition, and dung pats frequently remain largely intact for days following deposition, even during the moist summer months. The structure of these communities is variable over time.

CHAPTER 9

Dung Beetles in Tropical Savannas

Yves Cambefort

9.1. INTRODUCTION

Tropical grasslands, or savannas, comprise the biome with the most diverse mammalian fauna ever found on Earth (e.g., Bigalke 1972; McNaughton and Georgiadis 1986). Unfortunately, in all parts of the world except one, much of the fauna has become extinct in the latest geological times, during the Pleistocene extinctions (Axelrod 1967; Birks 1986; Owen-Smith 1987), possibly causing the extinction of many dung beetles. The only rich mammal fauna that continues to survive on savannas is the African one. Because dung beetles are closely associated with mammals, it is not surprising that they reach their peak in the tropical African savannas, with roughly seventy-five genera and fifteen hundred species of Scarabaeidae.

The taxonomy of African dung beetles is relatively well known, but only a few ecological studies have been completed so far. The CSIRO Dung Beetle Program yielded the first comprehensive results from the subtropical savannas of South Africa (Paschalidis 1974; Tribe 1976; Davis 1977; Stickler 1979; Bernon 1981; Doube, Chapter 8). The pioneering studies in tropical savannas are due to Kingston (1977; East Africa) and Walter (1978; Zaire), followed in western Africa by Cambefort (1980, 1982b, 1984). Tropical savannas in Asia have a rich assemblage of dung beetles, but little is known about them (Oppenheimer 1977; Oppenheimer and Begum 1978; Begum and Oppenheimer 1981; Mittal 1981a,b, 1986), and almost nothing is known about the situation in America (Kohlmann, Chapter 7). Dung beetles in Australian subtropical savannas are reviewed by Doube et al. in Chapter 15.

This chapter focuses on African savannas and is largely based on results from the Guinean savanna in the Ivory Coast (Cambefort 1980, 1982b, 1984). Following a description of the savanna environment (Section 9.2), I will briefly outline the structure of its dung insect communities (Section 9.3). I next describe the guilds of dung beetles present in tropical savannas (Section 9.4) and examine resource partitioning among the dozens of coexisting species (Section 9.5). The last two sections deal with spatial patterns (Section 9.6) and long-term temporal stability (Section 9.7).

9.2. BIOGEOGRAPHY OF AFRICAN SAVANNAS

The term "savanna" is used here for those tropical formations where (1) the grass stratum is continuous and dense, though occasionally possibly interrupted by trees and shrubs; (2) bush fires occur from time to time, usually once a year; and (3) the growth of vegetation is closely related to alternating wet and dry seasons (Bourlière and Hadley 1983). Tropical savannas have recently been the subject of a series of textbooks (UNESCO, 1979; Huntley and Walker 1982; Bourlière 1983; Cole 1986). Savannas are suitable for the cultivation of many important crops, and they are especially favorable for cattle breeding. On the other hand, tropical savannas are fragile ecosystems, and intensive agriculture and grazing are likely to lead to lateritization and desertification. Dung beetles, either native or introduced (Chapter 15), are an important component of the savanna ecosystem.

The origin of tropical savannas is in dispute. Climatic as well as biotic factors have most probably been involved, such as the influence of the pluvial and interpluvial periods of the Cenozoic, and the advance and retreat of tropical forests. During the Cenozoic, grass formations evolved between forests and (sub)deserts, probably at the expense of the former, and at least to some extent owing to the action of large herbivorous mammals (Proboscidians, Perissodactyles), which have played a significant role until recently (Cumming 1982). Perhaps the most important factor maintaining the present savannas is annual bush fires (in Africa: Monnier 1968; Trollope 1982; Gillon 1983; in South America: Coutinho 1982; in Australia: Lacey et al. 1982). It has been demonstrated in western Africa that humid (Guinean) savanna protected from the fire is likely to turn to evergreen forest (Vuattoux 1970, 1976; Devineau et al. 1984).

Africa has a concentric pattern of climatic zones from the equatorial humid climate to desert climates. Vegetation follows the same pattern (Fig. 9.1): a central nucleus of evergreen rain forest is surrounded by successive regions of humid savannas, dry savannas, and deserts. This zonation is especially clear in western Africa (Anonym 1979b; Lecordier 1974), where it is useful to distinguish between the southern "Guinean" zone and the northern "Sudanese" zone. The Guinean savanna is under the influence of the monsoon, and it is characterized by humid air, small range of temperatures, and an almost continuous supply of rain. There are four seasons (long rainy season, short dry season, short rainy season, and long dry season), but only the long dry season is truly dry. The Sudanese zone is under the influence of harmattan, a northern wind coming from the desert, and has only two seasons: a rainy season, which becomes shorter as one goes farther north, and a dry season. The Guinean zone supports rain forest only in its more humid, southern part, while the northern part is covered by Guinean savanna; the Sudanese zone is dominated by Sudanese savanna (Guillaumet and Adjanohoun 1971; Fig. 9.1).

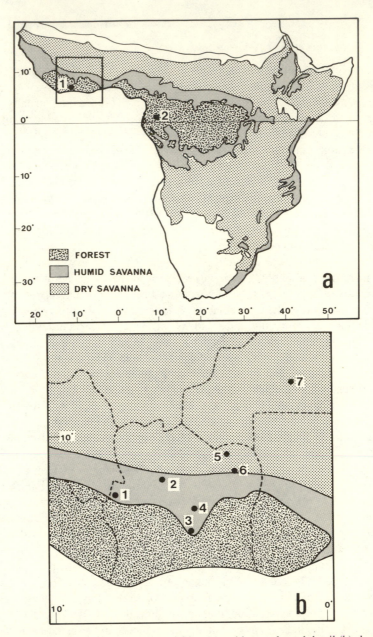

Fig. 9.1. Map of tropical and southern Africa (a), with an enlarged detail (b) showing the vegetation zones (simplified from White 1983) and the study localities for Chapters 9 and 11. The vegetation symbols are the same in the two maps (forest is evergreen rain forest, humid savanna includes savannas and woodlands of the Guinean type, and dry savanna includes savannas and woodlands of the Sudanian type). Localities in (a): 1, Taï (Ivory Coast); 2, Makokou (Gabon). Localities in (b): 1, Sipilou; 2, La Marahoue; 3, Lamto; 4, Abokouamekro; 5, Wango Fitini; 6, Kakpin (1–6 in Ivory Coast); 7, Pabre (Burkina Faso).

Bush fires mark the end of the long dry season, in January in Guinean savannas and in February in Sudanese savannas. Shortly after the fires, and even
before the first rains, grasses begin to grow, and they continue to grow until
fructification at the end of the year (Fournier 1982). At this point, all the stems
and leaves wither as the dry season proceeds, and the stage is set for another
round of fires. For convenience, it is assumed here that the dung beetle year
begins on February 1.

Dung beetle data are available from six localities, described in detail in
Cambefort (1984). The principal study site is Abokouamekro in the Ivory
Coast, situated in the Guinean savanna dominated by *Loudetia arundinacea*
(César 1977). The mammalian fauna consists of small mammals (Bourlière et
al. 1974) and cattle, with live weight around 140 kg per hectare, a value close
to the "natural" mammalian biomass (121 kg) in savannas, predicted from
rainfall (Coe et al. 1976; but see Bell 1982; East 1984). The second study
locality, Lamto, is close to Abokouamekro, but large mammals are practically
absent due to excessive hunting. In this well-known West African locality
(e.g., Lamotte 1978), the average temperature is 27.7°C and the average annual rainfall is 1,195 ± 234 mm (SD). The third study locality, Kakpin, is
situated in a national park, where the mammalian fauna is or recently was
close to its natural state (Anonym. 1979a).

The three localities of Lamto, Abokouamekro, and Kakpin, all situated in
the Guinean savanna with similar climatic conditions, allow comparisons between sites with the following kinds of mammalian faunas: (1) small mammals
present (mostly omnivores, including man) but no large, herbivorous mammals; (2) small mammals and cattle present; and (3) small, medium-sized and
large mammals present, including the elephant.

9.3. Dung Insect Communities in Savannas

It is useful to introduce a simple, functional classification of mammals based
on their dung (for the African mammals, see Dorst and Dandelot 1970). The
two basic types are the large herbivorous mammals, including buffalo and
cattle, and the small omnivorous mammals. The extreme, but important, example of the former is the elephant, which has fibrous dung with a large
amount of vegetable matter. Outside the study region, rhinos and horses (Perissodactyles) have almost the same type of dung (Tribe 1976), while cattle
dung is more fine-grained and probably richer in proteins. All the other mammals belong to the omnivore class in this classification. Many of them are
really herbivores (antelopes, duikers, etc.), but their dung is composed of
small or very small pellets, which are used completely differently by dung
beetles than is cattle or elephant dung. Some species, for example, the
Grimm's duiker, are true omnivores and include small animals in their diet.
All truly omnivorous mammals have dung richer in proteins than the dung of

large herbivores (Nibaruta et al. 1980), and their droppings are always small. For trapping dung beetles for the present study, the bait was usually human or cattle dung. These are not only typical examples of omnivore and herbivore dung, respectively, but they are also the two most widespread excrements of all (Halffter and Matthews 1966).

Mammal dung is a nutritionally rich resource that attracts many insect consumers and their arthropod predators (Hanski 1987a). The communities of dung insects have been well studied in South Africa (Bernon 1981), but no comprehensive data exist for the tropical African savannas. Some results from Lamto are given in Table 9.1. Scarabaeidae dung beetles strongly dominate the omnivore dung insect fauna, while Aphodiidae are scarce (they are more numerous in cattle dung). The numbers of species of Staphylinidae (35), Hydrophilidae (5), and Histeridae (14) are remarkably similar to the ones given by Bernon (1981): 34, 5, and 14, respectively, although he sampled cattle pats. Many Staphylinidae are predators (Hanski and Hammond 1986), as are most of the Histeridae and the larvae of Hydrophilidae. At Kakpin, some

TABLE 9.1

Insects caught in six pitfall traps baited with human dung in Lamto, Ivory Coast, in August 1987.

Order	Family	Number of Species	Number of Individuals
Beetles	Scarabaeidae	50	15,471
	Aphodiidae	4	5
	Hybosoridae	1	1
	Trogidae	1	17
	Staphylinidae	35	1,283
	Aleocharinae	10	513
	Euaesthetinae	1	1
	Oxytelinae	4	345
	Paederinae	1	3
	Piestinae	1	1
	Staphylininae*	13	321
	Tachyporinae*	2	9
	Xantholininae*	3	90
	Histeridae*	15	105
	Hydrophilidae	5	138
	Others	12	63
Diptera	Brachycera	46	3,650

Source: Cambefort (unpubl.)

Note: As pitfall traps do not collect flies satisfactorily, the figures for flies (Diptera) are not directly comparable with those for beetles (Coleoptera). Predators have been marked with an asterisk (*).

staphylinids, for example, the very large *Platydracus cantharophagus* ("beetle eater"), slaughter dung beetles in elephant dung pats. Judged by the remains of elytrae, legs, and other parts left in the pats, most of the prey species belong to small tunnelers (Oniticellini and Onthophagini), of which up to one fourth may be killed in some pats. At Lamto, large predatory insects are not abundant, probably because small omnivore droppings do not stay on the ground for very long and have the high concentrations of insects that occur in elephant and cattle pats. The extreme abundance of dung beetles reduces the chance of other insects breeding in dung pats, making fly larvae, for example, scarce in dung in western African savannas. In some dry regions, for example in Kenya, dung beetles are practically absent during the dry season (Fig. 8.7), when most of the elephant dung is used by termites (Kingston 1977).

Dung beetles play an important role in African savannas: Cambefort (1984) has estimated that beetles bury one metric ton of dung per hectare per year in West African savannas. Primary production is significantly higher in savannas with large herbivores and associated dung beetles than in savannas without mammals (Cambefort 1986).

9.4. THE GUILDS OF DUNG BEETLES

At Abokouamekro and in the rest of Africa, the two main guilds of dung beetles are the rollers and the tunnelers, while the dwellers (Aphodiinae) usually play an insignificant role. The rollers have about 350 species in African savannas, with 28 species occurring in the Ivory Coast, while the tunnelers have about 1,150 species, of which 179 occur in the Ivory Coast. (The Ivory Coast has about 50 species of dwellers: Bordat, 1983.) The rollers comprise the subfamily Scarabaeinae while the tunnelers make up the bulk of Coprinae. A few Coprinae do not dig tunnels by themselves but live as kleptoparasites of rollers or tunnelers (see below).

Rollers

The behavior of making and rolling dung balls is the most peculiar characteristic of Scarabaeinae dung beetles. Although rolling has probably appeared in some forest dung beetles (tribe Canthonini), it is now most widespread in the species occurring in open habitats, especially on sandy soils (Halffter and Matthews 1966).

In the Ivory Coast, individual rollers are very numerous (47% of all dung beetles in my study), but the number of species is relatively small (14% of the total; Table 9.2). Most species are small and diurnal, and belong to the two tribes of Sisyphini and Gymnopleurini. Some Sisyphini in particular are extremely abundant (Appendix B.9). It is remarkable that rollers are very scarce in systematic collections made in West Bengal (Oppenheimer 1977): two spe-

TABLE 9.2
Numbers of species, genera, and individuals in different guilds and tribes of dung beetles in the Ivory Coast.

Guilds	Tribe	Genera	Species	Common Species	Average Number of Individuals per Species
Rollers	Scarabaeini	2	4	1	72
	Gymnopleurini	3	8	6	1,345
	Canthonini	2	5	—	34
	Sisyphini	2	11	8	5,906
Tunnelers	Coprini	4	21	5	202
	Dichotomiini	3	6	—	13
	Onitini	2	14	1	34
	Oniticellini	6	18	7	1,082
	Onthophagini	11	101	30	455
Kleptoparasites	Dichotomiini	1	9	2	160
	Onthophagini	3	10	8	1,417
Total		39	207	68	783

Source: Data from Appendix B.9.

Note: "Common species" are the ones represented by more than 128 individuals in the pooled material.

cies of Gymnopleurini out of thirty-five species of dung beetles, representing a mere 1.5% of all specimens.

Sisyphus are active during most of the year and can use the excrements of all mammals, including carnivores such as the leopard, whose dung attracts almost no other dung beetles. Sisyphus also use the excrements of birds, such as the Guinea fowl, and of land tortoises and toads; some species even use carrion. Adults of the dominant species S. seminulum are active throughout the day, with a daily peak between 10 and 11 A.M. (Cambefort 1984). After emergence, the teneral adults make food balls for approximately 3 weeks. The individual fresh weight is 4.9 ± 1.5 mg (SD) during the feeding period, and the weight of the food ball is 28.8 ± 10.4 mg. By the beginning of the breeding period the individual fresh weight has risen to 8.4 ± 1.7 mg, and the balls, which are now rolled by pairs of beetles, have a fresh weight of 83.4 ± 20.3 mg. The nesting sequence, which consists of preparing the brood ball, constructing a nest, and laying an egg in the ball, can probably occur as often as every 3 days, as with some South African species (Paschalidis 1974). Because the female must copulate before each nesting sequence, a new pair is established each time. Sisyphus adults can live more than one year, and up to five generations may develop per year (Paschalidis 1974).

The *Neosisyphus* species are less abundant and more specialized than *Sisyphus*, the adults being attracted especially to cattle pats. They spend 24–48 hours in the pat, covered with the dung particles that make them almost invisible. They seem to feed directly on dung without making food balls. After mating in the pat, a pair of beetles rolls away a brood ball, which is buried very shallowly or may even be left at the soil surface after an egg is laid (Cambefort 1984). Some South African species attach the brood ball to a grass stem (Paschalidis 1974; see Fig. 17.2). Inside the ball, the larva becomes shriveled during the dry season, but it survives and resumes its development once the rains soften the ball.

The Gymnopleurini are the second most numerous rollers in the Ivory Coast, but have only eight species (Table 9.2). Most of them are diurnal and active in the beginning of the year, the first individuals emerging just after the bush fire. Most Gymnopleurini spend nearly 10 months in the soil, either as larvae or as adults. Although precise information is lacking, it is likely that there is only one generation per year, and the beetles can probably live for more than one year. The older, spent individuals are the first ones to emerge from the soil, before the beginning of the rains. These brightly colored beetles are very conspicuous and seem to be involved in mimicry complexes (see below).

In addition to these abundant taxa, there are other rollers worth mentioning. Most of the Canthonini (genus *Anachalcos*) are large, nocturnal species and make and roll relatively small balls. Another nocturnal species is the large *Scarabaeus goryi*. The genus *Kheper* includes some large species, diurnal or nocturnal, which prefer tree savannas or woodlands to open savannas (Edwards and Aschenborn 1988). In the Ivory Coast this genus is uncommon, but elsewhere it can dominate the dung beetle fauna, at least in biomass (e.g., in Transvaal: Stickler 1979; Chapter 8). The diurnal species spend up to one hour making a ball (Sato and Imamori 1986b), but the nocturnal ones, which are often involved in exceptionally severe competition (with large tunnelers; see below), can make a ball in a few minutes (Kingston 1977). In these species, the success of making and rolling balls depends on thoracic temperature and size (Bartholomew and Heinrich 1978; Heinrich and Bartholomew 1979a, b). *Kheper* shows remarkably intensive maternal care, and fecundity may be as low as one offspring per female per year (Edwards 1984; 1988a; Sato and Imamori 1986a,b, 1987, 1988).

Tunnelers

The pattern of species abundance in tunnelers is opposite of that in rollers: the tunnelers account for 77% of the species but only 43% of individuals (Table 9.2). In other words, there are many rare or relatively rare tunnelers. The Onthophagini, which tend numerically to dominate the dung beetle communities with Sisyphini, have a large number of species, but a species of Sisyphini is on average thirteen times more abundant than a species of Onthophagini.

The Oniticellini, another major tribe of tunnelers (Table 9.2), are less speciose but each species is more abundant, on average, than Onthophagini. Oniticellini generally prefer large dung pats, in which they may stay for a while, splitting and crumbling the pat rather than burying dung. The small *Drepanocerus* and the larger *Cyptochirus* use the same sort of camouflage as *Neosisyphus*: they are covered with dung particles and are difficult to see (Krikken 1983). Perhaps for this reason they were long considered to be very rare (e.g., Janssens 1953), whereas the *Drepanocerus* especially are among the most typical and abundant species in cattle pats in South Africa (Stickler 1979) as well as in western African savannas. Most Oniticellini bury only small quantities of dung, often very shallowly, and others adjust their nesting behavior according to climatic conditions (Rougon and Rougon 1982b). The "dweller" behavior in *Oniticellus* is evidence of the extreme development of the general tendency in this tribe to use large dung pats. They make their nests within pats rather than in the soil (Chapter 3).

Onthophagini and Oniticellini make up the guild of "small tunnelers," in which fecundity is relatively high, nest architecture is relatively simple, and maternal care is uncommon. The remaining tribes—the Coprini, Dichotomiini (*Heliocopris*) and Onitini—comprise the "large tunnelers," which have much reduced fecundity, elaborate nesting behavior, and often intensive maternal care. The small tunnelers produce generally more than one generation per year, while the large tunnelers probably only produce one generation each year.

Kleptoparasites

Some dung beetles do not dig or provision an underground nest but use the reserves secured by other species and the nests dug by them. The kleptoparasites account for nearly 10% of dung beetle species in the Ivory Coast (Table 9.2). I recognize two subguilds: the kleptoparasites of rollers and those of tunnelers.

The kleptoparasites of rollers are diurnal and many of them belong to the genus *Cleptocaccobius*, which are attracted more to dung balls that are being rolled than to dung pats. In West Africa, *Cleptocaccobius uniseries* is the most abundant species, probably using balls of both *Gymnopleurus* (in the beginning of the year) and *Neosisyphus* (at the end of the year). Other *Cleptocaccobius* are most abundant in the beginning of the year, coinciding with several *Gymnopleurus* species. The kleptoparasites of tunnelers are found in the buried reserves of large tunnelers. It is possible that some nonkleptoparasitic species become accidentally buried by the large tunnelers, but the true kleptoparasites are found in the host nests in large numbers and clearly more frequently than "accidental" species (Table 9.3). However, kleptoparasitism may have evolved from such accidents.

Mimicry

Arrow (1926, 1931) was the first to draw attention to apparent mimicry complexes involving African dung beetles in the tribes Gymnopleurini and Onthophagini, in which the *Pleuronitis* (Onitini) may also be involved. In the Ivory Coast, for example in Kakpin, a mimicry complex in a bright metallic blue species is apparent (Table 9.4). These species occur early in the year, and the complex is believed to involve both Müllerian and Batesian mimicry. The models are assumed to be the Gymnopleurini, which are exposed to predation while rolling balls. Although there is no direct evidence that these insects are unpalatable, Pluot-Sigwalt (1982, 1984) has demonstrated that they have exceptionally numerous pygidial glands that generally secrete repellent substances. In this case, the bright colors exhibited by these taxa may be interpreted as warning colors. Some diurnal birds (pied crows, Guinea fowls) and monkeys (baboons) are the usual predators of rollers (Paschalidis 1974; Tribe 1976). The three species of Gymnopleurini present at Kakpin are probably mutually mimetic, exhibiting the same blue color. The mimics (Onthophagini) do not have as many pygidial glands as the models. The mimics are associated with the models and usually have the same bright blue color, al-

TABLE 9.3
Dung beetles found in one nest of *Heliocopris antenor* (Abokouamekro, April 1981).

	Species	*Nest*	*HDT*	*CDT*
Presumed nonkleptoparasites				
Aphodiidae	*Notocaulus nigropiceus*	1	—	—
Scarabaeidae	*Drepanocerus endroedyi*	1	—	7
	Drepanocerus laticollis	1	19	142
	Caccobius auberti	3	1,771	149
Presumed kleptoparasites				
Aphodiidae	*Trichonotulus kindianus*	15	—	—
Scarabaeidae	*Onthophagus juvencus*	7	4	127
	Pedaria coprinarum	72	—	99
	Pedaria dedei	27	—	12
	Pedaria durandi	2	—	—
	Pedaria renwarti	1	91	65

Note: The table gives the numbers of individuals collected from the nest, which are compared with the numbers caught in human dung-baited (HDT) and cattle dung-baited pitfall traps (CDT).

TABLE 9.4
Models and mimics caught at Kakpin in six human-dung baited traps in May 1981.

Species	Color	Numbers
Presumed models		
Garreta nitens janthinus	blue	593
Gymnopleurus coerulescens	blue	168
Gymnopleurus puncticollis	blue	1,677
Presumed mimics		
Proagoderus auratus cyanesthes	blue	230
Proagoderus auratus auratus	coppery	1
Hyalonthophagus pseudoalcyon	blue	39
Phalops vanellus coeruleatus	blue	91
Phalops vanellus vanellus	coppery	2
Phalops batesi batesi	blue	2
Onthophagus cupreus cyanomelas	blue	54
Onthophagus cupreus cupreus	coppery	7

Source: Cambefort (1984).

though a few individuals are differently colored (Table 9.4). In other African localities, mimicry complexes have different colors. For example, a red complex occurs in northern Senegal, with three Gymnopleurini, four to five Onthophagini, and one *Pleuronitis*, while a green complex occurs in some East African localities.

Other possible mimicry complexes do not include Gymnopleurini. Holm and Kirsten (1979) have described a case of "speed mimicry" in large rollers, and Cambefort and Dudley (unpubl.) have observed in Malawi a case in which two large tunnelers, *Onitis dimidiatus* and *Diastellopalpus thomsoni*, are similarly bicolored (forebody bright metallic green, the rest of the body black).

9.5. RESOURCE PARTITIONING

Food Choice

Extensive trapping data are available for the two main types of dung, large herbivore dung (exemplified by cattle dung) and omnivore dung (human excrement). Most dung beetles, 72% of the species, are specialized and prefer either one or the other type of resource. In the rollers, only *Sisyphus goryi* and the *Neosisyphus* prefer cattle dung, which is favored by most large tunnelers, most Oniticellini, and by *Digitonthophagus* and *Milichus* species. Most of the remaining Onthophagini are more attracted to omnivore than herbivore dung. Like their host species, the kleptoparasites of rollers are more attracted to human dung, while the kleptoparasites of tunnelers are more attracted to herbivore dung. It is noteworthy that generally all the species in a genus show the same food selection, suggesting that food preference is a relatively conservative trait in these taxa.

Not only the quality of dung but also the size of the dung pat is of significance to dung beetles. The size is especially important for nesting: the female, with or without the help of the male, has to bury an amount of dung sufficient to make one or more brood balls. In the case of large tunnelers, especially the *Heliocopris* species, the quantity of dung needed is very large (Klemperer and Boulton 1976; Kingston and Coe 1977). The presence of large mammals thus has the greatest influence on dung beetles through larval requirements, while in the localities where they occur, adult *Heliocopris* also feed on the dung of small, omnivorous mammals. Elephant dung specialists can occasionally use a substitute when elephants have disappeared, but then the individuals remain distinctly smaller. This is the case at Pabré, where a species of *Heteronitis* occurs in donkey dung instead of elephant dung. One could expect a strong directional selection for smaller body size in such situations.

Small droppings of omnivorous mammals are not available to beetles for very long. When dung beetles are abundant and active, small droppings disappear in a few hours; and during the dry season, when beetles are less numerous, small droppings dry out quickly and attract only few dung beetles. In contrast, large herbivore dung pats, especially cattle pats, remain attractive to beetles for several days, though most dung beetles reach their highest numbers in the pats during the first day (Fig. 9.2). The numbers then gradually decline, and most species have left the pat by the fourth or even the third day. A few species are most numerous on the second day, and some *Pedaria* and especially *Drepanocerus* can stay in the larger pats for a longer period. The "dweller" *Oniticellus*, notably *O. formosus*, arrive at the pats later and make their nests in the largest remaining ones, in which they stay for several days.

The numbers of species attracted to human, buffalo, and elephant dung were about equal at Kakpin, around forty species in a random sample of three hundred individuals. In contrast, the diversity of beetles in traps baited with human dung in Lamto and in cattle dung pats in Abokouamekro was lower, thirty species in a random sample of three hundred individuals.

Some "dung beetles" are attracted to food sources other than dung, especially to carrion (Appendix B.9), though the African savannas appear to have no strictly necrophagous species. A peculiar case is *Onthophagus latigibber* and allied species, which are attracted almost exclusively to large milliped carcasses, which they bury like other species bury dung. In southern Africa, species of the *Scarabaeus*-like genus *Sceliages* also make brood balls of milliped carcasses (Bernon 1981). In the Ivory Coast, a few *Onthophagus* are attracted to rotten fruits as well as to carrion, and two peculiar Dichotomiini genera (*Coptorhina* and *Delopleurus*) have specialized on fungi, which they bury in their nests.

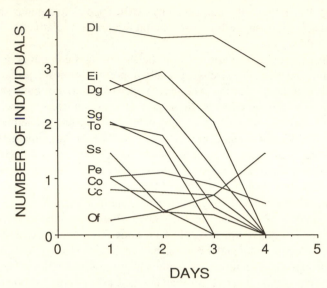

Fig. 9.2. Numbers of individuals in selected species of dung beetles in cattle pats plotted against the age of the pat. The vertical axis gives the logarithm of abundance. Species are: Dl, *Drepanocerus laticollis*; Ei, *Euoniticellus intermedius*; Dg, *Digitonthophagus gazella*; Sg, *Sisyphus goryi*; Ts, *Tiniocellus spinipes*; Ss, *Sisyphus seminulum*; Pc, *Pedaria coprinarum*; Co, *Copris orion*; Cc, *Catharsius crassicornis*; Of, *Oniticellus formosus*.

Seasonality

African savannas have a distinctly seasonal climate, with clear-cut rainy and dry seasons. Seasonality has a strong effect on dung beetle activity; for example, in Kenya a good correlation exists between the rains and the emergence of *Heliocopris* and other species (Kingston 1977; Kingston and Coe 1977; Fig. 8.7).

In the Guinean savannas of the Ivory Coast, which have two rainy seasons, the dung beetle community has two peaks of abundance: a very distinct one early in the year, soon after the first rains in February–March, and a second, less obvious one in October–November. These two peaks are due to two basic phenological types: species that occur only during a short period of time in the beginning of the year, and species that occur throughout the rainy season but often become more abundant toward the end of it. This dichotomy is especially clear in omnivore dung specialists, in which the first peak is more pronounced (Fig. 9.3). The species of the first phenological type are exemplified by the rollers *Gymnopleurus*, which are very numerous in the very beginning of the year and most probably produce only one generation per year. Another example is the small tunneler, *Metacatharsius*. The second phenological type

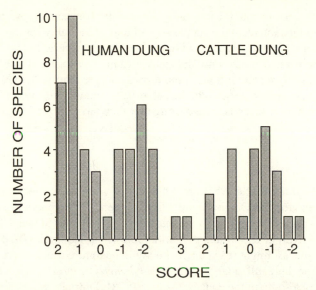

Fig. 9.3. Seasonality of dung beetles trapped in pitfall traps baited with human and cattle dung in Abokouamekro in 1980–81. This figure gives the results of a principal component analysis of the numbers of individuals in different species caught in six bimonthly trapping periods. The horizontal axis gives the loadings of the species along the first axis. A large score indicates early seasonal occurrence. Note the bimodal pattern in species caught in traps baited with human dung.

is exemplified by the most abundant *Sisyphus* species, *S. goryi* and *S. seminulum*, which may have up to five generations per year, and by the dominant small tunneler, *Onthophagus rufonotatus*. There is no reason to assume that the *Gymnopleurus* species have a lower potential rate of development than the *Sisyphus* species, and the observed difference in phenology may be due to interspecific competition (Hanski 1988): the *Gymnopleurus* species are probably inferior competitors compared to *Sisyphus*, but as the latter are abundant only after a significant amount of rain has fallen, *Gymnopleurus* may take advantage of their resistance to drought and occupy the less competitive few weeks in the beginning of the year. The same may be true about *Metacatharsius* versus most *Onthophagus* species.

In cattle dung, the seasonal pattern is less clear, perhaps due to less severe competition for this type of resource, and the second peak is now more distinct (Fig. 9.3). At Abokouamekro, *Catharsius crassicornis* is a dominant species and is most abundant in December–January, during the peak of its nesting activity. The nesting season may last until April, when individuals are attracted to human dung, perhaps to feed on this high-quality resource. Another abundant species, *Copris interioris*, has its peak abundance in April–May. These two large tunnelers have their peaks of activity in the beginning of the year, in contrast to many cattle dung specialists, for example *Neosisyphus*, which are more abundant during the second half of the rainy season.

In Sudanese localities such as Wango and Pabré, where there is only one rainy season, there is also only one peak of dung beetle abundance. *Gymnopleurus* and *Metacatharsius* are more dominant here than in the Guinean savannas. These genera, especially the latter, have many more species in the dry than in the humid savannas, suggesting that they have adapted to arid conditions.

Size and Diel Activity

Most of the dung beetle tribes are either entirely diurnal or nocturnal (Table 9.5). Except for the Onthophagini, diurnal species are generally small and nocturnal species are large (at Abokouamekro: 50 ± 86 mg (SD) and 311 ± 831 mg, respectively). The results of Bartholomew and Heinrich (1978) suggest that large species might reach a lethal body temperature if they were flying in full sunshine; but, still, a large roller, *Pachylomera femoralis*, is a definite day flier in Africa south of the equator (Walter 1978). Diurnal species do not generally fly all day, nor are nocturnal species active all night, but most species do have restricted flight periods (Fig. 9.4). Some diurnal species, such as *Sisyphus biarmatus*, are active only during the hottest part of the day, prefer the grass savanna (the most open environment), and are almost never found in bush or tree savanna. On the other hand, *S. gazanus*, which prefers relatively closed environments such as woodlands, is active in the morning and in the

TABLE 9.5

Numbers and biomasses of diurnal and nocturnal dung beetles in Abokouamekro, Ivory Coast (data from Appendix B.9).

Tribe	Number of Species		Number of Individuals		Biomass (g)	
	D	N	D	N	D	N
Scarabaeini	1	1	145	57	78	95
Gymnopleurini	5	1	3,868	209	387	24
Canthonini	—	2	—	11	—	7
Sisyphini	9	—	15,009	—	191	—
Coprini	—	12	—	1,242	—	511
Dichotomiini	1	6	8	1,159	*	98
Onitini	—	3	—	84	—	34
Oniticellini	10	—	10,717	—	187	—
Onthophagini	39	33	17,418	3,721	309	196
Percentage	53	47	88	12	54	46

Note: D, diurnal; N, nocturnal.

* <1.

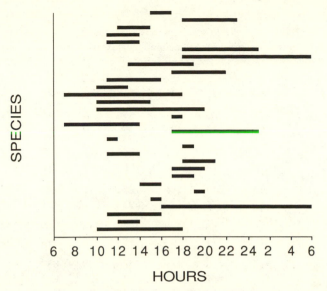

Fig. 9.4. Diel activity of thirty species of *Onthophagus* in tree savanna in northern Zaire, demonstrating great variation among species. The species are (from top to bottom): *aegrotus, altidorsis, bidentifrons, brazzavillianus, caelator, calchas, crantor, cyanochlorus, denudatus, duvivieri, euzeti, geminatus, grandidorsis, impressicollis, inflatus, kassaicus, lefiniensis, menkaoensis, merdrignaci, musculus, picatus, pisciphagus, pseudosanguineus, pullus, reticulatus, tuzetae, umbratus, variegranosus, willameorum,* and *xanthochlorus* (from Walter 1978).

afternoon. Many "nocturnal" dung beetles may in fact be "dusk and dawn" species (Tribe 1976).

Assuming that the difference in body size influences the degree of interspecific competition in a pair of species, we could expect that the most dominant species at a locality would be better spaced out in size than are species in a random selection of equally many species from the species pool (Hanski 1982). This is the case for the rollers in most of the localities for which data are available (Table 9.6). For example, the sizes of the eight most dominant species of rollers at Kakpin are remarkably well spaced out: 8, 17, 62, 130, 310, 540, 1,165, and 1,675 mg fresh weight. Knowing that competition is the key interspecific interaction in many rollers, these results strongly suggest that interspecific competition affects the abundance relations and community structure of rollers.

Soil

The type and especially the structure of the soil is crucially important for dung beetles that dig deeper or shallower burrows (e.g., Halffter and Matthews

TABLE 9.6
The observed size ratios in the 3, 4, . . . 9 most dominant roller species in five Ivory Coast savanna localities compared with a random expectation.

| Locality | Test | Species | | | | | | |
		3	*4*	*5*	*6*	*7*	*8*	*9*
Abokouamekro	Var	269	621	**56**	**54**	236	398	532
	Min	235	141	**35**	**10**	767	752	715
Sipilou	Var	306	339	**88**	192	603	418	571
	Min	**13**	**38**	**0**	35	592	492	383
Lamto	Var	615	161	127	**77**	330	407	326
	Min	473	512	592	514	427	298	209
Kakpin	Var	555	323	493	510	**50**	**33**	106
	Min	231	**93**	318	195	**98**	**39**	418
Wango	Var	459	489	358	633	830	769	747
	Min	**960**	**980**	701	821	857	550	346

Notes: The expected size ratios were calculated using equally many species drawn randomly from the local species pool. Dominance is defined as the abundance of the species times its weight. Two test statistics are used, the variance and the minimum value of log (x_i/x_{i-1}), where x_i is the size (fresh weight) of the *i*th most dominant species. The values in the table are the numbers of simulations out of 1,000 that gave a more extreme result than the observed one (significant results in boldface).

1966). In western Africa, most dung beetle species prefer a soft, sandy soil, but some taxa, for example *Onitis*, are more abundant on hard or gravel soils. At Abokouamekro, there are only three species of *Onitis*, all restricted to the hardest and driest places. The most abundant species in cattle dung is *O. alexis*, which also occurs during the dry season in semideserts (Chapter 13). At Wango, where the soil is gravelly, there are twelve species of *Onitis*. It seems that on favorable (sandy) soils, *Copris* are superior competitors to *Onitis* (both are large tunnelers), while the latter find a refuge on poor soils. This situation is somewhat similar to the apparent seasonal competition between *Gymnopleurus* and *Sisyphus*, the former perhaps being forced to use a short time period in the very beginning of the year because of competition.

9.6. SPATIAL PATTERNS

Local Distribution

At the local scale, several factors may be expected to influence the distribution of beetles among dung pats. The level of intraspecific aggregation among pats was measured by the index $J = (V/x - 1)/x$, where x and V are the average density and variance, respectively, of the species in a set of pats (Ives 1988a). The J index was calculated for six sets of similar cattle pats collected during different periods of the year (see legend to Table 9.8). Dung beetle tribe, diel activity, and mode of dung removal (roller versus tunneler) had no effect on aggregation. However, three factors were significantly correlated with the level of aggregation, which decreased with (1) the size and (2) the mean abundance of the species; and (3) cattle dung specialists were significantly less aggregated than generalist species (Table 9.7).

Degree of interspecific correlation in the distributions of species among pats was estimated using the index $C = cov/x_1 x_2$, where cov is the covariance between species 1 and 2, and x_1 and x_2 are the respective mean abundances (Ives 1988a). The C values were calculated for all pairs of species, with both species having average abundance greater than 1; the data are the six sets of dung pats referred to above. The results (Table 9.8) show a very clear pattern: (1) the greater the size difference between two species, the lower their spatial correlation (five significant results out of six); (2) nocturnal-diurnal species pairs always have lower spatial correlation than nocturnal-nocturnal or diurnal-diurnal pairs (four significant differences out of six); and (3) the level of correlation was always lower in two species belonging to different tribes than in species pairs from the same tribe (two significant differences out of six). Because the level of intraspecific aggregation was substantially higher than the level of interspecific aggregation (Table 9.8), the observed spatial patterns are expected to contribute toward coexistence of competitors (Hanski 1981, 1987b; Atkinson and Shorrocks 1981; Ives 1988a; Chapter 1: the variance-covariance model).

TABLE 9.7
Analysis of covariance for the level of intraspecific aggregation (J) among dung pats.

	F	P	Slope or Mean Value
Covariate			
Size	11.11	0.001	−0.074
Abundance	4.82	0.032	−0.080
Factor			
Specialization	17.56	0.000	specialists: 1.03
			generalists: 1.76

Notes: The data are the pooled results (70 J values) for 6 sets of cattle pats collected in different times of the year at Abokouamekro (see Table 9.8). The level of aggregation (log $[J + 1]$) is explained, in the model, by the logarithm of size, the logarithm of abundance, and the degree of specialization of the species (specialists are cattle dung specialists). Species with an average log abundance less than 0.5 individuals per pat are excluded. F is the F-statistic and P denotes the level of significance.

TABLE 9.8

Level of interspecific aggregation (C) in pairs of species in six sets of cattle pats at Abokouamekro.

Data Set	Size		Diel Activity				Tribe			J		C		n
	Coef.	P	0	1	2	P	0	1	P	x	med	x	med	
1	−0.046	0.007	−0.35	−1.59	0.49	0.000	0.29	−0.66	0.434	2.80	2.65	0.45	0.17	218
2	−0.074	0.016	0.27	−0.35	0.33	0.192	0.51	−0.17	0.076	3.14	2.33	0.68	0.03	127
3	−0.111	0.011	0.45	0.07	0.40	0.846	0.49	0.16	0.374	3.02	1.77	0.55	0.32	93
4	−0.083	0.000	0.41	−0.75	0.59	0.000	0.89	−0.38	0.000	3.70	2.57	0.63	−0.24	133
5	0.211	0.055	0.33	−0.40	0.82	0.000	1.22	−0.15	0.000	4.22	3.29	1.17	0.50	225
6	−0.059	0.000	0.54	−0.40	1.47	0.000	1.22	0.51	0.912	6.54	4.46	2.69	0.98	78

Notes: Data set 1: October 1980, 20 pats; 2: January 1981, 17; 3: February 1981, 25; 4: April 1981, 20; 5: June 1981, 15; 6: August 1981, 36. This table gives the results of an analysis of covariance, with C as a dependent variable, the logarithm of the size ratio as a covariate, and diel activity and tribe as two independent variables (*diel activity:* 0, nocturnal vs. nocturnal species pair; 1, nocturnal vs. diurnal species pair; and 2, diurnal vs. diurnal species pair; *tribe:* 0, two species belonging to the same tribe; and 1, two species belonging to two different tribes). The last five columns give the average (x) and median (*med*) values of J and C (defined in the text).

Large-Scale Spatial Distribution

The results of a factorial correspondence analysis of the material from the Ivory Coast revealed two principal factors affecting dung beetle distributions on a regional scale: the occurrence of large herbivorous mammals in the locality, and the type of dung available, essentially omnivore versus herbivore dung (Cambefort 1984). Certain other patterns are noteworthy in the results (Fig. 9.5). First, species belonging to the same tribe are usually closely associated, which parallels the result for the local scale (Table 9.8). Second, some between-tribe associations are evident, for example, between Oniticellini and Coprini. Although these are small diurnal (Oniticellini) versus large nocturnal tunnelers (Coprini), they are both attracted to large herbivore dung, leading to correlations at both local and regional scales. Another such association is be-

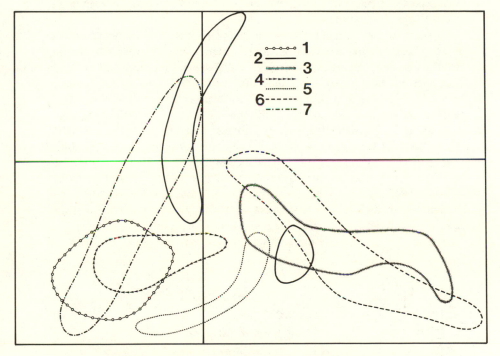

Fig. 9.5. A multivariate analysis of the occurrences of dung beetles at the trapping sites in the Ivory Coast. The horizontal axis is correlated with the occurrence of large mammals in the locality, while the vertical axis is related to the type of dung used by the species (herbivore versus omnivore dung). Species belonging to the same taxon have been encircled, and the symbols for individual species have been omitted. The taxa are: 1, Gymnopleurini (including Scarabaeini); 2, Sisyphini, divided into two parts, *Sisyphus* on the left and *Neosisyphus* on the right; 3, *Copris* and *Litocopris*; 4, *Catharsius* and *Metacatharsius*; 5, *Heliocopris*; 6, Oniticellini; 7, Onthophagini.

tween Gymnopleurini plus *Sisyphus* with most of Onthophagini, which re-flects the use of omnivore dung (but note that *Neosisyphus* are associated with Coprini and Oniticellini, which all use herbivore dung; Fig. 9.5).

The presence or absence of large mammals at a locality is reflected in the average size and species richness of dung beetles, both of which are highest in the locality with elephants (Fig. 9.6). A prime example is *Heliocopris*. No species of this genus has been found at Lamto in recent years. The two species found at Wango breed in cattle and buffalo dung, while at Kakpin two more species occur, breeding only in elephant dung. Kakpin also has *Heteronitis*, which are large elephant dung specialists. Incidentally, the estimated total number of species in Kakpin, about 150, is the same as reported by Edwards (1988a) for national parks in South Africa, in spite of a very small number of species (about 10) common to the two regions. Perhaps 150 is the level of local species richness of dung beetles that one could expect to occur in undis-turbed African savannas.

Apart from the correlation between the average size of mammals and the average size of dung beetles, what are the relationships between size and dis-tribution in dung beetles? Considering the 207 species present in the six study localities in the Ivory Coast (Appendix B.9), the following four patterns emerge: (1) There is a strong positive correlation between individual fresh weight and the average biomass of the species across the six sites: the larger the species, the more dominant in biomass it tends to be ($r = 0.614$; $P<0.001$). (2) There is a less significant negative correlation between individ-ual fresh weight and the average abundance of the species, the smaller species being, on average, more abundant ($r = -0.27$; $P<0.01$). (3) Another corre-lation exists between average local abundance and the number of localities in which the species is present ($r = 0.294$; $P<0.01$). This correlation establishes a relationship between local and regional levels: the more abundant a species is at the local level, the more widespread it is at the regional level (Hanski

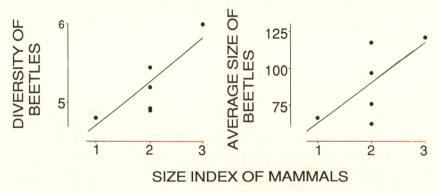

Fig. 9.6. Correlations between the average size and species richness of dung beetles and the average size of mammals in six savanna localities in the Ivory Coast.

1982). (4) Finally, the smaller the species, the greater the number of localities where it is present (r = −0.202; P<0.05).

9.7. LONG-TERM STABILITY

Dung beetles were trapped in exactly the same way using six traps baited with human dung in the same place at Lamto in August 1979 and in August 1987. Although the total numbers of species and individuals were somewhat higher in 1987 than in 1979 (50 and 15,471 versus 41 and 8,610), there is an excellent correlation between the relative abundances of the species in these two years (Fig. 9.7). The correlation is especially good for the rollers (r^2 = 0.95). The difference in the pooled abundance of beetles between 1979 and 1987 may be due to poor rains in August 1979 (17 mm), compared with August 1987 (133 mm; long-term average ± SD for August is 66±58 mm). Thus, in a locality that has remained essentially unchanged, there has been no significant variation in the species composition of dung beetles during eight years.

9.8. SUMMARY

In African savannas, the occurrence of dung beetles depends primarily on the rains. In the localities with two rain seasons (Guinean savanna), the fauna

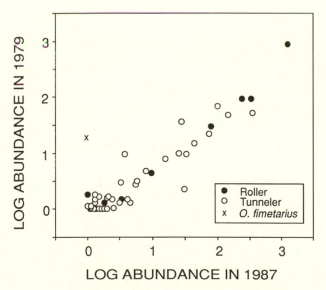

Fig. 9.7. Correlation between the numbers of dung beetles caught in August 1979 and August 1987 in the same place in Lamto, Ivory Coast, in pitfall traps baited with human dung. *Onthophagus fimetarius* was absent in 1987 but abundant in 1979.

consists of two phenological types, one with restricted occurrence early in the year and the other one with more continuous occurrence during the two rainy seasons. In the localities with one rainy season (Sudanese savanna), all the species have a more limited seasonal occurrence. The presence of large herbivorous mammals is another important factor, especially for the larger dung beetles. The average size of beetles and their species richness increase with the richness of the mammalian fauna. Some 150 species of dung beetles, including the largest ones (*Heliocopris*), occur in undisturbed savannas with elephants.

At the local level, differences in size, diel activity, and taxon (tribe) of pairs of dung beetles decrease their interspecific aggregation among droppings and hence facilitate the coexistence of the dozens of potential competitors. In the rollers, the most dominant species are better spaced out in size than expected by chance. Perhaps because of strong interspecific interactions, relative abundances of beetles have remained remarkably constant over eight years.

CHAPTER 10

Dung Beetles in Tropical Forests in South-East Asia

Ilkka Hanski and Jan Krikken

10.1. INTRODUCTION

Tropical forests in South-East Asia are fragmented among the thousands of smaller and larger, and more or less isolated, islands in the largest aggregation of islands in the world. The geological history of the region is not yet entirely understood (Whitmore 1987), though the major features were known already to Alfred Russell Wallace: tropical forests extend from the western Indo-Malayan region to the eastern New Guinean-Australian region; and while the large western islands on the Sunda Shelf have been repeatedly connected and disconnected to the mainland and to each other, many of the central islands are oceanic. These two geological features largely determine the kind of mammals that may be found in the rain forests in South-East Asia: only marsupials in the eastern region, and a decline eastward in the numbers of especially large eutherian mammals with distance from the Malay peninsula. On this basis we may expect quite different assemblages of dung beetles in different parts of the Indo-Australian archipelago, in spite of the fundamental uniformity of its tropical forests (Whitmore 1975).

Our knowledge of the ecology of dung beetles in this vast region, spanning some 6,000 km from west to east, is based on a number of shorter and longer field excursions by us and by others to the Malay peninsula, Sumatra, Borneo, Sulawesi, the Moluccas, Seram, and New Guinea. The most extensive fieldwork, covering one year, was carried out in North Sulawesi during the Project Wallace expedition in 1985 (Knight 1988). Although our understanding of the biogeography and distributional ecology of dung beetles in South-East Asia is now relatively good, more detailed biological and population ecological information is lacking. We hope that the broad patterns described in this chapter may provide a basis and stimulus for more fieldwork in this interesting area.

This chapter commences with a brief description of the insect communities that use dung and carrion in the rain forests in South-East Asia. We then outline the main biogeographic features of this exceptionally complex region (Section 10.3). Two aspects of population ecology, seasonality and the foraging behavior of beetles, are discussed in Section 10.4, while Section 10.5 is devoted to community patterns: resource partitioning, abundance distributions, elevational distributions on mountains, and distributions of beetles on islands.

10.2. DUNG AND CARRION INSECT COMMUNITIES

Table 10.1 lists the main taxa of insects in the dung communities of rain forests in South-East Asia, based mainly on studies conducted in Borneo and Sulawesi. A characteristic feature is the substantial overlap in species composition with the carrion insect community (Table 10.1). Many dung beetles (Scarabaeidae) are attracted to both dung and carrion, but it is not known whether they come only to feed at carcasses or whether they use carrion for breeding as well.

The carrion insect community is generally dominated by flies (especially Calliphoridae), but occasionally ants invade the carcass and keep out all other insects (I. Hanski, pers. obs.). Adult Hybosoridae are conspicuously numerous on carrion, and most montane localities have one species of *Nicrophorus* (Silphidae), which bury small carcasses. Both dung and carrion insect communities have a large number of predatory beetles, about which only one study has been published, from Sarawak in Borneo (Hanski and Hammond 1986).

TABLE 10.1
The main taxa of the dung and carrion insect communities in tropical forests in South-East Asia.

		Dung		Carrion	
Taxa	*Ecology*	Adults	Larvae	Adults	Larvae
Scarabaeidae					
Gymnopleurus	Large, diurnal rollers	Yes	Yes	Yes	?
Coprini	Large, nocturnal tunnelers	Yes	Yes	Yes	?
Onthophagus	Small, mostly diurnal	Yes	Yes	Yes	?
Other genera	Mixed	Yes	Yes	Yes	?
Hybosoridae	Nocturnal, larvae saprophagous?	No	No	Yes	No
Silphidae	Mostly montane, one species per locality	No	No	Yes	Yes
Staphylinidae	Coprophagous, necrophagous, and predatory taxa	Yes	Yes	Yes	Yes
Calliphoridae	Hanski (1983)	Yes	Yes	Yes	Yes
Formicidae	Erratic but may entirely exclude other taxa	Yes	No	Yes	No

This study reported nearly two hundred species of Staphylinidae, which is about the same number as found in temperate dung and carrion insect communities (Hanski and Koskela 1977). In the predatory staphylinids, the most common species in the lowland forests were relatively more, not less, abundant than their temperate counterparts, though there were also a larger number of rare species in Sarawak than in temperate samples, giving a markedly uneven distribution of species abundances (Hanski and Hammond 1986).

10.3. Biogeography

The geographical history of the Indo-Australian archipelago is characterized by extreme complexity (Whitmore 1987). Extensive land areas have existed since the early Tertiary, though the Australian and Laurasian land masses remained wide apart until some 15 million years BP. Quaternary changes in climate and vegetation are equally complex (Morley and Flenley 1987): almost everything—from contracting and expanding rain forests to savannas to glacial phenomena—seems to have occurred. The influence of man on the vegetation during the Holocene has further complicated the picture.

The earliest fossil records of mammals in the Indo-Australian archipelago are from the early Tertiary period, 40 million years BP, and they include anthracotheres, the ancestors of the hippopotamus family, from Timor and West Borneo (Von Koenigswald 1967). The better-known Plio-Pleistocene mammal faunas include a variety of large mammals, for example, *Mastodon*, *Stegodon*, *Elephas*, bovines, pigs, and deer. Indicators of lowland rain forest, for example, the orangutan and the gibbon, have been discovered in deposits dated from 30,000 to 100,000 BP. The history of the terrestrial faunas on the Indo-Australian islands is characterized by a very long mutual isolation of the Australian and Asian elements, followed by limited mixing in the contact zone, in Sulawesi and the Lesser Sunda Islands, since the Plio-Pleistocene times. There are certain puzzling observations, such as the Pleistocene-Holocene occurrence of elephant and rhino on Luzon, and generally the origin and dispersal of mammals throughout the archipelago remain highly controversial (but see Heaney 1986). Nowadays the larger dung producers in virgin forests are wild pigs and deer, with the elephant, rhino, tapir, and bovines occurring locally on the larger western islands.

The oldest records of dung beetles from the region date back to Linnaeus, who described a key species, the widespread large tunneler, *Catharsius molossus*, in 1758. Since the early part of this century the number of known species has steadily though slowly increased. Intensive trapping studies have been conducted only during the past 10 years, and today some 450 species of Scarabaeidae, most of which belong to *Onthophagus* (324 species), are known from the archipelago. Many of the smaller islands have endemic species, and because most of them have never been studied, there must still be a large number of undescribed species. We estimate that the total number of species in the Indo-Australian archipelago is likely to be between 1,000 and 2,000.

Similarities in the island faunas are relatively low. Sumatra and Borneo, with the most similar faunas of dung beetles, have 112 and 120 species, respectively, of which only 49 species occur on both islands. There are several apparently monophyletic groups of species in the archipelago, but their phylogeny is yet too imperfectly known to warrant any discussion of vicariance issues. A dendrogram of island similarities has three main clusters, corresponding to the Larger Sunda Islands, the Philippine Islands, and the "Wallacea," that is, Sulawesi and the Lesser Sunda Islands (Fig. 10.1). New Guinea has, as expected, a very distinct fauna. The major faunal boundary is the line of Weber rather than the line of Wallace; in Sulawesi, Oriental faunal elements are clearly dominant while characteristically Australian genera are absent. Some Australian canthonines, such as *Amphistomus* in Halmahera and *Tesserodon* in Seram, occur in the Moluccas, which have a mixed fauna, the Oriental species being represented by, for example, *Haroldius* in Seram.

Four major distribution patterns may be distinguished: (1) a widespread western element, almost certainly Holocene-recent, including many eurytopic and nonforest species (the forest species include *Catharsius molossus*, *Copris* species, *Liatongus venator*, and *Onthophagus rectecornutus*); (2) a Sundaland lowland rain forest element (including *Tiniocellus sarawacus* and many *Onthophagus* and *Phacosoma* species); (3) a Philippine element (mainly *Onthophagus*; data limited); and (4) an Australian-Papuasian element, present both in New Guinea and northern Australia (e.g., *Onthophagus atrox*).

The proportion of endemic species is lowest on the Malay mainland (11%) and highest in New Guinea (83%). Sulawesi has a large fauna and an exceptionally high degree of endemicity (75%, Fig. 10.1). However, the data basis

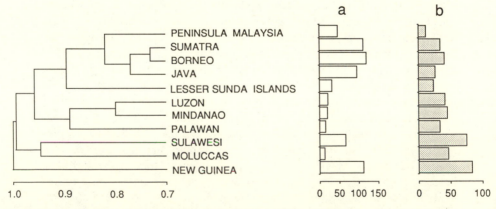

Fig. 10.1. A dendrogram of faunal similarities of Scarabaeidae dung beetles between the main islands in the Indo-Australian archipelago . The histograms on the right show the numbers of species (a) and the proportions of endemic species on different islands (b).

is still so patchy that these figures must be interpreted with caution; for example, the low figures for the two Philippine islands reflect, to an unknown extent, lack of modern fieldwork (the mammalian fauna on these islands shows a high level of endemicity; Heaney 1986). Among the Scarabaeidae dung beetles there is only one generic endemic in the Sundaland, the peculiar onthophagine *Cyobius wallacei*, possibly associated with ants (Krikken 1971; there is another species, *C. knighti*, from Sulawesi and Borneo). Thus, in spite of a very long existence of land in the area between continental Asia and Australia, there is no evidence of advanced post-Cretaceous evolution of dung beetles in this region. There is a large gap in the vertebrate fossil record before the Pliocene, and if this gap is indicative of absence of large herbivorous mammals, it may explain much of the apparently late evolution in the dung beetle fauna. In other scarabaeoid groups the situation is somewhat different, for example, there are several endemic cetoniid genera in Sulawesi (Krikken 1984). In mammals, of which most dung beetles are dependent, the number of endemic species is relatively low on the islands on the Sunda Shelf but the Philippine islands alone support at least seventeen endemic genera of rodents (Heaney 1986).

10.4. POPULATION ECOLOGY

Apart from short-term observations, no fieldwork has been conducted on the population ecology of dung beetles in the rain forests of South-East Asia. Next to nothing is known about their breeding biology, which may nonetheless be assumed to be similar to the breeding biology of dung beetles elsewhere in tropical forests (Chapters 11 and 12). This section is limited to seasonality and foraging behavior.

Seasonality

South-East Asia has a wide range of forests in terms of seasonality of rainfall, and in large areas there is no severe dry season at all (Whitmore 1975). The only long data series (10 months) available on the numbers of dung beetles comes from North Sulawesi, which has a seasonal climate but where the dry spells occur irregularly and where there is much geographical variation in seasonality within small areas (Paarmann and Stork 1987).

Beetles were trapped for 10 months with flight interception traps located in fixed positions. These traps consist of a vertical, black cloth, which the flying beetles cannot see in the relatively dark forest; beetles that happen to fly against the cloth drop down to collecting trays on both sides of the cloth (see Chapter 12 for possible differences among species in trappability; the trapping results are summarized in Appendix B.10.) All of the common species occurred throughout the year, with most species showing little variation in abundance from one month to another. In contrast, the uncommon species tended to show more variation as measured by the coefficient of variation (CV). All of the eleven species with average monthly samples greater than twenty had a CV less than 100, while eleven of the thirteen species with average monthly samples greater than one but less than twenty had a CV greater than 100 (P<0.001, Fisher's exact test). This result suggests that seasonal flight activity, whether involving some form of dormancy or not, is a less successful strategy than nonseasonal activity in this area. Assuming that the species compete with one another, which they most probably do, and that there is no predictable seasonality in the environment, selection should indeed favor nonseasonal activity: there is nothing to be gained by postponing reproduction. It is noteworthy that in other groups of insects using other resources, concurrent studies revealed various patterns of seasonality (Paarman and Stork 1987; P. M. Hammond, pers. comm.).

Foraging Behavior

Foraging beetles need to select the type of food they would use for feeding and breeding (Section 10.5), but they also need to make choices about the place, time, and mode of searching for food. Most species forage close to the ground level, where most of the resource is to be found, but some species also move in the canopy or are there exclusively. In North Sulawesi, *Onthophagus magnipygus* was caught only rarely at the ground level, but it was regularly caught in traps at heights from 5 to 20 m, where it may have been searching for the droppings of the Celebes monkey (I. Hanski, pers. obs.). In the same forest, the hugely abundant *Phaeochrous emarginatus* (Hybosoridae), whose breeding biology remains unknown, flies at all heights from the ground level to the upper canopy.

While moving close to the ground level, beetles may use several modes of foraging: they may fly around while searching for food (e.g., the large rollers *Paragymnopleurus*), or they may stay in one place waiting for the odor trails from fresh droppings to reach them. If they do the latter, they may stay on the ground or on leaves. Perching on leaves, usually some 50 cm above the ground level, is peculiar to many dung beetles in tropical forests. Many perching beetles seem to be foraging: they sit on a leaf with antennae outstretched, as if smelling the air currents; but there may be reasons other than foraging for perching (Howden and Nealis 1978; Howden and Young 1981; Hanski 1983; Young 1984; Chapters 11 and 12). In South-East Asia, it is bizarre that although perching is common in North Sulawesi (I. Hanski, pers. obs.; Table 10.2), it has not been observed at all, or it is extremely rare, in Borneo (I. Hanski, pers. obs.; J. Krikken, pers. obs.; P. M. Hammond, pers. comm.) and Seram (M. Brendell, pers. comm.). The explanation cannot be any important difference in forest structure, but may involve some unknown differences in resource supply.

Table 10.2 compares three sets of samples from North Sulawesi: beetles

TABLE 10.2

Comparing the species composition of dung and carrion beetles caught in pitfalls, in flight interception traps (FIT), and hand-collected while perching on leaves.

Species	Length (mm)	Perching		FIT	Pitfall (Dumoga)					
		A	B	A	HD	BC	RC	SF	LF	FR
O. aureopilosus	8	2		88	5		+			
O. mentaveiensis	7			176	1	2	6	3	4	1
O. aper	7	2	7	165	1	32	51	61	96	+
O. travestitus	6			31	1					
O. fulvus	5	4	3	22	1		+			
O. fuscostriatus	6	222	301	130	143	1	33	30	26	
O. hollowayi	4			6		+	+	1	1	
O. toraut	3	8	41	114	5	2	2	1	2	
O. sembeli	5			54	+		+	+		
O. sangirensis	6	6	7	299	+					+
O. biscrutator	6		1	83	+	4	6	7	17	+
O. spiculatus	4	3	2	1	1		+			
O. aereomaculatus	5			1	5	+	3	2	1	
O. magnipygus	6			7	+					
O. ribbei	10		1	15	15		1	+	5	
O. forsteni	8				1		+	+		
O. tumpah	3	10		3	1					
O. curvicarinatus	17				+					
Onthophagus sp.	7				+					
C. calvus	21				3		+	+		
C. saundersi	27				+					
C. macacus	11			1	14		2	+	+	
G. planus	19			7	7		+	+	+	
C. knighti	4			11						
Aphodius #3	4				+					
Aphodius #4	3				+					
Aphodius #5	6				6					
B. celebensis	9			129						+
P. emarginatus	11					15	6	13	247	
N. distinctus	25					1	1	1	1	

Notes: Data from lowland rain forest in North Sulawesi. The FIT results are the pooled results for February and March from Appendix B.10, the time of the year when the other two samples were collected. A and B are two localities, the Dumoga-Bone National Park and the Tangogo National Park, respectively, situated more than 100 km apart. The bait types are: HD, human dung; BC, bird meat; RC, rat meat; SF, small fish bait; LF large fish bait; and FR, fruit (usually mango).

+ <1.

collected perching on leaves, beetles caught in flight interception traps, and beetles caught in baited pitfall traps. It is evident that although ten species were collected while perching, this behavior is particularly characteristic for only one species (84% of observations), *Onthophagus fuscostriatus*, the dominant small coprophagous species in the forest (Table 10.2). Observations on perching in two forest areas separated by about 100 km gave very similar results (Table 10.2), indicating that there are species-specific differences in the rest of the species as well. In *Onthophagus*, the species that were seen perching were smaller, on average, than species not seen perching, though the difference was not significant (P<0.2). Perching and size are more strongly correlated if all species are considered, as the non-*Onthophagus* species do not perch and are generally larger than *Onthophagus*. In *Onthophagus*, the species that perched (10 species) were more numerous in the flight interception traps than were the species that did not perch (10 species; F = 4.73, P<0.04; log-transformed abundances), suggesting that perching and frequent flying are associated, rather than alternative, modes of foraging. There was no difference in the tendency to perch in coprophagous, necrophagous, and generalist species of *Onthophagus*.

The flight interception trap and pitfall trap catches were correlated but not very strongly; for example, in *Onthophagus* r^2 was only 0.25. Species that search for food other than that used as a bait in pitfall traps should be relatively more abundant in interception trap than in pitfalls. Thus *Bolbochromus celebensis* was very numerous in interception traps but was not caught in pitfalls. This species most likely feeds on subterranean fungi, like most other bolboceratine Geotrupidae (though the related amphi-Pacific *Bolbocerosoma* packs its burrows with litter; Howden 1955a). Judging by its morphology, *Cyobius knighti* is a myrmecophilous species (P. M. Hammond, pers. comm.) and was not caught in pitfalls, though, according to the interception trap results, it was relatively common in the forest. It is more surprising that some *Onthophagus* species, notably *O. sangirensis*, were exceedingly abundant in interception traps yet were rarely caught in pitfalls. Either there are unexpectedly large differences in species' flight activity, or there are unexpected differences in their food selection. The correct explanation cannot be decided with the data at hand. Two *Onthophagus* with a wide food selection, *O. mentaveiensis* and *O. aper*, were equally abundant in interception traps, though the latter species was by an order of magnitude more abundant in pitfalls, suggesting some difference in their foraging behavior.

10.5. COMMUNITY PATTERNS

Resource Partitioning

The food resources that are most important for dung beetles in rain forests in South-East Asia are dung and carrion, with decaying fruits and mushrooms being used by a smaller number of species. It is likely that many omnivorous species use different food material for adult feeding and for breeding, but we cannot distinguish between these two types of foraging with the data available.

In the lowland forests in North Sulawesi, most dung beetles (Scarabaeidae) were caught in largest numbers with omnivore (human) dung (21 species for which data are available), five species used carrion and/or fruits, and nine species were attracted to all bait types in relatively even numbers (Appendix B.10). In Sarawak, Hanski (1983) classified fifteen and twelve species as carrion and dung specialists, respectively, with seven species being considered as generalists. Such classifications are simplifications, and more detailed studies may reveal finer divisions of resource types. For example, *Onthophagus fuscostriatus*, the dominant small tunneler attracted to dung in North Sulawesi, was also caught in substantial numbers in traps baited with fish and mammalian (rat) meat, but hardly any individuals were caught in traps baited with bird meat (Appendix B.10). The size of the fish bait had a dramatic influence on the numbers of *Phaeochrous emarginatus*, the adults of which feed on carrion and prefer large carcasses.

Among the twenty *Onthophagus* species present in the lowland forests in North Sulawesi and for which good data are available, there were two coprophagous, two necrophagous, and three generalist species among the seven most abundant species caught in pitfalls. Among the thirteen rarer species, the respective figures were ten, three, and three species—in other words, most of the rarer species were coprophages. These figures suggest that the most abundant species in the local community have more diverse food habits than the less abundant ones (comparing the seven most abundant species versus the rest, and coprophagous versus noncoprophagous species; $P<0.06$, Fisher's exact test).

There are no extensive data on diel activity of dung beetles in the forests in South-East Asia, but the large majority of species appear to be diurnal, in contrast to other regions of tropical forest (Chapters 11 and 12). The only important group of nocturnal species is Coprini (*Catharsius* and *Copris*), which are large tunnelers.

Abundance Distributions

Fig. 10.2 shows two abundance distributions of dung beetles from lowland forests in Sarawak, Borneo, and from North Sulawesi. Both distributions are bimodal. No theoretical model has been proposed that would predict a bimodal abundance distribution (Engen 1978). It could be argued that the result from Borneo, based on trapping with baited pitfalls, is an artefact due to some species specializing on the types of resource used as baits, while other species might specialize on other resources and hence be caught in only small numbers. However, the result from Sulawesi is based on flight interception trap results, which should not depend on species' food selection, though they obviously depend on other aspects of foraging behavior (Table 10.2; Chapter 12). Alternatively, and more likely, the peak of rare species may be due to

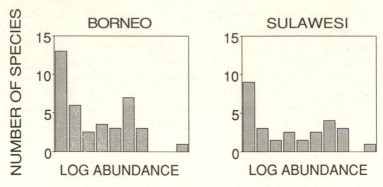

Fig. 10.2. Abundance distributions of dung beetles in lowland forests in Sarawak, Borneo (alluvial forest; from Hanski 1983) and in North Sulawesi. The results are based on pitfall (Borneo) and flight interception trapping (Sulawesi).

species that did not have a breeding population at the trapping site but were represented by dispersing individuals from elsewhere.

Complementary Abundance Changes

The large majority of dung beetles in South-East Asian forests are tunnelers. The only large rollers are *Paragymnopleurus*, present on the Sunda Islands and in Sulawesi, with typically one or two species being abundant in a local community. *Phacosoma*, small rollers, have many species in Borneo and Java, but little is known about their ecology, except that they show striking habitat differentiation where many species occur in the same region (Hanski 1983). *Paragymnopleurus* are diurnal, while the large tunnelers (*Copris* and *Catharsius*), with which they would be most likely to compete, are nocturnal. In the Dumoga-Bone Reserve in North Sulawesi, the only large (15–25 mm) dung beetles are *Paragymnopleurus planus, Copris saundersi*, and *C. calvus*. All three species occur up to 1,000 m elevation, but they showed a strikingly complementary pattern of abundance changes along an elevational transect (Fig. 10.3). Similar complementary abundance changes were found in the two species of *Onthophagus* that occur at high elevations on Mount Mulu in Sarawak (Fig. 10.3). The reasons for these patterns are not known, but it is likely that competition plays some role; all omnivore dung was rapidly removed in the forest in North Sulawesi, mostly by *Paragymnopleurus* and *Copris* (I. Hanski, pers. obs.).

Elevational Distributions on Mountains

Mutually exclusive elevational ranges in congeneric species of birds on tropical mountains have yielded convincing observational evidence for interspe-

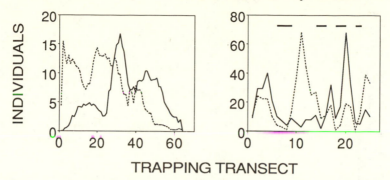

TRAPPING TRANSECT

Fig. 10.3. Complementary abundance changes along two trapping transects: *Copris* (*saundersi* and *calvus*; solid line) and *Paragymnopleurus sinuatus* (broken line) in North Sulawesi, and two species of *Onthophagus* at high elevations on Mount Mulu in Sarawak, Borneo (species 2 and 3 in Table 10.5).

cific competition structuring bird communities (Terborgh 1971, 1977, 1985; Diamond 1972, 1973, 1978; Mayr and Diamond 1976; Diamond and Marshall 1977). Table 10.3 shows elevational distributions of dung and carrion beetles and flies on Mount Mulu in Sarawak, Borneo. These data suggest that congeneric and/or ecologically closely related species have mutually exclusive elevational ranges more frequently than expected by chance (observed replacements 10.5, expected 4.2 ± 1.5 [SD]; statistical analysis in Hanski 1983). The reasons for little or no elevational overlap are not likely to be the same in all taxa, however. Hanski (1983) suggested that interspecific competition is likely to play a role in Hybosoridae (see below), while problems of species recognition and low fitness of hybrids may maintain exclusive ranges in the several closely related species of *Phacosoma*. Other cases of mutually exclusive elevational ranges in dung and carrion beetles in South-East Asia include *Nicrophorus*, *Copris*, and *Onthophagus* in Sulawesi (Hanski and Niemelä 1990) and *Liparochrus*, *Temnoplectron*, *Tesserodon*, and *Amphistomus* in New Guinea (Read 1976), though the latter results are only suggestive because of the relatively small number of elevations sampled.

Fig. 10.4 gives comparable trapping results from two contrasting mountains, the isolated peak of Mount Mulu in Sarawak, Borneo (2,370 m; Hanski 1983), and a lower peak (1,150 m) within a large mountain range in North Sulawesi (Hanski and Niemelä 1990). The number of species collected in the two localities, including the lowlands, was roughly the same, sixty to seventy species, but the two transects show a striking difference in the elevational change in species richness (Fig. 10.4). In Sulawesi, the number of species trapped per site remained practically constant, at an average of eighteen species, up to 800 m, and at 1,150 m there were still more than ten species per site (Fig. 10.4). At low elevations in Sarawak, species number per site was

TABLE 10.3

Elevational distribution of dung and carrion beetles and flies and their predators along a trapping transect on Mount Mulu in Sarawak, Borneo.

Altitude (meters)	Species																						
	1	2	3	4	5	6	7	8	9	10	11	12	13	14	15	16	17	18	19	20	21	22
90	1												117			5						
100							5						79			1						
105							11						75			1						
110	26			2			4			4			132			3						
130	13			5						122			74			10						
150	15			11						264			22			3						
150	1									39			1									
200	5			4						379			41			4						
300	4			2						218			46			5						
350	2			1			5			159			42			7						
500							3			108	1		9			3				4		
585	2						2			231			25			1				3		
670	1			1			2			177	3		4			1				2		
715	8				2		5			239			10			5				1		
765	9				3		2			310			4			4						
860	10				52		4			254	4		5	4		6				1		
895	8				6		5			207	9		5	2		8				6		
940	4				2		1			62	4		1	5			1			1		
1,020	4				5		5			36	21			6			3			1		
1,070	52				26		2	7		409	45			10			4	1		1		
1,130	9	1						61		54	69		1	18			7			7		
1,220	36					1		14		98				12			8	7				
1,275	28	4				1		28		105	1			12		1	11			11		
1,320	96	10	9			13				35				7			2	1	2	10		
1,420	32	24	29			1		9	1			2		9	3		3	2	14	26		
1,460	3	22	29					1	2								1	1	25	8		
1,510		22	40						6	1		1		8	3			3	14	2		
1,605		10	21						5					6	4		2	19	20	4		

Source: Hanski (1983).

Notes: The species are: Scarabaeidae: 1, *Onthophagus* sp. C; 2, *Onthophagus* sp. H; 3, *Phacosoma* sp. C; 4, *Phacosoma* sp. A; 5, *Phacosoma* sp. B; 6, *Phacosoma* sp. F; 7, *Phaeochroops gilleti*; 8, *P. acuticollis*; Silphidae: 9, *Nicrophorus podagricus*; Staphylinidae: 10, *Anotylus* sp. 12; 11, *Anotylus* sp. 28; 12, *Anotylus* sp. 11; 13, *Platydracus aeneipennis*; 14, *Philonthus* sp. 82; 15, *Philonthus* sp. 81; 16, *Philonthus* sp. 85; 17, *Philonthus* sp. 86; 18, *Philonthus* sp. 84; 19, *Philonthus* sp. 83; Calliphoridae: 20, *Lucilia porphyrina*; 21, *Calliphora atripalpis*; 22, *C. fulviceps*.

Elevation	1	2	3	7	9	13	14	16	17	18	20	21	22
1,640	4	10				1	9	13	1				1
1,650	3						13	4	4				2
1,650	1	4					7		2				
1,700	13	9	1				16		20				
1,700	36	6	5				7		52	1			
1,700	68	3	12				11		79				
1,700	44	8	1				11		31	3			
1,700	25	8	3	6			7		87				
1,700	27	11	4	3			7		47	5			
1,700	9	2					13		3	3			
1,700	11	11	2	1			5		8	2			
1,700	18	32	2				9		42	1			
1,700	1	9			1		3		4	2			
1,700	4	17	2				1		1	1			
1,700	19	68	5		8		16		1	1			
1,700	17	33	7		12		5		4				
1,750	1	6			1		1			1			
1,750	12	5			14		2		1	1			
1,810	69	15	6		12		14		29				
1,870	33	10	2		15		24		4	1			
2,000			1		3		3				24		
2,080	1				6		1		6	5			1
2,120					9		2		1	4	165		2
2,180					1						2		
2,200					1					1			1
2,240					5					5	90		1
2,310					9						18		8
2,365					4		4				1	16	2
2,370		4			4		4					18	5

BORNEO

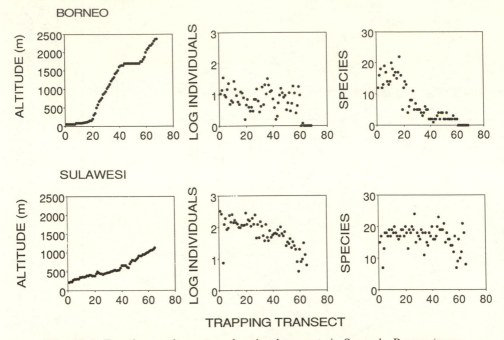

SULAWESI

TRAPPING TRANSECT

Fig. 10.4. Trapping results on two elevational transects in Sarawak, Borneo (upper panels), and in North Sulawesi (lower panels). The three panels give the elevations along the transects (on the left), the logarithm of the number of individuals (in the middle), and the number of species at each trapping site (on the right; trapping was conducted in the same way on both occasions; see Hanski 1983).

the same as in Sulawesi, between fifteen and twenty, but on Mount Mulu species number declined rapidly at elevations above 300 m; at 800 m only five to ten species were trapped per site, and at 1,150 m there were fewer than five species per site (Fig. 10.4). This difference is explicable by area (Mayr and Diamond 1976): the area of the montane habitat is more limited on an isolated mountain peak than in a comparable locality within a more extensive mountain range. In New Guinea, where montane forests cover huge areas, dung beetle diversity remains relatively constant up to 1,000 m (Read 1976). There are also clear differences in the proportions of montane species (>1,000 m) out of the regional species pools: 10% in Mulu in Sarawak (Hanski 1983), 20% in North Sulawesi, and nearly 50% in New Guinea (Read 1976).

The two transects show another and not less striking difference in the pooled density of beetles, but now the trend is reversed: steady decline of density with increasing elevation in Sulawesi, but not much change in Sarawak below the elevation of 1,700 m. This result is due to a density of beetles an order of magnitude higher in the lowland forest in Sulawesi than in Sarawak; at higher elevations the difference was smaller, and it disappeared at 1,000 m (Fig. 10.4). The difference may be due to a much higher density of especially small mammals in the lowland forests in North Sulawesi than in Sarawak (Appendix A.10; Hanski and Niemelä 1990), partly perhaps due to the practical absence of predatory mammals in Sulawesi.

Island Communities

The thousands of islands in the Indo-Australian archipelago varying in size and isolation provide a fascinating field for island biogeographic studies. In the case of dung beetles, an extra complication is the decline in the variety of larger mammals on the islands from west to east. The taxonomic mix of dung beetles becomes increasingly impoverished when one moves from west to east, as exemplified by dung beetles on the chain of islands in the Lesser Sunda Islands: Java (18 genera), Bali (9), Lombok (5), Sumbawa (6), Flores (5), Timor (2), and Wetar (2). The three genera reaching Timor and Wetar are *Onthophagus*, *Catharsius*, and *Onitis*.

Fig. 10.5 shows the species-area curve for thirteen islands ranging in size from about 5,000 square kilometers (Lombok) to more than 800,000 square kilometers (New Guinea). The slope of the regression line is very high (0.68) in comparison with most other published slopes (Connor and McCoy 1979). The corresponding slope for the mammals on six sets of islands in the Indo-Australian archipelago varied from 0.09 (Malay peninsula, mainland) to 0.44 (an isolated oceanic group of islands; Heaney 1986). The high slope is partly due to strikingly species-poor dung beetle assemblages on the relatively small (oceanic) islands; for example, only four species are known from Timor. Limited sampling is not likely to be the main explanation, because on the larger and better-known islands a substantial fraction of the known species can easily be collected from one site in a short period of time. Below, we shall examine in some detail the dung beetles of one relatively small oceanic island (Seram),

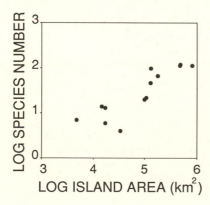

Fig. 10.5. Species-area curve for thirteen islands in the Indo-Australian archipelago. The parameter values of the regression line $\log S = \log C + z \log A$ are: $C = -1.89$ (SE $= 0.56$) and $z = 0.68$ (SE $= 0.11$). $r = 0.87$. (S and A are the species number and island area, respectively.)

as well as the dung beetles of New Guinea, a very large island but located far from the species pool of the Asian mainland.

Seram. This is the largest (ca 17,000 km²) island of the Moluccas, the Spice Islands, located in the "Wallacea," the transition zone between Asia and Australia. Seram still has a good forest cover. There are more than forty species of mammals, of which five are endemic; but all the native species are small in size (the deer *Cervus timorensis* and the wild pig *Sus scrofa* have been introduced by man).

Only six species of dung beetle are known from Seram (Table 10.4). All these species are small, 3–8 mm in length, which is not surprising since all native mammals are small as well. There are four *Onthophagus*, of which three are specialist coprophages apparently restricted to lowland forests, while the fourth one is a noncoprophagous, generalist species, distributed at least up to 1,000 m (Table 10.4). The two remaining species are *Haroldius* (lowlands) and *Tesserodon* (montane forests). A noteworthy detail in these results is that the regularly flooded alluvial forest had very few dung beetles (Table 10.4), possibly because beetles cannot breed in a flooded forest.

These results indicate very low species diversity, perhaps due to a small number of colonizations and low rate of local speciation (relatively small area). The four *Onthophagus* species belong to Papuasian species groups; one of them may originate from New Guinea, while the others are probably Moluccan endemics. *Haroldius* is an oriental genus while *Tesserodon* occurs only east from the Moluccas, in New Guinea and Australia.

New Guinea. Like Seram, New Guinea lacks large, native eutherian mam-

TABLE 10.4
Dung beetles of Seram.

Species	Length (mm)	FIT Traps A	FIT Traps B	Pitfall Traps						1,000 m
				Lowland Forests						
				HD	FI	ME	FR	FU	SA	
Onthophagus #1	8	23	6		3	6	5	1	1	5
Onthophagus #2	5	414	6	77	1					
Onthophagus #3	7	165	9	29						
Onthophagus #4	7	25	21	309	1					
Haroldius #1	5	5								
Tesserodon #1	3									6

Notes: Results based on trapping by M.J.D. Brendell and coworkers in the Manusela National Park in July–September 1987. Three flight interception traps (FIT) were operated in the lowland nonalluvial (A, 1 trap) and alluvial forests (B, 2 traps). Baited pitfall trapping was conducted in the lowland forests and at 1,000 m altitude. The bait types are: HD, human dung; FI, fish; ME, meat; FR, fruit; FU, fungi (in addition, one beetle was collected while sap feeding on a tree trunk, SA).

mals, while it has a rich fauna of smaller species and, in contrast to Seram, an enormous area (ca 800,000 km²). The number of dung beetles (Scarabaeidae) presently known from New Guinea is 112, which is comparable to the number known from Borneo (120) with similar area. As might be expected on the basis of entirely different mammal faunas, there is an important difference in the kinds of species that occur in Borneo and New Guinea. But on both islands *Onthophagus* is the numerically dominant group, with 70 species in Borneo and 85 species in New Guinea. Missing from New Guinea are the tribes of beetles with large species: Coprini and Gymnopleurini; large rollers of any sort are entirely absent. The largest species include some exceptionally large *Onthophagus* (up to 20 mm, in the *O. atrox* species group), apparently taking the place of the missing large tunnelers.

Nicrophorus

The mostly Holarctic genus *Nicrophorus*, the well-known burying beetles, is represented by several mostly montane species in the forests in South-East Asia. Typically, only one species occurs in one region, and most islands have just one species: Borneo (*podagricus*), Java (*insularis*), northern Philippines (*nepalensis*), Mindanao (*apo*), Flores (undescribed), and New Guinea (*heurni*). Three islands have two species, namely, Sumatra (*insularis* and undescribed), Sulawesi (*distinctus* and undescribed), and the Solomon Islands (*kieticus* and undescribed). In Sumatra the two species appear to occur in different parts of the island, and in Sulawesi they are elevationally separated (Hanski and Niemelä 1990). In contrast, in temperate regions several species often coexist in the same locality (Anderson 1982), though often there are differences in their habitat selection and seasonality (Pirone 1974; Shubeck 1976; Anderson 1982).

It is most interesting that this Holarctic genus has invaded the entire South-East Asian rain forest region, yet shows the minimum level of local species diversity. There exists no modern taxonomic study to indicate whether the occurrence of *Nicrophorus* in the archipelago is attributable to one or more invasions. Another interesting point is that, on most islands, *Nicrophorus* is entirely restricted to montane habitats, as might be expected for a Holarctic genus; yet in Sulawesi (and perhaps on the Solomon Islands) a lowland species is very abundant. We suggest that these observations indicate difficulty in invading the competitive lowland community, and competitive exclusion of all but one *Nicrophorus* locally.

Hybosoridae

Hybosoridae is a small family of fewer than two hundred species worldwide, with most species and genera confined to Australia and South-East Asia (Allsop 1984). Their biology is poorly known (Britton 1970). The species that occur in the rain forests in South-East Asia arrive in great numbers to feed on carrion at night (Kuijten 1978, 1981; Hanski, unpubl.), but the larvae are not necrophagous; they may be generalist saprophages living in the soil (Gardner 1935). In North Sulawesi, *Phaeochrous emarginatus* unexpectedly flew in large numbers to traps at all levels in the forest canopy (Hanski, unpubl.).

Table 10.5 summarizes the distribution of the species that are known from the Malay peninsula in the west to Sulawesi in the east (the Philippines and New Guinea are excluded because there is no ecological information for the respective species). These taxa include the widely distributed *Phaeochrous emarginatus* and ten species of *Phaeochroops* (2 or 3 rare species known only from type specimens have been omitted; Kuijten 1978, 1981).

Several details suggest that the Hybosoridae in the rain forests in South-East Asia have strong interspecific interactions and that two similar species do not generally coexist in the same habitat. First, the two smallest species of *Phaeochroops* have the widest distributions, together covering the entire region except for Sulawesi (no *Phaeochroops* occurs outside the Sunda Shelf; Kuijten 1981). They have been found together in Mount Leuser National Park in Sumatra, but with an elevational separation (Table 10.5). Remarkably, in Sumatra, *P. silphoides* occurred only above 800 m, whereas in Sarawak, where *P. rattus* was absent, *P. silphoides* was restricted to lowland forests (Hanski 1983). Second, the larger species of *Phaeochroops* occur on single islands (Table 10.5). Third, in Mount Mulu National Park in Sarawak, where four

TABLE 10.5
Distributional ecology of Hybosoridae in rain forests in South-East Asia.

Species	Length (mm)	Islands M S J B P S	Habitat and Elevation
P. silphoides	9.0	+ + +	<150 m in Borneo, 800–1,300 m in Sumatra
P. rattus	10.0	+ + +	<1,000 m in Sumatra
P. vulpecula	12.5	+	<600 m
P. lansbergei	12.5	+	Mostly mountains?
P. gilleti	12.5	+	Dominant in Sarawak <1,200 m
P. freenae	13.0	+	300 m, rare?
P. angulatus	13.5	+	
P. peninsularis	14.5	+	
P. acuticollis	15.0	+	1,200–1,500 m in Sarawak
P. gigas	15.5	+	Lowland heath forest in Sarawak
P. emarginatus	11.0	+ + + + + +	Restricted in Sumatra, absent in Sarawak, generalist and very abundant in Sulawesi

Note: The islands are: M, Malay Peninsula; S, Sumatra; J, Java; B, Borneo; P, Palawan; S, Sulawesi.

species of *Phaeochroops* occur, the three large ones show striking habitat and elevational segregations, and the only two species that occurred in the same habitat were a small (*P. silphoides*) and a large species (*P. gilleti*; Hanski 1983). The same was true in the Mount Leuser area in Sumatra, where *P. rattus* (small) and *P. vulpecula* (large) occurred together. Fourth, in Sumatra, with three species of *Phaeochroops*, *Phaeochrous emarginatus* was found abundantly only in a particular habitat (tree plantation at 150 m, no *Phaeochroops*). In Sarawak, with four species of *Phaeochroops*, it was not found at all, though the species is known from Borneo (Kuijten 1978); but in Sulawesi, with no *Phaeochroops*, *P. emarginatus* is exceedingly abundant in all forests from the lowlands up to 1,000 m (I. Hanski, pers. obs.).

Taken together, these observations are suggestive of resource competition and consequent ecological segregation of species (Hanski 1983). This hypothesis is plausible because Hybosoridae often attain extremely high densities. For example, in Sulawesi a pitfall trap baited with a 50 g piece of meat attracted up to 1,000 individuals of *P. emarginatus* in three days, with a total biomass of 50 g (Hanski, pers. obs.). The great abundance of adult Hybosoridae in carrion reminds one of the mass occurrences of some *Aphodius* species in the north temperate region, with adults feeding in cattle pats but larvae living as generalist saprophages in the soil (Chapter 5). The breeding biology of Hybosoridae in South-East Asia remains a challenge for further fieldwork.

10.6. CONCLUSIONS

The dung and carrion beetles of the Indo-Australian archipelago show considerable diversity of species and a high degree of species-level endemism on islands, but the fauna has very few phylogenetically strongly isolated endemics. There is a strong western dominance of widespread Oriental elements and a strong eastern dominance of Australian elements, with very limited trans-Wallacean transgression.

Many localities in South-East Asia have only a moderately seasonal climate. In North Sulawesi, the numerically dominant dung beetles showed completely aseasonal activity. There is substantial differentiation especially among the most abundant species in resource selection, with many carrion-feeding and some abundant fruit-feeding Scarabaeidae being characteristic for local communities. Complementary abundance changes, elevational replacements of congenerics on mountain slopes, and complementary distributions of species on islands are some of the observed patterns that can be attributed to interspecific competition. Competition is often severe: omnivore droppings are regularly removed in a matter of hours. The relatively small, oceanic islands in the Wallacea have strikingly species-poor assemblages of dung beetles, with the result that the species-area regression has an exceptionally steep slope, 0.68.

CHAPTER 11

Dung Beetles in Tropical Forests in Africa

Yves Cambefort and Philippe Walter

11.1. INTRODUCTION

When Halffter and Matthews reviewed the natural history of dung beetles in 1966, nothing was known about the tropical forest species outside the Neotropics. For the African species, all that existed a few years ago were some taxonomical and distributional records (e.g., Frey 1961; Balthasar 1967), and even today the ecology of African forest dung beetles remains poorly known. Studies have been conducted primarily in Zaire (Walter 1977, 1978, 1983), the Ivory Coast (Cambefort 1980, 1982b, 1984, 1985), and Gabon (Cambefort and Walter 1985; Walter 1984a,b, 1987), with little information available from Liberia (Hanski 1983) and Uganda (Nummelin and Hanski 1989). These studies have established that the dung beetle fauna in Afrotropical forests is clearly distinct from the savanna fauna, a situation typical for most groups of animals. However, contrary to the usual pattern in Africa, and contrary to the pattern in dung beetles in other tropical regions, forest dung beetles are less numerous in species and individuals than dung beetles in savannas.

African forest dung beetles are of special interest because of the richness of the Afrotropical mammalian fauna. A special feature is the presence of basically two kinds of dung beetles in African forests: species using the relatively small droppings of omnivorous mammals, and species exploiting the large droppings of the elephant and other large herbivores. The first kind of species are well represented elsewhere in the tropics (Chapter 10 and 12), but the second set of species is characteristic for Africa only. We have had an opportunity to study forest localities where elephants were abundant until recently, allowing a comparison between forest and savanna communities with qualitatively similar resource supplies. Africa is the only continent where such a comparison can be made.

This chapter first outlines the biogeography of African tropical forests, both in general terms and from the perspective of dung beetles in particular (Section 11.2). We then describe resource partitioning among species of dung beetles (Section 11.3) and patterns of species richness in their communities (Section 11.4), and in the concluding section we compare the forest and savanna dung beetles in Africa (Section 11.5).

11.2. BIOGEOGRAPHY

The evolution of African tropical forests in relation to its savannas is still in dispute (Monod 1957; Schnell 1976). Studies on water-level variation in African lakes (Servant and Servant-Vildary 1980) and on fossil pollen deposits (Maley and Livingstone 1983; Ritchie and Haynes 1987) have revealed that during and since the Pleistocene, the Afrotropical region has experienced successive pluvial and interpluvial periods (Chapter 4). The most recent pluvial period peaked in 9000 BP and ended by 5000 BP. During the pluvial periods, the rain forest has had a greater extent than it has today, and it used to consist of one continuous block as recently as 6000–5000 BP (Nagel 1986). At present, the African rain forest is divided by the Dahomey Gap (Booth 1954) into a smaller western block and a larger eastern block (Fig. 9.1).

Most of the information on African forest dung beetles is available from the Ivory Coast and Gabon, and especially from two localities, Taï National Park in the Ivory Coast (5°26′N, 6°55′W) and the Makokou Research Station in Gabon (0°34′N, 12°52′E). Both localities belong to the Guinean-Congolese botanical region, though Taï is located in the western forest block and Makokou in the eastern one.

Taï. The evergreen rain forest of Taï is located in a lowland region with an average altitude of 200 m. The climate is Guinean, with two rainy seasons in May–June and September–October; the annual rainfall is 1,777 mm, and the average monthly temperatures vary between 24.3°C and 27.6°C. Characteristic tree genera in the forest are *Diospyros* and *Eremospatha* (Guillaumet and Adjanohoun 1971), which grow on ferralitic clay-sand soils, with gravels on slopes (Collinet et al. 1984). The mammalian fauna is relatively rich, with fifty-five species recorded (Appendix A.11). Elephants were abundant until recently, with an average density of 0.5 and maximally up to two individuals per square kilometer (Merz 1982). At the time of the present study, the elephant biomass was estimated at 16–20 kg/ha. The density of other mammals is not known, but their biomass probably does not exceed 15–30 kg/ha (Appendix A.11).

Makokou. The station is situated at an average altitude of 500 m. The climate is strictly equatorial and of the "Ivindian" type (Saint-Vil 1977): there are four seasons, with average annual rainfall of 1,750 mm and relatively low average monthly temperatures, varying from 21.6°C to 25.3°C. The rainy seasons are in March–May and September–November. The soils are mostly xanthic ferralsols (Anonym 1974), and the vegetation is characterized by *Scyphocephalium ochocoa* and *Pycnanthus angolensis* (Caballe 1978). The mammalian fauna is better known than in Taï, with eighty-one species recorded (Appendix A.11). The main differences between the two areas are the presence of the forest buffalo and the pigmy hippo in Taï (absent from Makokou) and the gorilla in Makokou (absent from Taï). Elephants have been less abundant in Makokou than in Taï, probably making the mammalian biomass lower in that region.

Dung Beetles

African tropical forests have a rich fauna of dung beetles. Seventy-five species of Scarabaeidae and 15 species of Aphodiinae have been recorded from Taï, while the respective figures are 66 and 17 for Makokou (species lists in Appendix B.11). Altogether, there are 116 species of Scarabaeidae in the two localities, of which only 25 occur in both areas. The species abundance distribution is lognormal (Fig. 11.1).

Some of the species present in the two localities belong to exclusive forest genera not found in savannas: *Pseudopedaria*, *Lophodonitis*, *Mimonthophagus*, and *Tomogonus*. Other characteristic forest genera, such as *Paraphytus* and *Amietina*, are represented by different species in the two localities. *Allonitis* and *Drepanoplatynus* have not been collected at Makokou, but they are known to occur in the eastern forest range in Zaire (Simonis and Cambefort 1984). All these genera have only a few species, which are widely distributed in African rain forests. In Aphodiinae, the subgenera *Afrodiapterna* (two species) and *Colobopteridius* (one species) represent the same type of distribution. Except for *Paraphytus*, which lives in rotten trunks and cannot survive outside tropical forests, the forest genera are not clearly different in morphology, behavior, or ecology from their savanna relatives.

The rest of the species belong to genera that occur also in savannas. *Dias-*

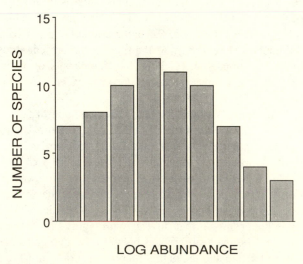

Fig. 11.1. Species abundance distribution of dung beetles in the Taï forest (data from Appendix B.11).

tellopalpus has eight species in the two localities, of which only two are common to both. This genus is endemic to the Afrotropical region, where it occurs mostly in forests but also in high plateaus from Guinea to Malawi. It never occurs in lowland savannas. The largest and most widespread genus of dung beetles in Africa is *Onthophagus*, which is as ubiquitous in forests as it is in savannas. However, in *Onthophagus* as well as in other genera, most species present at Taï and Makokou are exclusive forest dwellers, with only about ten savanna species occurring in clearings and on tracks, especially at Taï, which is located close to the savanna. These species are either common and widespread savanna species (e.g., *Tiniocellus spinipes*, *Onthophagus latigibber*, and *O. mucronatus*), tree-savanna species (e.g., *Sisyphus gazanus*), or elephant dung specialists (*Heliocopris haroldi* and *Onitis sphinx*).

The forest dung beetle communities are fragile, depending both on adequate tree cover and on the mammalian fauna. Extensive destruction of forest for timber, especially in the smaller western range, threatens the tropical forest fauna, though selective felling of trees may not be serious for dung beetles, provided the mammalian populations are not destroyed (Nummelin and Hanski 1989). Unfortunately, large mammals and especially elephants are heavily poached in African forests. As a consequence, the forest dung beetle fauna is rapidly vanishing in the western range.

11.3. Resource Partitioning

Food

Many species of dung beetles are specialized to utilize only one of the two main types of resource, the large dung pats of large herbivores or the smaller droppings of omnivorous mammals. The latter are relatively rich in nitrogen, while large herbivore dung is rich in carbohydrates (Chapter 2). The only true forest herbivore is the elephant. The pigmy hippo does not deposit its droppings but has the habit of dispersing them with its tail. Buffalo is not really a forest animal: it does not occur at all in Makokou, and in Taï buffalo herds occur mostly in clearings and on tracks. The smaller so-called herbivores, like duikers, have in fact a diverse diet, which includes leaves and fruits but also earthworms, snails, insects, eggs, etc. (e.g., Dubost 1984).

In Taï, with seventy-two species of dung beetles, there are more species strictly specializing on elephant (16) than omnivore dung (11), though the difference is not significant (Table 11.1). Elephant dung has many large nocturnal tunnelers, while all the omnivore dung specialists belong to *Onthophagus*. The latter genus includes only one elephant dung specialist, *O. vesanus*, which is distinct from all other described species (Balthasar 1967). Two other elephant dung specialists belong to strictly forest genera (*Mimonthophagus* and *Tomogonus*), while the remaining species are members of particular, relictlike species groups.

TABLE 11.1

Body weight (mg) of diurnal and nocturnal species of dung beetles specializing in human or elephant dung in Taï forest, Ivory Coast, and in all species of dung beetles from forest (Taï) and from savanna (Kakpin).

	Diurnal Species			Nocturnal Species				
	x	SD	n	x	SD	n	t-test	P
Human dung	16	10	5	55	20	6	− 4.1	0.003
Elephant dung	35	32	7	275	278	9	− 2.8	0.015
t-test and P	0.8	0.43		1.9	0.09			
Forest	49	78	30	250	505	42	− 4.4	0.001
Savanna	68	111	65	651	1374	67	− 5.7	0.001

Notes: A species was considered to be a specialist if more than 90% of specimens were collected from one type of dung with a total of at least 5 specimens caught; t-tests were run after logarithmic transformation.

Large herbivore dung seems always to be available in excess of the potential consumption by dung beetles, especially in the case of elephant dung pats, which may remain almost untouched for weeks. There is consequently relatively little competition, which explains the abundance of the dweller *Oniticellus pseudoplanatus*, an inferior competitor, in elephant dung pats in forests. The large amount of dung not used by dung beetles is occasionally used by other groups of beetles; for example, the cetonid *Campsiura trivittata* breeds in old elephant dung in Taï (Lumaret and Cambefort 1985). Such excess of the resource is not obvious in the case of omnivore droppings, which may disappear in less than an hour. Omnivore droppings attract a relatively much greater number of insects than elephant droppings, including both dung beetles and flies and their predators. Hanski (1983) found large numbers of calliphorid flies and hydrophilid beetles, their supposed predators, in the forests on Mount Nimba in Liberia. The hydrophilids are also frequent in traps baited with human dung in both Taï and Makokou. Hanski (1983) suggested that the abundance of calliphorid flies might be related to the scarcity of strictly necrophagous dung beetles in Afrotropical forests (Chapter 19).

Observations made in Taï (August 1980) provide quantitative figures for comparisons among the assemblages of beetles colonizing omnivore and herbivore dung. Human droppings of about 200 g typically yielded 1.5–3 g of beetles, while elephant droppings of about 2,500 g attracted 3–7 g of beetles. Human droppings were entirely buried, while only one fifth of the elephant dung was buried. In spite of this difference, species richness of dung beetles was the same in traps baited with human dung, in elephant dung pats, and in miscellaneous droppings (Appendix B.11). Species richness was significantly lower in buffalo dung, which is uncommon in forest.

Some species are attracted in small numbers to carrion (Appendix B.11). Of the nine necrophagous species in Taï, three species have been found only in carrion, one of them belonging to the genus *Amietina*, which is strictly necrophagous also in Makokou. The remaining six species are generalists.

One genus remains to be mentioned, *Paraphytus*, with three species in Taï and two species in Makokou (*P. bechynei* is common to both localities). These species live in the dead, rotten trunks of apparently many tree species. They are good examples of "soft saprophages" (Chapter 2), with adults probably feeding on the liquid part of rotten wood. Females construct their nests in the trunks, using decayed wood to make brood balls; one egg is laid per nest. The female stays with her offspring at least until pupation. In Taï, the three species do not occur together. One of them is a true corticolous species, occurring just under the bark, while the two other species live only in deep, rotten wood. Similar variation in microhabitat selection has been found in the family Passalidae: all the species live in dead trunks, the more flattened ones under the bark, the more convex ones in rotten wood (Reyes-Castillo, pers. comm.).

Temporal Patterns

Diel activity. The diel activity of dung beetles is not as well defined in African forests as in savannas. Some species are mostly diurnal but can fly at night, and vice versa, probably because in the dark forest even the most diurnal species have to be accustomed to low levels of light. Nevertheless, the taxa that are clearly either diurnal or nocturnal in savanna (Chapter 9) have the same diel activity also in forest: Sisyphini and Oniticellini are diurnal; Coprini, Dichotomiini, and Onitini are nocturnal. As in savanna, some Onthophagini are diurnal while others are nocturnal.

Seasonal activity. Beetles have been trapped six times during the year in Taï, where there are clear-cut wet and dry seasons. The dung beetle fauna is also seasonal (Appendix B.11), with species richness being highest in the beginning of the year (February–May) and total biomass peaking in August–November (Fig. 11.2). Numbers of species and individuals are lowest during the driest part of the year (December–January).

Vertical Distribution

Many mammals in tropical forests do not live in the forest floor but occur at various heights in the canopy. Dung beetles generally follow mammals, and they have done so up to the canopy in African rain forests. In Makokou, three species occurred commonly at 10–20 m above the ground in the rainy season, when only one individual was caught in traps set at the ground level (Walter 1984a; Table 11.2). Six months later, however, in the early dry season, two of the three species were not uncommon in ground traps (*Sisyphus arboreus*

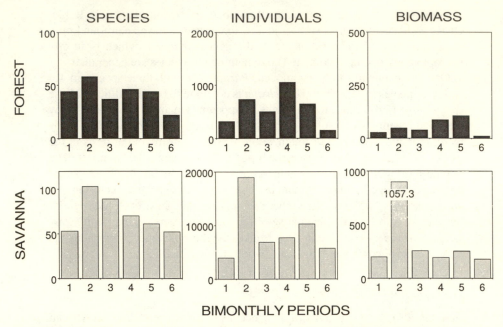

Fig. 11.2. Seasonality of dung beetles in a seasonal forest (Taï National Park) and in savanna (Abokouamekro) in the Ivory Coast. This figure shows seasonal changes in the number of species, numbers of individuals, and biomass of beetles (grams of fresh weight; data from Appendixes B.9 and B.11).

TABLE 11.2
Numbers of beetles collected in traps baited with human dung at various heights in forest canopy in Makokou, Gabon, in December 1981.

Species	Ground Level	3–5 m	10–20 m
Sisyphus arboreus	—	4	49
Onthophagus ahenomicans	—	1	3
Onthophagus laeviceps	1	16	46
Onthophagus mpassa	—	1	—
Onthophagus possoi	—	1	42

Source: Walter (1984a).
Note: Only species that were caught above the ground level have been included in this table.

and *Onthophagus laeviceps*). Unfortunately, we have no information on the biology and especially on the nesting behavior of the arboreal species. The *Sisyphus* were observed making and rolling balls in the canopy, on the platforms where the traps had been installed. Nesting could still take place in the soil, assuming that the beetles drop down with their balls, as reported for two

American rollers (Howden and Young 1981). Vadon (1947), working in a Madagascan forest, observed that the small roller *Arachnodes goudoti* was never attracted to traps at the ground level, but it was common in traps set on small trees, as low as 50 cm above the ground. In all cases observed, a beetle gathered a heap of dung and pushed it until it fell to the ground, where the ball was rolled as usual. *Onthophagus* do not make balls; they may construct their nests in the humus accumulated on large branches.

Another special behavior of forest dung beetles is "perching" on leaves: individuals of some species are occasionally found sitting on leaves of small underwood plants, at a height apparently related to their size (Howden and Nealis 1978). The perching behavior may have several functions, including thermoregulation (Young 1984; Chapter 12); but our observations from Taï and Makokou point to two additional explanations as well. First, some beetles seem to perch during foraging, instead of continuously flying, apparently waiting in one place for dung odors to reach them (Chapter 12). Such foraging behavior has been observed in rollers and small tunnelers, all of which are diurnal, and most of which are attracted to omnivore dung (Table 11.3). These beetles are wary and fly away fast when one attempts to catch them.

Other beetles perch during their period of rest, when they may be vulnerable

TABLE 11.3
Numbers of beetles caught perching on leaves in Taï (1980–81) and Makokou (1981–83) forests.

Species	Individuals	Ecological Parameters			Length (mm)
Neosisyphus angulicollis	1	HU	D	R	6.0
Sisyphus eburneus	9	HU	D	R	5.2
Sisyphus latus	20	HU	D	R	5.5
Amietina eburnea	1	CA	D	T	3.3
Onthophagus biplagiatus	5	EL	D	T	4.3
Onthophagus denticulatus	2	HU	D	T	5.8
Onthophagus fuscidorsis	4	HU	D	T	6.0
Onthophagus girardinae	6	HU	D	T	4.3
Onthophagus hilaris	10	MI	D	T	4.8
Onthophagus grassei	1	?	D	T	4.0
Onthophagus infaustus	1	?	D	T	4.3
Onthophagus justei	7	HU	D	T	4.4
Onthophagus laetus	1	?	D	T	3.4
Onthophagus laeviceps	3	HU	D	T	4.8
Onthophagus deplanatus	1	HU	N	T	6.8
Pseudopedaria grossa	4	HU	N	T	13.0

Notes: HU, EL, MI, and CA indicate that the species prefers human, elephant, or miscellaneous dung or carrion, respectively. D, diurnal; N, nocturnal; R, rollers; T, tunnelers.

to predation. Vadon (1947) reported that the diurnal *A. goudoti* firmly wedges itself in the angle between two branches for the night. We have observed this behavior in Taï and Makokou in the small tunneler *Onthophagus deplanatus* and the large tunneler *Pseudopedaria grossa*, both of which are nocturnal. The latter species was found sleeping on low plants at day time, wedged between a leaf and a stem. Such "rest perching" may occur at night in diurnal species, though we have not made any observations.

Size

Diurnal dung beetles are significantly smaller than nocturnal ones (means 49 and 250 mg, respectively, t = 4.4, P<0.001). In forest, the size difference between diurnal and nocturnal species cannot be explained by excessive elevation of body temperatures in large diurnal species (Chapter 3). Predation may play some role, though many of the insectivorous predators are themselves nocturnal. Omnivore dung specialists are smaller than elephant dung specialists, but the difference is significant only in nocturnal species (Table 11.1). No large species is found in the canopy, nor feeding on carrion, and only one was found perching on leaves. There is no relationship between the individual size and abundance in Taï, with small and large species being, on average, equally abundant. There is consequently a good relationship between the biomass of the species and its fresh weight: the heaviest species are the most dominant in biomass and generally also in resource use.

11.4. DUNG BEETLE COMMUNITIES

Species Richness

Table 11.4 compares the numbers of dung beetle species in five localities in the Ivory Coast and two in Gabon. In the localities with a rich mammalian fauna, and where elephants are still present, there is no significant impoverishment of the beetle fauna from the primary to the secondary forest (Table 11.4, localities b and c, both in Gabon; see also Nummelin and Hanski 1989). On the other hand, there is a clear decrease in species number when the forest becomes simplified in its structure and the mammalian fauna, either due to increasing altitude or decreasing area in the case of the "gallery" forest (Table 11.4, localities d–g). In the dry, deciduous forests in the northern Ivory Coast, there are only a few true forest dung beetles, while savanna species penetrate to these forests (Cambefort 1984).

The different dung beetle tribes are not equally affected by forest impoverishment. The Sisyphini and Onthophagini are always present, while the Dichotomiini (*Heliocopris*) and Oniticellini are more restricted to extensive forests with good mammalian fauna.

TABLE 11.4

The expected number of species in a sample of 300 individuals in seven localities and four different types of forest in West Africa (rarefaction, mean and standard deviation).

Rich Mammalian Fauna		Poor Mammalian Fauna		
1	2	2	3	4
48.76(a)	39.55(c)	27.13(d)	16*(f)	12.62(g)
±2.72	±1.85	±0.86		±1.41
43.83(b)		23.81(e)		
±2.62		±0.96		

Notes: The forest types are: 1, primary or old secondary forest; 2, secondary forest in patchy and/or exploited state; 3, montane forest; 4, riverine (gallery) secondary forest. The localities are: a, Taï, Ivory Coast; b, Makokou, Gabon; c, Nzoua-Meyang, Gabon; d, Sipilou, Ivory Coast; e, Yéalé, Ivory Coast; f, Mount Tonkoui (1,200 m), Ivory Coast; g, Lamto, Ivory Coast. See Appendix A.11 for the mammalian faunas.
* Sample of 297.

Dung Beetle Guilds

The four main types of dung beetles, the rollers, the tunnelers, the kleptoparasites, and the dwellers (Chapter 3), are all present in Afrotropical forests. The kleptoparasites are, however, very rare, and only one species of *Pedaria* has been collected in Makokou, where it probably exploits the brood balls of large *Catharsius* species.

Rollers. It is well known that the rollers as a group are more dominant in open habitats than in forests (Halffter and Matthews 1966). There are only five roller species in Taï and four in Makokou, and only one or two species are abundant. All the species are diurnal, and some of them can be found perching on leaves or occurring in the canopy (Section 11.3). Though the poor representation of rollers in forests is obvious in the Ivory Coast, it is difficult to explain. Rolling a ball across grass stems is more difficult than rolling balls on the almost bare forest floor. The efforts made in rolling may require high body temperature (Heinrich and Bartholomew 1979a,b), which is easier to achieve in open savanna than in forest. *Neosisyphus tai*, which occurs exclusively in deep-forest shade, is always very slow in its ball-rolling movements. *Sisyphus eburneus*, another forest species, is very fast-moving, both while rolling a ball and in flying away when disturbed from leaves where it was perching, in this case perhaps to increase its body temperature.

Tunnelers. There are sixty-seven species of tunnelers in Taï and sixty-two species in Makokou. Large tunnelers are more numerous in Makokou, probably because of the exceptionally rich mammalian fauna in this locality (Section 11.2). In Taï, the tunnelers include twenty-five diurnal and forty-two nocturnal species, of which the former are more abundant but have a lower average biomass per species than the nocturnal ones. Although the forest dung beetle fauna is impoverished in comparison with the savanna fauna, several pairs of seemingly very similar species occur in forest, especially in *Ontho-phagus* (e.g., *O. strictestriatus-tenuistriatus*, *fuscatus-fuscidorsis*, *hilaris-hilaridis*). A pair of similar large tunnelers occurs in elephant dung, *Copris camerunus* and *Onitis nemoralis*. The former species occurs under trees, the latter on tracks. This difference may be due to the washing of the shallow soils on the tracks, *Onitis* perhaps being restricted to the poorest sites, as in the savannas (Chapter 9).

Shallow soils undoubtedly explain why the nests of the tunnelers are noticeably shallow in African forests, even those of burrowers as skilled as *Heliocopris* and *Onitis*. Small tunnelers may be able to construct their small nests in the cracks between rocks and gravels, but the large ones do not have this possibility. The brood balls of *Copris* and *Heliocopris* are always covered with a layer of soil, which may have a defensive function. The large *Heliocopris* larvae are aggressive and try to bite an intruder, while the female in the nest stridulates energetically when disturbed. These features are more evident in the forest than in the savanna, perhaps due to a greater risk of predation in the forest.

Dwellers. Most Aphodiinae, the archetypal dwellers, are small in size. *Aphodius egregius* and *A. wittei*, two African forest species, are two examples of exceptionally large species, both measuring 18–20 mm. These species in fact dig burrows and function as tunnelers. Tegumentary expansions, or horns, which are frequent in Scarabaeidae but rare in Aphodiidae, are present in *A. renaudi* males. It appears that the Afrotropical forest *Aphodius* are unusually similar to Scarabaeidae in their biology.

11.5. FOREST-SAVANNA COMPARISON

Africa is the only continent with extensive tracts of both rain forest and savanna. In this section we compare the dung beetle faunas of these two biomes using data from two closely situated localities, Taï forest and Kakpin savanna (Chapter 9) in the Ivory Coast. Both study localities are situated within national parks with rich mammalian fauna. The results are summarized in Fig. 11.3 and Table 11.5.

Species richness is clearly lower in forest than in savanna, both in traps baited with human dung and in elephant dung pats, and regardless of whether calculated as the number of species per trap or per three hundred individuals (Fig. 11.3). The number of individuals per species is lower in forest (46) than in savanna (92), but the really striking difference is in beetle density: traps baited with human dung attract forty times more individuals in the savanna than in the forest, and elephant dung pats have six times more individuals.

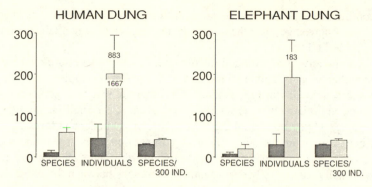

Fig. 11.3. Comparison of dung beetles attracted to human and elephant dung in forest (Taï) and savanna (Kakpin) in the Ivory Coast. The columns give the numbers of species and individuals, and the expected number of species in a sample of three hundred individuals. Dark columns are for forest, light columns for savanna (data from Appendixes B.9 and B.11).

TABLE 11.5
Comparison of dung beetle assemblages in tropical forest (Taï) and in savanna (Kakpin) in Ivory Coast.

	Forest			Savanna		
	S	N	B	S	N	B
Percentage of diurnal	42	62	25	49	87	45
Percentage of rollers	7	8	2	16	50	55
Percentage of large tunnelers*	22	8	62	24	9	31
Percentage of kleptoparasites	0	0	0	10	10	1

Notes: S, N, and B denote species, individuals, and biomass, respectively. See also Table 11.1 and Figs. 11.2 and 11.3.
* Longer than 13 mm.

(Fig. 11.3). The respective factors are six and three for the number of species. A complicating factor is that the radius of attraction of a trap or a pat may be somewhat greater in the open savanna than in the closed forest, but this difference can hardly explain more than a small part of the observed difference in density.

We do not have seasonal data for Kakpin, but we can compare Taï with another savanna locality, Abokouamekro (Fig. 11.2). Although the forest dung beetle community appears to be as clearly seasonal as the savanna community, the peak abundance of beetles occurs in the second half of the year in the forest, not in the first half as in the savanna. About half of the species are diurnal in both biomes, but in terms of individuals the diurnal ones dominate the community especially in the savanna (Table 11.5). The savanna species are larger, on average, than the forest species, especially among the nocturnal ones (Table 11.1).

Taxonomically, the forest and savanna beetle assemblages are rather similar. Scarabaeini and Canthonini are absent in our forest samples, but they are relatively scarce also in the savanna. The other rollers, Gymnopleurini and Sisyphini, are relatively more numerous in savanna, especially in numbers of individuals, while Oniticellini are relatively more abundant in forest, apparently benefiting from the relatively weak competition with large tunnelers. In both biomes, the Onthophagini is the dominant tribe, but even more so in the forest. Finally, there are two substantial differences between the two biomes in terms of the functional groups of beetles: the rollers are much more numerous in the savanna, especially in numbers of individuals; and the kleptoparasites are absent (Taï) or nearly absent (Makokou) in forest but represent 10% of the dung beetle fauna in savannas (Table 11.5).

These comparisons clearly indicate that Afrotropical savanna and forest dung beetles belong to the same taxonomic pool and that their differences are quantitative rather than qualitative. Generally, the savannas have a richer assemblage of dung beetles than the forests, and the density of beetles is much higher in the savannas. The oldest Scarabaeidae groups, Canthonini and Dichotomiini, have practically disappeared from both forests and savannas, being present only in the southernmost part of the continent (Chapter 4). The simplest explanation of the greater diversity and higher density of dung beetles in savannas is the greater availability of resources there than in forests. Some groups of species, especially the rollers Gymnopleurini and Sisyphini, have obviously diversified in savannas and secondarily penetrated to forests, but it is difficult to draw any similar conclusion about the other tribes.

CHAPTER 12

Dung Beetles in Tropical American Forests

Bruce D. Gill

12.1. INTRODUCTION

Central and South America enclose the largest expanse of tropical forests in the world, with a long evolutionary history and a rich fauna of dung beetles. Nearly all tropical American dung beetles belong to the family Scarabaeidae, with only a few species representing the Aphodiidae, Hybosoridae, and Trogidae.

Many groups of dung beetles in tropical America are still poorly known taxonomically, especially the genera with predominantly small species, such as *Ateuchus*, *Canthidium*, and *Uroxys*. Lack of revisionary work and the presence of many undescribed taxa makes identification difficult and has led some workers to study only the larger and more easily recognized species (Wille 1973; Wille et al. 1974; Janzen 1983a), while others have resorted to arbitrary numbering systems (Howden and Nealis 1975; Peck and Forsyth 1982). Although this latter approach has permitted the study of entire dung beetle assemblages, it has also made comparisons between different localities difficult if not impossible. To alleviate this problem and to facilitate ecological work, Howden and Young (1981) completed a taxonomic monograph on the Scarabaeidae of Panama. Much of this chapter is based on fieldwork conducted in Panama and Costa Rica.

The term "dung beetle" is taxonomically ambiguous since the habit of coprophagy is shared by taxa within several different lineages of the Scarabaeoidea (Chapter 2). On the other hand, many species of the true dung beetles (Scarabaeidae) do not feed on dung but use other resources such as carrion or decomposing fruit. Such diversity of feeding habits is especially characteristic of neotropical dung beetle assemblages and undoubtedly contributes to the high species richness in this region. Historical events such as the megafaunal extinctions of the Quaternary (Martin and Klein 1984) have probably favored dung beetles capable of exploiting alternative food resources. To better understand the modern dung beetle communities and the evolutionary history that has shaped the present fauna, it is important to consider all the species of Scarabaeidae regardless of their present feeding habits.

12.2. BIOGEOGRAPHY

Physical Setting

Lowland tropical forests covered much of South America by the close of the Cretaceous (Raven and Axelrod 1975). Miocene and Pliocene orogenies (Harrington 1962) fragmented lowland forests while giving rise to montane habitats. Other types of forest as well as savannas probably developed in response to changing patterns of rainfall due to changes in topographic relief. Orogenic activity was also responsible for completing the Central American land bridge 3 million years ago (Marshall et al. 1979), providing the first direct forest connection between the North and South American faunas.

Reduced rainfall resulting from lowered sea levels and lowered temperatures during the Pleistocene is thought to have shrunk the tropical forests into isolated refugia (Haffer 1969, 1974; Prance 1973, 1985). Pollen analyses from sites in South America (summarized in Absy 1985) indicate that savannas expanded several times, with forests presumably contracting to areas of higher rainfall. The existence of refugia has been used to explain speciation and distributional patterns in a variety of organisms, including birds (Haffer 1969), lizards (Vanzolini and Williams 1970) and butterflies (Brown 1982). Climatic changes may have also contributed to the extinction of many species of mammals (Patterson and Pascual 1968), giving rise to the current fauna, which is, in comparison with other continents, relatively poor in large mammals.

Within historical times, tropical forests have extended from approximately central Mexico to northern Argentina. Spanning 50 degrees of latitude and a wide range of elevations, these forests encompass an exceptionally rich assemblage of plant species. Forest classifications are traditionally based on relief, seasonality, biomass, and the presence of specific plant forms or taxa. For many groups of animals, including dung beetles, floristic parameters and plant biomass are perhaps less important than topography and seasonality in determining suitable habitats. Therefore, while discussing habitat associations of dung beetles, I shall use the Holdridge Life Zone system (Holdridge 1967; Holdridge et al. 1971).

The present status of tropical American forests is precarious. Most forested areas have been severely reduced in size during the past 50 years, and the remaining areas continue to be lost at a catastrophic rate (Caufield 1984; Myers 1984). Amazonia had a forest mosaic that covered much of its nearly 6 million square kilometers as late as the 1970s (Pires and Prance 1977, 1985), but vast regions have been destroyed in the last few years. The situation in Central America is no better (D'Arcy 1977): Janzen (1986) estimates that less than 2% of the original tropical dry forest of western Mesoamerica remains. The wholesale disappearance of forests is likely to cause the extinction of most dung beetle species along with many other forest animals, as shown by Howden and Nealis (1975) for dung beetles in clearings of wet forest at Leticia, Colombia.

Biogeographical Trends among
* Dung Beetles*

Three genera of Hybosoridae—*Anaides*, *Chaetodus*, and *Coilodes*—are occasionally found in dung and carrion in lowland and montane forests. The lethargic behavior of hybosorids (Young 1983) suggests that they are not successful competitors when Scarabaeidae are present. Among the Aphodiidae several species of *Aphodius* and *Ataenius* are frequently recorded from lowland forests (Howden and Nealis 1975; Peck and Forsyth 1982), and Aphodiidae are abundant in pastures and clearings both in lowland and montane habitats in Panama and Costa Rica. In forests the importance of Aphodiidae is, however, eclipsed by the much richer and more active fauna of Scarabaeidae, with about twelve hundred recognized species in tropical America (see Table 4.1). The true species diversity may be much greater since the largest tribe, the Dichotomiini, is also one of the least studied.

In the Onthophagini, represented by *Onthophagus*, species number decreases rapidly from Central to South America. Central American *Onthophagus* occupy a variety of forests from sea level to over 2,000 m, and they use a wide range of food resources. The Oniticellini has no species extending into the forests of Central America or South America, and of the nineteen species of Central American *Copris* (Coprini), only *C. incertus* has a range that extends to South America (Matthews 1962). The Central American *Copris* are primarily dung feeders found in pastures and in low and middle elevation forests.

The Phanaeini is a predominantly tropical group endemic to the New World (Edmonds 1972; Chapter 4). The approximately fifty species of *Phanaeus* are distributed from the eastern United States to Argentina, with about half of the species occurring in Mexico. Central and South American *Phanaeus* are diurnal dung feeders found in dry, moist, and wet forests; the dry forest species often venture to open pastures. *Coprophanaeus* is found from Mexico to Argentina but is most numerous in South America. The thirty or so described species are crepuscular or nocturnal and feed mainly on vertebrate carrion. The subgenus *Megaphanaeus* includes the largest New World dung beetles with lengths exceeding 50 mm. *Oxysternon* and *Sulcophanaeus* are diurnal dung feeders widely distributed in South American forests, with only a few species extending to Central America.

The Dichotomiini has twenty-three genera and nearly half of the dung beetle species in tropical America. *Dichotomius* includes medium-sized to large nocturnal dung feeders, which are often very abundant in forests. About half of the twelve Central American species are found in low- and middle-elevation moist forests, while the remainder occur in dry forests and pastures. *Ateuchus*, *Canthidium*, and *Uroxys* are ubiquitous members of the dung beetle assemblages in tropical America, with numerous species found over a wide range of forests and exhibiting a variety of feeding habits. Some of the larger species of *Uroxys* as well as some *Pedaridium* and *Trichillum* are phoretic upon sloths. Several species of *Ontherus* live in the nests of leaf-cutting ants.

The Eurysternini is restricted to the New World, with twenty-six species of *Eurysternus* (Jessop 1985; Martínez 1988). These species are mostly dung feeders and occur in a wide range of tropical forests from Mexico to Argentina. Another South American group is the Eucraniini, which is restricted to the arid and semiarid regions of Argentina. The Sisyphini is represented by two Mexican species of *Sisyphus* (Howden 1965), one of which (*S. mexicanus*) can be found in tropical dry forests as far south as Costa Rica.

Another exceptionally successful group of dung beetles in tropical America is the Canthonini, which is also relatively well studied with over three hundred described species. Of the twenty-eight currently recognized genera (Halffter and Edmonds 1982), about ten are monotypic and another ten have five or fewer species. The remaining genera include several that are predominantly West Indian (*Canthochilum*, *Pseudocanthon*, and *Canthonella*). The fourteen species of *Cryptocanthon* are widely distributed over Central and South America. *Cryptocanthon* and *Canthonella* include very small species, less than 3 mm in length, which are abundant in the deep litter of wet habitats, especially cloud forests. *Malagoniella* (9 species) and *Scybalocanthon* (16 species) are widely distributed in lower-elevation dry and moist forests of Central and South America. *Deltochilum* and *Canthon* are the largest genera of New World canthonines, with 76 and 148 species, respectively. *Deltochilum* is widely distributed over lowland and montane forests of tropical America, with large and small species feeding on carrion and dung. *Canthon* contains small to medium-sized species that are predominantly dung feeders, although a few have become carrion specialists. Most species of *Canthon* are restricted to forests below 500 or 1,000 m in elevation.

12.3. FOOD SELECTION AMONG TROPICAL AMERICAN DUNG BEETLES

Temporal and Spatial Availability of Food

The conditions of high temperature and humidity that are responsible for the luxuriance of plant life in tropical forests also favor the rapid breakdown of dung, carrion, and fruit. Successful scavengers must be able to locate and secure these resources as quickly as possible. The availability of fresh dung is more even in space and time than the availability of fresh carrion, and the dung of smaller mammals is more evenly spaced than the dung of larger mammals. Beetles specializing in small mammal dung may be expected to show the lowest level of aggregation in space. Locating carrion and larger dung pats necessitates a greater reliance on flight than locating the pellets of small mammals and invertebrates (see Section 12.4 on foraging behavior).

Dawn and dusk are the two periods when the defecation rate of mammals might be expected to peak due to a change in activity in both diurnal and nocturnal species. The activity patterns of many dung beetles seem to be geared to these periods (Section 12.5). Many types of dung are deposited in places that make it difficult for dung beetles to locate and use the resources. Tapirs and pacas show a strong tendency to defecate in water, while sloths descend to the ground to bury their dung. Unusually long periods of time between defecations in sloths impose problems that may have favored the development of phoresy in several species of *Uroxys* and *Trichillum* that specialize in sloth dung (Ratcliffe 1980). The dung of many arboreal mammals is released high in the canopy, with the result that much of it is available only to beetles that forage in the canopy (Section 12.5). Fossorial mammals may defecate within burrow systems, making their dung unavailable to most beetles.

Alternative Food Resources

Tropical American dung beetles utilize a variety of food resources other than mammalian dung. Among the records pertaining to the dung of nonmammalian vertebrates, Young (1981) recorded *Canthon moniliatus* and *Onthophagus sharpi* in traps baited with iguana and boa constrictor feces, and *O. sharpi*, *Ateuchus candezei*, *Canthon lamprimus*, *Uroxys gorgon*, and *U. micros* at toad feces on Barro Colorado Island. *Onthophagus nyctopus* and *Sulcophanaeus noctis* have been trapped with fresh *Ameiva* lizard dung in Costa Rica (Gill, unpubl.). Records of dung beetles using bird droppings and guano include *Anomiopus wittmeri* in northern Argentina (Martínez 1952), canthonines in Puerto Rico (Matthews 1965), *Onthophagus hopfneri* in Guanacaste, Costa Rica (Janzen 1983b), and *Canthidium ardens* pushing chicken droppings in western Panama (Gill, unpubl.).

Zonocopris gibbicollis is the only known example of a tropical American dung beetle that feeds exclusively on nonmammalian dung. Adults of this unusual canthonine are phoretic on large terrestrial snails (Arrow 1932) and presumably feed on the dung produced by their gastropod host (Martinez 1959). Larval habits of this species are still unknown. Snail excrements are also used by several canthonines in Puerto Rico (Matthews 1965), but none of these is restricted to this type of dung. The frass of large lepidopteran larvae is a potentially important resource for many of the smaller dung beetles, but this possibility has been little studied. Only two species, *U. gorgon* and *Dichotom-*

ius femoratus, have been reported on caterpillar droppings (Howden and Young 1981).

In addition to insect droppings, tropical forests produce a rain of plant debris that is exploited by some dung beetles. Mature and rotting fruits are particularly attractive to a variety of species (Table 12.1). Many of the species recorded at fruit are generalists and can also be found on dung and carrion, but a few species, such as *Onthophagus atriglabrus*, *O. belorhinus*, and *O. carpophilus*, may feed predominantly or perhaps exclusively on fruits. The genus *Bdelyrus* is occasionally collected in dung-baited pitfall traps, but their peculiar flattened shape appears to be an adaptation for living in terrestrial and epiphytic bromeliads (Pereira et al. 1960; Huijbregts 1984). These beetles presumably feed on the plant material or insect frass that accumulates in these plants. Debris piles of leaf-cutting ants are also used by dung beetles. Halffter and Matthews (1966) list fourteen neotropical species of Scarabaeidae from Attini nests, one of which, *Ontherus cephalotes*, has also been found in the fungus gardens and may feed on fungi. Fungal feeding is known for other species not associated with ants; for example, *Canthidium bokermanni* and *C. kelleri* were described by Martínez et al. (1964) from specimens taken from Hymenomycetes fungi in Brazil and Argentina. Other records of fungal feeding include generalist species such as *Canthon v. leechi*, *Deltochilum fuscocupreum*, *D. granulatum*, *Phanaeus daphnis*, and *P. endymion* (Halffter and Matthews 1966).

A final but significant category of alternative food resources for dung beetles is carrion. *Coprophanaeus*, *Deltochilum*, and *Canthon* have many species that are predominantly or exclusively necrophagous, but carrion-feeding has also been reported for *Agamopus*, *Ateuchus*, *Canthidium*, *Canthonella* (as *Ipsellisus*), *Copris*, *Dichotomius*, *Malagoniella*, *Onthophagus*, *Pedaridium*, and *Phanaeus* (Halffter and Matthews 1966). Necrophagy is not limited to vertebrate carcasses, as species of *Canthon* and *Deltochilum* are reported from dead insects and millipedes (Luederwaldt 1911; Pereira and Martínez 1956; Howden and Young 1981; Janzen 1983b). Several Brazilian species of *Canthon* are reported to kill reproductive *Atta* ants and to use them as carrion (Halffter and Matthews 1966). Halffter and Matthews (1966) also mention that *Deltochilum kolbei* may attack live millipedes. The broad range of genera involved indicates that carrion feeding is of major importance in tropical American dung beetles.

12.4. POPULATION ECOLOGY

Seasonality

Tropical insects exhibit great variability in their seasonality and year-to-year abundance changes (Wolda 1978a, 1983). Differences occur between species but also within species depending on local conditions. Variability in rainfall

TABLE 12.1
Tropical American dung beetles (Scarabaeidae) recorded from fruits.

Species	Locality	Type of Fruit	Ref
Ateuchus candezei	Panama	Palm	6
Ateuchus illaesum	W. Indies	?	1
Canthidium ardens	Panama	Palm	6
Canthidium barbacenicum	Brazil	Butia palm	11
Canthidium decoratum	Brazil	Butia palm	11
Canthidium elegantulum	Panama	Ficus	6
Canthidium nobilis	Brazil	Butia palm	11
Canthidium splendidum	Brazil	Butia palm	11
Canthidium tuberifrons	Panama	Solancus	3
Canthon angularis	Brazil	Butia palm	12
Canthon conformis	Brazil	Butia palm	12
Canthon latipes	Brazil	Butia palm	12
Canthon lituratus	Brazil	Butia palm	12
Canthon moniliatus	Panama	Entada	6
Canthon muticus	Brazil	Butia palm	12
Canthon opacus	Brazil	Butia palm	12
Canthon scrutator	Brazil	Butia palm	12
Copris lugubris	?	Banana, avocado	9
Dichotomius ascanius	Brazil	Guava, pineapple	7
Dichotomius ascanius piceus	Brazil	Coffee	11
Dichotomius glaucus	Brazil	Palm	8
Eurysternus plebejus	Panama	Gustavia	6
Onthophagus acuminatus	Panama	Virola	6
Onthophagus andersoni	Costa Rica	Citrus	13
Onthophagus atriglabrus	Costa Rica	Citrus, banana	13
Onthophagus belorhinus	Guatemala	Cacao	2
Onthophagus belorhinus	Mexico	Jackfruit	4
Onthophagus belorhinus	Panama	Solancus	3
Onthophagus nr. bidentatus	Ecuador	Banana	10
Onthophagus nr. canellinus	Ecuador	Banana	10
Onthophagus carpophilus	Mexico	Zapote	11
Onthophagus dicranius	Panama	Gustavia	6
Onthophagus mirabilis	Panama	Avocado	3
Onthophagus praecellens	Panama	Palm	6
Onthophagus praecellens	Costa Rica	Banana, jackfruit	3
Onthophagus sharpi	Panama	Palm, banana	6
Onthophagus sharpi	Panama	Gustavia, maguira	6
Phanaeus pyrois	Panama	Gustavia	6
Phanaeus apollinaris	Ecuador	Banana	10

References: 1, Arrow (1903); 2, Bates (1886); 3, Gill (1986); 4, Halffter and Matthews (1966); 5, Howden and Gill (1987); 6, Howden and Young (1981); 7, Luederwaldt (1914); 8, Luederwaldt (1931); 9, Matthews (1962); 10, Peck and Forsyth (1982); 11, Pereira and Halffter (1961); 12, Pereira and Martínez (1956); 13, Gill (unpubl.).

appears to be the prime factor influencing seasonality in tropical organisms (Wolda 1978b, 1982), and this seems to hold true for many dung beetles.

Janzen (1983a,b) found that most dung beetles were strongly seasonal in the tropical dry forest at Guanacaste, Costa Rica. The appearance of large species, such as *Dichotomius* and *Phanaeus*, coincided with the start of the rains in early May; fresh horse dung was rapidly degraded by these and many smaller species during the first month of the wet season. By the early dry season all of the larger beetles had disappeared leaving only the smallest species, such as *Onthophagus hopfneri*. The severity of the dry season in Guanacaste curtails the activity of most species (Janzen 1983a).

In the Brazilian state of Parana, Stumpf and coworkers (Stumpf 1986a,b; Stumpf et al. 1986a,b) found seasonal abundance changes in most dung beetle species. The numbers of *Ateuchus apicatus*, *A. mutilans*, and *Uroxys dilaticollis* correlated with rainfall, the minimum numbers occurring in February, the driest month of the study. In Panama, the large sloth beetle *Uroxys gorgon* is present throughout the year in blacklight samples on Barro Colorado Island (Wolda and Estribi 1985), where it shows a bimodal abundance pattern with increased numbers in the beginning and at the end of the wet season, and the lowest numbers coinciding with the dry season. Similar fluctuations in abundance were seen in *U. gorgon* populations at Fortuna, an area with high rainfall throughout the year (Wolda and Estribi 1985). Waage and Best (1985) found a similar pattern of seasonality for the sloth beetle *Uroxys besti* at Manaus, Brazil, where beetles were abundant on their *Bradypus* hosts in June and again in December. While the numbers of *U. besti* and the smaller sloth associate, *Trichillum adisi*, were not correlated with rainfall at Manaus, the numbers of *T. adisi* on sloths on a neighboring island showed a strong correlation with seasonal flooding (Waage and Best 1985). Beetles were rarely encountered on sloths during the period of inundation from May to August. Since sloths must defecate into water, beetles presumably underwent reproductive diapause or dispersed to areas of dry ground.

In contrast to the seasonal abundance changes observed for dung beetles in tropical dry and moist forests, Peck and Forsyth (1982) did not observe any overt seasonality among dung beetles in a tropical wet forest. Excluding rare species, the rank order of species' abundances remained constant over the wet-dry season transition at Rio Palenque in western Ecuador (the "dry" season at Rio Palenque being slightly drier than the wet season). Peck and Forsyth (1982) question the validity of any observations of seasonality based on pitfall trapping data by suggesting that differences in rainfall may limit available flight time for beetles. However, perching beetles (see below) do not seem to be perturbed by rain, and flight interception and pitfall traps baited during rain storms continue to collect beetles. But heavy rainfall can flood traps washing away specimens or the bait, while baits may dry out very quickly in the dry season, reducing their apparency or perhaps suitability to foraging beetles.

Apart from seasonality, beetle numbers vary from one year to another, though there is very little information on this topic. Howden and Nealis (1975) compared pitfall samples taken two years apart in a wet forest at Leticia, Colombia. Species composition and the relative abundances of the dominant species at carrion and dung were not greatly changed. Estimates of actual population densities are difficult to obtain since little is known about the distances traveled by beetles collected in traps. Wille et al. (1974) attempted to trap out the adult population of the large canthonine *Megathoposoma candezei* from a 3.8-hectare patch of isolated forest in Costa Rica. Fifty-six individuals were removed, giving a density of 15 beetles per hectare. Peck and Forsyth (1982) used a mark-recapture technique at Rio Palenque to estimate total population size for all dung beetle species in a forest of 80 hectares. Of 2,178 beetles marked and released over a 3-day period, only 41 were recovered in a recapture sample of 3,184 beetles, giving an estimate of over 2,000 beetles per hectare.

Foraging Modes and Dispersal

Dung beetles have evolved specialized foraging techniques to rapidly locate and utilize suitable food resources. Prompt detection is especially important in tropical forests, where the combination of high temperatures and rainfall hastens the deterioration of perishable materials such as fresh dung. All dung beetles rely upon olfactory sensillae in the antennae to detect airborne food odors. Arrival at suitable resources is generally accomplished by flight, with the exception of a few flightless species associated with montane habitats: *Cryptocanthon* species (Howden 1973, 1976) and *Onthophagus micropterus* (Zunino and Halffter 1981) among tropical American species.

Fully winged dung beetles detect food odors during cruising flights or while perching on vegetation, and most species exhibit a strong preference for one of these two basic foraging techniques. Cruising species can be loosely divided into fast and slow fliers. Fast fliers, such as the Phanaeini, are easily detected by the loud buzzing sound they produce during flight. These species may be flying close to the maximum range velocity (Pennycuick 1982) to cover as much terrain as possible during flight. *Oxysternon* and *Phanaeus* are generalist dung feeders that have flight periods lasting many hours per day. *Coprophanaeus* are carrion specialists that fly for a short time, usually less than 30 min, at dusk and dawn. It is tempting to speculate that carrion specialists have shorter flight periods because the temporal availability of fresh carrion is much less predictable than the supply of fresh dung: powered flight is not an efficient method of passing time while waiting for the appearance of fresh carrion. A short high-speed flight may be sufficient to determine whether a carcass is locally available on a given day.

Short periods of rapid flight are also seen in species other than carrion specialists. *Onthophagus belorhinus* and *O. mirabilis* have flight times of 20–40 min at dusk in the lower montane wet forests of western Panama (Gill, unpubl.). These species can be collected at a wide variety of resources that include dung, carrion, and fruit. The fact that the foraging periods of unrelated species (*Coprophanaeus* and *Onthophagus*) coincide at twilight may indicate that predation pressure on dung beetles is temporarily relaxed during these times.

Slow flight is typical for nocturnal species of all sizes and for some diurnal species. Slow fliers may be flying at the minimum power velocity (Pennycuick 1982) to reduce energy costs while maximizing the period of time they are airborne and searching for food. For many of the larger nocturnal species, for example *D. satanas*, slow flight also eliminates the production of buzzing sounds that might be picked up by foraging bats. Slow flight among diurnal species may be selected against by increased predation from aerial predators such as the asilid flies (Section 12.4). On Barro Colorado Island, only *Canthon c. sallei* and *C. moniliatus* appear to fly at the minimum power velocity. These two species are brightly colored and very conspicuous while they slowly meander over the ground at the height of 15–30 cm. *Canthon c. sallei* is known to produce a secretion that repels blowflies from its food (Bellés and Favila 1984). Beetles captured in flight have a distinctly unpleasant aroma (Gill, unpubl.), which may confer some form of protection from diurnal predators (Section 12.4). *Canthon moniliatus* may also be chemically protected, or it may be a Batesian mimic of *C. c. sallei*.

Like the slowly flying species, beetles that perch on vegetation may increase their foraging time without excessive energy expenditure. Perching beetles can be readily divided into two groups. Selective height perchers are found over a narrow range of heights above the ground. The average perching height for these species is usually between 10 and 50 cm above the ground and is strongly correlated with body length (Howden and Nealis 1978; Gill, unpubl.). The nonselective height perchers are found over a broad range of heights, and they apparently perch for a variety of reasons.

Selective height perchers are specialists on small, scattered resources, such as rodent droppings, which are potentially much more evenly dispersed in space and time than the dung of larger mammals. Certain dung beetles have adopted perching as a "sit and wait" tactic for locating these small resource patches. Since larger beetles require larger droppings, they presumably perch higher up to avoid being distracted by odors from resource units that are too small to be utilized. Field observations on Barro Colorado Island (BCI) suggest that beetles using this strategy may increase their chances of locating food by perching downwind of mammals.

Canthidium aurifex is active throughout the day on BCI (Fig. 12.1) with peak flight times in the beginning and at the end of the day. Between these two "rush hours," beetles spend most of their time perching on leaves about

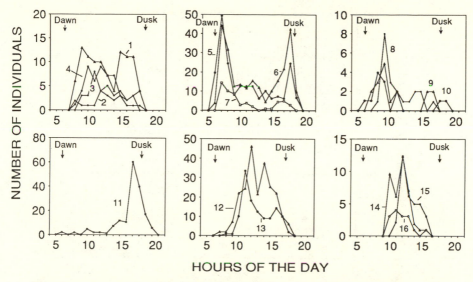

Fig. 12.1. Diel activity of selected species of dung beetles illustrating different patterns. Top left: examples of species showing eurythermic flight activity (1, *Canthon cyanellus sallei*; 2, *Canthon morsei*; 3, *Onthophagus crinitus panamensis*; 4, *Onthophagus sharpi*). Top center: examples of species showing bimodal flight activity (5, *Canthidium haroldi*; 6, *Canthidium aurifex*; 7, *Onthophagus acuminatus*). Top right: examples of species showing reduced flight activity at midday (8, *Onthophagus stockwelli*; 9, *Canthon moniliatus*; 10, *Anomiopus panamensis*). Bottom left: an example of species showing late afternoon flight activity (11, *Canthidium ardens*). Bottom center and bottom right: examples of species showing midday flight activity (12, *Canthon lamprimus*; 13, *Onthophagus coscineus*; 14, *Onthophagus lebasi*; 15, *Canthon septemmaculatus*; 16, *Eurysternus plebejus*).

25 cm above the ground (Fig. 12.2), with occasional flights to new perches. Seeding a study plot with human dung results in a noticeable increase in the numbers of beetles perching downwind. Large concentrations of perching beetles were often found on vegetation, although no dung was apparent in the vicinity. On one occasion several dozen *Canthidium aurifex* and *C. haroldi* were found on a bush immediately downwind from where an agouti had been sitting for 20 min. All of these beetles departed within some minutes of the agouti leaving. The absence of dung and urine indicated that the beetles were perching in response to the odor of the mammal, and were waiting for the appearance of food.

Nonselective height perchers may also use perching as a means of increasing foraging time while reducing the cost of flight, but these species are not specialists on ground-based resources. In contrast, most of these species use resources, such as monkey dung, that are distributed over a range of heights

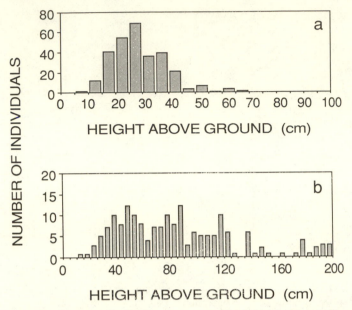

Fig. 12.2. Perching height distributions in two species of dung beetles on Barro Colorado Island, Panama. (a) *Canthidium aurifex*, a selective height percher (upper panel); and (b) *Canthon angustatus*, a nonselective height percher (lower panel).

above the ground. *Canthon angustatus* is a typical nonselective height percher that can be found on leaves up to several meters or more above the ground (Fig. 12.2). Since *C. angustatus* usually perches only in the vicinity of fresh howler monkey dung, perching beetles may be waiting for fresh dung to fall or the opportunity to steal the dung ball of another *C. angustatus*.

In addition to searching for food odors, some species of dung beetles use perching for thermoregulatory purposes. Young (1984) has reported that *Canthon septemmaculatus* preferentially perches on leaves that are located in sun flecks. *Eurysternus plebejus* also shows this behavior on BCI (Gill, unpubl.). By elevating their body temperatures, beetles may be more successful in foraging or in avoiding predators.

Differences in foraging behavior such as perching versus flying may affect the dispersal ability of a species. Among the species that rely upon flight, some of the fast fliers (*Phanaeus* and *Oxysternon*) are widely distributed over Central and northern South America. The slow-flying *Canthon cyanellus* also has good dispersal ability (different subspecies range from the United States to Venezuela), perhaps because it has a long period of daily flight activity (Section 12.5). Individual beetles may cover large distances; for example, *Oxysternon conspicillatus* flew at least 1 km in two days, and some *Onthophagus* were recovered at distances of 180 and 700 m after two days (Peck and Forsyth 1982).

Predation

Dung beetles fall prey to a wide variety of predators. Janzen (1983a, 1983c) describes graphically how *Eulissus chalybaeus*, a large staphylinid beetle, consumes small and medium-sized dung beetles (*Dichotomius yucatanus*, *Onthophagus championi*, and *Canthidium centrale*) in Guanacaste. The predatory beetles fly to fresh dung pats, where they await the arrival of their prey. In the lower montane wet forests in Costa Rica, another large staphylinid, *Leistotrophus versicolor*, is found on fresh dung where it feeds upon large blowflies (Gill, unpubl.) and probably also on small dung beetles. The tiger beetle *Megacephala affinis* is reported to prey on *Onthophagus landolti* and *O. marginicollis* from beneath cattle pats in a Panamanian pasture (Young 1980a).

The assassin bug *Apiomerus ochropterus* is a conspicuous predator on leaf surfaces on BCI (Gill, unpubl.), where it must frequently encounter the common leaf-perching dung beetles, *Canthidium ardens*, *C. aurifex*, and *Canthon angustatus*. When disturbed while perching, the typical defensive reaction of the first two species is to retract their appendages and to "sit tight." When confined in a jar with one *A. ochropterus*, such defense offered no protection, and the assassin bug readily consumed several beetles (Gill, unpubl.). In contrast, *Canthon angustatus* is chemically defended, and when disturbed while perching it assumes an aposomatic display stance with the head down, pygidium raised, and legs outstretched. When a single *C. angustatus* was confined in a jar with an *A. ochropterus*, the bug approached the dung beetle with its beak raised, ready to strike. Once the forelegs of the bug touched the beetle, the beetle instantly dropped into a display stance and released a strongly smelling odor that caused the normally staid bug to flip over backwards and to retreat to the far side of the jar. The release of this odor was an effective defense, as the bug showed no further interest in repeating the experience.

In a study of the prey captured by *Argiope argentata* spiders on BCI, Robinson and Robinson (1970) found that Scarabaeoidea constituted 1.3% of the total prey biomass over a 12-month period. Most of these were probably small dung beetles that are common in the area and likely to be caught in low-lying spider webs. Aerial-foraging robber flies prey on dung beetles. One example is *Senobasis corsair*, a common robber fly on BCI (Shelly 1985). T. Shelly (pers. comm.) recorded 178 captures of prey, 74% of which were beetles, and most of the latter were small metallic-colored dung beetles, including *Ateuchus aeneomicans*, *C. ardens*, and *C. aurifex* (Gill, unpubl.). Insectivorous bats may prey heavily on nocturnal dung beetles. Howden and Young (1981) report a specimen of *Onthophagus batesi* taken from the stomach of a bat in

Costa Rica. While data on stomach contents are difficult to obtain and destructive to bats, examination of feeding roosts (Dunkle and Belwood 1982) could yield quantitative estimates of bat predation on dung beetles. Beetle fragments from bat roosts on BCI (supplied by J. J. Belwood) contained an assortment of elytra belonging to species of *Copris* and *Dichotomius* (Gill, unpubl.).

12.5. DUNG BEETLE COMMUNITY ON BARRO COLORADO ISLAND, PANAMA

Barro Colorado Island is a 1,570-hectare reserve situated within Gatun Lake about 35 km northeast of Panama City. The island was created in 1914 during the formation of the Panama Canal and is administered by the Smithsonian Tropical Research Institute. With an elevational range of 20–170 m above sea level, BCI lies within the Tropical Moist Forest life zone (Holdridge 1967). The island is covered with a mixture of second growth and mature semideciduous lowland forest (Foster and Brokaw 1982), and it experiences a moderate dry season from late December to April. Heavy rains at the end of April signal the start of a 7-month rainy season. Additional information on the physical setting and ecology of BCI can be found in Leigh et al. (1982).

As a result of its isolation and protected status, BCI has maintained a rich mammalian fauna. Diversity and biomass estimates for many of the forty-two nonvolant mammals (Appendix A.12) recorded on the island compare favorably with other Neotropical lowland forests (Eisenberg and Thorington 1973; Eisenberg 1980; Glanz 1982). The dung beetle fauna on BCI was studied for a total of 13 weeks from April 27 to July 7, 1981; September 2 to 6, 1982; June 3 to 14 and September 5 to 12, 1983. These dates correspond to early and middle rainy seasons. Beetles were collected using flight interception traps, two sizes of dung-baited pitfall traps, and dung-baited soil traps. Fresh human dung was used in all baited traps. Details of trap design can be found in Bernon (1980), Peck and Davies (1980), and Peck and Howden (1984).

Species Composition and Abundances

Combined with data from Howden and Young (1981), a total of fifty-nine species of Scarabaeidae dung beetles are known to occur on BCI (Appendix B.12). Species composition and relative abundances during the 3-month field study varied in the results obtained by the different sampling methods used. In dung-baited traps, the size of the bait and the height of the trap clearly affected trap catches. The use of carrion or other type of dung would certainly have affected the species composition as well. To reduce some of the biases associated with baits, flight-interception traps were used to quantify relative abundances of species. Observations of beetles flying into these traps, however, revealed differences in flight behavior that affected trapping efficiency.

Small diurnal species (for example, *Canthidium*) fell into the collecting trough after colliding with the flight interception screen, but larger and more powerful fliers (for example, *Phanaeus*) often bounced off the screen and escaped, though some were collected as they attempted to fly beneath the obstruction. Large nocturnal species (for example, *Canthon aequinoctialis* and *Dichotomius satanas*) fly much slower than similar-sized diurnal beetles, and often hovered in front of the screen before changing direction and flying away. Contamination of the fluid in the collecting trough with dung from passing mammals or from a buildup of decomposing organic matter changed the flight interception trap to a large "baited" trap, in which *C. aequinoctialis* and *C. angustatus* were collected in large numbers.

Unusual behavior or feeding habits may explain the supposed rarity of many species. *Uroxys gorgon* should be abundant since it is phoretic on the three-toed sloth, *Bradypus infuscatus*, the most abundant mammal on BCI (Appendix A.12), but only two specimens were collected with dung-baited and flight-interception traps during three months. However, Wolda and Estribi (1985) collected more than twenty-five hundred *U. gorgon* on BCI with two black-light traps over a three-year period. In the case of *Ontherus brevipennis*, only two of the thirty-one specimens seen by Howden and Young (1981) had been collected since 1912, but twenty-three teneral and newly sclerotized adults were removed from their pupal chambers in the refuse dump of an *Atta colombica* nest on BCI. With such unusual nesting and, presumably, feeding habits, it is not surprising that *O. brevipennis* is rarely encountered. Two other apparently rare species are *Pedaridium bottimeri* and *P. brevisetosum*. These beetles are probably phoretic on sloths or other arboreal mammals.

Canthon juvencus, Canthon mutabilis, Canthon v. meridionalis, Dichotomius agenor, and *Uroxys microcularis* are abundant in the drier forests on some of the smaller islands in Gatun Lake and at Gamboa, on the neighboring mainland. On BCI such environmental conditions are restricted to some small clearings, which may limit the population sizes of these species. The two species of *Copris* recorded from BCI may be limited to the drier parts of the island.

Diel Activity

The dung beetle community on BCI is composed of thirty-three diurnal, twenty-four nocturnal, and two possibly auroral/crepuscular species. The preponderance of diurnal species contradicts the view of Halffter and Matthews (1966) that nocturnal species predominate in tropical American forests. Nocturnal species appear to be more abundant due to the great abundance of *Canthon aequinoctialis* (Appendix B.12), but it is difficult to say to what extent this result reflects the greater proclivity of nocturnal species to be caught in traps rather than their large population sizes.

Diurnal species display several distinctive patterns of flight activity. Constant levels of activity throughout the day are exhibited by *Canthon c. sallei*, *Canthon morsei*, *Onthophagus c. panamensis*, and *O. sharpi* (Fig. 12.1). These eurythermic species seem unaffected by diurnal temperature changes. Most species are more selective and show a peak of flight activity either in early morning, in late afternoon, or at midday. Species that remain inactive during the hottest hours of the day include *Canthidium aurifex*, *C. haroldi*, *Onthophagus acuminatus*, *O. panamensis*, *C. moniliatus*, and *Onthophagus stockwelli* (Fig. 12.1). *Canthidium ardens* has a long flight period but does not show peak activity until 2 hours before dusk (Fig. 12.1). In contrast to the previous species, *Canthon lamprimus*, *C. septemmaculatus*, *E. plebejus*, *Onthophagus coscineus*, and *O. lebasi* have shorter flight periods coinciding with the hottest hours of the day (Fig. 12.1). With the exception of *O. lebasi*, these species are commonly found in the drier forests and presumably have relatively high thermal preferences.

Nocturnal species on BCI concentrate their flight activity to the first few hours of darkness. Over a five-day period, flight interception traps caught thirty-four, seventeen, and three specimens of *C. aequinoctialis* during the first, second, and third hour of darkness. Additional specimens were seen flying in the hour before dawn.

Food Relocation

Dung beetles employ several techniques to sequester dung from droppings (Chapter 3). The most common one is the excavation of burrows beside or beneath droppings, which are then packed with dung. Beetles using this technique are the tunnelers (Chapter 3), represented by *Onthophagus* and many of the Coprini on BCI. Because they tunnel into the soil directly adjacent to their food, tunnelers are limited to resources that are located on substrates suitable for burrowing and nest construction. Several other relocation strategies appear to have developed from the basic tunneler technique.

Mexican species of *Phanaeus* are known to push fragments of dung overland before burying them (Halffter and Matthews 1966; Halffter and Lopez 1972; Halffter et al. 1974). On BCI three diurnal species, *Phanaeus howdeni*, *P. pyrois*, and *Sulcophanaeus cupricollis* exhibit similar behavior, often constructing burrows 5–20 cm away from the dropping. These beetles work in bisexual pairs, and while one beetle excavates the burrow, the other individual pushes fragments of dung to the burrow entrance. Such "head butting" is an above-ground extension of the behavior used by beetles to move dung through their underground tunnels. Halffter et al. (1974) point out that dung pushing reduces potential competition for nesting space among tunnelers. It also allows beetles to utilize dung that is located on unsuitable substrates, such as rocks, and may also be an effective defense against another group of dung beetles, the kleptoparasites (also called cleptocoprids; see Brussaard 1987).

Kleptoparasitism is a modified tunneler strategy, in which beetles remove dung from the subterranean supplies provisioned by other species, most often by large tunnelers. On BCI, *Onthophagus acuminatus* is a facultative klepto-parasite that provisions burrows like a typical tunneler at small droppings, but becomes a kleptoparasite at larger resource patches. Excavation of soil beneath dung pats of 200 cm³ revealed numerous *O. acuminatus* at 20–30 cm depth, stealing food from the large dung-filled burrows of *Dichotomius satanas*. Exclusion experiments were used to determine whether *O. acuminatus* was brought down in the dung by *D. satanas*. Placing coarse-mesh screen over fresh dung prevented colonization by *D. satanas* but had little effect on *O. acuminatus*, which tunneled deep into the soil and were found waiting in empty burrows. *Pedaridium pilosum* and *Scatimus ovatus* are two other klepto-parasites on BCI that rarely provision burrows by the tunneler method.

Another method of overland transport that does not require ball formation has developed in the coprine genus *Canthidium*. On BCI at least three species, *C. ardens*, *C. aurifex*, and *C. haroldi*, are accomplished pellet rollers that can rapidly disperse rodent droppings. The elongate shape and dry texture of rodent dung allows these beetles to run away from a pile of fresh dung with a single pellet rolling along on the front edge of the clypeus. The frontal tubercles and carinae that characterize these species probably assist in maintaining contact with and providing directional stability to the pellet during rolling. On rare occasions, beetles will attempt to roll fragments of human and howler monkey dung, but the amorphous shape and sticky texture of these types of dung prevent beetles from moving quickly. Species using pellet rolling utilize dung that is "packaged" for moving. This method of food relocation probably occurs in other groups of dung beetles that have specialized on rodent droppings.

The final strategy for transporting dung is the roller technique. *Canthon* and *Deltochilum* are the only dung beetles on BCI that exhibit typical ball-making and ball-rolling behaviors. Like head butters and pellet rollers, the true rollers avoid possible competition for nest sites by moving dung away from the site where it was deposited (Chapter 3).

Resource Utilization

Different types of dung are not evenly utilized by the different sorts of dung beetles on BCI. The howler monkey, *Alouatta palliata*, is the largest producer of mammalian dung on BCI (Appendix A.12). These diurnal primates are often found in fig trees, where they transform fruits into an aromatic yellowish-green dung, which is typically released high in the canopy, with the result that much of it lands on the lower vegetation as it cascades toward the ground. Defecation in the early morning by a troop of howlers attracts many species of dung beetles. *Canthon angustatus* is always the most abundant species. On one occasion, several thousand *C. angustatus* formed a dense cloud of beetles that hovered under a tree in which monkeys were defecating.

Howler dung that lands on vegetation is located by *C. angustatus* and transformed into balls. These balls are rolled off the vegetation, and they fall to the ground with the beetle riding on top. Rolling continues on the ground, where pair formation also takes place and intraspecific combat is common. Ball construction seems to be restricted to the canopy, however, as very few beetles were observed making balls of dung that had fallen directly to the ground.

The only other species on BCI that regularly removes dung from the surface of vegetation is *Canthon subhyalinus*. Though less abundant than *C. angustatus*, it is likewise found on leaves with fresh howler dung. Both species prefer dung that is located above ground level. A trap at 8 m above ground also collected several *C. aequinoctialis*, a species that feeds on human dung placed on branches and leaves but has not been seen to make balls above the ground level.

Howler dung that reaches the ground is quickly colonized by *P. howdeni*, *P. pyrois*, *S. cupricollis*, and many smaller species, including *Ateuchus aeneomicans*, *O. acuminatus*, *O. c. panamensis*, *O. sharpi*, and *O. stockwelli*. Numerous *C. lamprimus*, *C. aurifex*, and *C. haroldi* perch on vegetation surrounding howler dung, yet are rarely found on the ground in the dung. Howler dung that is defecated late in the afternoon is usually colonized by *Canthon aequinoctialis* and *Uroxys micros* after dark.

Capuchin monkeys, *Cebus capucinus*, also defecate in the canopy but their fecal masses are much smaller than those of the howlers and attract fewer beetles. Agouti, *Dasyprocta punctata*, are diurnal caviomorph rodents that produce elongate dung pellets. A pile of agouti pellets discovered in the late afternoon was in the process of being rapidly dispersed by eight *Canthidium aurifex* and six *C. haroldi*. The *Canthidium* species had no difficulty in rolling the pellets away. The large amount of agouti dung that is produced each year (Appendix A.12) may account for the abundance of *Canthidium aurifex* and *C. haroldi* on BCI (Appendix B.12).

12.6. SUMMARY

Tropical American forests support a rich and varied fauna of dung beetles that exhibit great diversity in feeding habits, including carrion- fruit, and detritus feeding. While many species are capable of utilizing a wide range of different types of dung, differences in the spatial and temporal availability of the dung types have led to feeding specializations among forest dung beetles. A case in point is the species that have evolved particular foraging behaviors to exploit monkey dung located high in the canopy. Other species have specialized in widely scattered ground-based resources, such as rodent pellets. Baited pitfall traps remove many of the extrinsic characteristics of different dung types and may give a rather distorted picture of local dung beetle assemblages. The dung beetle community on Barro Colorado Island has more diurnal than nocturnal species, although nocturnal species may be more readily collected by baited pitfall traps. A large number of questions remain concerning the taxonomy and ecology of tropical American dung beetles. We can only hope that enough forest will be saved to help answer some of these questions.

CHAPTER 13

Dung Beetles of the Sahel Region

Daniel Rougon and Christiane Rougon

13.1. INTRODUCTION

Sahel, "the edge of the desert" in Arabic, comprises a band of land in Africa running from the Atlantic Ocean to the Red Sea (Fig. 13.1). The Sahelian climate is distinguished by a long dry season from October to May, and by a short rainy season from June to September, during which pluvial agriculture is possible but risky. For the past twenty years, the Sahelian region has had exceptionally low rainfall, which, when combined with intensive use of the vulnerable soils by man, has given rise to extensive desertification.

Two groups of dung beetles have adapted to the contrasting seasons of the Sahel. Dung beetles of the dry season are represented by a small number of species, all of which are tunnelers. Dung beetles of the rainy season are more numerous in species and individuals, and they include rollers as well as tunnelers. Under the extreme environmental conditions of the Sahel, rainfall is the limiting factor, as clearly shown by a prominent increase in the numbers of dung beetles in the beginning of the rainy season.

In this chapter we focus on the dung beetles of the dry season, because, by their nesting behavior, they are actively involved in the fertilization of the Sahelian soils in regions where millet grass is cultivated. During the dry season, the Sahelian dung has a short "life" of nine days due to very rapid desiccation. The Sahelian dung beetles have evolved short life cycles, and they feature exceptional flexibility in their nesting behavior, which undoubtedly helps them survive the extreme conditions they face. This chapter is based mostly on our own studies conducted in Niger for ten years. No similar study has been carried out in any other arid environment.

13.2. BIOGEOGRAPHY OF THE SAHELIAN REGION

Characteristics of the Study Area

In Niger, the Sahelian region can be divided into two parts along the isohyet of 350 mm (Peyre de Fabrègues 1980): the northern "nomadic zone" with a primarily pastoral livelihood, and the southern "sedentary zone," where agriculture is practiced in the rainy season and cattle breeding during the rest of

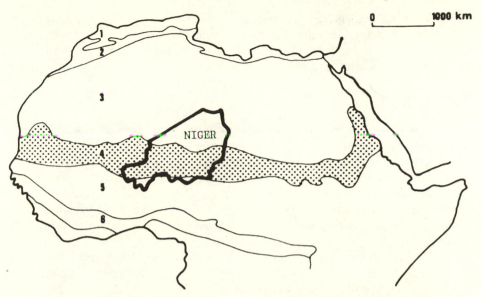

Fig. 13.1. The Sahelian region (dotted area; according to the African chorologic map of White 1965, 1976, simplified by Rougon). 1, Mediterranean endemism center; 2, Sub-Mediterranean transition zone; 3, Saharo-Sindian endemism center; 4, Saharo-Sudanian transition zone (Sahel); 5, Sudanian endemism center; 6, Sudano-Guineo-Congolese transition zone.

the year. Dung beetles have been studied in the southern area, at the university campus of Niamey.

The outstanding feature of the south Sahelian climate is dryness, attributable to low rainfall during one short rainy season from June to September, and to high temperatures, the annual average reaching 28–29°C. At the study site in Niamey, the average annual rainfall is 592 mm (SD 127 mm), and the ratio of the maximum over the minimum value is 21 for the years 1941–70. The Niamey climate is very warm with an average annual temperature of 29.1°C.

The south Sahelian area has tropical ferruginous soils, which are not at all or only a little washed, and are formed on sand or sandstone. The sandy texture of the soils is suitable only for the less demanding pluvial crops, such as millet grass and peanuts. The soil is very poor in organic carbon (C<0.2%), organic matter (M.O.<0.3%), and total nitrogen (N<0.01%), and it is slightly acidic (5.6<pH<6.2) and relatively desaturated (42%<S/T<66%).

After agriculture, cattle breeding is the most important economic activity in Niger. Cattle are kept primarily by the pastoral populations, the Peuls and the Tuareg. The zebu (humped ox, *Bos indicus*) dominates among the bovids. Constant moving of the herds of cattle is essential in the Sahel, allowing the use of the best grazing lands at the right time. In the beginning of the rainy season, stockbreeders leave the south and proceed to north, giving up the land to pluvial cultures. After the rainy season, herds of cattle are brought back to the south, where cattle are now fed with the straw of the millet grass, left in the fields to dry after the harvest.

Dung in the Sahelian Region

Because of the extreme climate, two microclimatic parameters are crucial for the Sahelian dung beetles: (1) the microclimate prevailing around droppings in April (dry and warm season) and in July (rainy season); and (2) changes in the temperature in the dung and soil over 24 hours. During the dry and warm season, air temperature reaches 44°C at 2 P.M.; relative humidity remains less than 25% from 11 A.M. to 8 P.M., and drops down to 10% or less from 1 P.M. to 6 P.M. Temperature in the dung and the underlying soil is highest in April, when it reaches 26°C at 8 A.M. and 48.5°C at 3 P.M. in droppings. In the soil, the thermal amplitude has a maximum of 32.3°C at the surface (between 7 A.M. and 2 P.M.), reduces to 16.4°C at a depth of 5.7 cm (between 7 A.M. and 4 P.M.), but is only 3.1°C at a depth of 45 cm (Fig. 13.2). Considering the average temperature (36.6 ± 5.9°C) and its amplitude (22.5°C) in droppings, zebu dung is relatively well buffered against ambient temperatures.

The main physico-chemical characteristics of Sahelian dung are the following:

Appearance. The dung consists of dissimilar, rough, and very fibrous com-

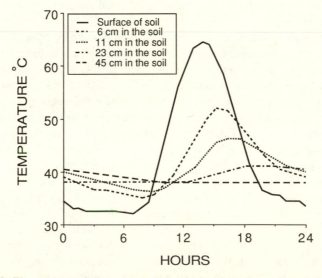

Fig. 13.2. Fluctuations of temperature during 24 hours on the surface and at different depths in the soil (Boudouresque and Rougon, unpubl., April 15, 1982).

ponents and has a less pasty consistency than in biomes with a less extreme climate, for example, in the Ivory Coast (Cambefort 1984).

Rate of production. A zebu weighing 250 kg consumes 6 kg of dry matter (straw of millet grass) per day and defecates 3 kg of dry fecal matter, or 10 kg of dung in fresh weight. A zebu defecates about ten times a day, giving an average weight of 1 kg per dung pat (Rougon 1987).

Humidity of dung. The average water content is $71 \pm 5\%$ (n = 88), which is much lower than in temperate (more than 82%; Frison 1967; During and Weeda 1973; Matthiessen and Hayles 1983) and in wet tropical regions (80%; Cambefort 1984). In the Sahel, dung loses all its water in nine days during the warm season in April; and after ten days, if not buried, the dung pat is like a stone. In the beginning and at the end of the rainy season in June and September, when it rains during the night, dung pats retain 45% of water after ten days.

Chemical composition. The Sahelian zebu dung is poor in nitrogen (about 1% of dry matter) and mineral elements, especially in calcium (0.05%). Two amino acids normally found in natural organic matters, b-alanine and Ornithine, are missing in zebu dung.

Availability of zebu dung to dung beetles. Following the harvest of millet grass, the most common crop in Niger (90% of arable land), an estimated 2.56 tons per hectare of straw remain in the fields and are available to zebu during the dry season. Digestibility of the straw is 50%, thus the potential quantity of dung is 1.26 tons per hectare per year (dry weight), or 4,400 droppings per hectare per year.

The Sahelian Dung Beetle Fauna

During the years 1978–79, 17,175 beetles were extracted from 240 zebu droppings (Table 13.1). Twenty families of beetles accounted for 93% of the total dung insect fauna. Among the beetles, Staphylinidae predominate (42% of individuals), followed by Histeridae (39%), Aphodiidae (11%), and Scarabaeidae (3%). Staphylinidae are represented by three subfamilies in the Sahel—Oxytelinae, Staphylininae, and Aleocharinae—of which the first has coprophagous species while the others are predators (Hanski and Koskela 1979). Two other beetle families, Lathridiidae and Anthicidae, are unusually well represented, while Hydrophilidae are relatively scarce (Table 13.1). A complete list of the dung beetles (Scarabaeidae and Aphodiidae), with ecological data, is given in Appendix B.13.

The activities of humans have affected dung beetles in two major ways in the Sahel. Since the frightful drought of 1968, the Sahelian governments have, for varied reasons, tried to settle the nomadic stockbreeders into specific areas. When the ancient practice of seminomadism disappears, large areas will remain ungrazed and the dung beetles will disappear. In addition, the increasing

TABLE 13.1
Numbers of beetles collected from 240 zebu droppings in Niamey, Niger, in 1978–79.

Family	Number of Species	Number of Individuals
Carabidae	4	30
Dytiscidae	1	2
Staphylinidae	11	7,186
Scydmaenidae	1	1
Hydrophilidae	4	205
Histeridae	8	6,771
Scarabaeidae	20	457
Aegialidae	1	10
Aphodiidae	20	1,953
Hybosoridae	1	13
Dynastidae	1	1
Cetonidae	1	1
Elateridae	1	2
Cucujidae	1	46
Lathridiidae	1	262
Coccinellidae	1	1
Anthicidae	1	221
Tenebrionidae	7	9
Chrysomelidae	1	3
Curculionidae	1	1
Total	87	17,175

use of airplane-dispersed insecticides in the fight against locusts and owlet moths and the use of nematicides against cattle parasites undoubtedly have harmful effects on dung beetles.

13.3. THE DUNG BEETLE COMMUNITY

Beetles were collected by depositing, once per month, 125 kg of fresh zebu dung in heaps of 1 kg. The 125 dung pats were placed in a line at one-meter intervals. Five pats and the underlying soil were removed after 1, 3, 5, and 10 days, and beetles were extracted by water flotation.

Seasonality

The numbers of predatory beetles, the Staphylinidae and Histeridae, peak in October. In November the numbers of insects in dung pats generally begin to decrease; they reach a minimum in the dry season, and begin to increase again in the beginning of the rainy season in June (Fig. 13.3). Scarabaeidae and

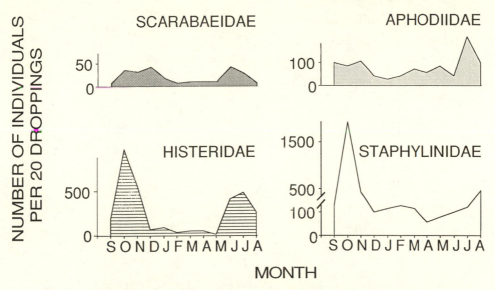

Fig. 13.3. Monthly changes in the numbers of individuals in four beetle families (from September 1978 to August 1979).

Aphodiidae show the same seasonal pattern, but have only a small decrease in the numbers during the dry season.

Food Web Relations in the Dung Insect Community

In July, freshly deposited dung pats attract many species of flies, mainly Sepsidae and Muscidae, seeking a place to deposit their eggs and larvae. Droppings attract flies for only two or three hours, however, because the rapidly developing hard crust prevents oviposition. Ovipositing flies are followed by coprophagous beetles, the Aphodiidae and Scarabaeidae, and by the first predators, the Staphylinidae and especially the Histeridae. Dung beetles reach their maximum numbers on the second (Aphodiidae) and third days (Scarabaeidae; Fig. 13.4). Scarabaeidae are entirely absent by the seventh day, while the numbers of Aphodiidae decrease more slowly. Dung beetles carry phoretic mites, which prey on the eggs and larvae of flies (Seymour 1980). Predatory beetles, the Staphylinidae and Histeridae, are the dominant beetles between the third and sixth days, reaching their peak abundance on the fourth day. Their main prey, fly larvae, reach their maximum numbers on the fifth and sixth days.

Beginning on the fifth day, a third wave of arthropods invades the droppings, which have, by then, lost their specific characteristics as a result of

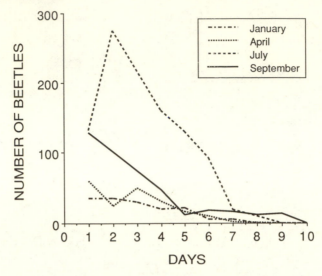

Fig. 13.4. Changes in the numbers of individuals of dung beetles (Scarabaeidae and Aphodiidae) according to the season and to the age of the dung pat.

desiccation and chemical changes. This third wave is comprised of species from the surface of the soil and from the soil itself: Entomobryidae, Isotomidae (Collembola), and Tenebrionidae, all of which are thought to be saprophagous; Formicidae, represented by the three genera *Pheidole* (granivorous and predaceous), *Monomorium* (omnivorous), and *Messor* (granivorous); and Carabidae and Araneidae, which are generalist predators.

In April, during the warm and dry season, the dung microhabitat features extreme climatic conditions, and the temporal sequence of colonization described above is greatly accelerated; the maximum numbers of dung beetles (four species of Scarabaeidae and *Nialus lividus*, Aphodiidae) are already reached during the first day (Fig. 13.4). The predatory Staphylinidae are dominant on the third day, while Histeridae are abundant from the third to the eighth days. From the third day onward, soil arthropods start to colonize the dung, which is rapidly drying out. Unlike in July, termites colonize droppings in great numbers in April, and they are by far the most abundant taxon beginning on the third day. Microorganisms do not appear to play an important role in the process of dung decomposition during the dry season, undoubtedly because of the rapid desiccation and hardening of the dung (see Merritt and Anderson 1977 for California).

13.4. POPULATION ECOLOGY OF DUNG BEETLES

Among the twenty species of Scarabaeidae collected in 1978–79 in Niamey, only five species accounted for more than 5% of the total number. Fig. 13.5

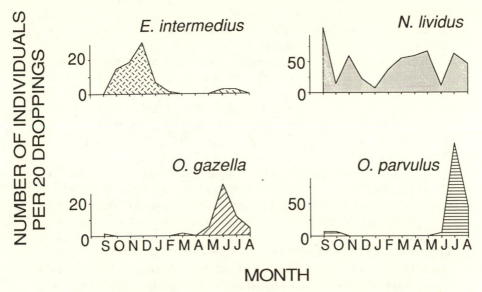

Fig. 13.5. Monthly changes in the numbers of four dominant species of dung beetles (from September 1978 to August 1979).

shows the monthly abundance changes in the two dominant species, *Euoniticellus intermedius* and *Digitonthophagus gazella*, which occur at different times of the year. Of the 1,953 individuals of Aphodiidae collected in 1978–79, only six out of twenty species were common, each representing more than 2% of the total number. Fig. 13.5 gives the monthly abundance changes in two species. *Nialus lividus* is the only Aphodiidae present in large numbers throughout the year, while *Orodalus parvulus* is a common species during the rainy season only.

*Adaptive Nesting Behavior in Sahelian
    Dung Beetles*

Studies on the nesting behavior of the most abundant species of Scarabaeidae, conducted both in situ and in the laboratory, have led to three main conclusions:

In the dry season on sandy soils, the galleries of dung beetles are shored up. In the Sahel, tunnelers shore up their shafts and the galleries leading to the nest with a stercorous matter, thus preventing them from caving in. Depending on the species, the thickness of the film of excrement used for this purpose varies from 1.1 mm in *Onitis alexis* to 1.5 mm in *Euoniticellus intermedius* and *Digitonthophagus gazella* (D. Rougon and C. Rougon 1980, 1982a,b; Rougon 1987).

Nests are stratified at different levels in the soil. The Sahelian dung beetles have evolved a stratified nesting behavior, which should decrease interspecific competition during the warm and dry season. The nests are tiered down to a depth of 50 cm beneath the soil surface.

Oniticellus formosus is a dweller active in the dry season. It makes spherical brood balls inside two or three cavities (5 or 6 cm in diameter), burrowed in large two-or three-day-old dung pats. Each cavity contains an average of nine cells (1.2 ± 0.1 cm in diameter and weighing 980 ± 120 mg, Fig. 13.6; Rougon and Rougon 1982b). The same type of nesting behavior has been described for this species in South Africa (Davis 1977; Bernon 1981). In the Sahel, the life cycle of *Oniticellus formosus* is 38 ± 2 days.

Three other common species—*Onitis alexis*, *Euoniticellus intermedius*, and *Digitonthophagus gazella*—build their nests at varying depths in the soil below the dropping. These three species bury a large amount of dung in a very short period of time; the dung will then be used to prepare the brood balls. *Euoniticellus intermedius* builds its nest only 3–8 cm beneath the soil surface. In the warm season the structure of the nest is of the "onitidian type," that is, access to the nest is by a shaft that penetrates the soil vertically for 3 cm, then branches off into two galleries parallel to the ground (Fig. 13.7). Along these galleries, six mounds of stercorous matter, each 3 cm wide, are deposited at a depth of 3—8 cm. Each mound is a miniature version of an *Onitis alexis* nest and is formed of four to six brood sausages, with a diameter of 7.5–8 mm, placed side by side (Rougon and Rougon 1982a). This compact structure protects the progeny from drought more effectively than would a nest in the shape of isolated ovoid cells. The life cycle of *Euoniticellus intermedius* is 28 days in the Sahel.

Digitonthophagus gazella constructs its nest at a slightly greater depth, 20–35 cm beneath the soil surface. The nest consists of a network of galleries with 13 ± 5 brood balls, each 2.6 ± 0.4 cm in length and 1.4 ± 0.3 cm in diameter (C. Rougon and D. Rougon 1980). The life cycle of this species is 41 days in the Sahel.

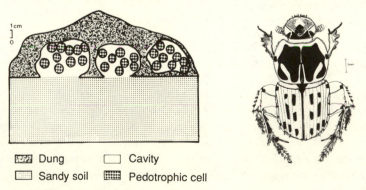

1 cm
]
0

| Dung | Cavity |
| Sandy soil | Pedotrophic cell |

Fig. 13.6. Nest of *Oniticellus formosus* Chevrolat.

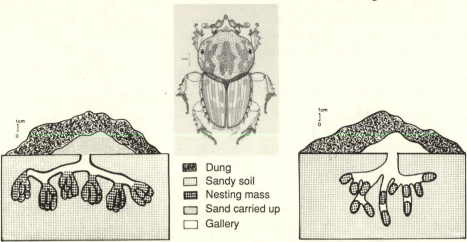

Dung
Sandy soil
Nesting mass
Sand carried up
Gallery

Fig. 13.7. Nest of *Euoniticellus intermedius* (Reiche) during the dry and warm season ("Onitidian type," on the left) and in the beginning of the rainy season ("Onthophagian type," on the right).

Onitis alexis, the largest dry-season dung beetle, builds a large nest at a depth between 28 and 46 cm. *Onitis alexis* always colonizes 1–48-hour-old dung at dusk. Its nest is reached via a perpendicular shaft, 1.6 cm in diameter and 30 cm in depth, which is shored up with a stercorous substance 1.1 mm thick. The nest has 204 ± 84 g of stercorous matter (after 4 days), comprising 7 ± 3 brood sausages parallel to each other or ramified and packed tightly together. Each roll contains two or three chambers with a single egg in each one (Rougon and Rougon 1982a). The life cycle of *Onitis alexis* is completed in 112 days on average, a very short period of time compared with the 2-year life cycle of *Onitis caffer* in South Africa (Oberholzer 1958). *Onitis alexis* plays a crucial role in dung burial in the Sahel: it buries, on average, a quarter of a dung pat weighing 1 kg.

The type of nest constructed depends on the season. *Euoniticellus intermedius* constructs two types of nest depending on the water content in the soil adjacent to the dung pat. In the warm and dry season, when the water content in the soil is less than 1%, *E. intermedius* builds a nest of the "onitidian type" (Fig. 13.7). In the rainy season, in contrast, the nest is of the "onthophagian type" (Fig. 13.7): access is by a shaft that opens to highly ramified galleries with one or several brood balls measuring 1.5 ± 0.2 cm by 0.8 ± 0.1 cm. During this season the higher water content in the soil (2–20%) prevents the balls from desiccating. Dung beetle nesting behavior is not fixed or unchanging; in semiarid regions it is flexible and adapted to the prevailing environmental conditions.

The remarkable flexible nesting behavior of *Euoniticellus intermedius* allows it to thrive all over Africa and may explain why this species has been so easy to introduce into other parts of the globe, for example, to Australia (Bornemissza 1976) and New Caledonia (Gutierrez et al. 1988; Chapter 16).

Kleptoparasitism in Aphodiidae

Four species of Aphodiidae—*Nialus bayeri*, *N. nigrita*, *N. lividus*, and *Mesontoplatys rougoni*—all apparently unable to survive the thermal and moisture conditions prevailing in droppings in the dry season, have become kleptoparasites of *O. alexis*, *E. intermedius*, and *D. gazella*. These Aphodiidae take advantage of the rapid (less than 48 hours) burial of dung by the larger beetles, and they ensure the development of their own progeny by using the subterranean dung reserves of the host species (D. Rougon and C. Rougon 1980, 1982a, 1983; Rougon 1987). Either the Scarabaeidae are passive carriers of the Aphodiidae adults when they descend to construct their nests, or the Aphodiidae actively seek out the dung buried by Scarabaeidae. Both processes probably occur.

The behavior of kleptoparasitic Aphodiidae has several distinctive features. The larvae move about actively within the nests and transport grains of sand, which obstruct the chambers where the larvae of *D. gazella*, *O. alexis*, and *E. intermedius* lodge. These grains of sand can damage the thin integuments of the host larvae, causing their death. The latter are often found desiccated in their chambers, reduced in size, and filled with grains of sand.

The larvae of Aphodiidae develop up to the third larval instar inside the host nest, but the pupal stage is spent outside the nest. *Nialus* larvae migrate in the third larval instar to the periphery of the nest of *O. alexis* or to the surrounding sand. The larvae construct small pupal cocoons composed of a mixture of sand grains and fragments of their excrement. The third instar larva of *M. rougoni* leaves the brood ball of *D. gazella* and burrows directly into the sand to form a small chamber 4 mm long and 2.5 mm wide, in which it pupates.

13.5. CONCLUSION

Ecological studies of the Sahelian dung beetles demonstrate that Scarabaeidae and Aphodiidae, though not represented by many species, have a relatively large biomass. To survive the extreme microclimate in zebu dung pats during the warm and dry season, dung beetles have evolved particular nesting behaviors: shoring up the galleries, stratification of the nests in the soil, flexible nest architecture, and kleptoparasitism.

By building shafts and subterranean galleries, dung beetles contribute to aeration of the soil and improve its structure and water circulation. Dung beetles in the Sahel play the role of earthworms in humid temperate and tropical biomes; the Sahelian soil, by its highly sandy texture and desiccation-prone structure, rules out the presence of earthworms. In the warm season, the abundant dung beetles play an essential role in dispersing organic matter over a wide area, thus preventing the Sahelian soil from becoming sterile. Field experiments (Rougon 1987) show that the nesting activity of dung beetles leads to a notable increase in the content of organic carbon ($\times$ 2), nitrogen ($\times$ 3), interchangeable bases ($\times$ 2), and humic acids in the soil. Without dung beetles, zebu dung becomes rapidly desiccated, and the potential organic enrichment to the soil is lost. Rougon (1987) showed that dung beetles are essential to the stability of the Sahelian ecosystem; hence it is necessary to preserve the association between agriculture and cattle rearing in the Sahel, and to enable the Peul cattle breeders to continue their seminomadic way of life.

CHAPTER 14

Montane Dung Beetles

Jean-Pierre Lumaret and
Nicole Stiernet

14.1. INTRODUCTION

Montane habitats have special ecological interest because of the predictable, systematic change in environmental conditions within short distances. The tree line comprises a natural ecological boundary between the lowlands and high altitudes, but its position depends on the montane system and the latitude. In the Alps, the tree line is located between 2,000 and 2,100 m on the southern and between 1,800 and 1,900 m on the northern slopes, while in the Himalayas it is at about 3,600 m (Mani 1968). Gams (1935) recognized four biotic zones above the tree line on the Central European mountains: the lower alpine zone, the upper alpine or grassy meadow zone, the subnival zone up to the closed meadows, and the nival zone. The three biotic zones below the tree line are the subalpine, the upper montane, and the lower montane zones. The elevational limits of the high montane habitats are not fixed, however, but they fluctuate in time, and the organisms have to respond to these variations. In the Alps, the snow line was located some 1,200 m below its present elevation at about 3,000 m during the Würm glaciation (20,000 BP). In the Himalayas, the snow line is at about 5,200 m.

Many dung beetles have been collected on high mountains all over the world, but ecological studies on them are scarce. Biogeographical data on montane dung beetles, including their elevational distributions, have been compiled and discussed by Horion (1950) for the Alps and the Carpathians, and by Reinig (1931), Mani and Singh (1961), Mittal (1981a), and Stebnicka (1986) for the Pamir, the Himalayas, and northwest India. Dung beetle distributions on mountains in Mongolia have been described by Nikolajev and Puntsagdulam (1984) and Grebenscikov (1985), and in Afghanistan and Iran by Petrovitz (1959, 1961, 1965, 1967) and Bortesi and Zunino (1974). Some ecological data on dung beetles on the Mediterranean mountains are included in Lumaret (1978), Martin-Piera (1982), Avila and Pascual (1981, 1986, 1987), Galante (1983) and Mesa (1985). Very little is known about dung beetles on tropical mountains (Hanski 1983; Morón and Terrón 1984; Halffter 1987).

In this chapter, we present an overview of the geographical and elevational distributions of dung beetles (Section 14.2) and describe the population ecology of high-altitude dung beetles, mostly *Aphodius*, in the Alps (Section 14.3 and 14.4). The rest of the chapter deals with montane dung beetle communities, species richness, and abundance distributions (Section 14.5). The focus of our discussion is the dung beetle community in the Swiss and French Alps, where recent quantitative studies have been carried out (Lumaret and Stiernet 1990).

14.2. DISTRIBUTION OF MONTANE DUNG BEETLES: AN OVERVIEW

The geographical distribution of montane dung beetles depends, like that of an insular fauna, on the present structure and the history of their discontinuous environment (Halffter 1987). Several examples of dung beetle distributions from various parts of the world demonstrate that mountain systems constitute both centers of differentiation and refuges for dung beetles. A case in point is *Agolius* and *Neagolius*, two alpine subgenera of *Aphodius*, which occur along the entire alpine arc and are distributed, respectively, from Spain to the Himalayas and from Spain to Central Asia. Most species and subspecies are endemics, patchily distributed and restricted to high altitudes, rarely collected below 1,800 m. Sexual dimorphism is often striking, and some females were first described as distinct species (Dellacasa 1983). An examination of morphological characters leads to the conclusion that *Agolius* and *Neagolius* are relicts with few affinities with other *Aphodius* (Baraud 1977).

The geographical isolation between the Pyrenees and the Alps, and between the different massifs of the Alps, has led to subspecific differentiation in many taxa. For example, *A. (Agolius) abdominalis* (=*A. mixtus*), a species widely distributed from Spain to Italy, is differentiated into four distinct subspecies. *Aphodius (Neagolius) schlumbergeri* is also divided into four subspecies, but numerous transitional forms occur in the Pyrenees between *A. schlumbergeri* s. str. and *A. schlumbergeri* ssp. *temperei*, including a cline in paramere shape (Baraud 1977). Apparently, *A. schlumbergeri* is in the active process of differentiation in the Pyrenees, both horizontally and vertically, while the two other subspecies in the Alps are more distinct because of their isolation.

Endemics may also differentiate from species with orophilous tendencies. In the western Alps, *Aphodius germandi* represents an alpine vicariant closely related to *A. obscurus*, a species widely distributed from Spain to Poland, Asia Minor, and the Caucasus. *Onthophagus baraudi* is the montane sister species of *O. joannae*, with a lowland to midaltitude distribution in western Europe. Another characteristic set of species in the Alps consists of species with boreo-alpine distribution. Most of them belong to the *Agoliinus* and *Agrilinus* subgenera of *Aphodius*, for example, *A. piceus*, *A. ater* (with several subspecies), *A. nemoralis*, *A. borealis*, and *A. tenellus*.

Halffter (1962, 1974, 1976, 1987) and Morón and Zaragoza (1976) have described distributional patterns in the montane dung beetles in Mexico. The mountains of western North America continue through the United States, Mexico, and Central America to merge with the northernmost spurs of the Andes. Mexico constitutes a crossing point where both Nearctic and Neotropical dung beetles may be found (Chapter 7). The occurrence of Nearctic species on the highest Mexican mountains is explicable by the recent transformation of a large part of the Mexican Altiplano to a desert; examples are *Ceratotrupes*, *Geotrupes*, *Copris*, *Onthophagus*, and *Aphodius*. A mixed, Nearctic–Paleo-American fauna dominates between 2,000 and 3,000 m, while all the dung beetles above 3,000 m are exclusively Nearctic (Halffter 1987). Species present in the alpine meadows between 2,980 and 3,050 m include *Geotrupes viridiobscura*, *G. sobrina*, *Copris armatus*, *Onthophagus chevrolati*, and *O. hippopotamus*.

Turning to the altitudes at which dung beetles occur on mountains, the maximum upper limit generally depends on the duration of snow cover and the length of the period when the average temperature exceeds 5°C. The maximum altitudes reached by dung beetles vary with latitude (Table 14.1). In the western Alps, *Aphodius* species are dominant and reach the permanent snow, whereas Onthophagini and Geotrupini do not exceed 2,400 m. In the high plateau of Iran and Afghanistan with its higher summer temperatures, Scarabaeidae are more numerous at high altitudes despite winter frost. In the Himalayas, several *Onthophagus* reach 5,000 m (Table 14.1).

14.3. BIOGEOGRAPHY OF THE ALPS

Lumaret and Stiernet (1984, unpubl.) have studied the ecology of dung beetles in the European Alps, in the French National Park of Vanoise, and in the Swiss National Park of Grisons. The Vanoise National Park (528 km²) lies at altitudes between 1,250 and 3,852 m. The relief is broken, with 107 peaks exceeding 3,000 m. Forty-five percent of the surface area belongs to the zone of vegetation mats and meadows, extending from the lowest limit of snow patches in July to the upper limit of animal life. Forests are very scarce (0.8%), while meadows and pastures account for 17% of the area. Human influence is negligible, except during summer, when sheep and cattle from adjacent valleys are grazed in the subalpine and alpine meadows up to 3,000 m. The park has a rich fauna of wild mammals, including wild goats (*Capra ibex*), chamois, marmots, and hares (Appendix A.14).

The Grisons National Park is smaller (170 km²) and stretches between 1,700 and 3,164 m in altitude. The abundance of wild mammals is high: 2,130 red deer, 1,230 chamois, 238 wild goats, and 57 roe deers were counted in the park in 1985. In addition, cattle, sheep, and a few horses (all outside the park limits) increase the abundance and diversity of dung in this area (Appendix A.14).

TABLE 14.1
High-altitude observations of dung beetles.

Locality (Latitude)	Altitude (meters)	Species
Swiss Alps, Grisons (46°30′N)	3,200	*A. abdominalis, A. obscurus*
	3,000	*Heptaulacus carinatus, A. alpinus, A. fimetarius*
French Alps, Vanoise (45°30′N)	3,000	*A. obscurus, A. germandi, A. alpinus, A. abdominalis*
	2,800	*A. satyrus, A. haemorrhoidalis, A. depressus, A. fimetarius*
	2,400	*Onthophagus fracticornis, O. baraudi*
Sierra Nevada, Spain (37°N)	2,600	*A. haemorrhoidalis, Geotrupes ibericus*
	2,500	*Sisyphus schaefferi, Gymnopleurus flagellatus, Euonthophagus gibbosus, O. fracticornis, O. vacca, A. scrutator, A. erraticus, A. fimetarius, A. sturmi, A. granarius*
	2,200	*Euoniticellus fulvus, E. amyntas, O. taurus, O. maki, O. similis*
Alai Valley, Pamir (37°N)	3,200	*G. impressus, A. nigrivitis, A. przewalskyi, A. vittatus mundus, A. distinctus, O. sibiricus*
Iran and Afghanistan (32°N)	3,200	*E. gibbosus, E. amyntas*
	3,000	*O. pygargus, O. turpidus*
	2,700	*O. taurus, Chironitis phoebus*
	2,450	*C. moeris, O. speculifer*
	2,200	*Scarabaeus carinatus*
	2,100	*Gymnopleurus flagellatus*
Sikkim, Tibet (30°N)	4,000–5,200	*O. cupreiceps*
	3,000–4,200	*O. tibetanus*
Western Himalaya (29°N)	4,200–5,000	6 *Aphodius* spp.
	3,600–4,200	18 *Aphodius* spp.
	3,100–3,600	19 *Aphodius* spp.
	2,400–3,100	26 *Aphodius* spp.
NW Himalaya (28°N)	3,660	*O. tibetanus*
	3,000–3,450	*Geotrupes kashmirensis*
	3,050	*E. gibbosus, O. sutlejensis*
Punjab (28°N)	3,800	*Caccobius denticollis*
Kashmir (26°N)	3,000	*Caccobius himalayanus*
	2,500–2,800	*Sisyphus indicus, Gymnopleurus opacus*
Colorado Mountains (38°N)	4,000	*Aphodius terminalis*
Mexican Altiplano (19°N)	3,000	*Geotrupes viridiobscura, G. sobrina, Copris armatus, O. chevrolati, O. hippopotamus*

Sources: Handschin (1963); Besuchet (1983); Lumaret and Stiernet (1990, unpubl.); Avila and Pascual (1987); Reinig (1931); Bortesi and Zunino (1974); Petrovitz (1961, 1965, 1967); Balthasar (1963); Stebnicka (1986); Mani and Singh (1961); Mani (1968); Halffter (1962, 1974, 1976); Morón and Zaragoza (1976).

The environmental conditions vary from the western to the eastern Alps. In the Vanoise, annual precipitation ranges between 860 mm (Bozel) and 1,100 mm (Pralognan) (Balseinte 1955). In winter, snow may lie 9 m thick on glaciers. Temperature decreases by about 5.5°C for an increase of 1,000 m in elevation. July and August are the warmest months, with maximum air temperatures ranging from 18°C (2,500 m) to 26°C (1,900 m) (Gensac 1978). Temperature fluctuations are great, and night frosts occur frequently, even in summer. In contrast, the soil temperature at a depth of 10 cm is more constant. In winter, the minimal soil temperatures scarcely decrease below 0°C, while in summer average temperatures are always 3°C higher in the soil than in the air, and exceed 20°C. The vegetative period (soil temperature exceeding 5°C) lasts for 5 months in the subalpine zone and for 3.5 months in the alpine zone (Gensac 1978).

In the Palearctic region, most of the alpine dung beetles belong to Aphodiidae, whereas on tropical mountains Geotrupidae and Scarabaeidae constitute a larger part of the fauna. This difference reflects the generally increasing dominance of Aphodiidae with increasing latitude (Chapter 4). For example, among the thirty-seven species recorded in the Vanoise National Park and its surroundings, Scarabaeidae account for only 18% of the species (six *Onthophagus* and one *Euoniticellus*) and Geotrupidae for 8% (*Geotrupes*, *Anoplotrupes*, and *Trypocopris*), while Aphodiidae include the remaining 73% (twenty-four *Aphodius*, two *Heptaulacus*, and one *Oxyomus*) (Lumaret and Stiernet, unpubl.). The results are very similar for the Grisons National Park, with thirty-six species (Handschin 1963; Besuchet 1983; Dethier 1985; Stiernet, unpubl.). Species richness in the Alps is much increased by the network of deep valleys, which make it easier for nonmontane, Medio-European species to penetrate the region. Many such species can be found on exposed slopes, for example, *Euoniticellus fulvus* and *Onthophagus joannae* in Vanoise up to 2,200 m, and *O. lemur* up to 1,900 m (Lumaret and Stiernet 1990).

14.4. POPULATION ECOLOGY OF HIGH-ALTITUDE DUNG BEETLES

This section is largely based on the results from the Vanoise National Park. Taking into account the relative abundances and seasonal dynamics of the twenty-four dung beetle species at the Vanoise study sites, three groups of species can be distinguished: (1) lower and upper montane dung beetles (two Geotrupidae, six Aphodiidae), which only exceptionally reach the subnival level; (2) midaltitude dung beetles (two *Onthophagus*, one *Heptaulacus*, and two *Aphodius*), all of which reach the subnival level (in this group *O. baraudi* is an endemic species); and (3) true high-altitude dung beetles, all *Aphodius*, with the elevational distribution often beginning in the lower montane or subalpine zone. Two subgroups of the high-altitude species may be distinguished: species that occur up to the upper alpine level (six species), and the ones that reach the subnival and nival levels (five species). Among these eleven typical high-altitude dung beetles, *A. (Neagolius) amblyodon* is an endemic and *A. (Agolius) abdominalis* is represented by its subspecies, *A. a. abdominalis*. *Aphodius germandi* is the high montane vicariant of *A. obscurus*, a polymorphic and dominant species widely distributed on exposed mountain pastures in central and southern Europe as far as the Caucasus.

Phenology of High-Altitude Dung Beetles

Many high montane species overwinter as adults, their emergence primarily depending on air and soil temperatures. Emergence typically begins when soil temperatures exceed 5°C, though a few adults often emerge earlier and can be seen walking on the still-frozen soil and snow (*A. fimetarius*, *A. obscurus*, *A. germandi*, *A. amblyodon*, and *A. abdominalis*). These species constitute the core of the spring community (Section 14.5). *Aphodius depressus* is a characteristic spring species, which may represent more than 60% of subalpine dung beetles at sites below 2,000 m. At even lower altitudes, *A. erraticus*, a species more resistant to high temperatures (Landin 1961), shows a similar phenology. *Aphodius fimetarius* is another characteristic species, with two generations per year, both males and females occurring from spring to the end of autumn in all types of dung. Under alpine conditions, only adults of this species hibernate, regardless of the altitude.

Other species spend winter as third instar larvae. On emergence the adults appear to be immature, with yellow or brownish appendices and tegument, as observed, for example, in *Aphodius haemorrhoidalis*, *A. rufipes*, and *A. satyrus*. *Aphodius haemorrhoidalis* is a typical summer species, which emerges in the beginning of June. It may be dominant at the lowest sites, and is always present in montane dung beetle communities. *Aphodius rufipes* emerges about one week later than *A. haemorrhoidalis*. It plays a significant role in the subalpine and lower alpine beetle communities because of its abundance and relatively large body size. *Aphodius satyrus*, an upper alpine and subnival species, may account for up to 40% of individuals in some summer dung beetle communities (e.g., Pont du Montet, 2,410 m). In the Vanoise National Park, *A. satyrus* occurs mostly at open sites that are not too dry, but in Grisons National Park, a more xeric region, this species is found mostly in forests. Finally, some montane dung beetles emerge only in the autumn, for example, *A. tenellus*, which may be dominant at the most humid and cold upper alpine sites.

Aphodius obscurus

Aphodius obscurus is a polymorphic species that reaches an altitude corresponding to the 0°C January isotherm (Lumaret 1978). It is often the dominant species, present in all montane dung beetle communities. In the upper alpine and subnival zones it may constitute more than 50% of all dung beetles; in Balcon du Montet, at 2,760 m, it comprised 96% of beetles in the autumn.

Aphodius obscurus is a spring and summer species at the lowest sites but it becomes a summer and autumn species at the upper ones. During the congregation period, *A. obscurus* oviposits under or in all types of dung, including manure at the resting places of cattle. Larvae may live inside large droppings, but they dig tunnels under small ones, which dry out quickly. Some other *Aphodius* species have similar habits under Mediterranean climatic conditions (Chapter 6). Pupation always takes place in the soil (2–6 cm depth).

At subalpine and lower alpine sites, *A. obscurus* has two generations per year, but above the upper alpine level, up to the subnival and nival levels, there is only one generation per year. The sex ratio is typically close to 50:50, but males are always in the majority in the beginning and at the end of the emergence period, with females dominant in July and August. Whatever the site, some individuals overwinter as adults (mostly males), while the rest overwinter as third instar larvae. The flexible phenology of *A. obscurus* may contribute to its dominance in montane dung beetle communities.

Aphodius amblyodon and A. abdominalis

Many species in the subgenus *Agolius* are saprophagous and breed in old dung pats and in decaying vegetable matter in the soil, though adult beetles can be found feeding in fresh cattle pats. *Aphodius amblyodon* and *A. abdominalis* are two examples, sometimes found together in the Vanoise (Table 14.2). Males are often observed walking on snow patches, perhaps attracted by reflected sunlight.

Aphodius amblyodon is a spring species, restricted to a few sites in the study region, though it may be abundant locally. The flight activity of males begins when large snow patches still cover the ground (June). Males are attracted to cattle pats in which they feed. Females, which are always short-winged, are never attracted to dung but remain in the soil. The ecology of *A. abdominalis* is similar. The density of this species may reach very high levels in the wettest places—for example, 19.9 ± 5.3 (SD) males per square meter at an altitude of 2,490 m in July 1987 (Lumaret and Stiernet, unpubl.). Like in *A. amblyodon*, very few females are attracted to dung, although they can fly. Most females remain in the soil, where they lay eggs in mold and decaying vegetable matter. Larvae develop in the soil and hibernate in the third instar. Larvae have been found in July under large stones deeply sunk into the soil and under old dung pats dropped in the previous autumn (Lumaret and Stiernet 1984).

Scarabaeidae

Few *Onthophagus* species reach high altitudes in the Alps, where the short summer does not permit them to complete their life cycle. Nonetheless, *O.*

TABLE 14.2
Elevational distribution of dung beetles in the Alps (Vanoise National Park).

Species	Site							
	1	*2*	*3*	*4*	*5*	*6*	*7*	*8*
Aphodius aestivalis	2							
Heptaulacus villosus	1	1						
A. putridus	142	1						
A. pusillus	1,001	97						
Geotrupes stercorarius	30	2	8					
A. rufus	155	3		1		1		
A. ater		45		1		1		
Anoplotrupes stercorosus	24	73		19		9		
Onthophagus baraudi	282	829				22		
A. corvinus	249	27	1			61		
A. erraticus	1,183	1,092	87		1	4		
O. fracticornis	439	1,074	160	1		120		
H. carinatus	367	229		4	10	2		
A. haemorrhoidalis	638	201	31	111	377	7	1	
A. fimetarius	269	156	615	23	20	30	3	
A. amblyodon			301	82				
A. depressus	3,774	147	60	230	85	67		
A. rufipes	464	85	488	61	10	2		
A. tenellus	17	28	571	146	95		2	
A. alpinus	40	118	702	861	971	867	168	1
A. satyrus	5	54	2	379	2,718	838	11	
A. germandi	6	45	2	3	20	1,061	141	23
A. abdominalis	1		5	17	77	145	154	
A. obscurus	873	2,119	891	1,644	3,081	17,674	5,637	92
Individuals	9,962	6,426	3,924	3,583	7,465	20,911	6,117	116
Number of species	22	21	15	16	12	17	8	3
Shannon-Weaver index	3.08	2.95	2.91	2.33	1.95	0.98	0.54	0.79
Equitability	0.69	0.67	0.75	0.58	0.54	0.24	0.18	0.49
Altitude (m)	1,750	1,960	2,000	2,230	2,410	2,440	2,760	2,960

Note: Sites: 1, Bessans; 2, Bonneval; 3, La Ramasse; 4, Grand Plan; 5, Pont du Montet; 6, Les Roches; 7, Balcon du Montet; 8, Col de Gontière.

baraudi and *O. fracticornis* occur in the Vanoise above the tree line. The former species emerges in spring, then disappears until the next year. *Onthophagus baraudi* probably hibernates in the adult stage, as very few immature individuals have been observed in spring. This species reaches the lower alpine zone (2,000 m), where it may represent 20% of the spring dung beetle community. *Onthophagus fracticornis* occurs at higher altitudes (up to 2,450 m, subnival zone), but mainly frequents the alpine zone. It has two peaks of activity, the first one corresponding to the spring emergence of overwintered adults, the other one being that of the new generation.

14.5. THE ALPINE DUNG BEETLE COMMUNITIES

The Dung Insect Community

On the European mountains, insects constitute an important food source for many predators. Large flying dung beetles, such as Geotrupidae, are taken by foxes around cattle pats, and moles dig tunnels below the pats to get to the insects. Ravens and choughs often scatter cattle dung and sheep droppings while searching for insects. Spiders catch small *Aphodius* that land on droppings.

Many types of droppings are present in the Alps. The colonization of cattle pats and red-deer droppings by dung insects has been studied in the Grisons (Stiernet, unpubl.). A large number of fresh cattle pats and red-deer droppings were placed simultaneously in the field and were subsequently sampled for thirty consecutive days. Staphylinidae were numerically dominant in deer droppings, while the species composition was more diverse in the larger cattle pats, with *Aphodius*, Hydrophilidae and Staphylinidae all well represented (Fig. 14.1). Flies were not scarce but the sampling method is not well suited for collecting them. When cattle pats and red-deer droppings become older, their attractiveness to insects decreases and most adult insects leave the droppings. At this stage *Aphodius* larvae are the dominant inhabitants of dung. On average, some 500 and 2,500 insects were attracted per kg (dry wt) of cattle dung and red-deer droppings, respectively, during the first nine days.

Species Richness of Dung Beetles

Eight sites ranging in altitude from 1,750 to 2,960 m were sampled in the Vanoise in 1985–87. At each site, ten dung-baited pitfall traps were set up, each consisting of a collecting pot covered with a screen supporting the bait, 200–250 g of fresh dung from the dominant livestock at the site (sheep droppings in subnival and nival zones, cattle pats at lower altitudes).

Species number varied from site to site but it remained high, around fifteen to twenty species, along the entire elevational gradient, excepting the highest sites, where species number was eight (2,760 m) and three (2,960 m; Table 14.2). The large number of species (seventeen) at site 6 (Les Roches) at 2,440 m, was more unexpected (Table 14.2), but this result is partly explained by the exceptional situation of this site, above site 2 (Bonneval). Most species at site 6 appeared to be immigrants, helped in their flight from the valley by air currents along the slopes of the mountain.

Density of dung beetles remained high at all altitudes, suggesting density compensation, with the species at the highest sites being more abundant, on

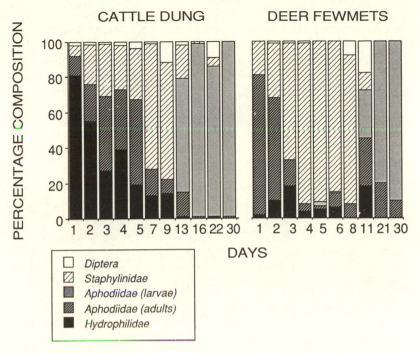

Fig. 14.1. Composition of the dung insect fauna in cattle pats and deer droppings varying in age.

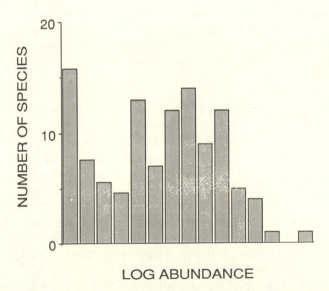

Fig. 14.2. Abundance distribution in the pooled material from the seven sites. The horizontal axis is logarithmic (the abundance classes are 1–2, 3–4, 5–8, etc.).

average, than species at lower altitudes. Hanski (1983) found a similar phenomenon on a tropical mountain in Borneo. A contributing factor may be decreasing competition between *Aphodius* and other coprophagous insects, flies, and Staphylinidae beetles becoming increasingly scarce with altitude. Only Hydrophilidae remain abundant along the entire gradient, occurring mainly in cattle pats. The marked decrease in the value of the equitability index with altitude (Table 14.2) indicates high degree of dominance by a few species at high altitudes, especially at the subnival and nival zones.

Abundance Distribution

Figure 14.2 shows the abundance distribution for the pooled material from sites 1 to 7. The abundance values are the numbers of beetles trapped at each site during one week per month from June to September (Table 14.2). The distribution is log-normal, apart from the peak of very rare species. Such bimodality suggests a natural dichotomy of species into the ones that have a breeding population at the site versus the others, the rare species, which are likely to be represented by immigrants from elsewhere.

Comparison between Communities at Different Elevations

Subalpine communities. Four successive waves of species replace each other from spring to autumn at the Bessans site, where eleven common species of the twenty-two local species form the core of the assemblage (Table 14.3). The percentage similarity of the species composition in successive months is only about 30%, indicating extensive seasonal turnover in species. Typically, each core group is comprised of species of different size. For example, in August the five core species have the following dry weights (mg): 1.3, 4.0, 4.6, 5.9, and 19.8.

The first wave of species in June consists mainly of three species—*Aphodius depressus*, *A. erraticus*, and *A. obscurus*—which represent 81% of the numbers and 85% of the biomass of beetles. The second wave in July also has three core species, of which only one was abundant in June (*A. erraticus*). This group accounts for 73% of beetles caught in July. Their numbers rapidly decline while the third wave of species emerges in August (Table 14.3), with twenty species and five core species, *A. rufipes*, *Heptaulacus carinatus*, *A. obscurus*, *A. haemorrhoidalis*, and *A. rufus* (79% of individuals and 80% of biomass). As most of the species active in August were already present in early spring, the summer community is stable in species composition in spite of great changes in dominance.

The autumn community appears suddenly (Table 14.3). Some of the autumn species were present at low density in summer, but their numbers rapidly

TABLE 14.3

Seasonal changes in the relative abundances of the core species in a subalpine community of dung beetles (Vanoise National Park in the French Alps, Bessans site, 1,750 m).

Species	June	July	August	September
A. depressus	60.6	2.8	2.7	2.2
A. erraticus	10.8	25.8	5.2	4.2
A. pusillus	8.1	30.3	1.3	0.3
A. haemorrhoidalis	3.0	16.8	10.6	2.4
A. obscurus	9.3	6.0	12.8	1.8
A. rufipes	0.2	3.6	24.0	2.8
H. carinatus	—	0.5	22.9	—
A. rufus	—	0.4	9.1	0.8
O. fracticornis	2.4	0.3	1.2	37.7
A. corvinus	0.8	0.8	2.0	21.7
A. fimetarius	0.8	0.3	4.5	20.1
Other species	4.0	12.4	3.7	6.0
Number of species	16	17	20	17

increase in September with the emergence of a new generation. Seventeen species were observed in September, but only three of them contributed to the core group: *Onthophagus fracticornis*, *Aphodius corvinus*, and *A. fimetarius* (80% of beetles). *Geotrupes stercorarius* constitutes an important functional species in September, with 55% of total biomass in spite of its low frequency (2.5%).

Lower alpine communities. Only three seasonal waves of species were recorded at the Bonneval site. Thirteen species are involved in the first wave with three core species. In July the core group consists of *Onthophagus baraudi* (15%), *O. fracticornis* (16%), *A. erraticus* (24%), and *A. obscurus* (24%). The autumn wave of species is relatively distinct as at lower altitudes, with eleven satellite and three core species: *O. fracticornis*, *A. obscurus*, and *A. fimetarius*.

Upper alpine, subnival, and nival communities. At mid- and high altitudes, dung beetles are scarce in spring (June) due to the late melting of snow, and the species number does not peak before late July. Species diversity then remains high until September, particularly in the nival communities. Two types of community organization can be distinguished in these communities: (1) species emerge in a rapid succession, but with little overlap between the peaks of emergence; the species tend to differ in size (upper alpine communities at La Ramasse and Grand Plan sites); (2) species emerge synchronously in summer, with prolonged activity in autumn (subnival and nival communities).

The species composition of the upper alpine communities does not change much from spring to early autumn, though the species composition may vary from one site to another. In La Ramasse, the early wave of dung beetles includes two dominant species, *A. amblyodon* and *A. obscurus*. These species are rapidly replaced in July and August by a stable community of eleven species, the three numerically dominant species representing 85% of the total. In September, this community collapses suddenly and the last wave of beetles emerges, dominated by *A. tenellus* and *A. fimetarius* (82% of total). The organization of the Grand Plan community at 2,230 m is similar to the previous one, in spite of the higher altitude, which delays the spring emergence of the earliest species. As soon as the snow has thawed, the summer wave suddenly emerges, with *A. obscurus* and *A. alpinus* being the two constantly dominant species, while a third one, *A. depressus*, is progressively replaced by the smaller *A. satyrus* with increasing altitude. The summer community disappears suddenly and is replaced by a new wave of dung beetles, among which *A. tenellus* is dominant.

Three main sites were studied in the subnival and nival zones. The thickness of the snow cover delays insect activity until the middle of June, sometimes later. The first dung beetles to emerge at the Pont du Montet site (2,410 m) are *A. obscurus* and *A. fimetarius*, represented by small numbers of individuals flying between large patches of snow. When the snow has completely disappeared, an explosive emergence of thousands of beetles can be observed within a few days (Appendix B.14). In early July, an average of 324 beetles was collected per pitfall trap in six days. Throughout the favorable period of activity, *A. obscurus* and *A. alpinus* (or *A. satyrus* depending on the site), form the core group. At the highest sites—for example, Balcon du Montet (2,760 m)—a community of six or seven species is dominated by *A. obscurus*, representing from 92% to 96% of individuals active at any time.

14.6. SUMMARY

At high altitudes, climatic constraints, in particular low temperatures and long-lasting snow cover, restrict the favorable season for dung beetles to a few months. High-montane dung beetles exhibit several adaptations to enable them to complete their life cycle in the short period of time available. One such adaptation is to be ready to rapidly exploit the resources when they become available in early summer. In many dung beetles this is achieved by overwintering in the adult stage. Species have typically very short breeding seasons, possibly because of competition with other common species. Short breeding seasons allow successive core species to develop during the favorable period, with those occurring at the same time being generally of different size. In spite of the harsh environment, the montane dung beetle communities appear to be highly structured, which may contribute to high species richness even at high altitudes, up to 2,200 m in the Alps.

CHAPTER 15

Native and Introduced Dung Beetles in Australia

B. M. Doube, A. Macqueen,

T. J. Ridsdill-Smith, and

T. A. Weir

15.1. INTRODUCTION

The Australasian land mass drifted apart from the rest of the Gondwana at the time when the dinosaurs were the dominant dung producers, 100 million years BP. During the past 25 million years, but excluding the last 200 years, the dominant herbivores and dung producers in Australia have been marsupial macropods with about fifty extant species (Tyndale-Biscoe 1971), together with the wombats in eastern Australia. The culmination point for dung beetles was the year 1788, when the arrival of Europeans and their livestock altered the herbivore complex dramatically and created an entirely new situation for dung beetles.

The current dung beetle fauna in Australia consists of two elements: the indigenous species, most of which are adapted to use marsupial dung (Matthews 1972, 1974, 1976), and a suite of species that have been introduced over the past two decades for the biological control of bovine dung and dung-breeding flies (Waterhouse 1974; Bornemissza 1976). The distribution of most indigenous species is given by Matthews (1972, 1974, 1976), but their biology and ecology are poorly known; they are of limited economic importance and are largely restricted to forest and woodland habitats. A few indigenous species are common in the open grasslands of southern Australia, where they use the dung of cattle, sheep, and horses. These beetles are probably now more abundant than they were prior to European settlement due to the increased supply of dung. In contrast, many of the introduced species are hugely abundant in open pastures, and some of these species have spread to most of Australia. Today, at least one introduced dung beetle occurs in practically every pasture in Australia.

In this chapter we review first the geographical distribution and habitat associations of the indigenous species (Section 15.2), then examine the attempt to establish an integrated complex of exotic dung beetles in Australian grasslands (Section 15.3), both in tropical and subtropical regions (Section 15.4) as well as in temperate regions (Section 15.5). The roles of biotic interactions and abiotic factors in determining the relative abundances of the species will be considered in Section 15.6. The invasion of dung beetle communities by new species is discussed in Section 15.7 in the light of the Australian experience.

15.2. THE INDIGENOUS FAUNA

There are 315 described species of indigenous dung beetles belonging to twenty genera in the family Scarabaeidae in Australia (Table 15.1). The indigenous fauna also includes 127 species of Aphodiidae and 40 species of Hy-

TABLE 15.1
Numbers of indigenous (N_{ind}) and introduced (N_{int}) dung beetle species in Australia and Tasmania.

Family Tribe Genus	N_{ind}	N_{int}	Tribe Genus	N_{ind}	N_{int}
Geotrupidae			Sisyphini		
Geotrupini			Neosisyphus	—	5 (3)
Geotrupes	—	1 (1)	Gymnopleurini		
Bolboceratini			Allogymnopleurus	—	1 (—)
Elephastomus	1	—	Canthonini		
Scarabaeidae			Canthon	—	1 (—)
Onthophagini			Aulacopris	3	—
Onthophagus	191*	10 (5)	Cephalodesmius	3	—
Onitini			Canthonosoma	3	—
Onitis	—	11 (6)	Amphistomus	18	—
Cheironitis	—	1 (—)	Labroma	3	—
Bubas	—	1 (1)	Mentophilus	2	—
Oniticellini			Coproecus	1	—
Liatongus	—	1 (1)	Tesserodon	8	—
Euoniticellus	—	4 (4)	Boletoscapter	2	—
Coprini			Aptenocanthon	3	—
Copris	—	6 (1)	Pseudignambia	2	—
Coptodactyla	11	—	Lepanus	21	—
Thyregis	4	—	Sauvagesinella	3	—
Dichotomini			Monoplistes	6	—
Demarziella	14	—	Diorygopyx	8	—
			Temnoplectron	10	—

Notes: Species are considered introduced if adults have been released in the field. The numbers in parentheses indicate the number of species known to have become established in Australia.

* Includes *O. depressus*, introduced accidentally from South Africa before 1900, and twenty species described by Storey and Weir (1989).

bosoridae, many of which are thought to be coprophages, as well as one coprophagous Geotrupidae (Carne 1965). Most species occur in the higher rainfall coastal regions, but a few species are present in the dry interior of the continent.

Some of the islands adjacent to the Australian mainland have simple dung beetle assemblages. Eight species of *Onthophagus* occur in Tasmania, which has been periodically isolated from the mainland, but only one of them is endemic (Matthews 1972). New Zealand has 14 species of Canthonini in two endemic genera as well as one introduced *Copris* (Matthews 1974). New Caledonia supports 20 species of Canthonini in eight endemic genera (Paulian 1986, 1987b), with one accidentally introduced *Onthophagus* and 3 other introduced species (Gutierrez et al. 1988). Vanuatu has 3 deliberately introduced species and one accidentally introduced *Copris* (Gutierrez et al. 1988). Norfolk Island has no endemic dung beetles but now supports an introduced fauna consisting of 3 species of *Onthophagus*, one species of *Euoniticellus*, and one species of *Onitis*. In contrast to these impoverished faunas on small islands, New Guinea and the offshore islands have 103 described species in nine genera in the tribes Onthophagini, Canthonini, and Coprini (Balthasar 1970; Paulian 1985; Krikken 1977; Weir, unpubl.). Seven of the genera and at least 7 species are shared with Australia, while a further 3 species have been introduced for the control of cattle dung.

Geographical Distributions

The major geographical constraint on beetle dispersal in Australia has been the central desert, which has isolated southwestern from eastern Australia. The different tribes of dung beetles differ markedly in their patterns of distribution over the Australian mainland. Onthophagini is represented by only one genus, *Onthophagus*, with 191 species (Table 15.1; Matthews 1972; Storey 1977; Storey and Weir 1989), most of which are restricted to the northern and eastern regions of Australia (see Fig. 12 in Matthews 1972). Eight of the 9 species in southwestern Australia are endemic.

The tribe Coprini includes two genera and 15 species, while Dichotomini has 14 species (Table 15.1; Matthews 1976; Matthews and Stebnicka 1986). *Demarziella* (14 species) and *Coptodactyla* (11 species) occur largely within several hundred kilometers of the northern and eastern seaboards of Australia. The genus *Thyregis* is represented by 3 species in southeastern Australia and one on the western coast.

The tribe Canthonini has sixteen indigenous genera with 96 species (Table 15.1; Matthews 1974; Storey 1984, 1986). Four genera (*Labroma*, *Coproecus*, *Mentophilus* and *Sauvagesinella*) are restricted to western Australia (WA), while most of the species in *Tesserodon* and *Monoplistes* occur in the summer rainfall regions of northern Australia. Three species of *Aulacopris* and *Lepanus* occur in Victoria with distributions extending to New South Wales (NSW). No indigenous Canthonini have been recorded in South Australia (SA).

Most of the species in the remaining genera fall into one of four broad distribution patterns: (1) coastal NSW and southern Queensland (*Diorygopyx*, *Cephalodesmius*, *Aulacopris*, and *Aptenocanthon*), with relict distributions in *Aulacopris* and *Aptenocanthon* in high-rainfall montane regions in north Queensland (Storey 1984, 1986); (2) Queensland (*Pseudignambia*, *Canthonosoma*, and *Boletoscapter*); (3) coastal Queensland and Northern Territory (NT) (*Temnoplectron*); and (4) widespread along the coast from NSW to NT (*Lepanus* and *Amphistomus*).

The Australian fauna shows a high level of endemicity at both generic and specific levels, which, together with the presence of archaic groups in the Coprini and Canthonini, gives the fauna a strongly insular aspect, unlike that of other regions of comparable size but reminiscent of the endemicity on some continental islands (Matthews 1976). One of the two genera of Coprini and eleven of the sixteen genera of Canthonini are restricted to Australia (Matthews 1974, 1976). Most of the Australian species of the cosmopolitan genus *Onthophagus* are endemic. Matthews (1972) considers that *Onthophagus* may have originated in Africa, where there are over 800 species, and to have entered Australia from the north (Chapter 4).

Habitat Selection, Food Preferences, and
Seasonal Activity

The majority of the Australian indigenous dung beetles are associated with woodland or forest (Matthews 1972, 1974, 1976; Donovan 1979). In north Queensland, Ferrar (1975) found that indigenous species were rare in cattle dung in open pastures, though there are some *Onthophagus* that occur in pastures and open savannas in northern (e.g., *O. capella*, *O. consentaneus*, and *O. rubrimaculatus*) and central Australia (e.g., *O. sloanei*: Matthiessen et al. 1986). In southern Australia, a few species (e.g., *O. ferox*, *O.capella*, *O. mniszechi*, *O. granulatus*, and *O. australis*) are common in open pastures as well as in woodlands, and in some years they become seasonally abundant and cause substantial disruption of cattle dung pats (Hughes 1975; Donovan 1979; Tyndale-Biscoe et al. 1981; Ridsdill-Smith and Hall 1984a). In contrast to Scarabaeidae, many indigenous species of Aphodiidae appear to be scarce in dense forest. For example, Williams (1979) and Williams and Williams (1982, 1983a, 1983b, 1983c, 1984) extensively surveyed the small wet forests of coastal NSW and found numerous species of Onthophagini and Canthonini but caught only small numbers of a few species of Aphodiidae and Hybosoridae. Ridsdill-Smith et al. (1983) found twenty to forty times more Aphodiidae in open forest (jarrah) and heath than in tall open forest (karri). In southwestern Australia, six species of *Aphodius* occur in pastures though only the accidentally introduced *A. pseudolividus* is abundant (Ridsdill-Smith and Hall 1984a).

The food preferences of the indigenous species have received little systematic attention, except for the work of Donovan (1979), who trapped six *Onthophagus* species using five different types of bait. He found that traps baited with human dung caught by far the largest numbers of beetles, while the other baits were less attractive, the order of decreasing preference being horse, sheep, cattle, and kangaroo dung. Matthews (1972, 1974, 1976) concluded that mammalian entrails are the most effective bait, human dung is less attractive, and herbivore dung is relatively ineffective. The large majority of species appear to be coprophagous, but other types of food are also consumed, and a small number of *Onthophagus* species are copro-necrophagous. Fourteen species of *Onthophagus* appear to be restricted to mushrooms, and one species feeds on fallen fruits (Matthews 1972; Storey 1977; Storey and Weir 1988). Ridsdill-Smith et al. (1983) noted that in southwestern Australia, species of *Onthophagus* and *Aphodius* prefer dung over carrion, while Canthonini are equally attracted to both types of bait.

Some *Onthophagus* species have morphological and behavioral adaptations that facilitate their access to fresh macropod dung. These beetles have prehensile tarsal claws by which they hold on to the fur of macropods, especially in the cloacal region. As a fresh dung pellet is extruded, the female beetle seizes it and falls to the ground with it. The pellet is then buried and provides food for one larva (Matthews 1972).

Seasonal changes in adult activity have been documented for a number of indigenous species (Tyndale-Biscoe et al. 1981; Ridsdill-Smith and Hall 1984a,b). These results conform to the usual pattern, with activity restricted to the moist seasons and being minimal during dry periods. Weir (unpubl.) found similar patterns among the Coprini in the forest regions around Cooktown in north Queensland, but there were some *Onthophagus* species that were most abundant during the dry season. Abundances can vary enormously from one year to another. For example, at Uriarra near Canberra, *O. granulatus* numbered from three hundred to three thousand beetles per trap in 1976–77, but during the next two years the catches rarely exceeded hundred individuals per trap. Such abundance fluctuations were explained by drought and associated poor-quality dung in critical times of the year (Tyndale-Biscoe et al. 1981).

The Effect of Human Colonization

Aboriginal colonization of Australia dates back at least 50,000 years, and it is apparent that the early colonists wrought substantial changes on the Australian landscape, primarily through the use of fire, hunting and the introduction of the dingo (Singh and Geissler 1985; Singh et al. 1981). Fire altered the patterns of vegetation distribution. The extinction of the large grazing and browsing herbivores (e.g., diprotodons and giant macropods) about 10,000 BP (Tyndale-Biscoe 1971; Owen-Smith 1987) reduced the types of dung available. These changes may have had severe consequences to indigenous dung beetles with narrow habitat and dung-type preferences.

The European colonization of Australia over the past two hundred years has caused equally or more dramatic changes to the landscape through the clearing of bushland for pasture, agriculture, and settlement, and through the introduction of exotic pasture plants and domestic mammals. The total area of many habitat types has severely contracted, thus restricting the distributions of their occupants, including the dung beetles. Large tracts of grazing land are now populated with cattle, horses, and sheep, but only a few species of indigenous dung beetles occupy this habitat. The CSIRO Dung Beetle Program is attempting to redress this ecological imbalance by importing exotic dung beetles that are adapted to use the dung of domestic cattle in open pastures.

15.3. THE AUSTRALIAN EXOTIC DUNG BEETLE FAUNA

Domestic animals were brought to Australia by the early European settlers. As animal numbers increased, a major problem arose with the accumulation of dung. Currently cattle are estimated to produce 350–400 million pats per day (Waterhouse 1974). In the absence of dung beetles, cattle pats foul valuable pasture for prolonged periods of time in the more intensively grazed areas, and the fiber and nutrients locked up in dung pats are not rapidly incorporated into the soil to maintain its fertility. Furthermore, two important fly pests in Australia breed almost exclusively in cattle dung: in tropical areas the blood-feeding buffalo fly, *Haematobia irritans exigua*, which infests cattle in large numbers; and in inland and southern areas the ubiquitous bush fly, *Musca vetustissima*, which pesters man and beast alike. The aim of the CSIRO Dung Beetle Program has been to establish in the Australian pastures a dung beetle fauna that would rapidly bury the dung of the domestic stock (mainly cattle), and thereby benefit pasture production and reduce the numbers of pest flies.

The selection procedure to secure the dung beetles has involved two phases. First, the climates of the Australian target areas were matched with those of the prospective donor regions overseas. The climatic maps of Walter and Lieth (1964) were used to identify regions with similar temperature and rainfall patterns (Fig. 15.1). Further considerations of the diversity of foreign dung beetle faunas led to the selection of parts of Africa and Europe as donor regions (Bornemissza 1979; Ridsdill-Smith and Kirk 1985; Kirk and Ridsdill-Smith 1986; Doube 1986). Tropical southern Africa was the chief source of species

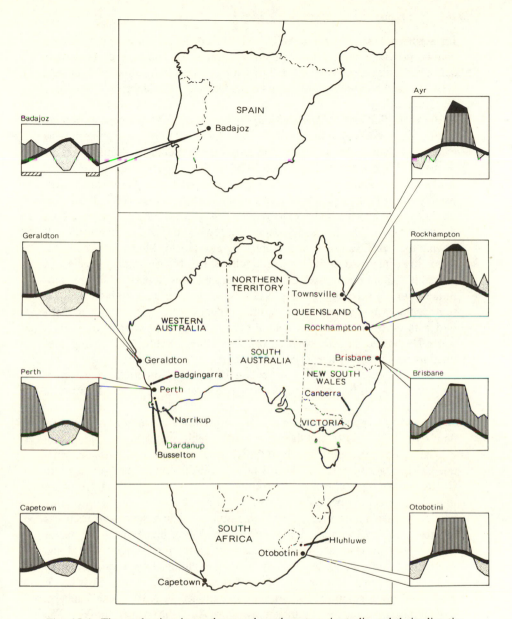

Fig. 15.1. The study sites in northern and southwestern Australia and their climatic diagrams. Overseas sites with homologous climates are shown for comparison. The climatic diagrams (after Walter and Lieth 1964) show mean monthly temperature (thick line) and rainfall on a common scale; dry periods (precipitation < evaporation) are shown as dotted areas and moist periods (precipitation > evaporation) are shown by vertical lines. Black areas denote wet months receiving rainfall in excess of 100 mm, on reduced scale (1 – 10).

for northern Australia, while the Cape Province (South Africa) and southern Europe provided species for southern Australia. The initial criteria used to select particular species were listed by Bornemissza (1976), and these criteria were progressively refined as a greater understanding of dung beetle biology was acquired.

Between 1968 and 1982, surface-sterilized eggs of fifty-two species of dung beetles were received at the CSIRO quarantine laboratories in Canberra. The eggs were placed into cavities in specially prepared dung balls, in which the beetles developed to the adult stage (Bornemissza 1976). Either the reared adults or their adult progeny were released in the field. The releases were usually conducted in such a way that possible interspecific competition was minimized. Where possible, individuals of the introduced species were confined in crude field cages and supplied with fresh dung for up to a week following their release to encourage breeding at one site. Most of the early releases in northern Australia involved multivoltine, high-fecundity species, of which 300–500 insectary-reared beetles were released at each of several sites in areas with a favorable climate. These species often built up large populations within a few years of release; their progeny were collected for redistribution to other areas to assist natural dispersal. A total of forty-one species was released and, to date, twenty-two species have established breeding populations in the field (Table 15.1). The reasons for the failure of establishment of some species are discussed in Section 15.7 (some imported species were not released at all because it proved impossible to rear them satisfactorily in the laboratory).

The early successful establishment of beetles has been summarized by Bornemissza (1976, 1979), and the progressive establishment and spread of species have been referred to by other CSIRO workers (papers by Wallace, Macqueen, Tyndale-Biscoe, Hughes, Doube, Ridsdill-Smith) and detailed in CSIRO annual reports. However, no comprehensive ecological analysis of the introductions has been published so far. Currently at least one exotic dung beetle occurs in almost every Australian pasture, and up to seven species occur together in a few localities. Some species occur over much of the continent, while others are restricted to the northern or southern regions. Tropical and subtropical study sites at which dung beetle populations have been monitored include Townsville, Rockhampton, and Brisbane in Queensland (Fig. 15.1), while temperate communities have been examined in NSW, near Canberra (cool temperate), and near Perth, southwestern Australia (winter rainfall). The next two sections examine the situations in Queensland and southwestern Australia in more detail.

15.4. BEETLES OF SUMMER RAINFALL REGIONS

The first species to be introduced to Australia were released in 1968 at several sites in coastal Queensland (Fig. 15.1), where the dung-breeding buffalo fly is a serious pest of cattle. Populations of these beetles became established in numerous localities, and they were monitored with pitfall trapping over a number of years at three of the original release sites near Townsville, Rockhampton, and Brisbane. Indigenous Australian species were scarce in open pastures, and relatively few were caught in the pitfalls. At Rockhampton in coastal central Queensland, seven introduced species have so far established breeding populations, of which six species were present in 1986 at the end of the monitoring period (Table 15.2).

Seasonal Abundance and Activity

The same broad pattern of seasonal activity has been apparent at each of the three monitoring localities, of which data are given for Rockhampton in Fig. 15.2. Adults of all species are relatively common during the warmer months of the year, and they are scarce during the cool, dry winter period, in July and August (Fig. 15.2). Two species, *Sisyphus (Neosisyphus) spinipes* and *Onitis viridulus*, emerge in large numbers earlier in the season than the others. *Onthophagus (Digitonthophagus) gazella*, *Euoniticellus intermedius*, and *S. spinipes* show typically one or more periods of exceptional abundance each season, when up to several thousand beetles may be caught per trap in twenty-four hours. Such mass occurrences, following the emergence of new adults, sometimes persist for several weeks. The patterns of seasonal change in these species have seldom been synchronous at the three localities, probably because of some differences in local rainfall patterns.

In coastal central Queensland, breeding by the tunnelers was studied by counting the number of broods excavated from beneath dung pats in the field. Breeding occurred primarily in spring, summer, and early autumn, but the species showed some differences in timing within breeding seasons (Fig. 15.2). *Euoniticellus intermedius* can breed throughout the year, albeit sporadically during the cooler months. The other species terminate breeding in autumn but the adults may survive for many weeks, depending on the severity of the winter. In mild winters some adults survive to breed in the following spring.

Larvae develop slowly to fully fed third instars during the winter and early spring, but they do not pupate until they receive an environmental stimulus. Some species (*S. spinipes* and *E. intermedius*) respond largely to increasing temperatures in the spring, since pupation and adult emergence can occur in the absence of rain. In others (e.g., *O. gazella*), the first emergence of new adults in the spring does not occur until there is adequate soil moisture (Macqueen, unpubl.).

Seasonal changes in beetle activity lead to corresponding changes in the level of dung dispersal, which has two components: a part of a pat is actually removed by beetles, while another part or all of the remainder is shredded into fragments but not removed (Fig. 15.3). Broad agreement was found between

TABLE 15.2
Abundance of introduced dung beetles at three localities in coastal Queensland.

Species	Year														
	71	72	73	74	75	76	77	78	79	80	81	82	83	84	85
Townsville															
O. gazella			61	53	40	27	28	29	12						
E. intermedius			—	1	2	40	21	16	5						
L. militaris			—	1	16	68	104	39	13						
O. sagittarius			—	—	—	*	1	2	1						
O. alexis			—	—	—	—	—	—	1						
Total			61	55	58	135	154	86	32						
Rockhampton															
O. gazella	46	28	50	70	81	50	48	146	30	90	137	81	99	98	242
O. sagittarius	3	3	37	30	22	11	6	12	3	3	2	1	1	3	3
L. militaris	—	—	2	6	11	29	15	66	12	26	94	5	96	46	77
E. intermedius	—	—	—	*	16	26	52	177	28	122	131	121	689	497	546
S. spinipes	—	—	—	*	7	19	17	246	19	39	36	6	792	1,208	490
O. viridulus	—	—	—	—	—	—	—	—	—	1	1	*	6	5	5
S. rubrus	—	—	—	—	—	—	—	—	—	—	—	—	—	—	5
Total	49	31	89	106	137	135	138	647	92	281	401	214	1,683	1,857	1,368
Brisbane															
O. gazella			191	113	90	82	179	52	126						
O. alexis			—	*	1	1	5	7	10						
E. intermedius			—	—	—	27	183	147	52						
S. spinipes			—	—	—	1	7	18	6						
Total			191	113	91	111	374	224	194						

Notes: The figures are mean numbers of beetles collected per trap in each year from September 1 to June 30. Species are listed in the order of establishment at the site.

* <1.

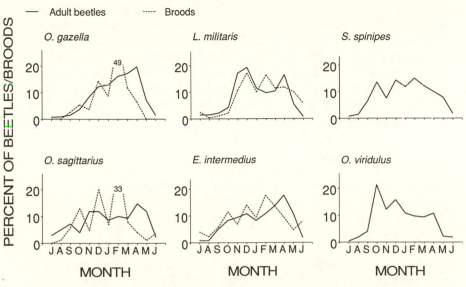

Fig. 15.2. Seasonal changes in abundance and breeding of dung beetles in grassland grazed by cattle at Rockhampton. The monthly average numbers per trap are expressed as percentages of the annual sum of the monthly averages for each of six species trapped in 1975–85 (solid line). The seasonal patterns of breeding in 1975–79 (dashed line) are based on the mean number of broods recovered per pat, expressed as percentages of the sum of the means (no equivalent data are available for *S. spinipes* and *O. viridulus*).

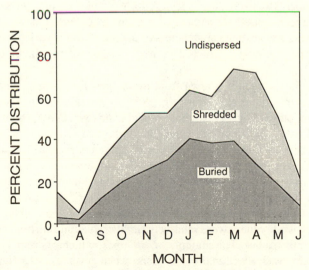

Fig. 15.3. Seasonal changes in the level of dispersal of dung pats at Rockhampton. This figure is based on results for eight years (1975–85).

the level of reproductive activity (numbers of brood balls produced) and the amount of dung dispersed, but there was no clear relationship between the numbers of beetles collected in baited traps and the level of dung burial or shredding (Doube, Giller, and Moola 1988). The proportion of the pat removed by beetles after seven days was independent of pat volume in the range from 0.5 to 3.0 liters, so that larger pats had a greater mass of dung remaining (Macqueen, unpubl.). This has important implications for the other members of the dung insect community.

Community Structure

The Australian experiment provides a unique opportunity to study the development of dung beetle assemblages by serial introduction of species. The observed abundance changes in the introduced species in three localities in coastal Queensland are shown in Table 15.2.

The most comprehensive data set, covering fourteen years, is available for Rockhampton. *Onthophagus gazella*, *Onthophagus (Digitonthophagus) sagittarius*, and *Liatongus militaris* were first released in 1968–69; *Euoniticellus intermedius* arrived by natural spread, *S. spinipes* was released in 1972, and *O. viridulus* in 1977. The first year of exceptional overall abundance occurred in 1978–79, following substantial rain in winter and spring of 1978. Beetles were also very abundant during the years 1983–86. However, the performance of different species has varied markedly, and there is no one pattern common to all species. Some species, for example, *O. gazella* and *L. militaris*, show no obvious trend over the study period, while *O. sagittarius* has declined markedly since 1978–79; other species, for example, *E. intermedius* and *S. spinipes*, have increased dramatically during the latter part of the study period. The rank order and absolute abundances of the species have varied widely over the seven-year period for which data are available for the three study sites (Table 15.2).

Two conclusions can be drawn from these data. First, some species have been consistently more abundant (e.g., *S. spinipes*, *E. intermedius*, and *O. gazella*) than others (e.g., *O. sagittarius* and *Onitis viridulus*), though no one species has been consistently dominant at all three localities during the years examined, despite substantial climatic and biological similarities among the sites (open grassland grazed by cattle on loam or clay soil with stocking rate ranging from 0.1 to 0.8 beasts per hectare). Second, there have been substantial year-to-year differences in the mean abundance of individual species as well as in the overall numbers of beetles. Some years were highly favorable for breeding while others were relatively unfavorable, and different years favored different species. For example, *S. spinipes*, which places its broods in grass tussocks, was scarce in drought years, when grass tussocks were grazed short and the broods were exposed to the sun and often trampled by cattle. In contrast, up to twelve thousand beetles were collected per pitfall trap in twenty-four hours in other years with high rainfall and/or low stocking rate, when the pastures grew tall and dense. In some years, *S. spinipes* has been scarce on overstocked, denuded pastures but abundant in adjacent, well-grassed pastures. At Rockhampton, *O. sagittarius* decreased in abundance following the establishment of additional species (Table 15.2). Tyndale-Biscoe (pers. comm.) observed a similar decline in the numbers of *E. intermedius* at Narrabri (NSW) following the establishment of *O. gazella*, and of *Onitis alexis* in the Araluen Valley (NSW) following the establishment of a number of species, including *Onthophagus binodis*. The explanation of these trends is not clear, but the declines of *O. sagittarius* at Rockhampton and of *E. intermedius* at Narrabri have been attributed to increased interspecific competition, especially with *O. gazella*.

Beetle Abundance in Africa and Australia

Because the nature of the habitat is a major determinant of dung beetle abundance (Nealis 1977; Doube 1983), comparisons of abundances between continents require that beetles are sampled in similar habitats. The data in Table 15.3 are based on beetles trapped during the active season over three years, in

TABLE 15.3

The abundance of five species of African dung beetles in grassland grazed by large bovids (pastoral—cattle; game reserve—Cape buffalo) in the Hluhluwe district of Natal, South Africa, and near Rockhampton, Australia.

Species	Africa						Australia		
	Hluhluwe Game Reserve			Hluhluwe Pastoral			Rockhampton Pastoral		
	1982	1983	1984	1982	1983	1984	1982	1983	1984
O. gazella	5	3	1	1	2	1	81	99	98
L. militaris	7	13	3	7	9	7	5	96	46
E. intermedius	13	14	1	18	28	4	121	689	497
S. spinipes	1	8	*	1	3	1	6	792	1,208
O. viridulus	—	—	*	—	—	*	—	6	5
Total	26	38	6	27	42	14	213	1,682	1,854
Percentage of all beetles trapped	44	11	12	43	20	25	100	100	100
Number of traps	108	81	42	108	81	42	78	78	72

Notes: Figures are mean numbers of beetles caught per trap during the period September 1 to June 30. Pitfall traps were baited with fresh cattle dung and the beetles were collected after 24 hours.

* <1.

three localities with similar climates and extensive tracts of clay-loam grassland grazed by large ruminants. The main result is that the introduced beetles at Rockhampton in Australia have been many times more abundant than populations of the same species near Hluhluwe, South Africa: *O. gazella* × 70, *Liatongus militaris* × 10, *E. intermedius* × 50, *S. spinipes* × 400, and *O. viridulus* × 390. The densities of these species in Hluhluwe were of the same order of magnitude than in other years and in other, comparable localities in South Africa (Doube, unpubl.). Even more striking than the differences in the abundances of these species is that the total biomass of all dung beetles was ten to one hundred times greater in Rockhampton than in Hluhluwe. Since the same type of bait and traps were used in both localities, we conclude that the introduced species breed much more successfully in coastal regions of northeastern Australia than they do in homologous areas of South Africa.

These differences are also reflected in the level of dung dispersal observed in the two localities. Doube, Giller, and Moola (1988) found that there was almost complete dispersal of dung at Rockhampton in the summer of 1981–82, whereas in parallel studies conducted in the Hluhluwe Game Reserve the level of dung dispersal was found to be 53% in grassveld and only 32% in bushveld. Additional unpublished data collected from 1980 to 1986 in Africa indicate that the level of dung dispersal observed in clay-loam soils inside and outside the game reserve is consistently lower than that observed in the northern Australian regions where exotic dung beetles are abundant. In contrast, the level of dung dispersal during summer is extremely high on deep sandy soils in the summer rainfall regions of South Africa (Doube, Giller, and Moola 1988). The reasons for the relatively low biomass of dung beetles in the African clay-loam systems are not known but may be related to predation on buried dung by termites (Chapter 8).

15.5. BEETLES OF WINTER RAINFALL REGIONS

One of the main aims of dung beetle introductions to southwestern Australia was to reduce the abundance of the pestilent bush fly. This fly shows a short period of rapid population increase in the spring (Matthiessen 1983), when few beetles are active at present (Ridsdill-Smith and Hall 1984a). The introduction of exotic dung beetles has increased the level of dung dispersal and has extended the time over which that dispersal occurs, particularly in summer and autumn, thus shortening the bush fly season substantially (Ridsdill-Smith and Matthiessen 1988).

Nine of the fourteen species introduced from southern Africa and Europe have become established so far. The first beetles were released in southwestern Australia in 1972. Three African species, *Onitis alexis*, *Euoniticellus intermedius*, and *Onthophagus binodis*, were present in substantial numbers by 1978, and the other species now established are *Onitis aygulus* and *Onitis caffer* from South Africa; *Onthophagus taurus* from Greece, Spain, Italy, and Turkey; *Euoniticellus pallipes* from Iran and Turkey; *Euoniticellus fulvus* from France; and *Bubas bison* from France and Spain. The distribution and basic biology of these species are described by Ridsdill-Smith et al. (1989). Of the nine established species, all except *B. bison* are also established in southeastern Australia. Introduced dung beetles now occur in most of the areas where cattle are present, though the distribution of some species is still very restricted (Ridsdill-Smith et al. 1989). The dung beetle fauna of this region has been increased substantially from eighteen to twenty-seven species, but Ridsdill-Smith and Matthiessen (1984) and Ridsdill-Smith and Kirk (1985) argue for the introduction of further species that would have their main period of activity in spring, to coincide with the critical fly-breeding season.

Seasonal Abundance and Activity

The seasonal occurrences of the native species and most of the introduced species are markedly different. The native species show a bimodal pattern of seasonal activity in the cooler regions but a unimodal pattern in the warmer regions. At Busselton, 200 km south of Perth, the endemic *Onthophagus ferox* shows a bimodal pattern with substantial adult activity in spring and autumn, when the soil is moist and the temperatures are sufficiently high; but its seasonal occurrence becomes unimodal in the warmer regions north of Busselton, where beetles are active throughout late autumn and winter (Table 15.4).

In contrast to the bimodal pattern in the native species, the introduced species *O. binodis* and *E. pallipes* are active during the dry summer and autumn months, and their seasonal activity is not affected by soil moisture, as the results are the same for localities where pastures are irrigated (Dardanup) and where they are not (Busselton; Table 15.4). Nonetheless, soil moisture is likely to affect the abundance of these species. Barkhouse and Ridsdill-Smith (1986) have shown that in dry, sandy soils breeding by *O. binodis* is restricted to the damp soil beneath the pat; hence at higher densities intraspecific competition for breeding space should be more intense in dry than in moist soil.

The situation in southwestern Australia bears a remarkable similarity to the situation in the Mediterranean regions of southern Africa (Davis 1987), where the dung beetle fauna consists of two elements. One is a group of southern endemic species, which are restricted to the winter rainfall regions. These species are active during autumn and spring but are inactive during the hot summer months, and they are largely restricted to regions of undisturbed natural vegetation. The second element, which includes *O. binodis*, occurs throughout the summer rainfall and winter rainfall regions, primarily in pastures; this is the dominant fauna in the pastures of the Cape Province during summer.

Other species introduced to southwestern Australia, such as *O. caffer* (Edwards 1986a,b) and *B. bison* (Kirk 1983), are univoltine, with adult activity

TABLE 15.4
Seasonal changes in abundance of the common dung beetles in two pastures and in the Badgingarra National Park, southwestern Australia, in 1983.

Species	Jan	Feb	Mar	Apr	May	Jun	Jul	Aug	Sep	Oct	Nov	Dec
Badgingarra												
O. ferox	—	—	—	—	51	19	7	13	3	—	1	—
O. ruficarpa	—	—	—	20	31	35	7	9	3	—	—	—
Aphodius sp. 81.168	—	—	—	—	1,365	2,233	9	87	3	—	—	—
Aphodius sp. 81.169	—	—	—	—	—	237	1	4	—	—	—	—
Dardanup												
O. ferox	—	—	—	—	—	—	—	1	1	—	—	—
O. binodis	109	751	50	47	78	15	1	43	90	113	185	1,006
E. pallipes	1	2	1	2	—	—	—	—	—	1	4	31
A. pseudolividus	44	106	50	62	19	3	—	8	36	1	7	32
Busselton												
O. ferox	2	4	—	13	285	123	2	2	4	39	8	12
O. binodis	475	1,271	143	361	296	19	1	119	3	18	142	131
E. pallipes	41	84	99	121	54	1	1	—	—	—	29	22
A. pseudolividus	102	33	157	131	29	8	1	1	7	1	3	9

Notes: Figures for Dardanup and Busselton are the numbers of individuals caught in pitfall traps baited with cattle dung set twice per month. Figures for the Badgingarra National Park are the numbers of individuals caught in six pitfall traps baited with human dung and set once per month.

in autumn and spring. In southwestern Australia, *O. caffer* and *B. bison* are present only in low numbers, but they appear to show the same phenology in Australia as in South Africa and Spain, respectively. As further univoltine species with restricted breeding seasons are added to the fauna of southwestern Australia (e.g., *Copris hispanus*; Ridsdill-Smith and Kirk 1985), the contrast in the seasonality of the native and introduced species will diminish.

Community Structure

The introduction of a suite of exotic dung beetles to southwestern Australia provides an opportunity to examine whether the introduction of new species affects the abundances of the species already present. However, until now most of the introduced species are active during seasons when the majority of the native species are inactive, and the introduced species occur in pastures where most native species are scarce or absent. It is therefore not surprising that while the numbers of the introduced beetles *O. binodis* and *E. pallipes* increased markedly in 1979–84, the abundances of the native species *O. ferox* and the cosmopolitan *A. pseudolividus* remained relatively constant (Ridsdill-Smith and Matthiessen 1988). However, as species with wet-season activity (e.g., *O. caffer*) spread and further species become established, competitive interactions between the native and introduced species may become more widespread. It is not known whether any of the introduced species will colonize forested regions, where most of the endemic species occur, and whether the introduced species would consume marsupial droppings, the preferred food of the native species.

The endemic dung beetle communities of the Mediterranean climate in southwestern and southeastern Australia are strikingly different (Ridsdill-Smith 1988). In southwestern Australia, there are ten species of Canthonini and seven species of Onthophagini, all but one of which are restricted to this region (Matthews 1972, 1974). In comparable climates in southeastern Australia, there are no Canthonini and nineteen species of Onthophagini, but only 25% of them have distributions restricted to Mediterranean climate regions (Matthews 1972). The two regions of Australia have thus developed distinctly different faunas, despite marked climatic similarities.

The indigenous dung beetle fauna of Australia is less rich in species than that of the Mediterranean climate regions of the Cape region of South Africa, and of southern Spain and southern France, but considerably richer than the Mediterranean climate areas of the west coasts of North America and South America (Ridsdill-Smith 1988). The most obvious difference is the higher biomass of spring-active coprine and onitine beetles in pastures in Spain and France (Ridsdill-Smith and Kirk 1985; Lumaret and Kirk 1987) than in southwestern Australia (Ridsdill-Smith and Hall 1984a).

15.6. DETERMINANTS OF RELATIVE ABUNDANCE

Most of the introduced dung beetles in Australia are adapted to cattle dung in open pastures, and they are less abundant in sheep pastures and in woodland and forest. Provided that a suitable habitat with adequate food resources is available, the geographical distribution of a species is determined by its climatic tolerances, with the caveat that natural climatic barriers, such as the central and southern deserts in Australia, may prevent species colonizing otherwise suitable regions. For example, it was necessary to introduce *Onthophagus taurus* and *O. binodis* separately to both southeastern and southwestern Australia, where they are now common (Henry 1983; Ridsdill-Smith et al. 1989); they are scarce or absent in the warmer subtropical summer rainfall regions and arid zones. Other species have broad climatic tolerances (e.g., *E. intermedius*) and have established themselves over most of the Australian mainland, including the arid zones (Matthiessen et al. 1986), and yet others remain restricted to summer rainfall regions (e.g., *S. spinipes* and *O. gazella*).

Studies conducted to date have been primarily concerned with species associated with cattle dung in open pastures. In the remainder of this section we consider the way species' life history characteristics, intraspecific and interspecific interactions, and dung type determine the abundance of the species and ultimately the structure of Australian dung beetle assemblages. Habitat characteristics are also important, especially vegetation cover and soil type, but these factors have already been discussed in the previous sections (see also Chapter 8).

Life-History Characteristics

The native and introduced dung beetles in Australia exemplify a variety of reproductive strategies. Some species, for example, the native beetle *Onthophagus granulatus* and the introduced *Onitis caffer* (established in NSW and WA), are univoltine in most circumstances. At Uriarra, near Canberra, *O. granulatus* emerges in summer and early autumn, but the newly emerged adults breed poorly or not at all because of the low quality of dung. The beetles breed later in the autumn, but larvae do not enter diapause and die during the winter (Tyndale-Biscoe et al. 1981). Adults overwinter in the soil and breed successfully in spring and early summer. Thus, while potentially multivoltine, the species has an annual life cycle over most of its range. *Onitis caffer* emerges and begins to breed in the autumn both in summer rainfall and winter rainfall regions, and breeding continues during winter and spring, temperature permitting. In the field, *O. caffer* requires ten to twelve months from oviposition to adult emergence, but a facultative larval diapause, induced in early summer, can delay the emergence of some individuals until the autumn one or two years later (Edwards 1986a,b).

Other species are multivoltine. For example, *Onitis alexis* has one or two generations per year in southeastern Australia, where it overwinters mostly as a larva in diapause induced by low temperatures (Tyndale-Biscoe 1985, 1988). The first annual generation develops without larval diapause if there is sufficient moisture, but in times of drought the larvae enter diapause (Tyndale-Biscoe 1988). In hot, humid areas this species has the potential for several generations per year (Tyndale-Biscoe 1985). Such a versatile life history, in combination with wide tolerance limits for temperature, enables the species to adapt to a variety of climates, and its distribution appears to be limited only by cold, wet winters (Tyndale-Biscoe 1985). Most of the successful and wide-spread species, such as *O. gazella* and *E. intermedius*, are multivoltine and have high fecundity and no diapause.

Dung beetles are generally most active during the seasons when humidity and temperatures are high, and they are relatively scarce during the dry and cool seasons. Hence beetles are abundant following the break of season in late spring in the summer rainfall regions, and following rain in desert environments (e.g., Alice Springs, central Australia: Matthiessen et al. 1986). Beetles are scarce during periods of summer drought and during the dry winter in the summer rainfall regions, during extended warm and dry periods in deserts (Matthiessen et al. 1986), and during the cold winter period in the winter rainfall regions.

Intraspecific and Interspecific Interactions

Intraspecific competition for dung has been demonstrated in laboratory and field experiments for a number of dung beetles that now occur in Australia (Ridsdill-Smith et al. 1982; Giller and Doube 1989). Results on the abundances of the native and introduced species in open pastures (papers by Ridsdill-Smith, Tyndale-Biscoe, Wallace, Donovan, Doube, Macqueen, and Fay) and observations on their dung burial behavior (Bornemissza 1969; Doube, Macqueen, and Fay 1988) clearly indicate that there are frequently times when the beetles have the potential to remove much more dung than is available. In these situations, intraspecific competition for dung or breeding sites must limit their reproductive success.

The consequences of intraspecific competition for dung vary with the dung-use strategy of the species. Ball rollers such as *S. spinipes* often convert a large portion of the dung pat to dung balls. When *S. spinipes* is abundant (thousands per pat), the entire dung pat is removed within one day, but only a small proportion of the beetles succeed in making a dung ball (Macqueen, unpubl.). In contrast, in some tunnelers increasing the numbers per pat above a critical density results in a progressive reduction in the level of dung burial and brood production. For example, Ridsdill-Smith et al. (1982) found that brood production by *O. binodis* increased with beetle density up to 20–30 beetles per liter of dung (up to one hundred brood balls per pat), but then decreased markedly until, at 320 beetles per liter, fewer than 5 brood balls were produced per litre of dung. Intraspecific competition becomes more intense as dung quality decreases (Ridsdill-Smith et al. 1986), and competition is increased by the presence of coprophagous fly larvae (Ridsdill-Smith et al. 1986) and other dung-inhabiting beetles (Giller and Doube 1989).

The breeding behavior of beetles can have a marked effect on their success in a competitive situation. For example, the roller *S. spinipes* may feed in the pat for up to 24 hours before removing a brood ball, whereas *Sisyphus (Neosisyphus) rubrus* constructs a ball within hours of arrival at the pat (Paschalidis 1974). In this respect, *S. rubrus* has a competitive advantage over those *S. spinipes* that spend a day feeding in the pat. However, *S. spinipes* places its brood balls above ground in grass tussocks, hence its reproductive success is relatively independent of soil characteristics, whereas *S. rubrus* buries its brood balls in the soil. *Sisyphus spinipes* has an advantage in heavy soil regions during dry periods. Both *Sisyphus* species have a competitive advantage over the slow-burying tunnelers, some of which (e.g., *O. viridulus*) do not begin dung burial until 4–6 days after arrival at the pat (Edwards and Aschenborn 1987; Doube, Macqueen, and Fay 1988). When beetles are abundant and dung is removed rapidly, there is diffuse competition between diurnal (e.g., *S. spinipes* and *E. intermedius*) and crepuscular/nocturnal beetles (e.g., *O. gazella* and *O. alexis*: Wallace and Tyndale-Biscoe 1983).

In conclusion, many introduced species are frequently so abundant that mutual interference and preemptive resource competition reduce their potential rate of reproduction, but so far only in habitats and during seasons when native species have always been scarce. As the introduced species disperse and mix with one another, levels of interspecific and intraspecific competition will increase.

Amphibians (for example, the introduced cane toad *Bufo marinus*), reptiles, and various insectivorous and omnivorous birds (e.g., ibis, crows, and magpies) and mammals (e.g., foxes) have been observed to eat dung beetles in Australia, but such predation appears to have little effect on beetle populations in most circumstances. There are no recorded parasites or diseases of dung beetles in Australia, but Ridsdill-Smith (unpubl.) has found nematodes and staphylinid beetles attacking the immature stages of beetles in the field.

Dung Quality

The reproductive performance of many species of dung beetle has been shown to vary with seasonal changes in dung quality (Tyndale-Biscoe et al. 1981; Matthiessen and Hayles 1983; Tyndale-Biscoe 1985; Ridsdill-Smith 1986; Macqueen et al. 1986). Reproductive performance is highest in dung produced by cattle feeding on actively growing pasture, and lowest in dung produced by cattle feeding on hayed-off pasture. In the summer rainfall region at Rockhampton, the breeding rate of *E. intermedius* is maximal in dung collected during summer and minimal in dung collected in winter (13 and 1–2 broods, respectively, per female per fortnight; Macqueen et al. 1986). In the winter rainfall regions (e.g., Perth, WA), brood production by *O. binodis* is maximal in spring dung (100 brood balls per pat) and lowest in summer dung (0 balls per pat; Ridsdill-Smith 1986). In southeastern Australia, the native *O. granulatus* produced five times more brood balls with high-quality than with low-quality dung (Tyndale-Biscoe et al. 1981). However, dung beetles, like other dung-breeding insects (Macqueen et al. 1986), do not have a standard response to changes in dung characteristics. For example, the reproductive performance of *Onitis alexis* is only little affected by seasonal changes in dung quality (Ridsdill-Smith 1986; Tyndale-Biscoe 1988).

15.7. DUNG BEETLE INVASIONS: THE AUSTRALIAN EXPERIENCE

The continuing introduction of dung beetles to Australia and their spread provide an opportunity to examine three recurring questions about the ecology of invasions: (1) Can predictions be made as to which species will be successful invaders? (2) Can predictions be made as to which communities will be most easily invaded? and (3) Can we predict the impact of the invaders on the communities they enter?

The Success of Invaders

The success of an invading species is measured by its numerical abundance and by its long-term persistence in the new environment. On this basis there are twenty-two successfully introduced dung beetle species in Australia, several species whose success is not yet certain, and less than twenty species that appear to have failed to establish. All the species that failed are known to occur in at least moderate numbers in those parts of Europe, Africa, or Asia with climates similar to the target regions in Australia.

The success and failure of the introduced species in northern Australia is being reviewed by Doube and Macqueen (unpubl.), who conclude that the keys to success have been climate matching, habitat matching, and our ability to manipulate the species' reproductive biology—for example, to control the diapause; release method and founder numbers are also thought to be important. Doube and Macqueen (unpubl.) relate African data on the relative abundances, habitat associations, and reproductive biology of the twenty-two established species to their success in coastal central Queensland. Only seven species were abundant and widespread within a decade of their release. These species are all relatively abundant in the Hluhluwe district of Natal, South Africa (Fig. 15.1), where they show a preference for grassveld but also occur on clay-loam soils in a pastoral environment (Table 15.5). The successful invaders are multivoltine, high-fecundity species. It seems likely that many K-selected species (Edwards 1988a) also have the potential to be successful invaders, but because of low numbers released in the field and low fecundity of these species, large populations would not appear for many years following their release. One such species, *Copris elphenor*, has become moderately abundant at one locality in Queensland ten years after it was released. In New Zealand, the introduced Mexican dung beetle, *Copris incertus*, was recorded as abundant in the early 1980s (Blank et al. 1983), nearly 30 years after its release in 1955–58 (Thomas 1960).

There is a variety of reasons for the failure of other species to establish in Australia. Some species were difficult to rear in the laboratory because of larval or adult diapause, and adults of uncertain quality were released in low numbers. Other species are now known to be strongly associated with sand-veld, bushveld, or the game reserve environment in Africa, and are considered to have failed to establish because their basic ecological requirements were not met; they were released in grassland on clay-loam soils with only cattle present. Holm and Wallace (1987) suggest that the use of superphosphate fertilizer on dairy properties reduces the probability of dung beetle establishment.

Invasion of Communities

The indigenous dung beetles are generally scarce in open pastures in Australia, hence the first introduced species encountered virtually no interspecific competition. Some introduced species occupy niches complementary to those of the indigenous species and are thereby partially separated in space and time from them; this is the case with the species introduced to Western Australia (Section 15.5). On the other hand, many of the introduced species, especially the ones introduced into Queensland, share the same pattern of seasonal activity (Fig. 15.2), and at Rockhampton it is common to find five to six introduced species in the same dung pat, resulting in obvious and intense competition between the species. The degree to which such competition inhibits the establishment of additional species is not known. In April 1985, about fourteen thousand *Sisyphus rubrus* were released at the Rockhampton study area, and the species has successfully established a breeding population in the presence of substantial numbers of the six species already present. The fact that local assemblages of dung beetles in subtropical Africa often contain fifty or more species, most of which are rare (Chapter 8), also suggests that competitive exclusion is unlikely to prevent establishment of further species at Rockhampton.

Table 15.5
Numerical abundance and habitat associations of 22 dung beetle species in the Hluhluwe district of Natal, South Africa, with their performance following release in the field in Queensland, Australia.

Species	Length (mm)	Habitat in Africa			Total Number	Year First Released
		Gr	Cl	Pa		
Abundant and widespread						
Digitonthophagus gazella	11	90	68	58	715	1968
Euoniticellus intermedius	8	79	68	49	9,980	1971
Liatongus militaris	9	84	85	38	2,852	1968
Neosisyphus spinipes	9	69	58	28	3,101	1972
Neosisyphus rubrus	8	92	96	45	6,585	1973
Onitis alexis	15	73	33	69	203	1972
Onitis viridulus†	16	—	77	—	65	1976
Established but not widespread						
Neosisyphus infuscatus	8	64	76	39	1,005	1976
Onitis pecuarius	16	—	—	—	*	1976
Copris elphenor	19	98	10	7	200	1988
Not known to be established						
Neosisyphus mirabilis	10	12	66	15	1,283	1972
Allogymnopleurus thalassinus	11	98	5	23	2,186	1979
Onthophagus binodis	10	—	—	—	*	1973
Onitis tortuosus	16	—	—	—	*	1976
Onitis westermanni	12	—	—	—	*	1977
Onitis caffer	15	43	43	64	14	1983
Onitis deceptor	19	81	2	50	235	1979
Onitis uncinatus	18	88	63	69	16	1979
Euoniticellus africanus	9	—	—	—	*	1973
Copris diversus	10	—	—	—	*	1976
Copris bootes	20	—	—	—	*	1977
Copris fallaciosus	21	79	47	68	34	1977

Source: Doube and Macqueen (unpubl.).

Notes: Beetles were trapped weekly at eight sites in Africa during fifteen months, from April 1982 until June 1983. Trapping sites were located in the following three pairs of contrasting habitats, with four sites per habitat: grassveld versus bushveld, clay-loam soil versus deep sand, and pastoral versus game-reserve environment. The figures shown (under habitat in Africa) are the percentages of individuals trapped in grassveld (Gr), on clay-loam soil (Cl), and in pastoral environments (Pa). The total number of beetles trapped in Africa is also given, as well as the year when the species was first released in Australia.

Of the abundant species, D. gazella and E. intermedius have reached their distribution limits while the other species are still spreading. The established but not widespread species are known from one locality only (N. infuscatus was recovered in 1982, it may be now extinct). Of the remaining species, which are believed to be extinct, O. binodis was recovered in 1975 and C. fallaciosus in 1978.

† Only three individuals were caught in Africa in 1982–83. The data presented are for pastoral grassveld on clay and sandy soil in 1983–84.

* These species are known to be present in the Natal region but were not trapped in 1983–84.

The Impact of Invaders

The aims of introducing dung beetles to Australia were discussed in Section 15.3. The most obvious result of successful introductions has been the significantly increased level of dung dispersal (burial and shredding) wherever the new species have become established. Secondary responses to increased dung dispersal—reduced numbers of dung-breeding flies and increased pasture productivity—have been less self-evident. However, there is clear evidence from field and laboratory studies that dung beetle activity reduces survival of flies in dung pats (Bornemissza 1970b; Blume et al. 1973; Moon et al. 1980; Ridsdill-Smith 1981; Wallace and Tyndale-Biscoe 1983; Ridsdill-Smith et al. 1986; Doube, Giller, and Moola 1988). Furthermore, at times, intense dung beetle activity appears to have suppressed regionally the level of abundance of the bush fly (Hughes et al. 1978; Ridsdill-Smith and Matthiessen 1988) and the buffalo fly (Macqueen and Doube, unpubl.). Dung burial by beetles increases the growth of pasture grasses (Bornemissza 1970a; Macqueen and Beirne 1975a; Fincher 1981), and beetle activity must improve pasture productivity to some extent.

The effect of the introduced dung beetles on the overall composition of the dung insect fauna has not been assessed in detail, but many workers studying pasture communities have shown that dung pats protected from dung beetles produce many more dung-breeding flies than pats not so protected, and the numbers of predatory insects are also decreased in beetle-attacked pats (Roth and Macqueen, unpubl.; Doube and Macqueen, unpubl.). Whether the same is true of forest and woodland habitats in Australia is not known, nor has the effect of the introduced beetles on the indigenous species been examined in detail. However, the introduced dung beetles now comprise a major component of many Australian dung insect communities, and in some regions (for example, in coastal Queensland and southwestern Australia), they have radically restructured the dung insect communities in open pastures.

PART THREE

Synthesis

THE ELEVEN CHAPTERS in Part 2 described regional dung beetle assemblages from northern temperate regions to tropical forests and savannas. There are obvious differences among the main biomes of the world in their beetle assemblages. Small species of dwellers (Aphodiinae) with high fecundity but low competitive ability predominate at high latitudes, where population densities in relation to resource availability are generally low. In contrast, the larger and competitively superior tunnelers and rollers (Scarabaeidae), with relatively low to extremely low fecundity, are numerically and functionally dominant at low latitudes, where populations are often resource limited. This is a beautiful example of r-species being replaced by K-species along a gradient of increasing competition. Although it may be too simplistic to ascribe the main pattern in dung beetle biogeography to competition, there is no doubt that competition has played a major role.

In the next four chapters we attempt to provide a synthesis of four topics: spatial patterns, competition, resource partitioning, and species diversity in dung beetles. The order of the four chapters could be different, but we begin with spatial patterns (Chapter 16) to emphasize habitat patchiness, the fundamental environmental context within which any consideration of dung beetle ecology must be based. (The next chapter, on competition, takes for granted some findings on aggregated spatial distributions of beetles between droppings, reviewed in Chapter 16.) In Chapter 16 we also examine albeit very briefly, regional population dynamics, and we review the data on dung beetle introductions.

Chapter 17 summarizes the existing evidence for competition in dung beetles and attempts to erect a conceptual framework about a competitive hierarchy. This chapter should help make sense of the variety of dung beetle communities described in Part 2 for different biomes and for different localities within biomes, varying in soil type, vegetation cover, mammalian species richness, and so on. We realize that the amount and kind of data presently available to formulate and test some of the ideas are very restricted, and while dealing with the most species-rich tropical communities, we are biased by our experience (YC) in West Africa. The situation is undoubtedly somewhat different, and perhaps more so than we now realize, in the more seasonal savannas in East Africa and in the subtropical savannas in southern Africa (Chapter 8). Still, it seems useful to look for broad generalizations, even if they need to be modified in the future; the alternative would be to succumb in the complexity of the dozens of special cases, and make endless qualifications.

Chapter 18 reviews resource partitioning in dung beetles. We do not aim at an exhaustive summary of the information presented in Part 2, but have rather restricted Chapter 18 to the main resource dimensions for dung beetles. We use as an example the community of dung beetles in West African savannas with more than one hundred coexisting species.

Chapter 19 summarizes patterns in species richness in dung beetles, with increasing latitude, between grasslands and forests, and how species number varies with varying mammalian species richness. This chapter also relates species richness to the three processes discussed in the previous chapters: competition, spatial aggregation, and resource partitioning.

The final chapter, Chapter 20 is devoted to recording some concluding thoughts about areas in dung beetle ecology in which further research would be well rewarded.

CHAPTER 16

Spatial Processes

Ilkka Hanski and
Yves Cambefort

INSECTS LIVING IN patchy habitats have typically aggregated spatial distributions; some habitat (resource) patches have large numbers of individuals while others, though they appear to be similar, have only a few individuals (dung: below; Hanski 1987b; carrion: Kneidel 1985; Hanski 1987b, 1987c; Ives 1988b; mushrooms: Shorrocks et al. 1984; Hanski 1989c; fallen fruits: Atkinson and Shorrocks 1984). Aggregated distributions in these insects is no surprise, as aggregated spatial distributions are also the rule in species living in nonpatchy habitats (Taylor et al. 1978), if any habitats are truly nonpatchy for insects. Optimal foraging theories (Krebs and Davies 1981; Stephens and Krebs 1986) predict that individuals using a patchy habitat should distribute themselves and their offspring in relation to the amount of resources in the patches. Parker's (1970, 1974, 1978) results on the yellow dung fly, *Scatophaga stercoraria*, is a putative example (but see Hanski 1980f; Curtsinger 1986), but generally the predictions of the optimal foraging theory are not supported by spatial distributions of insects in patchy habitats, most likely because the insects do not have the time and the information-processing capacity to achieve the predicted distribution (Hanski 1990). One could expect that as habitat patchiness increases and the durational stability of individual patches decreases, the level of spatial aggregation in the insect populations further increases.

Population ecologists are interested in aggregated spatial distributions for two different reasons. According to one view, by studying spatial distribution patterns we may learn something about the processes involved in the dynamics of populations and communities (Patil et al. 1971; Ord et al. 1980; Taylor 1986, and references therein). The other school of thought, while not deeming the causes of spatial distributions uninteresting, is more concerned with their population-dynamic consequences (Elton 1949; Hassell and May 1973, 1985; May 1978; Hassell 1978; Lloyd and White 1980; Hanski 1981, 1987b; Atkinson and Shorrocks 1981; Chesson and Murdoch 1986; Shorrocks and Rosewell 1987; Ives 1988b). Ultimately, it will be necessary to understand both the causes and the consequences of spatial distributions (e.g., Kareiva and Odell 1987), not least because they may be functionally related.

The population-dynamic consequences of aggregated spatial distributions to competing species were outlined in Chapter 1, where it was argued that intraspecific spatial aggregation is often much greater than interspecific aggregation. The net effect of aggregated distributions therefore is, more often than not, increased intraspecific competition in relation to interspecific competition (Chapter 1). Such a shift in the strengths of the two types of competition will generally facilitate coexistence of competitors. At this very general level, aggregation-mediated coexistence is comparable to coexistence based on resource partitioning (Ives 1988b), though the underlying mechanisms are fundamentally different. As with coexistence based on resource partitioning, the question is about the degree of intraspecific aggregation, not of its presence or absence. How much aggregation is enough to allow coexistence has been explored in the theoretical papers by Atkinson and Shorrocks (1981), Hanski (1981), Ives and May (1985), and Ives (1988a,b, 1990). A single magical figure does not exist, because the outcome of competition is affected by other factors besides spatial distributions, including resource partitioning. Therefore we cannot resolve the role of spatial aggregations in the dynamics of competitors once and for all, and for all systems. Our aim here (Section 16.1) is to document and compare the patterns of variance-covariance structure in some dung beetle assemblages. In addition, we will discuss the causes of aggregation in dung beetles, which, though complex and largely unknown, are worthy of further study.

Of the various approaches to the measurement of spatial aggregation (e.g., Southwood 1976; Pielou 1977), we have used two. To characterize the level of aggregation in a set of samples we use the following indices (Ives 1988b):

a measure of
intraspecific aggregation: $J_1 = \dfrac{v_1/x_1 - 1}{x_1}$
$$(16.1a)$$

a measure of
interspecific aggregation: $C_{1,2} = \dfrac{cov_{1,2}}{x_1 x_2} \cdot$
$$(16.1b)$$

J_1 measures the relative increase in the average number of conspecific individuals that an individual of species 1 must face above the number of conspecifics it would meet if the distribution were random (Ives 1988b). $C_{1,2}$ has an analogous interpretation for interspecific aggregation. These measures of aggregation have the advantage of being simple and having a straightforward ecological interpretation, but they are also useful because they are directly related to the theory of competition in patchy habitats (Ives 1988a,b).

Another widely used measure of aggregation is the slope b of the regression line of the logarithm of spatial variance against the logarithm of mean, which describes the change in variance with increasing mean, and may be used both in intraspecific and interspecific comparisons. The parameter b has been advocated as an important attribute of species' spatial behavior (Taylor 1961, 1986, and references therein; but also see Anderson et al. 1982; Hanski 1987d; Downing 1986; Thòrarinsson 1986). The value of b does not actually measure the level of aggregation as such, but it gives a rough measure of the degree of density dependence in the change of aggregation. If b equals 2, then no density-dependent processes have affected the spatial distribution during the time the average abundance has changed (Hanski 1987d; Anderson et al. 1982). On the other hand, the greater the role of (positively) density-dependent immigration to or emigration from habitat patches, the smaller the value of b, and it is always less than 2.

The spatial scale at which aggregated distributions of individuals between resource patches is of importance is the scale of a local population. Individuals that disperse form a pool from which colonizers to patches are drawn (Chapter 1). In Section 16.2 we turn to larger spatial scales. The distribution of species among "population sites" at the regional scale raises new and important questions: Are all species distributed among suitable sites in a similar fashion? How frequently is a species' presence at a site dependent on immigration from elsewhere, from larger and more permanent populations? How often do local populations go extinct? Do species occur at an extinction-immigration equilibrium at the regional scale, or is local equilibrium more typical? Our knowledge about dung beetles is far too scanty to attempt answering some of these questions. Still, it seems worth raising the issues here, if for no other reason than the dung beetles' exceptional suitability for such study. In many temperate regions, most dung beetles are entirely or mostly dependent on cattle dung. Pastures define the suitable population sites, at which resource supply for dung beetles is easily quantified. Dispersal between the sites is easy to measure, at least on a relative scale, because dung beetles are easy to attract to baited traps. Finally, spatial distributions of domestic mammals have changed dramatically in many countries, and continue to do so, providing interesting opportunities to study metapopulation dynamics—the dynamics in a set of local populations connected by dispersal—in a changing environment.

In Section 16.3 we move to a still larger spatial scale, and to a different type of problem. Dozens of species of dung beetles have been introduced, intentionally or otherwise, into islands and continents where they did not occur before. Some of the chapters in Part 2 have given examples of such introductions and presented analyses of special cases (see particularly Chapter 15 for the large-scale introduction of dung beetles to Australia). Section 16.3 provides a brief summary of dung beetle introductions. To where have dung beetles been introduced, and where from? What has been the success of introductions? What have been the consequences to native species? There is an increasing ecological interest in species introductions (e.g., Crawley 1987), to which dung beetle introductions add new data. It is also likely that more species of dung beetles will be introduced in the future to tackle the two major management problems that exist in this field, the accumulation of cattle dung in pastures where the native beetles are unable to efficiently bury the dung of the introduced cattle; and the often large populations of dung-breeding flies, which have ample opportunities to increase to a pest level in the absence of competitors and effective natural enemies.

16.1. SPATIAL AGGREGATION ACROSS RESOURCE PATCHES

Patterns of Intraspecific Aggregation

Let us start with two extreme examples. Macqueen and Doube (pers. comm.) trapped beetles in an Australian pasture with pitfalls baited with pats of homogenized cattle dung. The traps were located either in a grid with 5 m intervals, or along a transect line with 20 m intervals. The five introduced species—*Neosisyphus spinipes*, *Euoniticellus intermedius*, *Liatongus militaris*, *Digitonthophagus gazella*, and *Onthophagus sagittarius*—were the only abundant species in the pasture, and all had J values less than 0.67, not significantly different from zero. In other words, the distributions of these beetles in the traps were random. Doube in Chapter 8 refers to another study conducted in southern Africa with essentially the same result, that is, there was no aggregation of beetles among cattle pats located within short distances from each other (in this case, however, the average densities were so low that most distributions would appear random).

Kohlmann in Chapter 7 describes a case of the opposite extreme, where one species (*Ateuchus carolinae*) occurred in large numbers in some dung pats but was entirely absent in other, apparently similar, pats. Several other species, assumed to be competitively inferior to *A. carolinae*, were present in varying numbers in the pats in which *A. carolinae* was absent. Such an extreme pattern is unusual (we do not know of another example), and the reasons for the dominant species to concentrate in only a subset of dung pats remain unknown (Chapter 7). As this study was based on sampling naturally occurring dung pats, it is possible that some differences between the pats contributed to the observed pattern, even if similar-looking pats were collected.

The usual pattern of spatial distribution in dung beetles is between these two extremes. Beetles are generally not randomly distributed even among apparently similar droppings, yet the degree of aggregation is less than in Kohlmann's example, and the abundant species typically occur in varying numbers in all or most pats.

One factor that undoubtedly affects the level of aggregation is the average distance between the traps or dung pats, though short distances, as in Doube's studies, do not necessarily lead to random distributions. Hanski (1979, 1980b), working in southern England, found large variation in the numbers of beetles among traps located only 5–10 m from each other and baited with homogenized cattle dung. Two factors in this study were different than in Doube's: Doube's Australian and African studies were conducted in larger, more open areas; and the dung beetles in England (Aphodiinae) are a much smaller species than those in Australia and Africa (Scarabaeidae). The latter point is related to the continuum from the lottery dynamics to the variance-covariance dynamics described in Chapter 1: the smaller the species in relation to the size of the resource patch, the more toward the variance-covariance end of the continuum its dynamics are expected to be located (Chapter 1).

Cambefort's extensive material from West Africa (Chapter 9) provides the opportunity to examine intraspecific aggregation in a very species-rich assemblage and to compare the degree of aggregation among species with different foraging behaviors. In Chapter 9 Cambefort analyzed spatial distributions of beetles among similar-looking, naturally occurring cattle pats, and found the species to be more or less aggregated (Table 9.7). Three factors were related to the level of aggregation: aggregation decreased with the size and average abundance of the species, and cattle dung specialists were significantly less aggregated than generalist species. Here we complement these analyses with parameter estimates of the power function relationship between spatial variance and mean abundance (an interspecific regression). We can ask the question: Does the increase in variance with increasing mean abundance vary among different sorts of beetles?

Table 16.1 gives the results for six sets of cattle pats sampled at different times of the year in the same habitat. Two kinds of comparisons can be made: cattle dung specialists may be compared with the generalists, and species in four tribes representing different nesting behaviors may be compared with one another. In the first comparison, the specialist species tend to have smaller values of b than the generalists (F = 4.84, P<0.05; differences in mean abundance have no significant effect), suggesting that density-dependent immigration to and/or emigration from droppings is more characteristic for the specialists than for the generalists. Thus specialist species seem to be more sensitive than the generalists to intraspecific interactions in the use of the resource to which they have become specialized.

The comparison of the four tribes gives an even more striking result. We compared the b values in Table 16.1 using analysis of covariance, with the logarithm of mean abundance as a covariate and the tribe as the classifying factor. Both the mean and the tribe had highly significant effects (mean: F = 6.55, P<0.019; tribe: F = 11.38, P<0.0002). The higher the average abundance, the smaller the slope b, which is not unexpected, because density-dependent processes may be expected to occur especially in the most abundant species (see also Table 9.7). Sisyphini, the small diurnal rollers, had the highest slopes, while Coprini, large and mostly nocturnal tunnelers, tended to have the lowest slopes (Table 16.1). This result suggests that the large Coprini in-

TABLE 16.1
Parameter estimates of the power function relationship between the variance and the mean of dung beetles in cattle dung pats in West African savanna (Ivory Coast, Chapter 9).

Species	Data Set	a	b	r^2	Mean	Var	n
Cattle dung specialists	1	0.60	1.62	0.98	5.82	168.5	20
	2	0.46	1.65	0.94	6.28	174.9	19
	3	0.45	1.50	0.92	3.63	79.6	19
	4	0.59	1.63	0.98	3.93	164.3	22
	5	0.65	1.68	0.95	3.17	91.1	20
	6	0.49	1.52	0.96	1.27	18.0	20
Generalists	1	0.76	1.71	0.99	6.94	582.4	15
	2	0.77	1.76	0.97	1.55	33.4	16
	3	0.75	1.52	0.96	0.73	5.5	20
	4	0.89	1.76	0.97	2.13	333.6	30
	5	0.80	1.73	0.98	1.32	72.0	36
	6	1.04	1.76	0.97	1.31	65.3	20
Sisyphini (rollers)	1	0.47	2.39	0.93	2.96	59.6	4
	2	0.42	2.68	0.96	2.18	41.2	4
	3	0.70	1.63	0.99	1.56	13.8	3
	4	1.01	1.96	0.99	4.06	478.9	5
	5	0.42	2.68	0.96	2.18	41.2	4
	6	1.15	2.04	0.99	0.74	18.4	5
Coprini (large tunnelers)	1	0.46	1.62	0.98	1.07	5.7	5
	2	−0.04	0.85	0.87	0.75	0.6	4
	3	−0.28	0.77	0.61	0.43	3.7	5
	4	0.41	1.51	0.99	2.72	22.1	5
	5	0.50	1.39	0.96	0.67	3.0	7
	6	0.36	1.29	0.98	0.43	1.1	5
Oniticellini (small tunnelers)	1	0.63	1.65	0.98	11.35	489.9	9
	2	0.60	1.68	0.95	11.61	362.4	9
	3	0.56	1.58	0.98	6.79	172.3	8
	4	0.68	1.75	0.98	5.95	395.0	8
	5	0.66	1.57	0.96	6.70	199.5	7
	6	0.72	1.77	0.96	2.88	108.4	8
Onthophagini (small tunnelers)	1	0.71	1.66	0.99	8.52	662.5	11
	2	0.67	1.64	0.95	1.36	14.2	13
	3	0.72	1.45	0.96	1.17	11.6	17
	4	0.78	1.65	0.97	2.47	341.5	23
	5	0.79	1.71	0.98	1.49	90.1	27
	6	0.82	1.58	0.93	1.16	45.5	15

Notes: The six data sets are the same as in Table 9.9; number of pats varied from fifteen to thirty-six per occasion. Mean and Var give the overall mean and variance per species and per pat, respectively. n is the number of species.

terfere with one another and thus become distributed more evenly among the
dung pats than the small tunnelers, Onthophagini and Oniticellini, or the small
rollers, Sisyphini. Observational (Kingston and Coe 1977) and experimental
results (Giller and Doube 1989) on large tunnelers indicate that one pair of
beetles (*Heliocopris*) or a small number of beetles (*Catharsius*, *Copris*) may
use all or most of the resource in one dung pat. Relatively low abundance of
Coprini is undoubtedly a consequence of their large size and competition. Unfortunately, it is not possible to compare large rollers (Gymnopleurini and
Scarabaeini) with Coprini, because in the Ivory Coast the former tend to use
mostly omnivore dung (Chapter 9). The values of the *J* index (Table 9.7) suggested that the size of the beetle, not the tribe, significantly affects aggregation. As the size distributions of beetles in the four tribes are different, it is not
possible, with these data, to sort out the effects of size and tribe.

Patterns of Interspecific Aggregation

Interspecific aggregation measures the degree of covariation in the numbers of
two species across a set of habitat patches. As for intraspecific aggregation,
the studies by Macqueen and Doube (pers. comm.) and Kohlmann (Chapter
7) contribute two special cases. The five species in Australia that were not
intraspecifically aggregated did not show any interspecific aggregation either
(all *C* values less than 0.44), while Kohlmann's species represent another unusual case, in which the dominant *A. carolinae* versus the other species had
significantly negatively correlated distributions (Fig. 7.5). In this case it is
possible that the dominant *A. carolinae* somehow prevented other species
from immigrating to, or forced them to emigrate from, the pats where it was
abundant. Yasuda (1987) has found that the rate of emigration of *Liatongus
phanaeoides* is less dependent on the numbers of conspecifics in a dung pat
than on the numbers of another species, *Aphodius haroldianus*.

The usual pattern of interspecific aggregation is not negative association but
some—though usually far from complete—positive association between species. An important point is that intraspecific aggregation does not necessarily
lead to interspecific aggregation. Hanski (1986) showed that nine coexisting
Aphodius species had entirely uncorrelated distributions even if all the species
were intraspecifically aggregated (Fig. 5.2).

Causes of Spatial Aggregation

Aggregated distributions of insects among resource patches located in the
same habitat have two possible causes, heterogeneity among the resource
patches themselves and the movement behavior of insects. It is obvious to
anyone who has observed any insect species in the field that habitat quality
and resource quantity vary from one place (patch) to another, and such heter-

ogeneity undoubtedly affects the spatial distribution of species. Species with similar resource requirements are expected to be similarly affected, hence heterogeneity among resource patches is expected to increase both intraspecific and interspecific aggregation in a guild of similar species. However, while considering resource patches such as dung pats produced by the same species of mammal in one place at one time, it seems unreasonable to assume that differences in the quality of physically similar-looking droppings would be so significant that they alone would suffice to explain the observed spatial distributions of beetles. In any case, such differences must be inconsequential to the breeding success of dung beetles.

Now we turn from the environment to the species living in the environment. The more similar two species are in their morphology, behavior, and ecology, the more similar should be their patterns of colonization of a set of habitat patches, because they are similarly affected by the deterministic factors influencing colonization. In agreement with this argument, Hanski (1987b; see also Fig. 5.3) found that interspecific aggregation was significantly higher in guilds of dung-inhabiting beetles with similar biologies than in a guild consisting of biologically less similar species. Cambefort's (Chapter 9) results on West African dung beetles point to the same conclusion: interspecific aggregation was greater in diurnal-diurnal and nocturnal-nocturnal than in diurnal-nocturnal pairs of species; interspecific aggregation tended to be higher in species pairs of similar size than in pairs consisting of a small and a large species; and interspecific aggregation tended to be higher in two species from the same tribe than in two species from different tribes. Diel activity, size, and tribe will affect the probabilities of detecting and colonizing dung pats that are simultaneously available to beetles. For example, short-term variation in weather may change the environment (e.g., direction of wind) for beetles active at different times of the day and colonizing the same set of droppings. Beetles belonging to different tribes and of different size are likely to have many differences in their foraging behavior, tending to decrease interspecific aggregation. In this perspective, lack of interspecific aggregation may be a consequence of all kinds of differences between species, not only those that are directly related to resource use. The more similar two species are in their ecology, the greater not only the overlap in their resource use, but also the greater their spatial correlation across similar resource patches.

So far we have assumed that intraspecific aggregation is due to heterogeneity among resource patches, differences in the foraging behavior of different species, or to stochasticity in the colonization of patches by insects. Another candidate for intraspecific aggregation is pheromone communication between individuals of the same species. The first convincing demonstration of pheromone communication in dung beetles was due to Tribe (1975, 1976), working on the large African roller, *Kheper nigroaeneus*. In this species, both sexes have tegumentary glands and hairs used for secreting and disseminating the chemical, though only the male has been observed to adopt a special stance during pheromone emission, apparently to facilitate the dispersion of the chemical by wind (Neuhaus 1983). The product of the tegumentary glands emerges from minute pores as hollow, layered tubules impregnated with very small quantities of substances in which hexadecanoic acid, 2,6-dimethyl-2-heptenoic acid, and nerolidol are major components (Burger et al. 1983). The presence of similar morphological structures in many other dung beetles (Pluot-Sigwalt 1982, 1984, 1986, 1988; Houston 1986) suggests that pheromone secretion and communication may be common in these beetles, though in most cases, for example in *Onthophagus binodis*, the pheromones are most likely to be used for close-range species recognition and sexual attraction than for long-range attraction (Ridsdill-Smith 1990). A particular stance during pheromone emission similar to that described for *K. nigroaeneus* has been observed in *Garreta*, another African roller (Tribe 1976), as well as in the Neotropical roller *Canthon cyanellus* (Bellés and Favila 1984). In the latter species, a chemical that is spread on the dung ball appears to have a repelling function against *Calliphora* flies, potential competitors.

In summary, chemical communication with pheromones probably facilitates close-range individual, sex, and species recognition in many species of dung beetles, and chemicals may be used to mark balls, burrows, nests, and so on, but it is as yet unclear to what extent pheromone communication may affect the level of aggregation of beetles in dung pats. Our guess is that pheromones are not important in this respect, especially not in the species that occur in high densities.

Consequences of Aggregation

Our interest in spatial aggregation of dung beetles has been stimulated by the possibility that aggregated distributions facilitate coexistence of competitors (Chapter 1). Ives (1990) gives the following general result about the minimum effect of aggregation on coexistence of competitors. Let α_0 and β_0 be the threshold values of larval competition coefficients above which coexistence is impossible in randomly distributed species, and denote by α_a and β_a the corresponding values when species are aggregated. Ives (1990) shows that the minimum effect of aggregation on coexistence is given by

$$\alpha_a\beta_a > \alpha_0\beta_0 (J_1 + 1)(J_2 + 1)/(C_{1,2} + 1)^2, \tag{16.2}$$

where J_1, J_2 and $C_{1,2}$ are defined by Eqs. (16.1). To take an example, given the average values of J and C in Table 9.8, aggregation will allow two species to coexist even if their competition coefficients were six times greater than the maximum possible without aggregation. The interpretation of this result is straightforward for *Aphodius*, whose larvae move freely within a resource patch and interact with one another. In the case of tunnelers and rollers, the crucial interactions occur earlier, among adult beetles competing for the resource to provision their nests (Chapter 3). Nonetheless, the basic point remains essentially the same: aggregated interactions among competing individuals allow coexistence of competitors with greater competition coefficients than would be possible without aggregation.

The current theory deals with only two species, while in real dung beetle communities dozens of species may be using the same resource. It seems intuitively clear that the two-species model extends to many species, but where is the limit to guild size if coexistence is solely based on intraspecifically aggregated spatial distributions? Shorrocks and Rosewell (1986, 1988) have made a start in answering this question, and they reach the conclusion that in fungivorous *Drosophila* intraspecifically aggregated distributions would allow some five species to coexist. The expected number of coexisting species naturally depends on many factors, most notably on the degree of competition between the species and the degree of intraspecific and interspecific aggregation; the predicted guild size of five was based on empirical results on these and other parameters in fungivorous *Drosophila* (Shorrocks and Rosewell 1986; see Ståhls et al. 1989, for another guild of fungivorous flies). In dung beetle assemblages with dozens of species, there are typically some differences in species' selection of different dung type, soil type, and vegetation cover; in their diel activity, seasonality, and nesting behavior; and in their size and its correlates (Chapter 18). Not all these differences entirely eliminate competition between pairs of species, but they have a potentially important effect in facilitating coexistence. Unfortunately, because of the multitude of species in many dung beetle assemblages with various kinds of ecological differences, it is not generally possible to delineate guilds of like species; competition and other forms of interspecific interactions are diffuse. Examples of sets of species with no known ecological differences include *Proagoderus lanista*, *P. tersidorsis*, and *P. quadrituber* in South Africa (Doube 1987) and several pairs of *Onthophagus* in the forests in West Africa (Chapter 11).

In Chapter 1 we described two types of dynamics in species competing for nonrenewable resources in patchy environments: the lottery and the variance-covariance dynamics. Here we showed that dung beetles generally exhibit aggregated spatial distributions, but also that different types of beetles show consistent differences in the level of intraspecific aggregation. Species that are large in relation to the size of resource patches are expected to obey the lottery dynamics and to show less intraspecific aggregation than small species (Chapter 1), which indeed is the pattern found in dung beetles (Fig. 16.1 and analysis of Table 16.1 above). Especially in the ecology of dung beetle assemblages consisting of small species, intraspecific and interspecific aggregation are critical ingredients, though this is only one factor among many factors that need to be considered for a comprehensive understanding of patterns of coexistence.

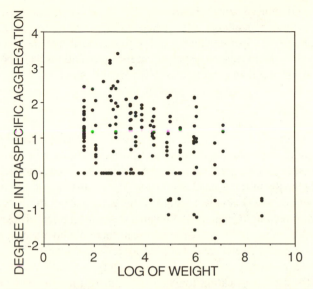

Fig. 16.1. The relationship between intraspecific aggregation and the logarithm of beetle weight in tunnelers in West Africa (same data as in Table 9.8). Aggregation on the vertical axis is measured by log $(J + 1)$. The slope of the regression line differs highly significantly from zero ($t = 5.33$; $P<0.001$; $r = -0.37$).

16.2. METAPOPULATION DYNAMICS

Neighboring regions located only a few kilometers apart may have strikingly different communities of dung beetles. Chapter 6 gives one example from southern Europe, where two closely situated sites had little overlap in species composition. In this case there were differences in soil type due to different parent rocks, probably explaining much of the observed difference in species composition (Chapter 6). Restricting our attention to regions without apparent climatic, edaphic, or other environmental differences of significance to dung beetles, we may ask about the patterns of occurrence of species at sets of similar sites. Following Levins (1969, 1970), population ecologists use the term "metapopulation" to describe systems of local populations connected by dispersal. What is the role of metapopulation dynamics in dung beetles?

The metapopulation scale is not very important for the locally abundant and regionally common species, which are present at all sites most of the time and are affected mostly by processes within populations. The metapopulation scale is important for the less common species, with local populations occasionally going extinct and new ones being established by immigration from other populations. A metapopulation may persist only if the rate of establishment of new local populations exceeds the rate of local extinctions when the number of local populations is small (Levins 1969). How correlated the extinction events are across populations also makes a difference (Quinn and Hastings 1987; Gilpin 1990). Species that have a high dispersal rate and uncorrelated local dynamics should have more persistent metapopulations than species with a low rate of dispersal and highly correlated local dynamics (Hanski 1989b).

Dispersal

The flight-oogenesis hypothesis (Johnson 1969) suggests that females, before commencing reproduction, enter a distinct phase of dispersal. Although valid for some insect species, the flight-oogenesis hypothesis does not account for dispersal behavior in dung beetles (Hanski 1979). Instead of a conspicuous dispersal period, newly emerged dung beetles enter a shorter or longer period of maturation feeding (Chapter 3). Because the consumption of resources by dung beetles does not affect resource renewal, there are no reasons for immature beetles to move elsewhere to feed before reproduction.

The species of dung beetles with the most highly evolved breeding behavior remain in the same nest caring for their offspring during the entire breeding season (Halffter and Edmonds 1982). Such species probably have a low dispersal rate. In contrast, other species need to fly between egglaying. *Aphodius* represents this latter type, and it is noteworthy that mature females are relatively more frequent among long-distance dispersers than immature beetles (Hanski 1980d). Mature females may benefit of movements in two ways: by depositing their eggs in several dung pats in which the success of their offspring varies independently (''spreading of the risk''), and by possibly locating an area where the density of the species is low. In *Aphodius* in southern England, between-pasture variance in numbers decreased during the adult flight season in most species, and the species with the highest frequency of long-distance movements tended to have the smallest between-pasture variance in numbers (Hanski 1979, 1986). Such observations suggest that the rate of dispersal between populations is high.

The maximum rates of dispersal can be examined in species that have been introduced into areas where they did not occur before. Table 16.2 gives estimates for two medium-sized tunnelers, *Digitonthophagus gazella* and *Onthophagus taurus*. The observed dispersal rates are surprisingly similar and very high, from 50 to 130 km per year. These values may not be representative for most dung beetles, as Onthophagini seem to be exceptionally good dispersers (Section 16.3). It is also possible that the estimates in Table 16.2 are inflated by human transport of beetles on cattle trucks, for instance.

In summary, large, low-fecundity dung beetles are expected to have low rates of dispersal, but many small species may have very high rates of dispersal. Data on the rate of dispersal of the large species, especially, are badly needed.

TABLE 16.2
Rate of dispersal of introduced dung beetles from the initial point of introduction.

Species	Locality	Rate of Spread, km/yr	Ref
Digitonthophagus gazella	Australia	50–80 (12)	1
Digitonthophagus gazella	North America	58 (12)	2
Onthophagus taurus	North America	129 (7)	3

References: 1, Seymour (1980); 2, Chapter 7; 3, Fincher et al. (1983).

Note: The figures in parentheses give the number of years over which the rate has been calculated.

Extinction

Extinction of local populations of insects is notoriously difficult to prove. The task is easier when the time span of observations is increased. Pleistocene extinctions of dung beetles can be attributed to varying climates and a dramatically declined resource supply (Chapter 5). Regional extinctions of species during the past 100 years have been explained by reduced availability of resources (Chapters 5, 6, 9, and 11). In some cases the key may be the reduced availability of resources in microenvironments particularly favorable for the species—for example, the right kind of dung on the right kind of soil.

Metapopulation persistence depends on the average rate of local extinctions and on the degree of correlation of local extinction events. Hanski (1979, 1986) calculated that if the average density of *Aphodius* in pastures in England were less than 0.5 beetles per cattle pat, many females would remain unfertilized, and hence the population growth rate would be decreased (the Allee effect). The bimodal distributions of log abundances that are frequently observed in dung beetles (Fig. 16.2) may result from such a critical density in local populations; many or most of the very rare species are likely to be represented by dispersers from elsewhere (Chapter 5 presents evidence supporting this hypothesis for temperate *Aphodius*). Bimodal abundance distributions are uncommon in the literature (Williams 1964), but this may only reflect the difficulties of sampling most communities of insects as exhaustively as one may sample dung beetle communities.

There are little data to examine the degree of correlation in local dynamics, and even less in local extinction events. However, one may guess that the large tropical dung beetles whose emergence is critically dependent on sufficient rainfall (Chapters 8 and 9 and references therein) have highly correlated dynamics over large areas. These species are also expected to have low dispersal rates (see above), which is a combination not favorable for metapopulation persistence. To make things worse, many of the largest dung beetles, such as *Heliocopris*, *Heteronitis*, and *Kheper*, are more or less dependent on

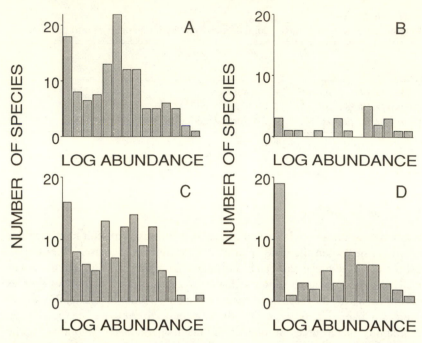

Fig. 16.2. Four species abundance distributions in local communities of dung beetles. Localities: A, tropical savanna (West Africa, Chapter 9); B, north temperate pasture (England, Chapter 5); C, montane pasture (the Alps, Chapter 14); D, tropical forest (Barro Colorado Island, Panama, Chapter 12). Note the tendency toward bimodality in these distributions.

elephant dung, which is a declining resource type in many areas. And finally, the game reserves in Africa, where most elephants now occur, are becoming increasingly isolated habitat islands, further decreasing the chances of meta-population survival.

In other dung beetles, local populations have less correlated dynamics than is probably the case with large African rollers and tunnelers. Fig. 16.3 shows that there is substantial between-population variation in the seasonal flight activity of temperate *Aphodius*. Since breeding success in insects depends much on weather conditions, the kinds of seasonal differences in adult flight activity shown in Fig. 16.3 are bound to lead to substantial variation in breeding success and hence in local dynamics between populations.

To summarize these considerations on dispersal and local extinction in dung beetles, we propose the working hypothesis that the large, low-fecundity species living in seasonal environments in the tropics have a low dispersal rate and highly correlated local dynamics, while many other species, for example, temperate *Aphodius*, have the opposite characteristics, making their metapopulations more persistent.

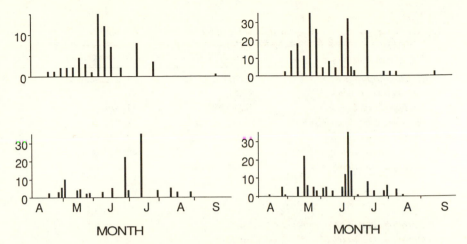

Fig. 16.3. Simultaneous pitfall trap catches of *Aphodius ater* in four pastures situated within 5 km from each other in southern England. The columns give the numbers of individuals caught. Note the marked differences among the pastures (from Hanski 1980e).

16.3. DUNG BEETLE INTRODUCTIONS

Over long periods of time, species may greatly expand their geographical ranges. According to Matthews (1972), the present fauna of 260 species of *Onthophagus* in Australia originates from at least thirty-four independent invasions of that continent. Long-distance dispersal of dung beetles is likely to be facilitated by man, leading to accidental introductions of species to new regions and continents. Paulian (1961) suggests that *Digitonthophagus gazella* was accidentally introduced into Madagascar with cattle at some point between 2,000 and 300 years BP (this very vagile species is also one of the most successfully introduced species when the introductions are intentional; see below). Table 5.2 summarizes the accidental introductions of *Aphodius* to North America, while Table 16.3 gives such information for Scarabaeidae. Except for the spread of *Euoniticellus cubiensis* from Cuba to Florida, all these species involve Onthophagini, which is another testimony to the exceptionally good powers of dispersal in this cosmopolitan tribe with more than 2000 species (see also Table 4.1 and Chapter 10 for the occurrence of *Onthophagus* on small islands in South-East Asia).

During the past decades, about one hundred species of dung beetles have been intentionally introduced into Australia, New Caledonia, North America, Puerto Rico, and Hawaii. Of these, about forty species have become established so far (Table 16.4). The two aims of dung beetle introductions have been improvement of dung removal from and enhancement of nutrient cycling in pastures, and decrease in the amount of resources available to dung-breed-

TABLE 16.3
Accidental introductions of Scarabaeidae dung beetles.

Species	Introduced from	Introduced to	Ref
Digitonthophagus gazella	Tropical Africa	Madagascar	1
Oniticellus cubiensis	Cuba	Florida	2
Onthophagus batesi	Salvador	Martinique	3
Onthophagus bituberculatus	West Africa	Martinique	3
Onthophagus cervus	India	Mauritius	4
Onthophagus consentaneus	Australia	New Caledonia	5
Onthophagus depressus	South Africa	Madagascar	1
		Mauritius	4
		USA	6
		Australia	7
Onthophagus nuchicornis	Europe	N America	6
Onthophagus taurus	Europe	USA	8
Onthophagus unifasciatus	India	Mauritius	4

References: 1, Paulian (1960); 2, Woodruff (1973); 3, Matthews (1966); 4, Paulian (1961); 5, Gutierrez et al. (1988); 6, Howden and Cartwright (1963); 7, Matthews (1972); 8, Fincher and Woodruff (1975).

TABLE 16.4
Intentional introductions of dung beetles.

Introduced to	Introduced from	N_{int}	Years	N_{est}	Ref
Australia	S Africa	44	1968–82	16	1
	Europe	10	1972–82	5	1
	Tropical Asia	1	1968	1	1
New Caledonia	S Africa	4	1978	3	2
N America	Europe	1	1982	?	3
	S Africa	3	1972–77	3	3
	Asia	1	1978–81	1	3
	S America	1	1981	1	3
Puerto Rico	Various	8	1920–30	0	3
Hawaii	Various	26	1906–83	10	3

References: 1, Chapter 15; 2, Gutierrez et al. (1988); 3, Fincher (1986).
Note: N_{int} is the number of species introduced, while the next columns give the years of introduction and the number of species known to have become established, N_{est}.

ing flies, which in many areas have become a serious nuisance to domestic animals and even to humans (Waterhouse 1974; Bornemissza 1979; Doube 1986). It is not our purpose here to try to assess how successful the introductions have been in achieving these aims. Chapter 15 reviews the experiences from the Australian introductions, which have been the most extensive introductions of dung beetles anywhere. The current consensus is that the introduced dung beetles have made a major contribution toward solving the problem of dung accumulation, but the success in controlling the populations of dung-breeding flies has been more variable. In Australia, dung beetles have not made much impact on the populations of the buffalo fly, but large populations of the introduced *Onthophagus binodis* have essentially halved the bush fly season (Chapter 15; Ridsdill-Smith and Matthiessen 1988). Introductions to other regions have given similar results (Patterson and Rutz 1986). Solving the fly problem would apparently require a nearly complete and rapid removal of dung (Palmer and Bay 1984; Hanski 1987a), which is not likely to occur except under exceptional circumstances (Chapters 8, 9, 15, 17). Following the early suggestions by Campbell (1938) and Bornemissza (1968, 1976), more attention has been given recently to the ideas of introducing some of the predators and parasitoids of dung-breeding flies instead of dung beetles, the competitors (Wallace and Holm 1983; several chapters in Patterson and Rutz 1986). Two apparently successful cases are the introduction of a parasitic wasp to Mauritius, where it caused a drastic decline in the abundance of the dung-breeding stable fly (Greathead and Monty 1982), and the control in Fiji of dung-breeding flies by *Hister chinensis* (Legner 1986). Other experiences have not been equally encouraging, however (Patterson and Rutz 1986).

From Where Have Dung Beetles Been Introduced, and to Where?

In the geographical regions where large herbivorous mammals are abundant, they are trailed by large numbers of specialized dung beetles, searching for droppings with some particular characteristics, for example, large size and a high carbohydrate/nitrogen ratio. In contrast, in regions lacking indigenous large herbivorous mammals, the local dung beetle fauna is unlikely to have many species that use their excrements efficiently. The pioneering idea of Bornemissza (1960) was to introduce dung beetles into Australia from areas with similar climates and where cattle, or closely allied large herbivorous ungulates, occur naturally. Nowadays, such areas are primarily Africa and, to a lesser extent, tropical Asia, but also Europe and the Middle East (temperate and Mediterranean climates). The areas where cattle have been introduced but which lack a dung beetle fauna that is adapted to use the droppings of large herbivores are Australia and many oceanic islands, as well as parts of the United States (Table 16.4).

Four Very Successful Species

For a dung beetle to be considered for introductions, it must have two basic characteristics: it must be efficient in burying cattle dung, and it must have the potential to become numerous and widespread. In the beginning of the Australian introduction program these requirements were clearly understood, but the role of habitat specificity in particular was underestimated, and the biology of beetles in their native countries was poorly known. As many as fifty-five dung beetle species were shipped to Australia, and forty-one species were released in the field (Table 16.4 and Chapter 15). Of these, twenty-two species have become established, but only five to seven species are presently really common and widespread in (North) Australia, though others are multiplying rapidly (B. M. Doube, pers. comm.).

For closer scrutiny we have selected four species of dung beetles, which have been introduced frequently and are considered to be among the most successful ones in terms of burying large quantities of dung. These species are *Neosisyphus spinipes* (Sisyphini), *Onitis alexis* (Onitini), *Euoniticellus intermedius* (Oniticellini), and *Digitonthophagus gazella* (Onthophagini) (Fig. 16.4). The key ecological characteristics of these species are given in Table 16.5.

One's first impression is that the four species comprise a surprisingly diverse assemblage. They belong to four different tribes of dung beetles. Though most of them are widespread, *N. spinipes* has a relatively small geographical range in southern Africa. The four species include two small tunnel-

TABLE 16.5
Ecological characteristics of four frequently and successfully introduced species of dung beetles.

Species Native biome and geographical distribution	L	R	T	N	D	A	F
Digitonthophagus gazella Tropical savanna Africa, Asia	11	15	30	ST	N	60?	90
Euoniticellus intermedius Tropical savanna Africa	9	12	28	ST	D	45	120
Onitis alexis Tropical-subtropical grasslands Africa, Mediterranean region	16	15	60	LT	N	60	130
Neosisyphus spinipes Tropical savanna Southern Africa	9	1	52	SR	D	100	44

Notes: The following data are given: L, length in mm; R, approximate geographical range in millions of square kilometers; T, developmental time in days from egg to adult; N, type of dung beetle (L, large; S, small; R, roller; T, tunneler); D, diel activity (D, diurnal; N, nocturnal); A, average longevity in days; F, mean lifetime fecundity under laboratory conditions.

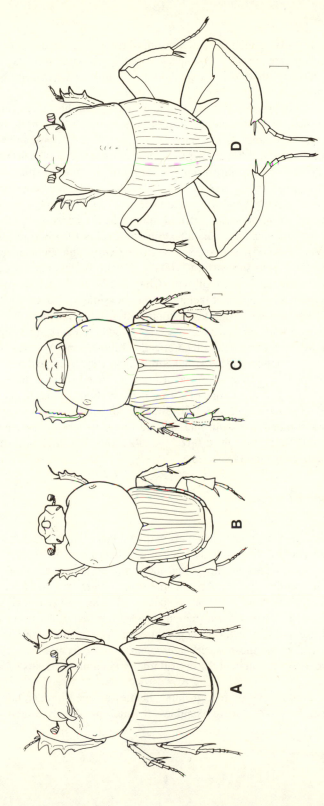

Fig. 16.4. Four frequently and successfully introduced species of dung beetles: A, *Digitonthophagus gazella* (Onthophagini); B, *Euoniticellus intermedius* (Oniticellini); C, *Onitis alexis* (Onitini); D, *Neosisyphus spinipes* (Sisyphini). The scale is 1 mm. See Table 16.5 for the ecological characteristics of these species.

ers, one (relatively) large tunneler, and one small roller; the large rollers and very large tunnelers are missing. Two species are diurnal while the other two are nocturnal. Clearly, there is not a single type of dung beetle that has been successfully introduced.

Nonetheless, there are also important similarities among the successful invaders. All the species are of medium size, from 9 to 16 mm in length, have relatively short times of immature development, from 28 to 60 days, and have high fecundity, from 50 to 130 eggs per female. A closer look at the Australian introductions (Table 15.5) confirms that small and medium-sized species are the ones that have been introduced successfully (ANOVA, $P<0.10$; comparing the lengths of the "abundant and widespread species" with the others in Table 15.5). Among the twenty-two species in Table 15.5, the successful species appear to be more abundant than the others in the locality of origin in South Africa (ANOVA, $P<0.05$), and species with relatively high abundance on clay-loam soils have been successful (ANOVA, $P<0.05$). The latter species are probably not very sensitive to soil type (Chapter 17). With so few species and so many potentially important factors, it is impossible to deduce whether all the factors would make a significant contribution independently, but the main point is that certain attributes make beetles good colonizers. The importance of fecundity is not surprising: theoretically, high-fecundity (r-selected) species make good colonizers (MacArthur and Wilson 1967). Large dung beetles, which are missing among the successfully introduced species, have extremely low fecundity, down to one offspring per breeding season in, for example, *Kheper nigroaeneus* (Table 3.2). Such low-fecundity species have a necessarily low population growth rate and are likely to have a low rate of dispersal (Section 16.2), though in favorable situations they may ultimately outcompete the high-fecundity species.

Effect of Introduced Species on
  Native Species

Not many observations have been made about the impact of introduced species on native dung beetles. In some cases not much interaction is expected to occur as the exotic species are introduced into communities that entirely or virtually lack any similar species. An example is the winter rainfall regions in Australia (Chapter 15), where the native species have similar seasonality to that observed in the Mediterranean region: beetles are active in spring and autumn, when the soil is moist and temperatures are neither too high nor too low. In contrast, the introduced species *Onthophagus binodis* and *Euoniticellus pallipes* are active throughout the dry summer and autumn months (Chapter 15).

Sequential introductions of foreign dung beetles in the summer rainfall regions in Australia provide some examples in which previously established species have greatly decreased in abundance when more species were added. Perhaps the best example is *Onthophagus sagittarius*, which was abundant at a locality in Rockhampton in 1973 to 1979, when the only other abundant species was the introduced *Digitonthophagus gazella* (Table 15.2). In the early 1980s, three other introduced species became very abundant (*L. militaris*, *E. intermedius*, and *N. spinipes*), and the concurrent decline in the density of *O. sagittarius* to about 10% of its abundance in the 1970s was probably due to interspecific competition (Chapter 15). Another example is the decline of *Euoniticellus intermedius* following the establishment of *Digitonthophagus gazella* at another locality in Australia (Chapter 15).

Digitonthophagus gazella has been identified as a superior competitor in the man-made assemblages of dung beetles in Australia. In North America, *D. gazella* and some other introduced species have invaded relatively rich assemblages of native dung beetles. Howden and Scholtz (1986) resampled a locality studied ten years earlier by Nealis (1977). The pooled density of beetles, as estimated by pitfall trapping conducted in the same way on both occasions, was twice as high in the earlier study. But in view of the effect of weather conditions and other such factors on trap catches, the difference does not imply any great change in beetle density. During the ten years, the study locality was invaded by *D. gazella*, which now accounted for 24% of all individuals. Another native species, *Onthophagus alluvius*, had similarly increased from 0% to 25% relative abundance in ten years, while the relative abundance of *Onthophagus pennsylvanicus* had dramatically decreased, from 55% to 12%. *Onthophagus alluvius* may have benefited from a gradual change in vegetation, but the earlier dominant *O. pennsylvanicus* and some other species probably declined because of competition with *D. gazella* and *O. alluvius* (Howden and Scholtz 1986). Note that Cambefort (Chapter 9) found no great changes in the relative abundances of dung beetles over a period of eight years at a locality in the Ivory Coast where no species had naturally or otherwise invaded. On the other hand, Doube (1987) reported substantial changes in relative abundances of species over a period of five years in a locality in southern Africa. Stability of both indigenous and introduced populations of dung beetles is a topic in need of further study (Chapter 20).

16.4. SUMMARY

In this chapter we have reviewed three topics and three spatial scales in population dynamics: aggregation of beetles across habitat (resource) patches; dynamics of local populations connected by dispersal (metapopulation dynamics); and dynamics of species introduced from one continent to another.

Small-scale spatial aggregation, such as occurs in dung beetles between dung pats, generally has a stabilizing effect on single-species dynamics, and if intraspecific aggregation exceeds interspecific aggregation, small-scale aggregation tends to facilitate coexistence of competing species. There are varying but generally high levels of intraspecific aggregation especially in small species of dung beetles, and we have discussed the causes of intraspecific and interspecific aggregation. Interspecific aggregation is lower in pairs of species belonging to different tribes than in those belonging to the same tribe, and in diurnal-nocturnal pairs of species than in diurnal-diurnal or nocturnal-nocturnal pairs. It decreases with increasing size difference between two species. These and other differences not directly related to resource use may nonetheless affect spatial variances and covariances and thereby the outcome of competition.

We have examined dispersal and extinction in dung beetles and suggested the working hypothesis that the large, low-fecundity species living in seasonal environments in the tropics have a low dispersal rate and highly correlated local dynamics, while other species—for example, temperate *Aphodius*—have the opposite characteristics, making their metapopulations more persistent.

Some one hundred species of dung beetles have been introduced into Australia, New Caledonia, North America, Puerto Rico, and Hawaii. Of these, some forty species have become established so far. Several of the introduced species in Australia have become hugely abundant, demonstrating clear density compensation when compared with the species-rich assemblages in their native countries (Chapter 15). The most successful species include a variety of dung beetles in terms of diel activity, food relocation technique, and the tribe, but all are medium-sized, high-fecundity species with a high population growth rate and good dispersal ability, up to 100 km per year.

CHAPTER 17

Competition in Dung Beetles

Ilkka Hanski and Yves Cambefort

INTRASPECIFIC AND interspecific competition occur at least occasionally in nearly all communities of dung beetles (see most chapters in Part 2). In some situations—for example, in African savannas on sandy soils in the rainy season (Chapters 8 and 9)—competition is severe and undoubtedly greatly influences the structure of the communities. Unfortunately, there is a lack of rigorous experimental work to demonstrate the degree of competition in different kinds of dung beetle communities, and an even more acute lack of results on the consequences of competition vis-à-vis the dynamics of populations and communities. Lack of relevant empirical studies, not lack of competition, explains why dung beetles have not featured in the reviews of the incidence of competition in natural populations (Connell 1983; Schoener 1983). Even at relatively low densities, when not all resources are rapidly removed, competition may nonetheless be the only type of interspecific interaction of any importance; other forms of interaction are simply uncommon in dung beetles (Chapter 1).

Two kinds of evidence can be put forward for competition in dung beetles, and our aim in this chapter is to collate a range of pertinent observations. First, for the rollers in particular, there are innumerable anecdotal observations, though from a relatively small number of biomes, of direct interference competition both within and between species, of individuals attempting to steal dung balls from one another. Halffter and Matthews (1966) have suggested that interference competition (combat) is restricted to rollers, but the lack of numerous observations for dwellers and tunnelers, which hide mostly in dung pats and in the soil underneath, may well be due to difficulty of observation. Second, numerous observations have been made of preemptive resource competition in dwellers (in *Aphodius*, not in *Oniticellus*), tunnelers, and rollers: entire droppings are removed or shredded by beetles, sometimes in a matter of minutes. We supplement observations on preemptive resource competition with records of high rates of immigration of beetles to dung-baited traps, indicating the potentially large numbers of beetles attracted to single resource patches.

Dung beetles provide intriguing examples of how species' life histories have profound consequences on population dynamics and on interactions between two or more species. Recall the resources required by the three types of dung beetles—the dwellers, the tunnelers, and the rollers (Chapter 3). The dwellers really consist of two distinct subtypes. In *Aphodius*, adults and larvae, both of which live freely in droppings, may compete for both space and food, while in *Oniticellus*, which construct nests in droppings, there is less chance of larval competition, each larva being confined to its individual brood ball. In tunnelers, the larvae do not generally compete with one another, for the same reason as in *Oniticellus*, but the adult beetles compete both for food and for space in the soil below the dropping. Finally, in the rollers, competition for space in the soil is eliminated by the transport of dung balls away from the food source, but adult competition for food remains and is frequently amplified by direct interference (as it may be in many tunnelers). Indeed, it is paradoxical that competition appears to be most frequent and most severe in the rollers, where only one form of competition—adult competition for food—is possible, and least frequent and intense in the dwellers (*Aphodius*), in which both adults and larvae may compete both for space and food. This tentative generalization is not meant to imply that some tunnelers, especially the largest species, may not be engaged in equally or even more competitive interactions than most rollers.

Differences in the behavior and life history of dwellers, tunnelers, and rollers, and other less striking differences among species in the same functional group, give rise to definite competitive asymmetries among different kinds of beetles. These asymmetries create a more or less predictable competitive hierarchy, which should allow us to build a theory of community structure in dung beetles, based on the behavior and life history of the species. We will make a start toward building such a theory in Section 17.2.

17.1. TYPES AND INCIDENCES OF COMPETITION

Interference Competition in Rollers

With the total number of only 1,000 + described species, of which a mere 350 species belong to modern tribes (Chapter 4), the rollers still present the most obvious examples of competition in dung beetles. Both exploitative and interference competition are commonplace in rollers. Indeed, it may not be a coincidence that the number of extant rollers is small, and that guilds of locally coexisting rollers are also small, typically fewer than 10 species, and maximally about 20. Rollers may not do well in coexisting with one another because of their biology (below) and because of the frequently severe competition.

Tables 17.1 and 17.2 summarize observations on intraspecific and interspecific fights for dung balls in rollers. Casual observations are numerous, but unfortunately more detailed studies are rare (fine examples are Matthews 1963, and Heinrich and Bartholomew 1979a). Nonetheless, wherever rollers

TABLE 17.1

Observations on intraspecific interference competition in dung beetles.

Species	Length (mm)	Fr	Dung Type	Locality	Biome	Ref
Rollers						
Sisyphus biarmatus	9	3	human	W Africa	grassland	1
Sisyphus seminulum	5	3	human	W Africa	grassland	1
Sisyphus seminulum	4	2	cattle	S Africa	grassland	2
Sisyphus schaefferi	9	?	various	Europe	grassland	3
Neosisyphus paschalidisae	11	2	cattle	W Africa	grassland	1
Neosisyphus spinipes	9	2	cattle	Australia	grassland	2
Gymnopleurus coerulescens	9	3	human	W Africa	grassland	1
Gymnopleurus puncticollis	11	3	human	W Africa	grassland	8
Gymnopleurus geoffroyi	13	?	various	Europe	grassland	3
Paragymnopleurus sinuatus	18	?	not given	Asia	grassland	4
Paragymnopleurus maurus	14	3	human	SE Asia	forest	5
Allogymnopleurus thalassinus	12	2	cattle	S Africa	grassland	2
Kheper nigroaeneus	30	2	cattle	S Africa	grassland	2
Scarabaeus semipunctatus	20	?	cattle	Europe	grassland	6
Canthon pilularius	15	3	cattle	N America	grassland	7
Kheper laevistriatus	37	3	elephant	E Africa	grassland	8
Canthon angustatus	10	3	human	S America	forest	9
Tunnelers						
Canthidium spp.	7	3	human	S America	forest	9
Coprophanaeus ensifer	45	3	carrion	S America	forest	10
Phanaeus tridens	15	2	cattle	C America	grassland	11
Heliocopris dilloni	50	1	elephant	E Africa	grassland	12
Oxysternon conspicillatus	25	3	human	S America	forest	9
Dwellers						
Aphodius rufipes	11	2	cattle	Europe	grassland	13
Aphodius spp.	7	2	cattle	Europe	grassland	14

References: 1, Cambefort (1984); 2, Doube (pers. comm.); 3, Prasse (1958); 4, Honda (1927); 5, C. Malumphy (pers. comm.); 6, Goggio (1926) and Heymons and von Lengerken (1929); 7, Matthews (1963); 8, Bartholomew and Heinrich (1978); 9, Peck and Forsyth (1982); 10, Otronen (1988); 11, Chapter 7; 12, Kingston and Coe (1977); 13, Holter (1979); 14, Chapter 5.

Note: Fr denotes the frequency of observations: 1, isolated observations; 2, has been observed many times; 3, regularly observed.

are a conspicuous element in dung beetle communities, interference competition is regularly observed. Interference has been observed both in open grasslands and in forests, and in species using omnivore dung and in species using the larger droppings of herbivore dung (Table 17.1). Interference occurs in small and in large species of rollers (Table 17.1).

The elegant study by Heinrich and Bartholomew (1979a) demonstrated that

TABLE 17.2
Observations on interspecific interference competition in dung beetles.

Dominant (Subordinate) Species	Length (mm)	Fr	Dung Type	Locality	Ref
Gymnopleurus coerulescens	10	1	human	West Africa	1
(Sisyphus biarmatus)	9				
Kheper lamarcki	35	2	cattle	South Africa	2
(Pachylomera femoralis)	40				
Kheper laevistriatus	37	3	elephant	East Africa	3
(Kheper platynotus)	32				
Canthon septemmaculatus	10	3	carrion	Central America	4
(Canthon cyanellus)	8				
Paragymnopleurus maurus	14	2	human	SE Asia	5
(Sisyphus thoracicus)	6				
Copris lugubris	16	1	cattle	Central America	6
(Phanaeus tridens)	15				
Neosisyphus spinipes	9	2	cattle	South Africa	7
N. infuscatus	8				
Canthon angustatus	11	2	human	South America	8
(Canthidium sp.)	7				
Oxysternon conspicillatus	26	2	human	South America	8
(Canthon angustatus)	11				

References: 1, Cambefort (1984); 2, Bernon (1981); 3, Heinrich and Bartholomew (1979a); 4, Howden and Young (1981); 5, C. Malumphy (pers. comm.); 6, Chapter 7; 7, Paschalidis (1974); 8, Peck and Forsyth (1982).

Notes: Each observation is about two species. The subordinate species (if any) is given in parentheses. Fr is the frequency of observations, as in Table 17.1.

the outcome of combat in the large African roller *Kheper laevistriatus* depends on body size but even more so on thoracic temperature (Heinrich and Bartholomew 1979a; Table 17.3). A fight for a dung ball was typically initiated immediately after a beetle had arrived at a dung pat, apparently to take full advantage of the elevated body temperature attained during the flight. *Kheper laevistriatus* may also maintain endothermically elevated thoracic temperature while making a dung ball. Heinrich and Bartholomew (1979a) show that the rate of ball construction is related to thoracic temperature, hence high thoracic temperature helps a beetle in fierce preemptive resource competition as well as in interference.

Interference occurs frequently among different species of rollers (Table 17.2). As one could expect, it has commonly been observed that larger species

TABLE 17.3

Weight and thoracic temperature of *Kheper laevistriatus* involved in contests over dung balls.

	Body Weight (g)			Temperature °C		
	x	SE	n	x	SE	n
Winners	3.52	0.07	118	38.7	0.23	120
Losers	3.22	0.06	118	35.2	0.34	120
t-value	3.15			8.65		
P	0.001			0.001		

Source: Heinrich and Bartholomew (1979a).

are competitively superior to smaller ones (Table 17.2). Young (1978), working in Panama, demonstrated in a laboratory experiment a linear competitive hierarchy among nine diurnal and seven nocturnal species; larger species were competitively dominant over smaller ones, and rollers (e.g., *Canthon*) were competitively superior to "butters" (e.g., *Phanaeus*), species which push a piece of dung or carrion some distance from the food source before burying it (Chapter 12). Very small balls are not worth stealing, however, hence interspecific interference occurs only between species that are not too dissimilar in size. The idea of limiting similarity in the body size of coexisting rollers is supported by guilds of rollers with unexpectedly well spaced-out body sizes (Section 18.2). Occasionally more than two species may compete for the same dung ball. J.-P. Lumaret (pers. comm.) has observed in Corsica a fight involving *Scarabaeus typhon*, *S. sacer*, and *S. laticollis*. The ball was prepared by *S. typhon*, which was first unsuccessfully attacked by *S. sacer*. When it was later attacked by *S. laticollis*, the same individual of *S. sacer* returned and all three beetles fought each other. *Scarabaeus typhon* remained the owner of that ball.

Other Cases of Interference

Fighting for dung balls by rollers is the most conspicuous kind of interference in dung beetles, but other forms of interference occur between tunnelers and dwellers and, in the case of nest parasitic dung beetles (kleptoparasites), between two very different kinds of beetles. Interference may increase the rate of emigration from a dung pat, as shown by Holter (1979) for the dweller *Aphodius rufipes*, which colonizes cattle pats in Europe, and by Yasuda (1987) for three species of dung beetles in Japan, *Aphodius haroldianus*, *Onthophagus lenzii*, and *Liatongus phanaeoides*. Laboratory experiments have shown that the rate of oviposition per female decreases with increasing density of beetles in *A. rufipes* (Holter 1979), *Onthophagus nuchicornis* (Macqueen and

Beirne 1975b), *O. binodis* (Ridsdill-Smith et al. 1982), *Euoniticellus intermedius* (Hughes et al. 1978), and undoubtedly in many other species. Laboratory studies are useful in demonstrating what may happen when beetle density is very high, but even more valuable would be results from natural populations.

Exploitative Competition

Table 17.4 gives examples of situations in which entire dung pats are rapidly removed by beetles, and where there consequently is no doubt that competition occurs and affects the populations of beetles. Individual pairs of large tunnelers commonly bury up to 500 g of dung (Edwards 1986a; Edwards and Aschenborn 1987; Doube, Giller, and Moola 1988), hence even small numbers of these beetles in the same dung pat are likely to compete with one another (Fig. 17.1). As with interference competition, we must emphasize that preemptive resource competition occurs regularly, though often seasonally, in many dung beetle communities (Kingston 1977; Peck and Forsyth 1982; Heinrich and Bartholomew 1979a; Chapters 8–12, 15), while it is more irregular in other communities. Most observations on preemptive resource competition are from tropical savannas and forests, and the beetles concerned are either rollers or tunnelers.

Table 17.5 gives some examples of the rate of immigration of beetles to dung-baited traps. Such trapping results indicate how many beetles would

TABLE 17.4

Preemptive resource competition in dung beetles: Examples of situations in which all dung is rapidly used by dung beetles.

Dominant Taxa	Dung Type	Fr	Locality	Biome	Ref
Heliocopris antenor	cattle	2	West Africa	grassland	1
Heliocopris dilloni	elephant	1	East Africa	grassland	6
Heliocopris spp.	elephant	2	South Africa	grassland	2
Large Coprinae	cattle	2	South Africa	grassland	2
Large Coprinae	cattle	2	South Africa	grassland	2
Large rollers	wildebeest	2	South Africa	grassland	2
Large rollers	elephant	3	East Africa	grassland	3
Gymnopleurus spp.	omnivore	3	West Africa	grassland	1
Sisyphus, Onthophagus	omnivore	3	West Africa	grassland	1
Paragymnopleurus, Copris	omnivore	3	SE Asia	forest	4
Canthon, Oxysternon	omnivore	3	South America	forest	5

References: 1, Cambefort (pers. obs.); 2, Doube (pers. comm.); 3, Kingston (1977); 4, Hanski (pers. obs.); 5, Peck and Forsyth (1982); 6, Kingston and Coe (1977).

Notes: Generally more than one species is involved. Competition is typically seasonal is grasslands but may be aseasonal in tropical forests. Fr is frequency of observations, as in Table 17.1.

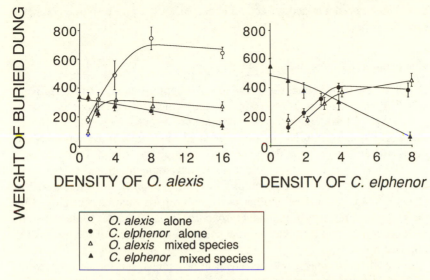

WEIGHT OF BURIED DUNG

DENSITY OF *O. alexis*

DENSITY OF *C. elphenor*

○ *O. alexis* alone
● *C. elphenor* alone
△ *O. alexis* mixed species
▲ *C. elphenor* mixed species

Fig. 17.1. Field competition experiment between *Onitis alexis* and *Copris elphenor*. The horizontal axis gives the number of pairs of one of the species, while the second species was either absent or had a constant density (two pairs of *Copris*, four pairs of *Onitis*). The vertical axis gives the total amount of dung buried (from Giller and Doube 1989).

TABLE 17.5
Maximal rates of immigration to dung-baited pitfall traps.

Dung Type	Bait Size	Trapping Period	Individuals Caught	Species Caught	Locality	Ref
Human	60 ml	1 day	1,000–4,000	25–37	West Africa	1
Cattle	600 g	1 day	1,000–3,000	20–40	South Africa	2
Cattle	1,000 g	1 day	50–500	10–15	West Africa	1
Elephant	500 g	15 min	4,000	?	East Africa	3

Reference: 1, Cambefort (pers. obs.); 2 Doube (pers. comm.); 3, Heinrich and Bartholomew (179b).

Notes: Results are given per trap. All examples are from tropical or subtropical savannas.

have arrived at a dropping if the arriving beetles would not deplete the resources. If the number of arriving beetles is greater than the number sufficient to remove all of the resource, then there clearly is potential for resource competition. Although trapping results must be interpreted with caution, as they undoubtedly eliminate many biological idiosyncrasies among the species, the numbers of beetles actually trapped are often so enormous—by an order of magnitude or more greater than needed to remove all of the resource—that there can be no doubt about frequent occurrence of severe resource competition.

Competition in Droppings: The Dwellers

Dwellers, the dung beetles that breed inside droppings, are typically relatively small in comparison with the dung pats they colonize, and single beetles or their offspring have little effect on the amount of resource in the pat. However, the shear numbers of *Aphodius* are occasionally so large that the hundreds and thousands of beetles break dung pats into small pieces, which dry up quickly and become entirely unsuitable for larval development. Some of the best-known cases of mass occurrence in temperate *Aphodius* are summarized in Chapter 5. Larval competition has not been properly studied; but judged by the space requirements of individual *Aphodius* larvae (Chapter 5), larval competition may be commonplace even at densities where most of the resource is not used up.

Some *Aphodius* species do not breed in droppings, but the larvae feed on decaying vegetable matter in the soil. As the larval resource in these species may be very abundant in comparison with the availability of dung, it is not surprising that the adults can be exceedingly numerous in droppings (Chapter 5). It may even be that populations of these species are limited by adult food availability. Another similar case is carrion-feeding Hybosoridae in tropical forests in South-East Asia: adult beetles often feed in hundreds and sometimes in thousands per carcass, but the larvae use some other resource (Chapter 10).

In conclusion, competition occurs in temperate *Aphodius*, but only occasionally are the numbers of beetles so large that their effect on dung pats is as obvious as in the case of many tunnelers and rollers. More subtle effects include density-dependent rates of emigration and oviposition (Holter 1979; Yasuda 1987) and competition among larvae (Landin 1961). In contrast to *Aphodius* with free-living larvae, *Oniticellus*, the dwellers that construct nests in dung pats, are never hugely abundant. *Oniticellus* are tropical and subtropical species, and often very much affected by the competitively superior tunnelers and rollers (Section 17.2).

Competition for Space in Tunnelers

Tunnelers have two essential requirements for successful breeding: food for the larvae and a space in the soil, below the dropping, to construct their nest. Both resources may be limiting and may be competed for (Fig. 3.2).

Different species use different depths in the soil below the dropping for nest construction (see Plate 3.3), and such interspecific differences, whether ultimately due to competition or something else—for example, body size—will decrease interspecific competition relative to intraspecific competition for space. Rougon and Rougon (Chapter 13) suggest that the stratified nest construction by the few species of dung beetles in the Sahel significantly decreases competition among species. There is also extensive variation in the location of the nest in the soil depending, for example, on soil moisture (Edwards and Aschenborn 1987) and soil type (Chapter 13), to which different species may have different responses.

The great diversity of nest architecture in Scarabaeidae dung beetles (Halffter and Edmonds 1982; Chapter 3) is related to the tendency of increasing parental care and decreasing fecundity in the most evolved taxa, but competition for space may be another cause for the diverse nest types. The species of tunnelers that construct and provision their nests fast (especially Coprini) need to ensure that other beetles do not disturb their nests; these species tend to construct their nests deep in the soil. Other species, for example many *Onitis* (Edwards and Aschenborn 1987), use older droppings which were not entirely removed by the faster and competitively superior species, and these species may construct their nests shallowly in the soil or, in the case of the dweller *Oniticellus*, inside droppings (Fig. 13.6).

Density Compensation

Interspecific competition is generally expected to decrease the average abundance of competing species, though in multispecies communities indirect interactions between species may complicate matters (Bender et al. 1984). Density compensation refers to an increase in the average abundance of competitors with decreasing species number while resource availability remains constant.

The few species of introduced dung beetles that now occur in the pastures in Australia may reach densities by two orders of magnitude higher than the same species have in climatically and edaphically comparable areas in South Africa, their native country (Chapter 15). Part of the difference may only be due to the higher food availability per hectare in the pastures in Australia than in the game parks in Africa, yet it is difficult to accept that this or some other trapping bias would provide the entire explanation. Surprisingly, even the pooled density of beetles may be much higher in a species-poor Australian assemblage of introduced species (ca 5 species) than in a species-rich African assemblage (ca 100 species; Chapter 15). In theory, such a difference could be due to excess density compensation (Case et al. 1979), which has been suggested to occur in some imported dung beetle assemblages (Wallace and Tyndale-Biscoe 1983). However, since not all of the dung is used by beetles on clay-loam soils in Africa (Chapter 8), as is often the case in Australia (Chapter 15; B. M. Doube, pers. comm.), it is more likely that the overall difference in abundance is due to some subtle differences in the soil type, known to have a major impact on dung beetle abundance (Chapter 8), and possibly to predation of the buried dung by termites in Africa (Chapter 15).

Another case of apparent but not proven density compensation occurs on mountains, where species richness declines with altitude (Hanski and Niemelä 1990). The average density of montane species has been found to be higher than the average density of lowland species both on tropical (Chapter 10; Hanski 1989a; Hanski and Niemelä 1990) and temperate mountains (Chapter 14). As resource availability is likely to decrease with increasing elevation, this result could well be due to density compensation.

17.2. FROM POPULATION BIOLOGY TO COMMUNITY STRUCTURE

Population Dynamic Consequences of Beetles' Behavior

Throughout this book we have employed two classifications to organize our thinking: (1) dung beetles have been divided into three main functional groups—the dwellers, the tunnelers, and the rollers (Chapter 3); and (2) we have suggested that their competitive dynamics span a continuum from lottery dynamics to variance-covariance dynamics (Chapter 1). It is now time to couple these two classifications and to examine the consequences of the different types of behavior to the population dynamics of dung beetles and to the structure of their communities.

The key difference between lottery dynamics and variance-covariance dynamics is the pattern of resource division and hence the reproductive success among breeding individuals (Chapter 1). In the lottery model, the individuals that manage to breed are selected randomly from the pool of individuals, and their reproductive success is relatively even. In the variance-covariance model, all individuals may attempt to breed, but they have very unequal success because of intraspecific aggregation between resource patches. The variance-covariance model may involve scramble competition in contrast to contest competition in the lottery model, but this is not essential; the key point is intraspecifically aggregated competitive interactions among individuals. What matters, then, are the factors that affect the distribution of competitive interactions and resource division among individuals. The significant interactions in dung beetles occur at droppings, and the key processes are the arrival of beetles at droppings, possible density-dependent emigration, and the speed by which beetles are able to secure sufficient resources for their offspring.

The order of arrival of beetles may be expected to be more or less random when comparing species with similar resource requirements and similar modes of foraging (not forgetting that many assemblages of dung beetles include species varying in foraging behavior: Chapters 10 and 12). On the other hand, some droppings are always easier to detect than others; hence the overall rate of arrival varies between droppings. The dwellers, of which *Aphodius* is the main example, represent one extreme. These are relatively small beetles in comparison with the size of the herbivore droppings they most frequently colonize. The larvae of *Aphodius* take some 5 days to hatch (Ridsdill-Smith 1990); hence the offspring initially have little effect on resources in the dung pat, and the colonization phase may last for many days (Hanski 1980d). During this period many females oviposit in droppings, and between-dropping differences in the numbers of eggs and small larvae accumulate. Ovipositing females are unlikely to be able to assess the density of eggs already laid, and only if the instantaneous density of adult beetles in dung pats becomes high, density-dependent emigration and oviposition may occur (Holter 1979; Yasuda 1987), decreasing between-dropping variance in the numbers of beetles and their progeny. In summary, relatively small size and the breeding behavior characteristic for dwellers easily lead to variance-covariance dynamics in their populations.

The other extreme is represented by many rollers, especially the species in the tribes Gymnopleurini, Scarabaeini, and Sisyphini. Many rollers prefer relatively small omnivore droppings (Chapter 18), increasing the ratio of the size of the beetle to the size of the resource patch, a key parameter of lottery versus variance-covariance dynamics (Chapter 1). Rollers are often so abundant that all the available resource is removed rapidly, on a first-arrived, first-served basis. Ideally, all the first n beetles to arrive gain enough resources to make a dung ball with enough food for one larva, at which point all the resource is used up and no extra beetles arrive at the dropping. The resource is divided relatively evenly among the n beetles that are drawn more or less randomly from the individual pool. This is an example of lottery competition (Chapter 1).

In practice, the division of resources is often not so orderly. As it takes some time to make a dung ball (below), large numbers of beetles may accumulate before the whole resource is exhausted, and interference can be severe. Large species are typically superior in interference competition to small ones (Table 17.2), which may give rise to a bias toward large size in the set of species that manages to breed. Large numbers of beetles may also disrupt the dropping to the extent that fewer than n females manage to breed. For example, the small (introduced) roller, *Neosisyphus spinipes*, may be so abundant in pastures in Australia, with thousands of beetles per pat, that only a small proportion of beetles is able to make a brood ball, and much of the resource is "wasted" (Chapter 15; *Neosisyphus* may be especially sensitive to such interference because it is slow in making a ball). However, this situation does not invalidate the lottery model. Other dung beetles may also interfere with ball making. Heinrich and Bartholomew (1979a) describe an extreme case from East Africa, where tens of thousands of beetles aggregate in elephant dung pats during the peak dung beetle season, and the large roller *Kheper laevistriatus* is often unable to make a ball because other beetles (tunnelers) have crumbled the dung pat so badly.

Most rollers, especially the larger species, have low to very low fecundity, down to one offspring per breeding (rainy) season in, for example, *Kheper nigroaeneus* and *Circellium bacchus* (Table 3.2). Generally, only one egg is laid per dung ball, and only one ball is obtained from a particular dropping (Halffter and Matthews 1966; Halffter 1977; Halffter and Edmonds 1982). If the female prepares several nests during the breeding season, she needs to search for another dropping after the previous nest has been completed. Laying single eggs per (resource obtained from one) dropping decreases intraspecific spatial aggregation in comparison with species that lay larger clutches (Atkinson and Shorrocks 1984). The large *Kheper platynotus* makes only one nest per breeding season, into which a pair of beetles rolls an exceptionally large ball of elephant dung, with a diameter up to 9 cm. Depending on the size of the ball, the female prepares, in the nest, from one to four brood balls, each of which receives one egg (Sato and Imamori 1986a,b, 1987, 1988). Most frequently the female makes two brood balls (Sato and Imamori 1987), which may indicate that competition is often so severe that beetles are not able to make balls as large as they could roll and bury, and consequently females may not be able to realize their maximal fecundity.

If the dwellers and many rollers represent the two end points of the continuum from variance-covariance to lottery dynamics, the situation is more complicated with the tunnelers. Small tunnelers, such as the approximately 2000 species of *Onthophagus*, may be expected to have similar levels of aggregation between droppings as *Aphodius*. Dung pats are generally large in comparison with the size of these beetles, and large numbers may accumulate in or under single droppings. Laboratory experiments have shown that *Onthophagus binodis*, *O. ferox*, and *O. gazella* rear fewer offspring per female with increasing beetle density at dung pats (Ridsdill-Smith 1990; Ridsdill-Smith et al. 1982; Lee and Peng 1982), indicating that the reproductive success of beetles present in the most crowded droppings is impaired. In contrast to the small species, one dung pat is often occupied by just one pair of the largest tunnelers, and any other arriving beetles fly away. Examples include *Heliocopris* (Kingston and Coe 1977) and large Coprini (Ridsdill-Smith 1990). Thus the size of the tunneler in relation to the size of the dropping is expected to be a major determinant of the type of dynamics, with small species approaching the variance-covariance dynamics but large ones better fitting the assumptions of the lottery model. Field observations support this hypothesis, showing that small species have significantly more aggregated distributions than large species (Fig. 16.1).

There is much variation in the breeding behavior of tunnelers, which, apart from the size of the beetle, will influence the type of competitive dynamics. For example, the time spent inside droppings before completing nest construction varies greatly (Edwards and Aschenborn 1987; Doube, Giller, and Moola 1988). Giller and Doube (1989) studied intraspecific and interspecific competition in two genera of tunnelers, *Copris* and *Onitis*, in southern Africa. While *Copris* start to bury dung quickly after arrival at a dropping, *Onitis* typically provision their nests over many days (possible reasons for this difference are discussed in Chapter 20). In both genera, increasing the density from one to eight pairs per cattle pat (1.5 kg) decreased the amount of dung buried per pair and hence the expected production of offspring per female (Fig. 17.1). In the large *Catharsius tricornutus*, the amount of dung buried increased with increasing numbers of beetles, but there was no increase in the total amount of dung that was successfully transformed into brood masses (Giller and Doube 1989), indicating severe competition between pairs of beetles at densities beyond one pair of beetles per pat. Unfortunately, we do not know how many pairs of beetles would stay breeding at one dropping in the field, and how many beetles would emigrate, and at what point, in search of a less crowded pat. Clearly, if the "extra" beetles emigrate (which is more likely in *Copris*) and the rest achieve their maximal fecundity, the situation approaches the lottery dynamics as, in rollers; whereas if all or most beetles remain to breed even if their reproductive success is impaired by competition (more likely in *Onitis*), the dynamics are more toward the variance-covariance end of the continuum.

To summarize this section, dwellers and small tunnelers fit the assumptions of the variance-covariance model, while rollers and fast-burying large tunnelers are expected to obey the lottery dynamics.

Two Trade-offs Affecting Rollers

Rolling dung balls creates two related trade-offs in rollers: (1) a trade-off between a morphology well suited for making and rolling balls and a morphology suited for burrowing; and (2) a trade-off between the ease with which a ball is rolled and the size of the ball, the latter being related to the amount of resource the ball contains for the offspring.

Fig. 17.2 illustrates the legs of representative rollers and tunnelers. The legs of tunnelers are relatively short and thick, and the beetle possesses other morphological adaptations for improved digging ability: thickened exocuticle and other strengthening devices, and specialized structures in the head and front tibiae for moving soil (Halffter and Edmonds 1982). In contrast, the legs of rollers, particularly the hind tibiae, are narrow and elongated, apparently selected for ball construction and rolling. *Neosisyphus* is an extreme example with its exceedingly elongated legs (Fig. 17.2). Since the rollers avoid competition for space below the dropping, there may be no extreme need for them

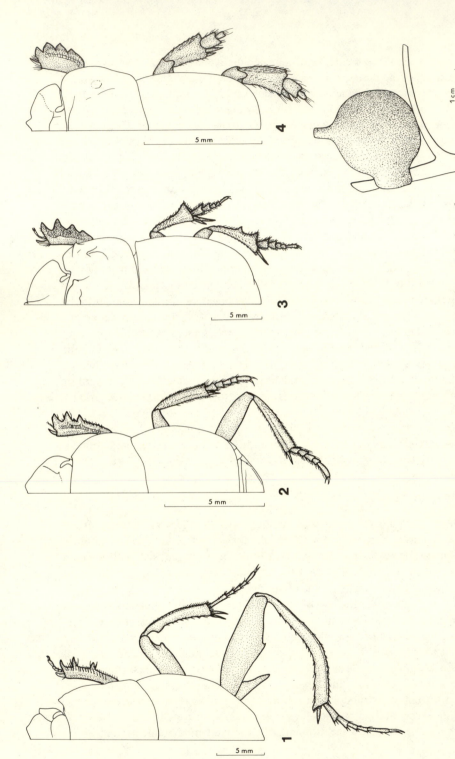

Fig. 17.2. A brood ball of *Neosisyphus spinipes* cemented onto a twig (from Paschalidis 1974), and the legs of two rollers (Scarabaeinae) and two tunnelers (Coprinae), ranging from an "extreme roller" (number 1) to an "extreme tunneler" (number 4). The species are: 1, *Neosisyphus spinipes* (Sisyphini); 2, *Canthon indigaceus* (Canthonini); 3, *Copris interioris* (Coprini); and 4, *Dendropaemon planus* (Phanaeini) (the scale is 5 mm).

to construct the nest in a hurry, apart from possible risk of predation during ball rolling (about which little is known). The rollers can select a favorable spot for burrowing, but there must be soils that are so difficult to penetrate for dung beetles generally that many rollers have great difficulty in making a burrow (Fig. 8.6). The nests of rollers are often shallower than the nests of comparable tunnelers, though the depth varies much between species (Edwards and Aschenborn 1987) and depends, for example, on soil moisture (Rougon and Rougon 1983; Edwards 1986a; Chapter 13). To take an example, the large tunneler *Heliocopris dilloni* typically constructs its breeding chamber at the depth of one meter in East Africa (Kingston and Coe 1977), while the syntopic large rollers, *Kheper platynotus* and *K. aegyptiorum*, dig their breeding chambers at depths of 30–40 cm and 20 cm, respectively (Sato and Imamori 1986a,b). As a testable working hypothesis, we suggest that the faster a roller is in making and rolling balls, the worse it is in digging burrows.

The size of the dung ball is related to the size of the beetle that has prepared it (Table 17.6; Halffter and Matthews 1966). Balls made by single beetles for feeding tend to be smaller than brood balls, which are often rolled by a pair of beetles (Table 17.6). The food balls are also made in a shorter period of time than brood balls; for example, in *Canthon pilularius* the average time to make a food ball is 15 min, whereas brood balls take more than 30 min to prepare (Matthews 1963). The brood balls may be so large that single beetles have difficulty in rolling them (e.g., *Kheper platynotus*; Sato and Imamori 1987).

The ratio of the weight of the ball to the weight of the beetle varies by the order of magnitude shown in the data in Table 17.6, from 6 in *Sisyphus* to 79 in *Kheper platynosus* (see also the photographs in Chapter 3). Some of this variation is probably due to the difference between food balls and brood balls; but even allowing for this difference, it remains true that some species make relatively much larger balls than others and are hence able to provide their offspring proportionately more of the resource. A comparison of the species that use omnivore and large herbivore dung reveals that the former tend to have smaller balls in relation to their size than the latter (Table 17.6). Because omnivore dung is potentially of higher quality than herbivore dung (Chapter 3), the species using omnivore dung may possibly compensate for the relatively small amount of resource they are able to provide their offspring by providing higher quality. These figures may be further compared with the amount of dung that tunnelers provision per offspring (Fig. 17.3). There is no significant difference in this respect between the tunnelers and the rollers that use herbivore dung, though the rollers tend to provision somewhat less food per offspring than the tunnelers; but both of these groups provide significantly more food per offspring than do rollers using omnivore dung. The amount of dung used for breeding per dung pat and per pair of beetles is, however, more than an order of magnitude greater in tunnelers than in rollers (Doube, unpubl.), independent of beetle size, because the nests of tunnelers contain many brood balls and offspring.

TABLE 17.6
Relationship between body size and ball size in rollers (fresh weight in g).

Species	Ball	Body Weight		Ball Weight		Ratio	Ref
		Mean	SD	Mean	SD		
Scarabaeus goryi	S	1.7	0.58	19.9		12	1
Scarabaeus sacer		2.0		30.0		15	2
Kheper platynotus	S	1.8		24.0		13	3
Kheper platynotus	P	1.8		142.0		79	3
Allogymnopleurus youngai	P	0.17		1.9		11	1
Gymnopleurus coerulescens	P	0.062	0.021	0.89	0.23	14	1
Garreta nitens	P	0.31	0.069	4.1	0.70	13	1
Anachalcos convexus	S	1.2		10.6		9	1
Anachalcos cupreus	S	1.9		11.9		6	1
Sisyphus biarmatus	P	0.046	0.006	0.61	0.24	13	1
Sisyphus seminulum	S	0.005	0.002	0.029	0.010	6	1
Sisyphus seminulum	P	0.008	0.002	0.083	0.020	10	1
Canthon pilularius	P	0.3		4.2		14	4
Neosisyphus paschalidisae	P	0.104	0.010	3.7	1.0	36	1
Neosisyphus barbarossa		0.093		1.2		13	5
Neosisyphus calcaratus		0.065		0.56		9	5
Neosisyphus infuscatus		0.074		1.19		16	5
Neosisyphus fortuitus		0.141		2.36		17	5
Neosisyphus rubrus		0.074		1.50		20	5
Neosisyphus mirabilis		0.122		2.00		16	5
Neosisyphus spinipes		0.079		2.57		33	5

References: 1, Cambefort (pers. obs.); 2, Marsch (1982); 3, Sato and Imamori (1987); 4, Matthews (1963); 5, Paschalidis (1974).

Notes: S and P indicate that the ball has been made by a single beetle or a pair of beetles, respectively. Cambefort's (pers. obs.) results are based on weight measurements; the other values have been estimated from known ball diameter and by comparing, if necessary, the beetle to other species with known fresh weight.

The data in Fig. 17.3 suggest another constraint on rollers: if a species is not able to make an exceptionally large ball, it may be restricted to use omnivore dung, which is of relatively high quality. A related factor is the time needed to construct the ball. The *Sisyphus* species, which use omnivore dung, make their balls very quickly (in minutes) in comparison with the *Neosisyphus* species, which use herbivore dung. The latter species have highly evolved camouflage behavior (Chapter 9), which presumably decreases the risk of predation while the beetles are feeding in dung pats without making food balls and while they are slowly making their brood balls. There is much variation in the ratio of the ball size to the beetle size in *Neosisyphus* (Table 17.6). Whether a part of this variation is related to differences in food choice is not known. The two South African species that definitely use cattle dung, *N. spi-*

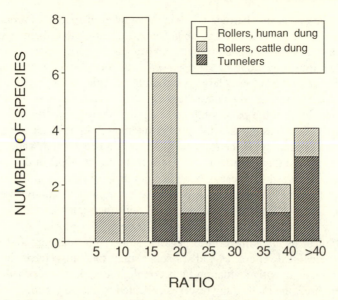

Fig. 17.3. Ratio of the amount of dung provided per offspring to the size of the beetle in rollers using omnivore dung, in rollers using herbivore dung and in tunnelers. Data for rollers are from Table 17.6, data for tunnelers are from the following sources: Kingston and Coe (1977), Klemperer (1981, 1982a, 1982b), Klemperer and Boulton (1976), Kirk (1983), Rougon and Rougon (1980), Cambefort (1984, unpubl.), and Ridsdill-Smith et al. (1982). Rollers using omnivore dung have significantly smaller ratios than rollers and tunnelers using herbivore dung (P<0.001).

nipes and *N. rubrus*, make relatively large balls (Table 17.6), for which reason they were considered suitable for introduction into Australia (Paschalidis 1974).

Competitive Hierarchy

It is clear that those species of rollers and tunnelers that are able to remove the resource to their underground nests rapidly are competitively superior to the dwellers, which can breed successfully only in droppings that remain relatively intact for several weeks. If the numbers of tunnelers and rollers are large, the dwellers have no chance of breeding (unless they adopt the klepto-parasitic life-style; see below and Chapter 4).

This obvious competitive asymmetry between the three functional groups explains well a major trend in the geographical distribution of dung beetles: the dwellers typically comprise an insignificant component in those tropical dung beetle assemblages in which rollers and tunnelers dominate and competition is severe (Chapters 4 and 19). It is noteworthy that the conspicuous decline with decreasing latitude is not in the number of species of Aphodiinae (Table 19.3) but in their average abundance, indicating that there is nothing intrinsically adverse in all subtropical and tropical environments for Aphodiinae. When the competitive pressure is weak, dwellers can do well at low latitudes. Bernon (1981) found large populations of Aphodiinae in cattle pats in South Africa in winter, when other beetles were relatively scarce. At this time Aphodiinae bred in the pats, while in summer, when competition from tunnelers and rollers was intense, the smaller number of Aphodiinae bred, atypically for the subfamily, in the soil below the pat. Cambefort and Walter (Chapter 11) compared dung beetle assemblages that exploit elephant dung pats in savannas and forests in West Africa. The numbers of beetles in forests were relatively low, and elephant dung pats were not removed rapidly like in the savanna. The dweller *Oniticellus pseudoplanatus* was abundant in the forest, apparently benefiting from weak interspecific competition.

The typical dwellers have little negative impact on tunnelers and rollers. Some species of dwellers, as well as some small tunnelers, have, however, reversed the competitive situation by becoming kleptoparasites of rollers and tunnelers, breeding in the dung reserves collected by the larger species in their nests. The offspring of the host species is assumed to be killed, though there are not many positive observations from the field (Chapter 13). Kleptoparasitic dung beetles are known from temperate regions (Hammond 1976), but they are especially common in subtropical and tropical grasslands (Chapter 9) and even in arid regions (Chapter 13). In the species-rich beetle assemblage in West African savannas, about 10% of all dung beetle species are kleptoparasites (Chapter 9).

While the interaction between nonkleptoparasitic dwellers and other beetles is clearly asymmetric, it is less straightforward to draw conclusions about possible asymmetries in the interactions between the rollers and the tunnelers. In situations where preemptive resource competition is severe, the crucial question centers on who manages to remove the resource fastest. The removal time has two major components: the time it takes to locate and arrive at a dropping, and the time it takes to remove the resource from the reach of other beetles.

There is no reason to assume that either rollers or tunnelers should be inherently faster at locating droppings; both are often extremely fast. In tropical forests in Ecuador, Peck and Forsyth (1982) found that human dung was typically discovered in less than one minute. Smaller species were the first to arrive, but this may simply reflect their greater numbers. An important consideration, however, is that although most rollers are typically diurnal, most of the large and potentially competitively dominant tunnelers are nocturnal (Chapters 8 and 18). If the rate of resource removal is rapid, differences in diel activity give a definite competitive advantage to both diurnal and nocturnal species at droppings that were deposited during the day and the night, respectively (Chapter 12).

As to the process of food removal from droppings to underground nests, the rollers seem to have an advantage over those tunnelers that are active at the same time of day. Rollers can start preparing a dung ball immediately after landing at a dropping, and once the ball has been made the equivalent amount of resource is away from the tunnelers, though not necessarily from other rollers, which may attempt to steal the ball (Section 17.1). The time needed to prepare a ball for transport varies widely, from less than 2 minutes (*Kheper laevistriatus*; Bartholomew and Heinrich 1978) to more than an hour (*Neosisyphus* species; Cambefort, pers. obs.). A typical figure for many species is 15–30 min (Marsch 1982: *Scarabaeus sacer*, 32 min; Peck and Forsyth 1982: *Canthon angustatus*, 4–30 min; Matthews 1963: *Canthon pilularius*, >30 min; Sato and Imamori 1986b: diurnal *Kheper*, < 1 hour).

The tunnelers are not ready to start removing dung to their burrow before the burrow has been constructed. Peck and Forsyth (1982) found that *Oxysternon conspicillatus* took only 15–20 min to dig a burrow. We would guess that most tunnelers are not so fast, but quantitative figures are not available. Some large tunnelers, for example, *Heliocopris* species, first construct a rough gallery into which dung is taken, before constructing the final nest, most probably to speed up dung removal from the dung pat. Another trick used by the largest species of *Heliocopris* is to rapidly cover the entire dropping with soil, a process that is faster than removing the entire dropping underground (Cambefort, pers. obs.). The time needed to cover the dropping may be equivalent to the time it takes to make a ball; hence these large tunnelers may be comparable in speed to large rollers. A female *Heliocopris dilloni* completed provisioning her nest in 6 hours (Kingston and Coe 1977).

Doube, Giller, and Moola (1988) make a distinction between two patterns of dung burial in tunnelers. Coprini (*Copris*, *Catharsius*, *Metacatharsius*) typically complete dung burial within 1–2 days, while many Onitini (*Onitis*) bury dung for many days, up to 6 weeks in *O. caffer* (Edwards 1986a). The first type of beetles, the fast-burying tunnelers, obviously have a competitive advantage over the second type, the slow-burying tunnelers.

In conclusion, the above considerations suggest the following generalized competitive hierarchy in dung beetles: (1) rollers and fast-burying tunnelers, of which many rollers may often be marginally dominant; but because there tends to be a difference in the diel activity of the two kinds of beetles, the possible small difference is not likely to be important in the field: the first beetles to arrive have a competitive advantage; (2) slow-burying tunnelers and small tunnelers; and (3) dwellers. However, there are two further factors that enter the picture and need to be considered before the outcome of competition in the field can be understood. The first complication is an interaction between competitive ability and soil type, the second one is an interaction between competitive ability and fecundity. Both interactions facilitate the coexistence of inferior competitors with the top competitors.

*Interaction between Competitive Ability
and Soil Type*

Soil type has only indirect effects on the majority of dwellers, which spend their entire life cycle in droppings; but it is of decisive importance to tunnelers and rollers, which construct underground burrows and nest chambers for their offspring. The tunnelers are, as their name implies, specialist burrowers, with a morphology clearly adapted for digging (Fig. 17.2). The rollers, as we have discussed above, have a dilemma: they have two tasks that clearly call for different sorts of morphologies—namely, carving and rolling a ball and making a burrow for it. The difference in the morphology and, by implication, in the burrowing ability of the two groups of beetles suggests that the rollers suffer more than the tunnelers from soils that are difficult to dig in. The good competitive ability of the fast-burying tunnelers is based on speed of burrowing, which must be affected by soil type; hence one may expect an interaction between competitive ability and sensitivity to soil type also among the tunnelers.

Figure 17.4 illustrates our working hypothesis about a negative relationship between species' intrinsic competitive ability and their sensitivity to the soil

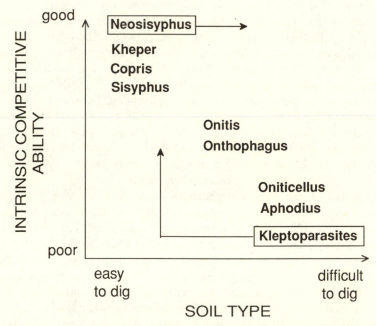

Fig. 17.4. A schematic illustration of the assumed negative relationship between the intrinsic competitive ability and the sensitivity to the soil type in dung beetles. Many superior competitors are restricted to soils that are easy to dig, while inferior competitors find a refuge in soils that are difficult to dig. See the text for discussion.

type. The rollers and the tunnelers with the highest intrinsic competitive ability are expected to dominate where the soil is most favorable for burrowing; the dwellers are predicted to dominate where burrowing is difficult, or nesting in the soil is impossible for other reasons; and the remaining species occupy the middle ground.

Some beetles at the two extremes of the scheme in Fig. 17.4 have escaped the constraints imposed by soil type and low intrinsic competitive ability, respectively. Some *Neosisyphus* species, which have the longest legs in dung beetles but which are clearly poorly suited for burrowing, have adopted an alternative nesting behavior: they attach their brood balls to grass stems (*N. spinipes*; Fig. 17.2) or simply leave the ball on the soil surface after laying an egg in it (Paschalidis 1974; Cambefort 1984), thus avoiding nest construction entirely. *Neosisyphus spinipes*, originally an African species, does well in Australia in situations where the grass is not too much affected by drought or intensive grazing (Chapter 15). At the other extreme shown in Fig. 17.4, some dwellers and small tunnelers have escaped their low status in the competitive hierarchy by becoming kleptoparasites, using the dung reserves in the nests of competitively superior species.

The following competition model modified from Lande (1987; see also Hanski 1991) illustrates the population-dynamic consequences of the trade-off between good competitive ability and elevated sensitivity to soil type. Within any region, there is likely to be small-scale variation in soil characteristics, such that some dung pats are located on more favorable spots for digging by beetles than others. Let us consider two species: a competitively superior species 1, which is sensitive to soil type; and a competitively inferior species 2, which is less affected by soil type. For simplicity, let us assume that each female breeds in one dung pat at most (as many dung beetles do). Each female is able to search for m pats, but she perishes without breeding if the first m pats to be located are on unsuitable soil or, in the case of species 2, they are occupied by a superior competitor. For simplicity, let us assume that the dung pats may be located on three kinds of soil patches: the worst patches (fraction u_1 of all patches) are unsuitable for both species; the best patches (fraction h) are acceptable for both species; and the remaining patches (fraction u_2) are suitable for species 2 but unsuitable for species 1. Competition is entirely asymmetric: species 2 can establish itself and breed in a dung pat only in the absence of species 1, while species 1 is not affected by species 2. If p_1 is the fraction of pats occupied by species 1 ($0 \leq p_1 \leq 1$), at equilibrium (Lande 1987),

$$(1 - (u_1 + u_2 + p_1 h)^m) R_0' = 1, \tag{17.1}$$

and

$$p_1 = 1 - (1 - k_1)/h, \tag{17.2}$$

where R_0' is the net lifetime production of female offspring per female, conditional on the mother's finding a suitable pat for breeding (for details, see Lande 1987), and $k_1 = (1 - 1/R_0')^{1/m}$. The equilibrium fraction of sites occupied by species 2 is

$$p_2 = 1 - (1 - k_2)/(u_2 + (1 - p_1)h). \tag{17.3}$$

Fig. 17.5 shows that, depending on the structure of the environment—in other words, on the values of u_1, u_2, and h—and depending on fecundity, longevity (R_0'), and the foraging ability (m) of the two species, neither species can maintain a population in the environment, species 1 or species 2 will occur alone, or the two species may coexist. The model could be extended to multispecies assemblages, where the species composition would change with changing habitat quality (soil type, in this case). The point is that the best competitors are expected to dominate on the soils that are easist to dig, whereas the inferior competitors are expected to dominate on harder soils.

The best data set available to test this prediction comes from the subtropical savannas in South Africa, where Doube (Chapter 8) conducted a factorial trapping experiment involving the soil type (deep sand versus clay-loam), vegetation cover (grassveld versus bushveld), and type of habitat (pasture versus game reserve). The species were divided into five functional types, as explained by Doube (Chapter 8): large rollers, small rollers, fast-burying (large) tunnelers, and slow-burying large and slow-burying small tunnelers (omitting the dwellers, which were uncommon). Table 17.7 gives the results of an analysis of variance for the pooled biomass of beetles in the five functional groups. Vegetation cover (grassveld or bushveld) had no significant effect on beetle biomass, though there are individual species that strongly prefer one or the other of the two habitats (Chapter 8). The remaining three factors had a sig-

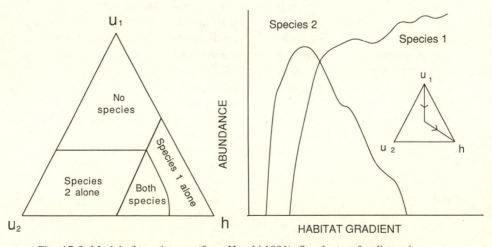

Fig. 17.5. Model of coexistence (from Hanski 1991). See the text for discussion.

TABLE 17.7
Analysis of variance for the pooled biomass of dung beetles in south African savanna.

Factor	df	SS/1,000	F	P
Functional group of beetles	4	320	6.50	0.0011
Soil type	1	90	7.35	0.0122
Habitat type	1	139	11.36	0.0025
Functional group × soil	4	504	10.26	0.0001
Functional group × habitat	4	108	2.19	0.1005
Habitat type × soil	1	44	3.55	0.0718
Residual	24	295		
Total	39	1,499		

Source: Data from Table 8.3.

Note: Vegetation cover had no significant effect on beetle biomass and was omitted from the analysis.

nificant effect (Table 17.7); but the most interesting point is that, of the two-factor interactions, only the soil type versus functional group interaction was significant, and very strongly so (Table 17.7). The biomass of large rollers (superior competitors) was an order of magnitude higher on sandy soil than on clay-loam, while the slow-burying tunnelers (inferior competitors) showed the opposite pattern, having a biomass twice as high on clay-loam soil as on deep sand (see Table 8.3). It is difficult to see anything intrinsically unfavorable in the sandy soils for the tunnelers; it is more likely that their numbers are suppressed by the large numbers of competitively superior large rollers on sandy soils. The biomass of the fast-burying tunnelers was lower on clay-loam than on sandy soil, though the difference was not as great as in large rollers. The fast-burying and slow-burying tunnelers are differently affected by soil type, and Doube's (Chapter 8) distinction between the two groups is an important one.

Interaction between Competitive Ability and Fecundity

Large rollers and the largest and competitively most dominant tunnelers have very low fecundity, down to the absolute minimum of one offspring per breeding season in, for example, *Kheper nigroaeneus*, a dominant species in the dung beetle assemblage studied by Edwards (1988a,b) in southern Africa. Many large rollers and tunnelers produce around five offspring per breeding season (Table 3.2), though this total may be achieved in different ways: some species (typically rollers) produce a series of clutches of only one from a series of dung pats, while others (typically tunnelers) produce a clutch of, say, five from a single dung pat. The competitively dominant fast-burying tunnelers have lower fecundity than the competitively inferior slow-burying tunnelers—around five offspring per breeding occasion in the former, versus twenty to forty offspring per breeding season in the latter (Doube, Giller, and Moola 1988, and references therein). Small tunnelers have even higher lifetime fecundity—for example, up to one hundred offspring or more in *Onthophagus binodis* (in the laboratory; Ridsdill-Smith et al. 1982). In *Aphodius* lifetime fecundity is also typically high, up to fifty offspring or more (Table 3.2).

Large rollers and fast-burying tunnelers are superior competitors because of their ability to remove dung to their burrows quickly. To ensure that their nests remain undisturbed by other, later-arriving beetles, the nest must be constructed deep in the soil (tunnelers) or away from the food source (rollers). Large size is an advantage in digging deep tunnels fast; but as the size of the beetle becomes large compared to the size of the dropping, the number of offspring that may be produced becomes necessarily small, creating a selection pressure toward a high level of parental care. Interference competition among rollers (Section 17.1) selects for larger size, though many small rollers are also highly efficient competitors (e.g., Sisyphini). Small rollers typically construct many nests in rapid succession during the breeding season; but many large rollers stay in the first nest caring for their offspring, and they have consequently very low fecundity. Large rollers may often have difficulty in finding a dung pat where a good brood ball could be prepared (Heinrich and Bartholomew 1979a), which may have selected for increased parental care. Good competitive ability, low fecundity, and high level of parental care constitute a single syndrome of large dung beetles.

The inferior competitors can breed successfully only in droppings that for some reason were not much used by the superior competitors; but the inferior competitors, which are typically smaller than the superior ones, may produce a large number of offspring in one such dropping. One mechanism allowing the coexistence of the two kinds of beetles is a variation of the one described by Kotler and Brown (1988) for desert rodents: due to trade-offs in travel costs and foraging efficiency, some species may specialize on rich resource patches, while others use low-quality patches more efficiently. The dung beetle case is very asymmetric, however, and if all or most fresh dung pats were suitable for and were detected in time by the superior competitors, the inferior competitors would be excluded from the community. In practice, the superior competitors are not likely to use all dung pats. These species search for fresh dung, and some pats may remain little used while fresh—for example, because of temporarily unfavorable weather for foraging. Other pats may be located in microhabitats that are unsuitable for the superior competitors—for example, pats located on hard soil. The important point is that the relatively small size and high fecundity of the inferior competitors help them survive even in a highly competitive situation. Even a single dung pat not used by the superior competitors may produce dozens of individuals of an inferior competitor.

17.3. SUMMARY

Intraspecific and interspecific competition occur occasionally in nearly all dung beetle communities, and in some communities, especially in tropical grasslands and forests, competition is constantly or seasonally severe: droppings are removed by beetles in a matter of hours. Competition may occur between adults and larvae, and beetles may compete for food and for space, the latter either in droppings or in the soil underneath, where tunnelers construct their nests. The kind of competition that may occur is dependent on the breeding behavior of the species. Paradoxically, competition is most obvious in the species in which only adult beetles compete only for food (rollers), while competition is generally weakest in the species in which both adults and larvae may compete for both food and space (dwellers). We have given examples of various types of competition.

We discussed a competitive hierarchy in dung beetles, in which good competitors are able rapidly to remove food from the reach of other beetles. The top competitors are rollers and the large, fast-burying tunnelers; the weakest competitors are the dwellers. Competition in the rollers and large tunnelers is of the lottery type, while the variance-covariance model gives a better description of competition in small tunnelers and dwellers. In rollers, the two tasks of making and rolling balls and digging burrows in the soil call for different types of morphology, and we may assume that rollers are generally more inept than tunnelers in making burrows. The good competitive ability of fast-burying tunnelers is based on the speed at which dung can be buried, and the speed must be affected by soil characteristics. We therefore propose that there exists a negative relationship between the competitive ability and the sensitivity of the species to soil characteristics. This hypothesis predicts a shift from the dominance of rollers and fast-burying tunnelers on sandy soils (easy to dig) to the dominance of slow-burying tunnelers and dwellers on more difficult soil types for dung beetles. Quantitative data from African savannas and numerous other observations support this prediction. Small-scale heterogeneity in soil characteristics may provide spatial refuges for the inferior competitors, while dung pats not detected rapidly by the top competitors may be used efficiently by inferior competitors with relatively high fecundity.

CHAPTER 18

Resource Partitioning

Ilkka Hanski and
Yves Cambefort

IN CHAPTER 16 we emphasized the role of spatial aggregation in dung beetle ecology, and how intraspecific aggregation between droppings may promote coexistence especially in dwellers and small tunnelers, the two groups of dung beetles with the largest numbers of species (Chapter 4). This perspective was inspired by the fact that most dung beetles use essentially the same resource, which they find in the same place. Hence classical resource partitioning seems a priori an insufficient explanation of coexistence of many similar species frequently engaged in severe competition (Chapter 17). Nonetheless, it must be recognized, on balance, that there are various kinds of ecological differences among cooccurring species of dung beetles, many of which may facilitate coexistence. For instance, there are subtle and not so subtle differences in the type of dung used, and how it is used, by different species; there are differences in diel activity and seasonality; and there are differences in habitat selection at small and large spatial scales. Not only in dung beetles but in all animals in general coexistence of competitors in natural communities is affected by many factors. Various kinds of nonequilibrium mechanisms of coexistence have been much explored recently (Chesson and Case 1986), but as emphasized by Chesson and Huntly (1988), the nonequilibrium mechanisms do not wipe out resource partitioning and other equilibrium mechanisms—the challenge is to develop an integrated theory embracing both equilibrium and nonequilibrium mechanisms.

Schoener's (1974) analysis of resource partitioning in animals led to the conclusion that "habitat dimensions are important more often than food-type dimensions, which are important more often than temporal dimensions." Dung beetles fit the first part of Schoener's generalization: the contrast between grassland and forest assemblages is indeed great in tropical biomes where most species are found (Chapters 11 and 15). What seems to be special about dung beetles is the critical importance of temporal dimensions.

In this chapter we discuss resource partitioning in dung beetles based on information in Part 2 of this book and in the literature. We focus on four sets of dimensions. First, the kind (quality) and size (quantity) of resource patches selected by coexisting species: What do they explain about dung beetle guilds and communities? Second, dung beetles span four orders of magnitude in size, and size differences may affect their competitive interactions. Third, dung beetles show habitat selection at the scale of a single dropping and at larger spatial scales, the latter often based on soil type and vegetation cover. And fourth, most dung beetle assemblages have well-defined guilds of diurnal and nocturnal species using the same resources, and in many biomes coexisting species show striking phenological differences not explicable by specialization to different resources. In the final section of this chapter we turn to multidimensional resource partitioning via an example from West African savannas, which has one of the most species-rich communities of dung beetles found anywhere. The main conclusion we draw from this analysis is that the four dimensions discussed in the earlier sections collapse into two composite dimensions: size of the species is correlated with diel activity, and food selection is related to seasonality. The dominant species in the community are distributed nonrandomly in the space defined by these composite dimensions.

18.1. SELECTION OF RESOURCE TYPE

The two main types of food resource used by dung beetles are (large) herbivore dung and omnivore dung. In tropical forests in particular, many dung beetles are attracted to both dung and carrion, or even exclusively to carrion, for reasons to be discussed below. Carnivore dung attracts relatively few dung beetles, which comprise a mixture of species that colonize dung and carrion (Hanski 1987a), undoubtedly reflecting differences and similarities in the chemical composition of these resource types. A small number of species, mostly found in tropical forests, has specialized on very particular resources, or has a very particular foraging behavior. A broad generalization, though largely based on experience from Africa, is that rollers tend to use relatively more omnivore than herbivore dung, while many large tunnelers use herbivore dung exclusively. We have suggested in Chapter 17 that the limited amount of food that the rollers are able to secure in a transportable ball may bias their food selection toward high-quality omnivore dung, while the large size of many tunnelers makes breeding impossible except under the large droppings of large herbivores.

The extensive data from the Ivory Coast in West Africa (Chapters 9 and 11) allow a quantitative comparison of the numbers of omnivore and herbivore dung specialists and of generalist species. If a specialist is arbitrarily defined as a species with more than 90% of individuals (see Appendixes B.9 and B.11) caught from one food type, the figures for savanna are forty-three omnivore dung specialists, sixteen herbivore dung specialists, and thirty-one generalists, while the respective figures for rain forest are eighteen, eighteen, and two (rare species with fewer than ten individuals trapped have been excluded). These figures demonstrate that most species use primarily one or the other of the two main types of dung.

Species Specializing in Large Herbivore Dung

The most obvious group of species specializing in large herbivore dung consists of the large, nocturnal tunnelers, especially Coprini, though herbivore dung is strongly selected also by Oniticellini, which are mostly small, diurnal species, and by some large (*Kheper*) and small rollers (*Neosisyphus*). The comprehensive data sets from subtropical and tropical savannas in Africa give a more detailed picture of food selection in large assemblages of beetles and demonstrate that there are species specializing in herbivore dung in nearly all groups of species (Appendix B.9), suggesting frequent divergence in food selection in many taxa. The largest dung beetles may use various kinds of dung for their own nutrition, but during the breeding season they are constrained to use large herbivore dung because of the large quantity of food needed for their larvae. For example, the large *Heliocopris* species bury from 1.8 to 2.6 kg of dung for brood formation (Klemperer and Boulton 1976; Kingston and Coe 1977), and the largest species can breed only under elephant dung pats (e.g., *Heliocopris dilloni* and *H. colossus*; Kingston and Coe 1977).

Species Specializing in Omnivore Dung

While many species strictly specialize in large herbivore dung, there is generally a less strong preference by beetles for omnivore dung. Some species show seasonal shifts in the type of resource used. In West African savannas, many *Pedaria*, *Caccobius*, and *Cleptocaccobius* species select omnivore dung early in the season, soon after the rains have started, but later on shift to herbivore dung (Chapter 9; Appendix B.9). Young (1978) has reported seasonal shifts from coprophagy to necrophagy in dung beetles on Barro Colorado Island in Panama. It is possible that such shifts are related to varying food selection between feeding (immature) and breeding (mature) beetles. Mobile adult beetles may search for the nitrogen-rich omnivore dung for their own nutrition, even if they provision their nests with the often more abundant and carbohydrate-rich herbivore dung. An analogous difference in the nutrition of adults and larvae has been commonplace in the evolution of coprophagy in beetles (Chapter 2). It is also possible that shifts in food selection are related to seasonally varying quality of herbivore dung. Kingston (1977) found that the nitrogen content of elephant dung varied seasonally from 1% to 5%, being highest in the middle of the rainy season (see Chapter 15 for studies conducted in Australia). Still missing are accurate field observations on the kind of dung used for feeding and breeding by different sorts of dung beetles.

Resource Selection in Tropical Forests

One of the tenets about animal communities in tropical forests is that they consist of a very large number of highly specialized species (MacArthur 1972; Pianka 1974; Pielou 1975). This is not so with tropical forest dung beetles. In Africa, there are more species of dung beetles in savannas than in forests (Chapter 11), presumably because resource availability to dung beetles is generally higher in savannas, and has been so for a very long time. Second, many tropical forest dung beetles have a broad diet, one that is broader than the diet of their relatives in savannas. Dozens or even some hundreds of tropical forest species worldwide are attracted to both dung and carrion, though it is not known whether they use both resources for breeding as well as for feeding. A parallel shift from a relatively narrow food selection in temperate species to a wide food selection in tropical forest species has been documented for bark and ambrosia beetles, which Beaver (1979) attributes to rarity of most host trees in tropical forests: high degree of specialization would be disadvantageous because of difficulty in locating a particular host species. This quantity hypothesis may be contrasted with a quality hypothesis: there may be some qualitative differences in resource characteristics in tropical forests facilitating or favoring the widening of food selection.

Availability of resources to dung beetles in most tropical forests is relatively low because of the generally low density and biomass of large mammals and low availability of any particular type of dung (Hanski 1989a). Halffter and Matthews (1966) attribute the shift from coprophagy to necrophagy in many South American dung beetles to the general predominance of forests in tropical South America (true until recently), the absence of large numbers of large herbivores in the forests (and on grasslands; Janzen 1983b), and to the relative scarcity of necrophagous insects, potential competitors to necrophagous dung beetles, in South America. This quantity hypothesis may explain why necrophagous dung beetles are significantly less numerous in Africa (Chapter 11) than in South-East Asia (Chapter 10) and South America (Chapters 12, 19). Until recently, large herbivores have been present in relatively large numbers in African forests, where a large number of dung beetles has specialized to use their dung instead of that of omnivorous mammals (Chapter 11; see figures in the beginning of this section).

Two forms of the quality hypothesis are plausible. From a nutritional point of view, carrion is more similar to omnivore dung than to large herbivore dung (Hanski 1987a), and as most tropical forest mammals are omnivores, especially in South-East Asia and South America, tropical forest dung beetles may be expected to use carrion more often than the savanna species. Second, it is possible that many species consume carrion only as adult food while still provisioning their nests with dung. In this case, carrion would represent a high-quality (nitrogen-rich) resource of which the mobile adults may take advantage. We have suggested in Chapter 2 that coprophagy has evolved from sa-prophagy via several steps, the first one being a shift in adult diet to high-quality mammalian dung from the relatively low-quality vegetable diet. In an analogous manner, tropical forest dung beetles may be in the process of mov-ing from dung to carrion.

Extreme Feeding Specializations

Some dung beetles have become extremely specialized either in their mode of foraging or in food selection. To the former category belong the tropical forest species that forage in the canopy (Chapters 10 to 12). Some species may even breed in the canopy, possibly using the soil that accumulates on large branches. In the other extreme, many dung beetles live in the burrows and nests of mammals, especially in arid environments (Chapter 5), while other species are found in caves and in ant nests (Halffter and Matthews 1966).

The constant race to get to the resource first has led to some striking adap-tations. In South America several species of small dung beetles (*Trichillum*, *Pedaridium*, *Uroxys*) live in the pelage of sloths (Ratcliffe 1980; Howden and Young 1981; Chapter 12), and they probably move to the excrement as soon as it is produced. Since sloths bury their dung (Ratcliffe 1980), these species may be the only ones to have access to this resource. Some Australian *On-thophagus* (*Macropocopris*) hold onto the fur of macropods with their prehen-sile tarsal claws (Chapter 15). The female seizes the fresh dung pellet, falls with it to the ground, buries the pellet, and lays one egg in it (Matthews 1972). *Onthophagus bifasciatus*, *O. unifasciatus*, and *Caccobius vulcanus* in India have taken the next logical step and occasionally colonize the resource before it actually exists, by entering the human intestine (Halffter and Matthews 1966; Perez-Inigo 1971). Other forms of extreme food specialization in dung beetles are described by Halffter and Matthews (1966) and by Gill (Chapter 12).

18.2. SIZE DISTRIBUTIONS

Size Differences and Local Coexistence

Size variation in dung beetles covers four orders of magnitude, from the small-est species of *Panelus* (about 1 mg) to the largest *Heliocopris* species, weigh-ing more than 10 g of fresh weight, or four times the size of the smallest mammals, such as the smallest shrews. Size has far-ranging consequences to all aspects of animal ecology (Peters 1983; Schmidt-Nielsen 1984). Size dif-ferences in species using the same resource are often attributed to past com-petition, or to size-dependent assembly of local communities from a regional species pool (MacArthur 1972; Case 1983; Rummel and Roughgarden 1985;

Roughgarden et al. 1987). Many dung beetle communities are intensely competitive (Chapter 17), and several mechanisms related to size may facilitate coexistence:

1. Size may affect food selection. This happens especially in the case of the largest species, which are dependent on the large droppings of the largest herbivorous mammals. The asymmetry between small and large species is reversed when compared with predators (Wilson 1975), since in dung beetles the largest, not the smallest, species have the narrowest range of resources available to them.
2. Size may affect the way a particular dropping is used. For example, Holter (1982) demonstrated differences in the parts of cattle pats used by different species of *Aphodius*, and such differences are likely to be size-dependent. In tunnelers, the use of space in the soil below the dropping must depend on the size of the beetle, at least if very small and very large species are compared.
3. Size has special importance in rollers, because of frequent interference competition and because of the relationship between the size of the beetle and the size of the dung ball (Chapter 17).
4. Our findings (Chapters 9 and 16) about decreasing spatial covariance with increasing size difference in a pair of species suggest yet another, more subtle mechanism by which size differences may facilitate coexistence, by decreasing interspecific aggregation in relation to intraspecific aggregation (Chapter 1).

While examining the spacing of species in size or along some resource dimension, it is informative to compare the most abundant or dominant species with the rest of the species in the community. Species-rich assemblages of invertebrates are likely to be mixtures of species in terms of their status at a particular site (Pulliam 1988): some species have locally breeding populations, while others are represented by ''sink'' populations, maintained by dispersal from elsewhere. Many of the rarer species may not have local populations at all (Chapters 5 and 16); hence size may be irrelevant to their presence or absence in the community, or it may be more related to their powers of dispersal than to their competitive ability. Among the locally breeding populations, interspecific interactions are most frequent among the most abundant species. Thus if coexistence and dominance of competing species is dependent on their size relative to the size of other species, we may expect the most dominant species to be more evenly spaced out in size than a random selection of species (Hanski 1982). This approach avoids many of the difficulties of deciding what is a sensible ''species pool'' to which a particular community should be compared (Harvey et al. 1983).

Several chapters in Part 2 reported on dung beetle assemblages in which cooccurring species are well separated in size. In three cases, in *Aphodius* in northern temperate regions (Chapter 5), in *Phanaeus* in subtropical North America (Chapter 7), and in rollers in African savannas (Chapter 9), the most dominant species in abundance or biomass were better spaced out in size than were species in a random selection of equally many species from the community. Dominance in these assemblages is thus dependent on the species' size relations. Of the above mechanisms, numbers 2 and 4 are likely to be important in *Aphodius*, numbers 1 and 4 in *Phanaeus*, and numbers 1, 3, and 4 in the rollers.

Comparison between Africa and
  South America

Africa and South America have large dung beetle faunas with little taxonomic overlap (Table 4.1). In the rollers, South America is dominated by the old tribe of Canthonini, followed by the endemic Eucraniini and Eurysterniini, while the African communities are dominated by Scarabaeini, Gymnopleurini, and Sisyphini. In the tunnelers, South American communities have large numbers of Dichotomiini, Phanaeini, and Onthophagini, whereas the bulk of the African tunnelers belong to Onthophagini, Coprini, Onitini, and Oniticellini.

In spite of these taxonomic differences, size distributions of rollers and tunnelers are strikingly similar on the two continents (Fig. 18.1). In the rollers, there is a clear distinction between small and large species on both continents, though the small rollers tend to be even smaller in Africa (Fig. 18.1). In the tunnelers, the size distributions of genera have identical parameter values on the two continents (Fig. 18.1). This distribution is not much affected by the very large African *Heliocopris* species. Similar large coprophagous species may have existed in South America before the extinction of the Pleistocene megafauna (Janzen 1983b). The largest extant South American species (*Coprophanaeus*) are necrophagous.

18.3. TEMPORAL RESOURCE PARTITIONING

Diel Activity

Dung beetle communities often consist of about equally large guilds of diurnal and nocturnal species using the same resource. Many species have short diel flight periods (Figs. 9.4 and 10.2); for example, many "nocturnal" species are in fact crepuscular and may fly during less than one hour per 24 hours (Chapter 8). Differences in diel activity are critical in dung beetles for three reasons.

First, if there is interference competition, which can take place only among individuals active at the same time, differences in diel activity increase the relative strength of intraspecific competition. Interference competition is common in dung beetles (Chapter 17). Second, competition is occasionally so severe in dung beetles, especially in tropical savannas and forests and among the species that use the relatively small droppings of omnivorous mammals, that all or most of the resource is removed within a few hours (Chapter 17). In

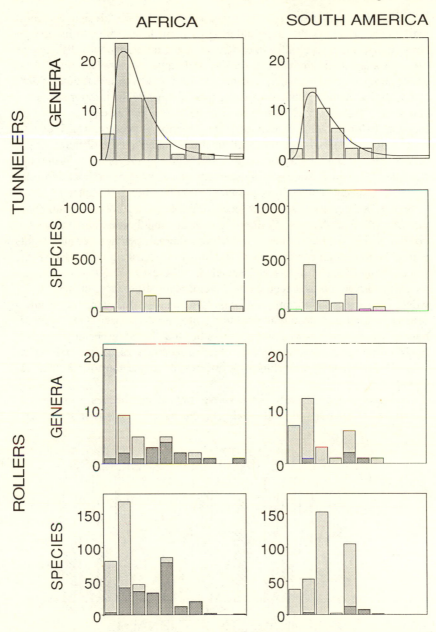

Fig. 18.1. Size distributions of genera and species of rollers and tunnelers in Africa and South America (data from Plates 4.1 to 4.8). The species distribution was calculated by multiplying the genus distribution by the number of species in the genus. In the rollers, Canthonini are shown with a lighter shading (note the major difference between Africa and South America).

this case differences in diel activity give a species a definite advantage in competition for droppings that appear during or immediately prior to its time of activity. Species are typically either diurnal or nocturnal, undoubtedly because different adaptations are required for day and night activity. Therefore, a superior competitor cannot exclude a species that flies at a different time of day, provided that enough resources become available while the inferior species is active, and that the latter is able to remove resources fast enough to a safe location, before the time of activity of the superior competitor. Third, even if beetles do not manage to remove all of the resource within 12 hours or so, differences in diel activity decrease interspecific aggregation in droppings (Chapter 16), which decreases the strength of interspecific competition relative to intraspecific competition and facilitates coexistence (Chapter 1).

Figure 18.2 compares sets of co-occurring diurnal and nocturnal dung beetles from different biomes. Northern temperate dung beetle communities are dominated by *Aphodius*, most of which are diurnal, perhaps because nights tend to be too cold for flight activity (Landin 1961; Koskela 1979). However, since *Aphodius* are dwellers and spend days or even weeks in droppings, where the larvae complete their entire development, differences in diel activity would not be very significant in *Aphodius*, in any case. In contrast, in tropical biomes diurnal and nocturnal (including crepuscular) species are approximately equally numerous, though often either the diurnal (especially in grasslands) or the nocturnal guild (in tropical forests) dominates in terms of biomass (Fig. 18.2), probably reflecting diel differences in the rate of resource renewal.

Diel activity is a relatively conservative trait in dung beetles, as most species in most tribes are either diurnal or nocturnal/crepuscular. In Africa and

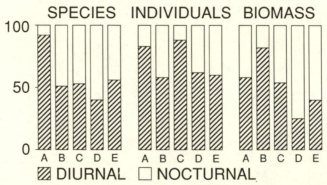

Fig. 18.2. Proportions of diurnal and nocturnal dung beetles in terms of species, individuals, and biomass in five assemblages: A, north temperate pasture (Chapter 5; 25 species); B, subtropical savanna (Chapter 8; 120 species); C, tropical savanna (Chapter 9; 100 species); D, tropical forest in Africa (Chapter 11; 75 species); E, tropical forest in Panama (Chapter 12; 59 species).

probably elsewhere as well, all Sisyphini and Oniticellini are diurnal, while the clear majority of Coprini are nocturnal. Gymnopleurini are mostly diurnal, though there are some nocturnal species of *Garreta*, and even one *Gymnopleurus* (*profanus*) is nocturnal in West Africa. The largest rollers, *Kheper* and *Scarabaeus*, have both diurnal and nocturnal species. Most tropical Onitini are nocturnal, but *Pleuronitis* appears to be diurnal, and so are many temperate species (Doube, pers. comm.). It is tempting to speculate that the exceptional species in largely diurnal or nocturnal tribes have shifted their diel activity because of interspecific competition with congeneric or ecologically similar species. The two most species-rich genera of dung beetles, *Onthophagus* and *Aphodius*, are notable for a variety of diel activities represented by different species (Fig. 9.4). Kohlmann (Chapter 7) gives an example of a species (*Megathoposoma candezei*) that is diurnal in some regions but nocturnal in others.

Doube (Chapter 8) and Cambefort (Chapter 9) have emphasized a broad difference in diel activity among members of the functional groups of dung beetles. As a rule, large tunnelers are nocturnal, while most rollers tend to be diurnal; small tunnelers include both diurnal and nocturnal species. Recalling that, as another broad generalization, rollers tend to use more omnivore dung and large tunnelers use primarily herbivore dung, the difference in diel activity between the two groups has two possible explanations. Either something in the roller behavior is better adapted to daylight conditions and something in the tunneler behavior is adapted to darkness; or the difference in diel activity reflects a difference in diel changes in the rate of production of omnivore and herbivore dung. Among herbivorous mammals, the larger species need to spend more time foraging than smaller species, up to 75% of a twenty-four-hour day in the African elephant, with the result that the ratio of daytime foraging to nighttime foraging decreases with body size (Owen-Smith 1988). Elephants tend to show three peaks of activity: in the early morning, in late afternoon, and around midnight (Owen-Smith 1988), and as the largest amount of dung is voided at the end of the feeding period (Tribe 1976), it seems probable that more fresh elephant dung becomes available to nocturnal/crepuscular species than to diurnal species. Many dung beetles have "dusk and dawn" flight activity, which may coincide with the peak production of dung by diurnal and nocturnal mammals, respectively (Chapter 12).

Seasonality

Seasonality in insects is generally controlled by three factors: resource availability, temperature, and rainfall (Wolda 1988). In dung beetles, rate of resource renewal varies little if at all seasonally, though resource availability and quality may still show seasonal variation. During hot and dry weather, dung becomes quickly unsuitable for most beetles (Chapter 13), hence resource availability as experienced by beetles may vary with varying temperature and rainfall. Studies conducted in Australia have demonstrated that variation in dung quality affects the breeding success of many dung-breeding flies and beetles (Hughes and Walker 1970; Matthiessen 1982; Macqueen et al. 1986; Ridsdill-Smith 1986, 1990), though other species are little affected (Chapter 15).

All three possible combinations of temperature and rainfall (soil moisture) as limiting factors are well documented for dung beetles. In northern temperate (Chapter 5) and montane regions (Chapter 14), temperature is the key factor restricting dung beetle development. In subtropical (Chapter 8) and tropical (Chapter 9) grasslands, rainfall demarcates the dung beetle seasons, though the proximate factor is soil moisture rather than atmospheric humidity. Some species take advantage of riverine or otherwise moist soils during the dry season (Cambefort, pers. obs.). Both temperature and rainfall are critical in the Mediterranean climates (Chapters 6 and 15), where winter is cold but midsummer is very dry, and dung beetle activity is concentrated in spring and autumn. Some species have, however, adopted an exceptional phenology. The dweller *Aphodius constans*, inferior in competition to rollers and tunnelers, breeds during the cold winter months in southern Europe, when the superior competitors are dormant (Chapter 6). Other species occur throughout the dry summer period (Lumaret and Kirk 1987; Ridsdill-Smith and Matthiessen 1988).

The shortest favorable seasons for dung beetles occur at high altitudes on mountains (Chapter 14) and in strongly seasonal tropical grasslands, for example, in East Africa (Kingston 1977; Fig. 8.7). In the dry season in savannas, dung dries out quickly (Chapter 13), and the soils may become so hard that tunneling becomes difficult or impossible (Chapter 8). Beetles can survive in the soil during long periods, perhaps even several years, awaiting the heavy rains to soften the soil and their brood balls (Kingston 1977; Kingston and Coe 1977). The large *Heliocopris dilloni* emerges in the evening of the second to the fifth day after more than 100 mm of rain has fallen (Kingston and Coe 1977). In the more humid Guinean savannas in West Africa, where the soil never dries out completely, some species appear before the first rains, taking advantage of the relative abundance of dung at that time of the year. Cambefort (Chapter 9) found that one-year-old individuals of *Gymnopleurus coerulescens* were the first to emerge, often preceding the beginning of the rains, while the young individuals emerged after the first rains. *Neosisyphus spinipes* attaches its brood balls to grass tussocks and is hence less affected by soil moisture. In Australia it emerges regularly at the same time of the year regardless of rainfall (Chapter 15).

Other temperate and tropical biomes have long, favorable seasons for dung beetles. In temperate regions, coexisting *Aphodius* typically show a clear seasonal succession of species, which is especially striking in the dominant species (Chapters 5 and 14), suggesting that seasonality has been molded by interspecific competition. In the West African savannas where two rainy seasons are separated by a short dry season, conditions are favorable for dung beetles for most of the year (Chapter 9). Some species breed continuously and may produce several generations per year, up to five generations in *Sisyphus* in southern Africa (Paschalidis 1974), with numbers increasing throughout the breeding season. Other species—for example, most Gymnopleurini—only breed in the very beginning of the first rainy season and spend the rest of the year in the soil.

Seasonal dormancy (often diapause) is expected to occur whenever the cost of dormancy is less than the (negative) net benefit of remaining active. The literature on insect dormancy is primarily concerned with abiotic environmental factors, such as temperature and humidity (e.g., Tauber et al. 1986; Danks 1987); but in taxa such as dung beetles, in which competition is occasionally severe, it is likely that intraspecific and interspecific interactions will also affect the costs and benefits of dormancy versus activity (Hanski 1988). Whenever the degree of competition varies seasonally, some species (inferior competitors) may do better by becoming dormant during the period of most severe competition. It is possible that *Gymnopleurus* in West African savannas spend most of the year in the soil to avoid competition with *Sisyphus* in the late rainy season.

It is important to recognize that competition alone is not sufficient to make dormancy, or "escape in time," advantageous—variation in the degree of competition is essential. In nonseasonal environments, dormancy is not expected to occur, regardless of competition, since an individual may gain nothing by postponing development. In tropical forests in Sulawesi, Indonesia, the most abundant species occurred at remarkably constant numbers throughout the year, while the species with more seasonal occurrence were, on average, significantly less common, even during the peak of their seasonal occurrence (Chapter 10). Peck and Forsyth (1982) have reported an almost aseasonal activity of dung beetles in an Ecuadorian rain forest with no severe dry season. In a more seasonal forest in Panama, most dung beetles occur throughout the year (31% of species) or are more abundant in the rainy season (41%), but a few species are restricted to the dry season (Howden and Young 1981; Chapter 12). In the clearly seasonal forests in the Ivory Coast, dung beetle activity broadly follows the seasonal pattern of rainfall (Chapter 11).

18.4. SPATIAL RESOURCE PARTITIONING: HABITAT SELECTION

Dung beetles may show habitat selection at two spatial scales: at the scale of a single dropping and its immediate surroundings, and at the macrohabitat scale, where the two most frequently studied factors are soil type and vegetation cover.

Microhabitat Selection

Resource partitioning is conceptually related to interspecific competition: different species are assumed to use different resources because of past or present competition (MacArthur 1972). Microhabitat selection in dung beetles is not, however, a powerful mechanism for facilitating coexistence. Whenever competition is severe, the entire resource patch is consumed rapidly, and the critical question is who is fastest, not, for example, which part of the dropping some species will use first. Severe resource competition, approaching the lottery type described in Chapter 1, is characteristic for many tropical communities (Chapter 17).

In other communities preemptive resource competition is not common, and much of the resource remains unused for several days or even weeks. Different species may now be found in different parts of the dropping, apparently using it in different ways. Holter (1982) divided the northern European *Aphodius* into top specialists, bottom specialists, periphery specialists, and generalists, on the basis of which part of dung pats the species were most frequently found in. In southern Africa, different species of *Sisyphus* and *Neosisyphus* tend to use different parts of cattle dung pats (Paschalidis 1974). *Sisyphus sordidus* and *S. seminulum* make their food and brood balls at the edge of the pat or in crevices on its surface; *Neosisyphus infuscatus* and *N. mirabilis* are found under or in the edge of the pat; and *N. fortuitus*, *N. rubrus*, and *N. spinipes* utilize the more central parts of dung pats. Such microhabitat selection, whatever its causes, increases intraspecific interactions and therefore amplifies the effects of spatial aggregation of beetles between droppings (Chapter 16). Intraspecific aggregation between and within droppings, apart from increasing intraspecific competition in common species, may be important in facilitating mate finding in rare species (Holter 1982).

Tunnelers and rollers construct their nests in the soil below or some distance away from the dropping. There are clear differences among different species in nest location and architecture (Halffter and Edmonds 1982; Edwards and Aschenborn 1987). Such differences, which comprise another element of resource partitioning at the within-patch scale, can be important even if preemptive resource competition is severe. The major behavioral difference between the rollers and the tunnelers is best understood as a form of resource partitioning at this scale: rollers largely avoid competition for space in the soil by rolling their share of the resource some distance away from the food source.

Macrohabitat Selection

The contrast in the species composition of dung beetles in open grasslands and forests varies with latitude. Table 18.1 gives three examples from north temperate, south temperate, and tropical biomes. In north temperate regions, forests are generally poor in dung beetles, though there are some continental

TABLE 18.1

Comparison of open grassland and forest assemblages of dung beetles in northern temperate (Appendix B.5), southern temperate (Table 6.5), and tropical biomes (Appendixes B.9 and B.11).

Biome	Grassland Species	Forest Species	Comments on Forest Species
N temperate	26	6	No exclusive forest species
S temperate	34	5	No exclusive forest species
Tropical	123	75	Only exclusive forest species

differences; for example, North America seems to have more forest species of *Aphodius* than Europe (Chapter 5). Most of the species that can be found in woodlands in England are common pasture species (Appendix B.5). South temperate forests have more dung beetles than north temperate forests, but most of them are still ubiquitous species present both in open habitats and in forests, and typically they are more abundant in open habitats (Chapter 6). From subtropical Costa Rica, Janzen (1983b) has reported some of the dominant species from both pastures and forests. Finally, in the tropics, both open grasslands and forests have a rich fauna of dung beetles, and there is practically no overlap in species composition (Chapter 11). The general pattern thus is a change from much overlap in species composition between the grassland and forest assemblages at high latitudes to no such overlap at low latitudes. These changes undoubtedly reflect changes in resource availability in forests and grasslands with decreasing latitude: tropical forests have more resources for dung beetles than temperate forests. The contrast between forest and grassland habitats may be greater at low than at high latitudes, but one may also speculate that the higher diversity of beetles at low latitudes has led to a greater degree of habitat specialization. Several chapters in Part 2 have discussed in detail the effect of vegetation cover on dung beetles in particular biomes (see especially Chapters 5, 6, 8, 11, and 15). Other studies on macrohabitat selection in dung beetles include Koskela (1972), Howden and Nealis (1975), Nealis (1977), Doube (1983), and Fay (1986).

Soil type has both direct and indirect effects on dung beetles. Soil type may indirectly affect all beetles by changing the conditions in droppings. Waterlogged soils are an example of habitats that are generally poor for all dung beetles. Lumaret and Kirk (Chapter 6) contrast two kinds of pastures in southern France on compact limestone and on a flinty limestone with marls. The latter site was poorly drained and occasionally flooded, and had very low density of dung beetles, apparently because of high mortality of beetles and their larvae in the soil. Apart from the indirect effects, soil type is of little importance to dwellers, which spend most of their life cycle inside droppings. In contrast, soil type is of decisive importance to tunnelers and rollers, which construct underground burrows and nest chambers in the soil. Several of the chapters in Part 2 have documented changes in dung beetle assemblages from one soil type to another, and in Chapter 17 we outlined the expected effects of soil type on different kinds of species.

Doube's (Chapter 8) data from southern Africa allow a quantitative assessment of the distribution of 120 species of dung beetles in four soil types. In a principal component analysis, more than two thirds of the variance in beetle numbers is accounted for by the first component, reflecting general abundance differences between the species (Table 18.2). The second component (20% of variance) describes a contrast between sandy soils versus loam and clay soils, while the third component (5% of variance) reveals a contrast between the latter two soil types. Many species are more or less restricted to particular soil types, as discussed in Chapter 8. After the effect of the first principal component is factored out, the species that are located farther away from the origin of the plane of the second and third components have higher biomass than species located in the center (Table 18.3). This result indicates that the dominant species tend to be specialists on particular soil types. Regional species richness of dung beetles may be greatly enhanced by variation in soil type.

TABLE 18.2
Principal component analysis of the occurrence of dung beetles in four soil types in southern Africa.

Soil Type	PC1	PC2	PC3	PC4
Deep sand	−0.40	−0.81	−0.13	0.41
Duplex (sand over clay)	−0.56	−0.14	0.04	−0.81
Clay	−0.52	0.32	0.71	0.34
Skeletal loam	−0.50	0.48	−0.69	0.23
Percentage of variance	73	20	5	2

Source: Data from Appendix B.8.

TABLE 18.3
Unweighted least squares linear regression of the logarithm of beetle biomass against PC1 and the distance from the origin of the plane of PC2 and PC3.

Variable	Coefficient	SE	t	P
Constant	1.78	0.12	14.57	<0.0001
PC1	−0.35	0.05	−7.17	<0.0001
Distance	0.72	0.13	5.67	<0.0001

Notes: Principal components from the analysis in Table 18.2. $n = 120$, overall $F = 68.51$, $r^2 = 0.54$.

18.5. MULTIDIMENSIONAL RESOURCE PARTITIONING: AN EXAMPLE FROM WEST AFRICAN SAVANNA

Four Primary Dimensions Collapsed into Two

Dung beetles do not divide resources along one resource dimension at one time but, like the populations of most animals and plants, populations of dung beetles are affected by several dimensions simultaneously. In this section we examine multidimensional resource partitioning with one particular example, the dung beetles of West African savannas, described in detail in Chapter 9. This assemblage has 119 species (Appendix B.9; there are also four species recorded from carrion or vegetable matter).

Figure 18.3 presents the distributions of the 119 species along the four dimensions we discussed separately in the previous sections: type of resource (omnivore versus herbivore dung), size of the beetle, diel activity, and seasonality. Most species in this assemblage show a clear preference for either omnivore or herbivore dung (Fig. 18.3). The size distribution is not normalized

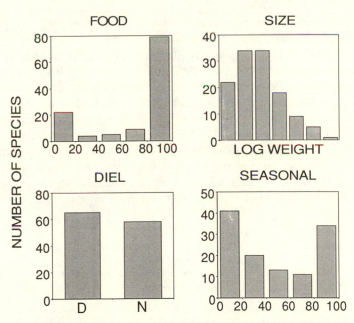

Fig. 18.3. The four panels give data on food selection (the horizontal axis gives proportion of individuals attracted to omnivore dung), size (logarithmic transformation), diel activity (diurnal versus nocturnal species), and seasonality (the horizontal axis gives proportion of individuals caught in April–May) in dung beetles in West African savanna.

by logarithmic transformation, but it remains skewed toward large size (Fig. 18.3). Diel activity is recorded only as diurnal or nocturnal, though both groups include species with more restricted times of activity. Many species have an "early" seasonal occurrence at the study site in the Ivory Coast, being active primarily or exclusively in April–May, following the beginning of the rainy season, while other species have a longer or later seasonal occurrence (see also Fig. 9.3).

Data on the positions of the 119 species along the four dimensions were subjected to a principal component analysis to find out the dimensionality of the community. The first two principal components accounted for most of variance, 40% and 31%, respectively, because of two correlations between the original variables: nocturnal species tend to be larger than diurnal ones, and while most omnivore dung specialists have an early seasonal occurrence, the herbivore dung specialists tend to have a late seasonal occurrence (Table 18.4). The 119 species are well spaced out on the plane defined by the first two principal components (Fig. 18.4).

Nonrandom Community Structure

The competition-centered community ecology (MacArthur 1972; May 1973; Roughgarden 1979) is much concerned with the distribution of species along resource dimensions, and ecologists often examine whether the spacing of the species along one or more such dimensions is more even than expected by chance (Schoener 1974), which is often taken as evidence for past or present competition. Overdispersed niches have been documented for a variety of species, for example, for lizards (Pianka 1975; Schoener 1974), granivorous desert rodents (Bowers and Brown 1982), insectivorous mammals (Hanski 1991) and intestinal helmints (Bush and Holmes 1986). For the reasons explained in the section on size distributions (Section 18.2), we have compared here the spacing of the most dominant species in the community along resource dimensions, with the spacing of all species on average. Dominance has been mea-

TABLE 18.4
Correlations between the first two principal components (Fig. 18.4) and the four resource dimensions in the West African dung beetle assemblage.

Resource Dimension	Small Values In . . .	PC1	PC2
Beetle size	Small species	−0.52	−0.47
Food type	Omnivore dung specialists	0.54	−0.43
Diel activity	Diurnal species	−0.51	−0.48
Seasonality	Early season species	0.42	−0.61
Percentage of variance		40	31

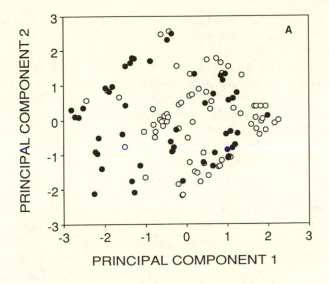

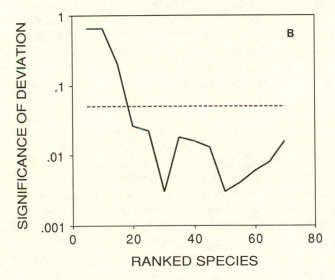

Fig. 18.4. (A) Loadings of the 119 species of dung beetles along the first two principal components in an analysis based on the variables shown in Fig. 18.3. For correlations of the components with the original variables, see Table 18.4. The fifty most dominant species (largest biomass) have been indicated with a black dot. (B) The probability that the *x* most dominant species (horizontal axis) have the same average distance from the origin in panel (A) than equally many randomly selected species.

sured by biomass, the individual weight multiplied by numerical abundance, because of huge size differences among the 119 species, from 3 to 5,800 mg in fresh weight.

Our approach is to compare different subsets of the most dominant species with a randomly selected group of equally many species from the total of 119 species. It is apparent that the dominant species have a highly nonrandom distribution in the plane of the first two principal components: the dominant species are located farther away from the origin than are all species on average, while many of the rarer species are clustered in the center (Fig. 18.4). Furthermore, the average distance between two dominant species is significantly greater than the average distance between two uncommon species (the pairwise distance among the 59 most dominant and the 60 least dominant species was less than 2.5 in 57% and 78% of pairs, respectively; the difference is highly significant). The pattern in Fig. 18.4 is just the opposite to the one described as centrifugal community structure by Rosenzweig and Abramsky (1986), in which the dominant species occupy the central positions in habitat space (Fig. 18.4 is not, of course, a description only of habitat selection).

Two conclusions may be drawn from these results. First, certain combinations of ecological parameters are more frequent among the dominant species than could be expected by chance. Thus in the early season omnivore-dung guild, relatively large diurnal rollers (Gymnopleurini) are dominant (10 of 14 dominant species), while the late season omnivore dung guild is dominated by small diurnal and large nocturnal species (8 and 7 of 17 dominant species, respectively). The late season herbivore guild is dominated by large diurnal and nocturnal species (5 and 10 of 18 dominant species). Almost missing completely are early season species specializing in herbivore dung, regardless of diel activity and size. Although absent in West African savannas in the Ivory Coast, such species occur in localities with a short rainy season, for example, in the drier parts of East Africa (Kingston and Coe 1977) and in southern Africa (Doube, pers. comm.), where dung beetle activity is severely constrained by soil moisture (Fig. 8.7). Our second conclusion is that the distribution of the dominant species along the two composite dimensions in Fig. 18.4 is less clustered than the distribution of species with lower biomass and hence with smaller impact in resource use. We suggest that this pattern is due to the frequently severe competition in this community of dung beetles.

18.6. Summary

This chapter has examined partitioning of four kinds of ecological dimensions by dung beetles: the type of resource, the size of beetle, habitat selection, and temporal dimensions. The two main resource types are omnivore and herbivore dung, which are colonized by relatively distinct species assemblages. In tropical forests, many dung beetles use carrion and decaying fruits. The size of dung beetles varies by more than four orders of magnitude, and in several assemblages the most dominant species in abundance or biomass are more evenly spaced out in size than are all species on average. Partitioning of space for breeding in the soil below and in the vicinity of droppings is the key form of small-scale spatial resource partitioning, while at larger spatial scales habitat selection with respect to soil type and vegetation cover is commonplace. The species-rich tropical communities have typically about equally large guilds of diurnal and nocturnal species using the same resource. Dung beetles provide examples of many types of seasonality to entirely aseasonal dynamics in some tropical forests.

In a West African savanna community, which represents one of the most species-rich dung beetle assemblages, with 119 species, the dominant species in biomass have a nonrandom distribution among all the species with respect to composite resource dimensions (principal component analysis). Certain combinations of ecological parameters are more frequent among the dominant species than could be expected by chance, but the distribution of the dominant species along the composite dimensions is also significantly less clustered than the distribution of the uncommon species.

CHAPTER 19

Species Richness

Ilkka Hanski and
Yves Cambefort

LOCAL COMMUNITIES of dung beetles may have dozens of species using the same resource (Chapters 6, 8–12). In the previous chapters we have discussed two mechanisms that may allow their coexistence in spite of frequently severe competition (Chapter 17), namely, aggregated spatial distributions (Chapter 16) and resource partitioning, including interspecific differences in foraging and breeding behavior (Chapter 18). Ultimately, one would like to have a theory capable of predicting patterns in species number and other such general attributes of communities. We do not have such a theory, only some building blocks that may prove useful for constructing the theory. Our purpose in this chapter is to summarize global patterns in species richness of dung beetles and to relate trends in species number to these building blocks: varying level of interspecific competition, aggregated spatial distributions across habitat patches, and resource partitioning.

One particularly interesting case is the insect community inhabiting cattle dung pats, a resource now universally available from north temperate to tropical biomes. In some regions, cattle dung has been present for long periods of time. In other regions it complements or replaces other, similar types of dung; and in still other regions cattle pats represent an entirely new kind of resource for insects. A comparative study of the insect communities exploiting this single resource in different parts of the world provides some interesting opportunities for a community ecologist.

19.1. PATTERNS IN SPECIES RICHNESS

Latitude and Altitude

The common latitudinal pattern in nearly all taxa is an increase in species number with decreasing latitude. Table 19.1 compares local dung beetle assemblages in open grasslands, from northern temperate pastures (61°N) to tropical savannas. Species belonging to the three functional groups of dwellers, tunnelers, and rollers have been treated separately. The dwellers, largely consisting of *Aphodius*, are the most widespread group of dung beetles, with no great change in species number with latitude, at least not in terms of species

TABLE 19.1
Latitudinal change in the number of species and the relative (pooled) abundance of the three functional groups of dung beetles in open grasslands.

Locality	Latitude	Species			Rel. Abundance			Ref
		D	T	R	D	T	R	
North temperate								
Finland	61°N	18	1	—	100	+	—	1
N England	54°N	14	—	—	100	—	—	2
S England	52°N	23	2	—	100	+	—	3
South temperate								
S France	44°N	14	16	4	5	86	9	4
Bulgaria	42°N	22	14	—	26	74	—	5
Tropical								
S Africa	25°S	28	68	12	53	39	8	6
Ivory Coast	6°N	27	104	19	+	64	36	7
Zaire	3°S	29	72	13	?	?	?	8
Kenya*	2°S	8	75	18	5	92	3	9

References: 1, Hanski and Koskela (1977); 2, White (1960); 3, Chapter 5; 4, Chapter 6; 5, Breymeyer and Zacharieva-Stoilova (1975); 6, Bernon (1981); 7, Chapter 9; 8, Walter (1978); 9, Kingston (1977).
Note: D, dwellers; T, tunnelers; R, rollers.
* Elephant dung only.

number in local communities. The rollers represent the other extreme, with substantial numbers of species being found only in subtropical and tropical communities. The tunnelers show a more steady increase in species number from northern temperate to tropical communities (Table 19.1).

The pattern is different in numbers of individuals, which reflect better than species number the interactions among the three functional groups. The northern temperate communities are entirely dominated by *Aphodius* (dwellers), with only a few species of tunnelers (*Geotrupes, Onthophagus*) and no rollers. The southern temperate communities present a mixture of species, with tunnelers generally being the dominant group. The tropical communities are dominated either by tunnelers or by rollers, depending on season (Chapter 9: rollers are often most abundant soon after the start of rains), soil type (Chapter 8: rollers are most abundant on soils that are easy to dig), and dung type (Chapter 9: many rollers specialize in omnivore dung). A detailed picture of the distribution of beetle biomass among the functional groups in south African habitats is given in Table 8.3.

Table 19.2 gives some examples of numbers of coexisting species in local communities, expressed as the expected number of species in a random sample

TABLE 19.2

Species richness of dung beetles in different biomes and habitats (species in the families Scarabaeidae, Aphodiidae, Geotrupidae, and Hybosoridae).

Biome and Habitat	Species Number (±SD)	Ref
N temperate pasture	13.5 ± 0.8	1
S temperate forest	16.3 ± 1.6	2
S temperate pasture	19.6 ± 1.5	3
Tropical primary forest rich in mammals	48.7 ± 2.7	4
Tropical secondary forest rich in mammals	39.6 ± 1.9	4
Tropical secondary forest poor in mammals	27.1 ± 0.9	4
Tropical montane forest	16.0	4
Tropical gallery forest	12.6 ± 1.4	4
Tropical savanna with elephants	50.4 ± 3.4	5
Tropical savanna with cattle	47.3 ± 3.2	5
Tropical savanna without large mammals	29.7 ± 2.6	5

References: 1, Chapter 5; 2, Lumaret (unpub.); 3, Chapter 6; 4, Chapter 11; 5, Chapter 9.

Note: Species richness was estimated by rarefaction (Simberloff 1979) and is expressed as the expected number of species in a sample of 300 individuals.

of three hundred individuals. These figures range from about ten in north temperate pastures to about fifty in tropical savannas in Africa.

Species number decreases with increasing altitude. The highest recorded altitudes for dung beetles are from the largest mountain ranges, between 4,000 and 5,000 m in the Himalayas (Chapter 14). The decline in species number with increasing altitude is slower on larger mountain ranges than on isolated mountain peaks (Chapter 10), undoubtedly because of an area effect (MacArthur and Wilson 1967; Mayr and Diamond 1976).

Mammalian Species Richness

Three aspects of mammalian species richness have direct consequences for dung beetles. The general abundance of mammals sets the level of resource availability to dung beetles. The range of different kinds of mammals determines the range of dung types available to beetles, while the size of mammals is of special importance to the largest species of dung beetles, which are more or less dependent on large droppings for breeding. These three aspects of mammalian communities are often related to one another. The African game parks are favorable for dung beetles in all three respects, which undoubtedly goes a long way toward explaining why they have more species of dung beetles than any other ecosystem.

In West African savannas, both the diversity of dung beetles and their average size are highest at localities with a rich fauna of mammals, including the elephant, and lowest at localities with only small mammals and man (Fig. 9.6). The expected number of species in a sample of three hundred individuals was fifty in savanna with elephants, forty-seven in savanna with cattle, but only thirty in savanna without large mammals (Table 19.2). Similarly, African secondary forests with a rich mammalian fauna had forty species among three hundred individuals, while forests with a poor mammalian fauna had only twenty-seven species (Table 19.2). On the other hand, it is noteworthy that both in the data in Table 19.2 and in Table 8.3 for southern African localities, grasslands with only cattle and grasslands with wild large herbivores have similar levels of species richness. These results suggest that abundance of large herbivore dung is more important for high species richness than the range of dung types available (the pastoral regions probably have some omnivore dung to support species specializing in this dung type).

The largest species of dung beetles are dependent on the dung of the largest herbivores. The extreme example consists of the one hundred species or so that use exclusively or primarily elephant dung to provision their nests (Cambefort, an unpublished estimate). When elephants and other large herbivores go extinct, the largest dung beetles are doomed. This is probably what has happened to many *Heliocopris* species. *Heliocopris* is an old genus with a much more extensive geographical distribution in the past than at present (today it is mostly restricted to Africa). For example, *Heliocopris* is known from the Japanese Miocene (Fugiyama 1968), and there are old and a few recent records from Borneo, Java, and Sumatra, which had or still have (remnant) populations of elephants. Janzen (1983b) suggests that many large dung beetles may have gone extinct in South America with the extinction of the Pleistocene megafauna: ground sloths, gomphotheres, glyptodonts, horses, and so forth. Africa is the last continent that still has a rich mammalian megafauna (Fig. 19.1), and a rich fauna of very large dung beetles (Plates 4.7 and 4.8); but locally and regionally the largest beetles are threatened also in Africa.

Some large species of *Heteronitis* have successfully bred in donkey dung in the absence of elephant dung, but the size of individuals reared with the relatively small donkey droppings is smaller than the size of individuals bred with elephant droppings (Chapter 9). Disappearance of elephants creates a strong directional selection toward smaller size in the largest dung beetles and, incidentally, provides interesting material for the study of the evolution of body size.

An interesting example of the reverse situation is presented by Sulawesi, which has large herbivores (forest buffalo) but no species of *Catharsius* and *Synapsis*, the largest South-East Asian tunnelers. Sulawesi has many species of *Copris*, however, and the largest *Copris* in Sulawesi are larger than the congeneric species in Borneo, where there are *Catharsius* and *Synapsis*. Large

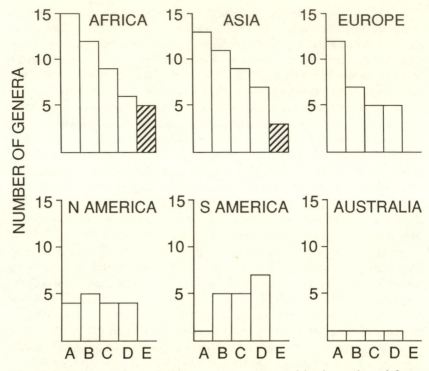

Fig. 19.1. Numbers of megaherbivore genera represented in six continental faunas during different periods of time: A, Pliocene; B, Early Pleistocene; C, Mid-Pleistocene; D, Late Pleistocene; E, Recent (from Owen-Smith 1988).

species of *Copris* have probably evolved in Sulawesi to occupy the niche of the largest tunnelers.

Comparison of Tropical Forest Regions

An interesting continental comparison can be made in the species inhabiting tropical forests in South-East Asia, Africa, and South America. There are obvious differences in the mammalian faunas between the continents, especially in the absence of large herbivores in South America and most of South-East Asia; but if the comparisons are restricted to species using omnivore and small herbivore dung (most species), there are no qualitative differences in resource availability.

Tropical forests in South-East Asia, Africa, and South America typically have 50 to 60 species of dung beetles. Isolated fragments of forest (Peck and Forsyth 1982; Nummelin and Hanski 1989), riverine forests, and forests with clearly impoverished mammalian fauna (Table 11.4) have fewer species. Local species richness in extensive tracts of forest is surprisingly constant and apparently little related to the size of the continental species pool. Thus the total number of dung beetles (Scarabaeidae) recorded for Borneo is 120, of which some 60 were found in a comprehensive survey of the Mount Mulu National Park in Sarawak (Hanski 1983). South America has about 1,250 species, of which some 55 were found in a locality studied in Colombia (Howden and Nealis 1975). Clearly, not all South American species can be expected to occur in Colombia, but the result remains nonetheless striking and suggests an upper limit to the number of locally co-occurring species.

In terms of the kinds of species present in tropical forests, there is relatively little difference between South American and South-East Asian communities, while the African forest communities are different from both (Table 19.3). This divergence is not unexpected, because while the South American and South-East Asian dung beetle faunas evolved on forested continents, the African fauna has fundamentally originated in savannas. In the taxonomic composition, canthonine rollers are well represented in South America, are quite numerous in some regions in South-East Asia, but entirely absent or very rare in Africa. Rollers on the whole are represented by only a few uncommon species in African forests (Chapter 11), while typically in South-East Asia one or two large rollers (*Paragymnopleurus*) play an important role in local commu-

TABLE 19.3

Tropical forest dung beetle assemblages in South-East Asia, Africa, and South America.

Category	SE Asia		Africa		S America
	1	2	3	4	5
Rollers	13	4	5	4	21
Tunnelers	48	37	70	62	35
Dwellers	1	4	15	17	3
Diurnal species	?	?	30	28	33
Nocturnal species	?	?	45	38	26
Unknown	62	45	15	17	—
Coprophagous species	8	25	84	81	34
Necrophagous species	12	5	3	2	—
Generalists & others	13	9	3	—	16
Unknown	29	6	—	—	9
Perching on leaves	—	10	10	7	15

Note: The figures in the table are numbers of species. Localities and data from: 1, Sarawak, Borneo (Hanski 1983); 2, North Sulawesi (Chapter 10); 3, Taï forest, Ivory Coast (Chapter 11); 4. Makokou, Gabon (Chapter 11); 5, Barro Colorado Island, Panama (Chapter 12).

nities, and South American forests have typically several abundant species of Canthon (Chapter 12).

The most striking difference among the continents is in food selection. In South America and South-East Asia, many species use both dung and carrion or are entirely restricted to carrion, whereas the vast majority of African forest species are coprophagous (Table 19.3: comparing localities 2, 3, and 5, for which best data are available, $\chi^2 = 20.6$, $df = 2$, P<0.001). The possible causes of this difference were discussed in Chapter 18.

19.2. THE INSECT COMMUNITY IN CATTLE DUNG

Cattle dung is currently available in large quantities in open grasslands virtually everywhere in the world, allowing comparisons between species assemblages that were drawn from different ecological and evolutionary backgrounds to the same resource.

The three main kinds of insects in cattle dung are dung beetles, dung flies, and predatory beetles (Chapter 1). The interaction between dung beetles and dung flies is competitive and typically asymmetric. Dung beetle activity increases the mortality of fly eggs and larvae (Tyndale-Biscoe and Hughes 1969; Bornemissza 1970b; Blume et al. 1973; Sands and Hughes 1976; Doube 1986; Fay and Doube 1987), and where tunnelers and rollers are the dominant dung beetles (subtropical and tropical grasslands), dung-breeding flies have a reduced chance to breed because of resource preemption by beetles. However, the experiences with introduced dung beetles from Australia (Doube 1986; Chapter 15) and the southern United States (Palmer and Bay 1984; Legner 1986) indicate that a nearly complete removal of dung by beetles is necessary to suppress the density of dung-breeding flies to a low level. The interaction between Aphodius (dwellers) and dung-breeding flies is more symmetric: fly larvae can complete their development by the time Aphodius larvae hatch in droppings, but the immature stages of flies are vulnerable to disturbance by adult beetles. Low or only moderately high abundance of dung-breeding flies in many temperate regions, where Aphodius dominate among dung beetles, cannot be attributed to competition with beetles, but it is known that predatory beetles often inflict heavy mortality on fly eggs and larvae (Hammer 1941; Macqueen and Beirne 1975c; Valiela 1969; Olechowicz 1974; Roth et al. 1983; Doube, Macqueen, and Fay 1988).

The often large numbers of small Oxytelinae (Staphylinidae) beetles in cattle pats (Koskela 1972; Hanski and Koskela 1977) probably do not have a significant impact on either dung beetles nor on dung flies. Their feeding habits are not well known, but they use some decomposing material that is present in droppings.

Table 19.4 describes the species composition of the beetle community in-

TABLE 19.4

Numbers of species of beetles inhabiting cattle dung pats in grassland habitats in different parts of the world.

	Temperate		Subtropical				Tropical	
	Eur	Ame	Aus	Afr	Afr	Afr	Afr	Ame
Taxa	1	2	3	4	5	6	7	8
Coprophages/Saprophages								
Aphodiidae	18	18	3	28	43	20	4	2
Geotrupidae	1	1	—	—	—	—	—	—
Hydrophilidae	16	11	3	5	13	4	5	—
Scarabaeidae	—	4	16	80	147	20	64	4
Staphylinidae	15	1	6	3	19	?	4	—
Predators/Parasitoids								
Carabidae	5	9	—	—	—	4	—	2
Histeridae	4	4	3	14	21	8	15	—
Staphylinidae	119	30	10	31	79	11	31	2

Note: Localities and data for *North temperate*: 1, South Finland (Hanski and Koskela 1977; includes forest species); 2, Minnesota, USA (Cervenka 1986); *Subtropical Australia*: 3, Rockhampton (Chapter 8); *Subtropical Africa*: 4, South Africa (Bernon 1981); 5, South Africa (Davis et al. 1988; Chapter 8, regional study, includes woodlands); 6, Niger (Chapter 13, semi-desert); *Tropical Africa*: 7, Ivory Coast (Chapter 9; figures for families other than Scarabaeidae are from a short-term study using pitfalls baited with human dung); *Tropical America*: 8, French Guiana (Y. Cambefort, unpubl.).

habiting cattle pats in different parts of the world. Comparable data are available from Europe, North America, Africa, South America, and Australia, which differ in terms of climate, original and modified habitats suitable for large herbivorous mammals, the species pool of large herbivores and dung beetles, and geographical history. Paralleling the extent of natural grasslands, the native fauna of large herbivorous mammals is very rich in Africa but poor in South America. Continentwide comparisons reveal certain patterns in the beetle community inhabiting cattle pats (Table 19.4).

First, there are temperate regions, including most of Europe, where an extremely species-rich community of beetles and flies (Skidmore 1985; Chapter 5) colonizes cattle pats, with up to two hundred species being present at one locality, including dozens of predatory beetles that are microhabitat generalists. The characteristic feature of this community is the great diversity of both dung-breeding flies (Table 5.1) and predatory beetles. Staphylinidae appear to be more diverse in Europe than in North America (Table 19.4; for North America see also Harris and Blume 1986), but this may be partly due to the better-known European fauna. The numerical and functional importance of dung beetles increases with decreasing latitude, with a shift from the dominance of *Aphodius* to the dominance of tunnelers and rollers (Chapter 6; Table 19.1).

Second, subtropical and tropical grasslands in Africa have another species-rich community, but with a different species composition from that in temperate regions. Large dung beetles, that is, tunnelers and rollers, often strongly dominate the community (Chapters 8, 9, and 17), and whenever they are so abundant that most or all of the resource is rapidly removed underground, the rest of the insect community has only a limited chance to breed (Table 9.1). Of the predatory beetles, Histeridae are well represented but Staphylinidae are locally only equally or less diverse than in Europe, probably reflecting the competitive effect of large dung beetles in the tropics on dung-breeding flies, the main prey of predatory staphylinids (Koskela and Hanski 1977).

Third, in contrast to the two previous types of communities, many parts of the world have a species-poor community of beetles colonizing cattle pats, exemplified by French Guiana in South America and Australia (Table 19.4). These regions have lacked an indigenous fauna of large herbivores in open grasslands for long periods of time. Many tropical and subtropical regions have had an extensive forest cover until recently, and as there is hardly any overlap in species composition between tropical grassland and forest dung beetles (Section 18.4; Chapter 11), a practical dung beetle vacuum is created when the forest is cleared (Howden and Nealis 1975). In Australia, the new superabundant resource of cattle dung was colonized by two species of dung-breeding flies, the buffalo fly *Haematobia irritans* and the bush fly *Musca vetustissima* (Chapter 15), and in many regions these species increased, in the absence of effective competitors and natural enemies, to very high densities (Hughes et al. 1972; Doube 1986).

Fig. 19.2 is a sketch of the relative abundances of the three main groups of insects found in cattle pats—dung beetles, dung-breeding flies, and predatory beetles—on three continents representing three different community structures. The key interaction in Europe is predation by staphylinid and, to a lesser extent, histerid beetles on dung-breeding flies. In many parts of Africa the key process is preemptive resource competition, and the community is dominated by large dung beetles. In many parts of Australia, the bush fly and the buffalo fly increased to their environmental carrying capacity in the absence of effective predators and competitors.

It is instructive to consider the slowness with which native species have colonized the new resource. In eastern North America, cattle have been present for more than four hundred years and in Australia for two hundred years, yet on both continents the dung beetle communities in pastures are greatly or entirely dominated by intentionally (Australia, Chapter 15) or accidentally (North America, Chapter 5) introduced dung beetles. On both continents, forest and woodland habitats have scores of native species using resources, such as deer dung in North America (*Aphodius*) and marsupial pellets in Australia

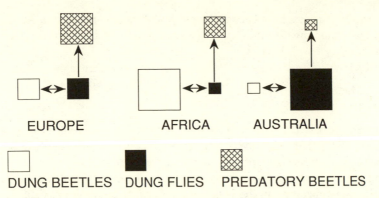

EUROPE   AFRICA AUSTRALIA

☐   ■   ▨

DUNG BEETLES DUNG FLIES PREDATORY BEETLES

Fig. 19.2. A schematic illustration of the relative abundance of dung beetles, dung-breeding flies and their predators in European pastures, African savannas, and Australia (the latter before the introduction of foreign dung beetles). The double-pointed arrows indicate competitive interaction, the single-pointed arrows indicate predator-prey interaction.

(*Onthophagus*); but these species have generally not been able to move to pastures to take advantage of the new resource, even if congeneric species in other parts of the world occur abundantly in cattle dung in pastures.

Colonization of the new resource in open grasslands by native forest species may be hindered by two factors: type of resource and difference in climate between forests and open grasslands. The former is unlikely to be of great significance, because cattle dung in forests is readily colonized by many native species. Microclimatic conditions in droppings in open pastures are probably so different from the conditions in forests that species are unable to make the shift even in hundreds of generations. The Rougons' (Chapter 13) results from the Sahel report truly extreme temperatures (up to 48°C) in droppings in open habitats. Landin's (1961) measurements from southern Sweden indicate that even in northern temperate regions, temperatures in droppings may reach levels that are lethal to forest species, and he concluded that the distribution of *Aphodius* "in different habitats does not depend on the kind of dung, but on climatic factors." Others have drawn the same conclusion about other species of dung beetle (Halffter and Matthews 1966; Gordon and Cartwright 1974; Fincher et al. 1970; Oppenheimer 1977; Chapter 18).

In summary, the insect community colonizing cattle dung is an exceptionally interesting system in which to study community organization because of the different mixtures of species and interactions that prevail in different parts of the world. The data basis is yet scanty, and there is a lack of studies in which both flies and beetles have been studied in any detail. Some of the key questions awaiting further study are: Does preemptive resource competition by beetles suppress the diversity of flies and their predators in African grasslands? Does the species number of dung-breeding flies decrease in regions where a few species have become hugely abundant, for example in Australia? Does predation mediate local coexistence of dozens of dung-breeding flies in Europe?

19.3. MECHANISMS AFFECTING SPECIES RICHNESS

Our purpose here is to return to the three processes that were discussed in the three previous chapters, and to relate these processes to the observed patterns in species richness. Interspecific competition (Chapter 17) is often severe in dung beetles, and makes coexistence of similar species more difficult. Of the mechanisms operating in the opposite direction, facilitating coexistence and maintaining species richness in local communities, we have reviewed spatial aggregation (Chapter 16) and resource partitioning (Chapter 18).

Competition and Species Richness

Two tribes of dung beetles, Canthonini (rollers) and Dichotomiini (tunnelers), are considered to be older than the others (Chapter 4). Canthonini are well represented in South America, Madagascar, and Australia, while Dichotomiini are really dominant only in South America (Table 4.1). In Africa, where the number of extant dung beetles is highest, the old tribes have relict distributions (Chapters 4 and 9), suggesting that they are in the process of being replaced by the more modern tribes.

Fig. 19.3 gives the numbers of dung beetle species in South America, Africa, and Australia, separately for the old (Canthonini and Dichotomiini) and more modern tribes, and separately for the four functional groups of small rollers, large rollers, small tunnelers, and large tunnelers (Plates 4.1 to 4.8). The species are relatively evenly divided among the old and more modern tribes, and among the four functional groups (Table 19.5); but there is a significant interaction between the age of the tribe and the functional group, and a nearly significant interaction between the age of the tribe and the continent (Fig. 19.3, Table 19.5). These interactions support the idea that the more modern tribes are replacing the older tribes of Canthonini and Dichotomiini.

One might expect that species number is smaller in communities in which competition is regularly severe than in communities in which competition is weak or nonexistent. This is not generally the case in dung beetles. The most competitive dung beetle assemblages occur in tropical forests and in subtropical and tropical grasslands (Chapter 17), but these biomes also have the most species-rich communities of dung beetles. In extensive tracts of tropical forest on all three continents, one locality has some 50 to 60 species of dung beetles (Section 19.1), while comprehensive surveys of subtropical and tropical grasslands in Africa have uncovered some 120 to 140 species per locality in West

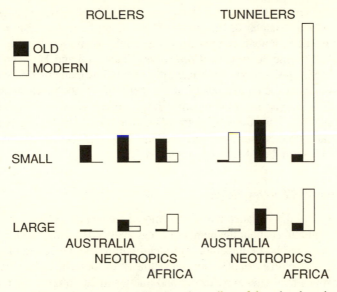

Fig. 19.3. Numbers of species in old and modern tribes of dung beetles, shown separately for small and large rollers and tunnelers in Australia, Neotropics, and Africa (modern species include the ''intermediate'' taxa in Chapter 4). For a statistical analysis, see Table 19.5.

TABLE 19.5
Analysis of variance for the numbers of different kinds of dung beetles on different continents.

Source	df	SS	F	P
Continent (A)	2	6.01	12.79	0.007
Functional group (B)	3	2.48	3.52	0.089
Old vs modern (C)	1	0.03	0.14	0.718
A*B	6	1.28	0.91	0.543
A*C	2	1.84	3.92	0.082
B*C	3	4.43	6.29	0.028
A*B*C	6	1.41		
Total	23	17.48		
Grand total	1	67.99		

Notes: The data are illustrated in Fig. 19.3. The continents are Australia, Neotropics, and Africa, the functional groups are small and large tunnelers and rollers, and the third factor is old versus modern tribes of dung beetles.

Africa (Chapter 9), East Africa (Kingston 1977; Cambefort, unpubl.), and southern Africa (Chapter 8; Edwards 1988a). In northern temperate regions, where competition among dung beetles is generally weak, local communities have typically some 20 species of *Aphodius* (Chapters 5 and 14). The latitudinal change in species diversity is naturally affected by many factors and may tell us little about the consequences of competition. Of the mechanisms that may facilitate coexistence of competing dung beetles, we have examined aggregated spatial distributions (Chapter 16) and resource partitioning (Chapter 18). We shall now consider the likely contributions of these mechanisms to species richness in dung beetle communities.

Aggregation and Species Richness

The lottery and the variance-covariance models reviewed in Chapter 1 yield different predictions about the number of regionally coexisting species: in the lottery model, the best competitor will ideally dominate the community, and in practice we would expect low species richness. In contrast, variance-covariance dynamics facilitate coexistence, and many similar competitors may coexist in spite of competition. Of the functional groups of dung beetles (Chapter 3), dwellers and small tunnelers best fit the assumptions of the variance-covariance model, while rollers and large tunnelers have dynamics approaching the lottery model (Chapter 17). If everything else were equal, we would expect lowest species richness in rollers and large tunnelers, and highest species richness in dwellers and small tunnelers.

At the global level, there are some 1,650 species of *Aphodius*, which comprise the bulk of the dwellers. However, competition is not generally strong in northern temperate communities (Chapter 17), where *Aphodius* dominate (Table 19.1), and it would be misleading to compare them with the mostly tropical rollers and tunnelers.

In the old tribes of rollers (Canthonini) and tunnelers (Dichotomiini), there are about equal numbers of species, about 800 species in both; but in the modern tribes the tunnelers outnumber the rollers by an order of magnitude, with about 3,500 versus 350 species. Thus, while the modern tribes have been replacing the old ones in Africa and Eurasia, species number has apparently decreased in the rollers but substantially increased in the tunnelers. The difference is particularly striking in small species (Fig. 19.3): modern small rollers, although now entirely dominant in most of Africa, number only about a third of the extant species of Canthonini, the old rollers, while in small tunnelers there are twenty times more species in modern than in old tribes. This difference is consistent with the prediction based on lottery dynamics in rollers and variance-covariance dynamics in small tunnelers.

Turning to a smaller spatial scale, local communities in Africa have typically five to ten times more species of tunnelers than rollers (Table 19.1). Fig.

19.4 shows the rank abundance distributions for the rollers and the tunnelers in six West African localities, expressed both in numbers of individuals and in biomass (the latter reflects better than the former the impact of the species in resource use; data from Appendix B.9). It is apparent that while in rollers a few species strongly dominate local communities, the distribution of species abundances is more even in the tunnelers. Furthermore, there are guilds of apparently very similar yet coexisting tunnelers, while coexisting rollers show striking size differences (Chapter 18), supporting the hypothesis that two or more similar tunnelers but not rollers may coexist in the same community. It is not clear why the old tribes of tunnelers and rollers show a different pattern. It would be instructive to compare the population biology of representative species of old and modern rollers. The fact that the old tribes in Africa have largely been replaced by the modern ones, and now remain primarily in southern Africa and Madagascar, suggests that in Africa members of the modern tribes are better competitors than species belonging to the old ones.

In conclusion, the difference in species richness of small rollers and small tunnelers is consistent with the prediction based on the difference in their pop-

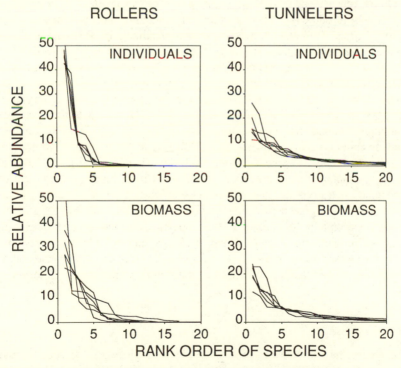

Fig. 19.4. Rank abundance lists for rollers and tunnelers in West African savannas in terms of numbers of individuals and biomass (data from Appendix B.9).

ulation dynamics, with rollers fitting the lottery model of competition better, while small tunnelers fit the variance-covariance model better.

Resource Partitioning and
Species Richness

A possible explanation for the observed trends in species richness (Section 19.1) is resource partitioning: species number is higher where the range of resources is greater, or where for some other reason the means of dividing the resources are better. Of the trends in species richness described in Section 19.1, the increase in species number with decreasing latitude, which is especially dramatic in forest species, is most probably related to resource partitioning. One obvious example is the presence of about equally large guilds of diurnal and nocturnal species in tropical savannas and forests, and the huge size variation of tropical dung beetles in comparison with temperate beetles (Chapter 18). Different soil types have different but overlapping assemblages of dung beetles (Chapters 6, 8, and 18), which increases regional species richness.

On the other hand, there is no strong evidence to suggest that partitioning of different resource types is important in maintaining species richness, apart from the broad difference between species specializing in omnivore and herbivore dung, and the very specialized species living in tropical forests. In southern Africa (Chapter 8) and West Africa (Chapter 9), the game parks that have a range of wild herbivores and the pastoral systems that have only cattle show similar levels of species richness (previous section). The contrasts in dung beetle diversity between African savannas and forests with rich and poor mammalian faunas (previous section) may have more to do with the amount than diversity of dung. The recent declines in regional species number in northern temperate (Chapter 5), southern temperate (Chapter 6), and tropical biomes (Chapter 9) may have more to do with the changing management of cattle and other domestic mammals than with changes in their relative numbers. Intensively grazed pastures may be generally unfavorable for rollers, which may explain the decline and disappearance of many large species in southern Europe in this century (Chapter 6). The interaction between breeding behavior and sensitivity to soil type (Section 17.2) suggests that species number declines when the environment becomes more homogeneous in terms of soil type.

19.4. SUMMARY

This chapter has summarized patterns in species richness of dung beetles and the mechanisms underlying these patterns. Local species number of dung beetles in open grasslands increases from about 20 in northern temperate pastures to about 120 in tropical grasslands (10 and 50 species in a random sample of 300 individuals, respectively). The species composition changes dramatically with latitude, from near complete dominance of *Aphodius* (dwellers) in northern temperate regions, to a varying mixture of tunnelers and rollers in tropical biomes. The number of species of dwellers does not decrease with decreasing latitude; but their decreasing relative abundance is due to competition with tunnelers and rollers, and possibly to a faster rate of drying out of droppings in tropical than temperate climates, making the dweller life history inferior in the tropics. Species richness of dung beetles is generally related to mammalian species richness, but cattle dung alone appears to support as many species in tropical grasslands as a range of wild herbivore dung. Cattle dung is presently available in large quantities all over the world, allowing comparisons among species assemblages that were drawn from different ecological and evolutionary backgrounds to essentially the same resource. Three types of insect communities occur in cattle dung: (1) a species-rich community of dung beetles (mostly *Aphodius*), dung-breeding flies, and predatory beetles in northern temperate regions; (2) a species-rich community dominated by dung beetles (tunnelers and rollers) in grasslands in Africa; and (3) species-poor communities in regions lacking native large herbivores and hence have only few insects adapted to use cattle dung. In some of the latter regions, for example in Australia, some dung-breeding flies have increased enormously in the absence of effective competitors and natural enemies.

Patterns in global species richness suggest that the modern tribes have negatively affected species richness in the oldest tribes, Canthonini and Dichotomiini, which are now well represented only in South America, Madagascar, and Australia. However, the most competitive extant dung beetle assemblages in tropical grasslands have also the highest numbers of species, up to 120 or more. Two mechanisms that contribute to coexistence of large numbers of competitors are aggregated spatial distributions and resource partitioning.

CHAPTER 20

Epilogue

Ilkka Hanski

DUNG BEETLE ecology has reached an exciting stage. New data are accumulating and new ideas are being contemplated. An ecological context is emerging into which old and new pieces of natural history observations fit. Yet so little work has been done in population and community ecology of dung beetles that it is easier to ask new questions than to review old ones. While assimilating the data and the ideas presented in this volume, my two recurrent feelings were that we are still in the dawn of unraveling many of the more significant ecological questions about dung beetles, but also that many of them are tractable and are just a matter of doing the right thing in the right place. I have compiled below a list of seven observations, concepts, and ideas that are worthy of further study.

1. COMPETITION IN THE DUNG INSECT COMMUNITY

The key interspecific interaction discussed throughout this book is competition. Although there is no doubt that competition occurs in many dung beetle populations (Chapter 17) and has probably played a significant role in the assembly of local communities (Chapter 18), there is a need for experimental field studies that quantify the population-ecological consequences of competition, and a need for experiments that deal with entire dung insect communities. The balance between the three main components of these communities, consisting of dung beetles, dung-breeding flies, and predatory beetles, varies from one biome to another (Section 19.2), but the direct and indirect interactions among them remain little studied (for some interesting examples see Valiela 1969, 1974; Doube, Macqueen, and Fay 1988). A better understanding of these interactions has great practical value in the biological control of pestilent dung-breeding flies (Doube 1986; Legner 1986; Chapter 15).

Considering dung beetles in particular, it would be informative to conduct experiments in which certain types of beetles were excluded from droppings, for example, by taking advantage of the striking differences in beetles' diel activity. Introductions of foreign dung beetles offer unique opportunities to conduct interesting experiments and to make valuable observations. The species that have been most successfully introduced so far are relatively small, multivoltine, high-fecundity species (Chapters 15 and 16), while many natural dung beetle communities in subtropical (Chapter 8) and tropical grasslands (Chapter 9) are dominated by large, univoltine, low-fecundity species with highly evolved parental care—the classical K-species. What happens when the latter species are introduced into new areas and have had sufficient time to become abundant? Do they outcompete or, more likely, significantly suppress the abundance of the more r-type species introduced earlier? Some observations from Australia and New Zealand demonstrate that it takes dozens of years before the large, low-fecundity species become abundant (Chapter 15).

2. TYPE I AND TYPE II DUNG BEETLES

To be a successful competitor a dung beetle needs to possess two traits: it must be able to locate a resource patch quickly, before other beetles have had a chance to remove all of it, and it must itself be able to remove, to a safe location, a sufficient amount of resource for breeding. Paradoxically, some beetles appear to do just the opposite: they arrive late at droppings and provision their underground nests slowly, or they breed in droppings, remaining vulnerable to disturbance by other insects. I shall denote the two kinds of beetles as Type I and Type II species, respectively. The other classifications of dung beetles described by Doube (Chapter 8) and Cambefort (Chapter 9) are refinements of the basic dichotomy between the Type I and Type II beetles as defined in Table 20.1.

Type I beetles are superior competitors in preemptive resource competition, and their dynamics often fit the lottery model (Chapter 1). They are adapted to use fresh dung as fast as possible, while Type II species are adapted to use older dung. The distinction between Type I and Type II species is related to the idea of species being either efficient in locating resource patches but inefficient in exploiting them, or vice versa (Kotler and Brown 1988). Small, high-quality omnivore droppings are often entirely dominated by Type I beetles, whereas the larger herbivore droppings have typically both Type I and Type II beetles. Most rollers and the fast-burying tunnelers (Chapter 8) are Type I species, while the slow-burying and many small tunnelers, dwellers, and kleptoparasites are Type II species.

Type I species are superior competitors because they are fast in securing resources for their exclusive use. Type I beetles are often but not always large, while Type II beetles are typically small or relatively small. In tunnelers, large size allows fast burrowing and fast removal of large quantities of dung underground. In rollers, large size is an advantage in interference competition (Section 17.1). The cost of large size is reduced fecundity (Section 17.2). Many Type I beetles invest much time and energy in constructing their nests in places where disturbance by later-arriving beetles is not likely. Rollers bury their brood balls some distance away from the food source and effectively

TABLE 20.1
A simple classification of dung beetles into two basic types.

Attribute	Type I Beetles	Type II Beetles
Represented by	Most rollers; fast-burying tunnelers	Slow-burying tunnelers; dwellers; kleptoparasites
Type of resource used	Omnivore and herbivore dung; species are adapted to use fresh dung	Herbivore dung; species are adapted to use older dung
Arrival at resource patch	Fast	Often slow; beetles locate the resource patches not used by Type I species
Rate of exploitation	Maximally fast	Slow but perhaps more efficient
Competitive ability	Good	Poor
Type of population dynamics	Lottery	Variance-covariance
Fecundity	Low in large species, high in small species	Generally high

eliminate any disturbance, but the cost is lost contact with the food source. Type I tunnelers may minimize disturbance by digging deep burrows below the dropping. We may expect a negative relationship between the depth of the nest in the soil and the speed of nest construction. Such a difference is obvious between the fast-burying tunnelers (Type I species) and dwellers (Type II species), but studies on different species of tunnelers would be especially interesting (some supporting observations are described in Chapter 6 on *Onthophagus* and by Edwards and Aschenborn 1987 on *Onitis*). Other open questions include: Why are some Type I beetles very large and have extremely low fecundity and extensive maternal care (e.g., *Kheper*, *Heliocopris*), while other Type I species are small and have high fecundity and no or only limited maternal care (e.g., *Sisyphus*, *Digitonthophagus gazella*)? And are Type II beetles more efficient in using the resources than Type I beetles?

3. SOIL TYPE AND DUNG BEETLES

Both rollers and tunnelers are more or less affected by soil characteristics, and, as suggested in Chapter 17, the best competitors, the Type I beetles (Table 20.1), have the disadvantage of being generally more affected by adverse soil type than the inferior competitors. The weakest competitors, the dwellers, are the least affected by soil type, since their entire life cycle is completed inside the dung pat.

The trapping studies by Doube in southern Africa have produced interesting results on the effect of soil type on different functional groups of dung beetles (Chapter 8). More such observations are badly needed. It would be interesting to obtain results not only on a large spatial scale (as in Table 8.3) but also on a small scale, comparing the use of dung pats located on patches of different types of soil. Small-scale patchiness in soil type may be an important factor maintaining high species richness in the most competitive dung beetle communities by preventing Type I beetles from dominating the community. Informative experiments to test this hypothesis would be straightforward to conduct.

4. AGGREGATED SPATIAL DISTRIBUTIONS

One of the major organizing concepts in this book has been the distinction between two kinds of competitive dynamics: the lottery and the variance-covariance dynamics (Chapter 1). In reality there is not a sharp distinction between the two but rather a continuum, and the question is where, along this continuum, different kinds of dung beetles are located. In general, large and Type I beetles fit the lottery model better, while small and Type II beetles fit the variance-covariance model better (Chapter 16). Note that variation in the quality of the dung pat or the soil below it may also generate intraspecific aggregation (point 3 above), but such aggregation is conceptually different from the one considered here, which occurs independently of any variation in the quality of resource patches and is not due to adaptive behavior, as the first kind of aggregation may be. In practice, however, the two kinds of aggregation may be compounded.

Different chapters in this book have reported contradictory results on the likely importance of aggregated distributions in dung beetles. Doube (Chapter 8) concludes that such aggregation is unlikely to be of any significance in South African grasslands, while Cambefort (Chapter 9; see also Chapter 16) finds much aggregation, and interspecific patterns related to aggregation, in West African savannas. One reason for these conflicting results is probably a difference in the spatial scale in the two studies, the replicate dung pats being located only meters apart in Doube's study but dozens of meters apart in Cambefort's study. Doube (Chapter 8) is concerned with large tunnelers, while Cambefort's (Chapter 9) analysis includes many kinds of species. As explained above and in Chapter 16, aggregation is not expected to be an important factor in the largest tunnelers, whose dynamics approach the lottery model.

Another question is what is causing the observed aggregation of species? One possible answer is that aggregation reflects differences in the detectability of dung pats; another possibility is that aggregation is due to the behavior of the beetles colonizing the pats. These two mechanisms may both contribute to aggregètion, but note that the difference between large versus small species, and between Type I versus Type II species, supports the second hypothesis. Consistent patterns of interspecific aggregation (Fig. 5.3) also support the behavior and biology of species as determinants of aggregation. Experimental studies are needed to elucidate the causes of intraspecific aggregation.

5. POPULATION BIOLOGY OF ROLLERS

Rollers present many fascinating and little-explored problems in population biology. Questions about breeding behavior, parental cooperation, sexual selection, and so on are largely beyond the scope of this book (for a fine example, see studies on *Kheper nigroaeneus* by Edwards 1988a,b; Edwards and Aschenborn 1988, 1989). Interesting population ecological questions focus on the kinds of balls made by rollers. Do beetles generally make balls as large as is economically feasible? Does the type of dung affect ball size, as suggested in Chapter 17? A comparative study of the ability of rollers to prepare and roll dung balls versus the ability to dig burrows would be most rewarding, and should not be too difficult to carry out, as both the type of dung and the kind of soil offered to beetles may be experimentally varied. Also of interest are the differences and similarities among rollers belonging to the old and modern tribes (Chapter 4). There are only a few small, modern rollers, most of which are very abundant, and a larger number of mostly uncommon old, small rollers. Does this contrast reflect a difference in their population dynamics? For example, do the modern but not the old small rollers fit the lottery model of competition? The pattern between the old and modern species is essentially reversed in large rollers, as there are very few old, large rollers (Fig. 19.3). It is possible that the old, large rollers have mostly gone extinct.

6. PREDATION ON DUNG BEETLES

While competition is the dominant interspecific interaction in dung beetles, it would indeed be surprising if predation were as insignificant as the meager information now available suggests. One could imagine that large dung beetles that often occur in huge concentrations would present an excellent food source for many predators. Many of the patterns examined in the previous chapters may be related to predation, including the cases of probable mimicry in African rollers (Chapter 9) and the generally nocturnal habits of large tunnelers. Kingston and Coe (1977) have suggested that predation by the ratel *Mellivora capensis* may explain the great depth in the soil, down to 120 cm, at which the huge *Heliocopris dilloni* excavates its breeding chambers. Kleptoparasites are often abundant in the nests of *Heliocopris* (Table 9.3). It would be interesting to know whether kleptoparasitism varies with the depth of the nest and, more generally, what features of dung beetle breeding behavior are possibly affected by kleptoparasitism. Predation might be expected to be heavy on adult rollers, because most of them are diurnal and very conspicuous while rolling dung balls. Systematic observations would be valuable, provided that the observer takes care not to scare off the potential predators. Unfortunately, in many regions of the world most of the original predators may have already gone extinct or are scarce.

7. TEMPORAL STABILITY OF COMMUNITIES

Several chapters in this book have examined temporal stability of dung beetle populations (Chapters 5–9, 15). Doube (Chapter 8) and Cambefort (Chapter 9) reached opposite conclusions about beetle assemblages in African subtropical and tropical grasslands. According to Doube (1987), there is great variation in relative abundances of species, even in the relative abundances of different functional groups of beetles (Chapter 8), while Cambefort (Fig. 9.7) found striking constancy of relative abundances over a period of eight years. The reason for the disparate results can hardly be due to differences between subtropical and tropical grasslands. A possible explanation, however, is that Doube's community of beetles using herbivore dung on clay-loam soils in a game reserve has many Type II species (Table 20.1) and is apparently much less competitive (Chapter 8) than the community of largely Type I beetles using omnivore dung in Cambefort's study. These results suggest that the structure of the most competitive (Type I) dung beetle communities is more constant than the structure of communities not so strongly and consistently affected by competition, consequently including many Type II species. Cambefort's results are in agreement with the classical competition theory, which assumes a stable equilibrium point (Chesson and Case 1986). Temporal variation of populations on clay-loam soils leads to discontinuous competition, which may facilitate coexistence, assuming that different species have some differences in their responses to fluctuating density-independent mortality or whatever factors are promoting instability. Surprisingly, the two communities that have apparently different sorts of mechanisms allowing coexistence are still about equally rich in species.

Appendix A

APPENDIX A gives data on resource availability at the study localities that are dealt with in detail in the chapters in Part 2, and for which Appendix B gives lists of species of dung beetles and ecological information. The data in this appendix are not equally comprehensive for all localities, nor can the data be presented in a uniform format. Appendix numbers refer to corresponding chapters.

APPENDIX A.5

Oxford, England, north temperate region.

The large herbivores consist of cattle in the pastures and deer in the woodland (Hanski 1979).

APPENDIX A.6

Southern France, south temperate region.

In the small area of the Garrigue north of Montpellier where most of the ecological studies on dung beetles have been conducted, some 6,700 hectares are used for sheep grazing, of which about one half is used intensively. The number of ewes present in 1982 was 4,410. The sheep were distributed among twelve flocks, three of which were less than 150 head in size and three more than 450 head. A herd of cattle (100 animals) was established in the same area in 1984.

APPENDIX A.7

Santa Cruz Acatlan, Mexico, subtropical North America.

Most of the resource available to beetles consisted of cattle, horse, and human dung, but it was not possible to quantify their respective availabilities.

APPENDIX A.8

Mkuzi Game Reserve, southern Africa, subtropical savannas.

The topography, vegetation cover, and fauna of the Mkuzi Game Reserve have been described in Edwards and Aschenborn (1988) and Osberg (1988). The size of the reserve is 25,000 ha. It is predominantly flat, consisting of open bushveld with small areas of open grassveld and dense bush. Soil types range from deep sand to heavy clay. The mammal fauna is diverse and it included, in April 1984, 8,100 impalas (45 kg), 820 zebras (310 kg), 180 giraffes (1,200 kg), 40 white rhinoceros (1,800 kg), 100 black rhinoceros (1,100 kg), 1,400 wildebeests (220 kg), 680 kudus (145 kg), 3,100 nyalas (85 kg), and 480 warthogs (66 kg), as well as some smaller antelopes, baboons (23 kg), vervet monkeys (5 kg), small mammals, and a few hyenas (75 kg), cheetahs

(40 kg), and jackals (7 kg). Most species occur throughout the reserve. Zebras and rhinoceros produce coarse-textured herbivore dung. Wildebeest dung is similar to that of small cattle. Giraffes and antelopes produce pelleted dung.

TABLE A.8
Mammals in Mkuzi Game Reserve.

Type of Mammals	Biomass (kg/ha)	Dung Production* (kg/ha/yr)
Large herbivores (>200 kg)	38.4	750
Medium-sized herbivores (>20 kg)	30.3	400
Small mammals (<2 kg)	unknown	

* Rough estimate.

APPENDIX A.9

Ivory Coast, West Africa, tropical savannas.

TABLE A.9
Mammals in Ivory Coast tropical savannas.

Locality	Type of Mammals	Biomass (kg/ha)	Dung Production (kg/ha/yr)
Lamto	Large herbivores	2	40
	Medium-sized mammals	20	300
Abokouamekro	Large herbivores	140	2,900
	Medium-sized mammals	20	300
Kakpin	Large herbivores	50	1,000
	Medium-sized mammals	35	525

Note: No data available for small mammals.

APPENDIX A.10

Dumoga-Bone National Park, North Sulawesi, tropical forests in South-East Asia.

Larger mammals are represented by the abundant crested macaque, *Macaca nigra*; the Sulawesi wild pig, *Sus celebensis*; the babirusa, *Babyrousa babyrussa*; the Sulawesi dwarfed buffalo (anoa), *Bubalus depressicornis*; and the introduced deer, *Cervus timorensis* (Musser 1987). Small and medium-sized mammals are very abundant. Durden (1986) collected ca 15 species of rats within a small area of lowland forest, obtaining the rough density estimate of 20 individuals per 100 trap-nights for the small and medium-sized mammals (Durden, pers. comm.). This is a very high density for tropical forests. The only native carnivore is the civet, *Macrogalidia musschenbroeki* (Musser 1987). Small lizards, *Sphenomorphus amabilis* (Müller) and *S. temmincki* (Duméril and Bibron), are conspicuously common in the forest floor.

APPENDIX A.11

Taï, Lamto (Ivory Coast) and Makokou (Gabon), tropical forests in Africa.

TABLE A.11
Numbers of mammalian species and their estimated biomasses in three forest localities.

Type of Mammals	Taï	Makokou	Lamto
Primates	11	17	4
Rodents	10	34	10
Pholidota	3	3	3
Carnivora	13	13	3
Proboscidians	1	1	0
Hyracoidea	1	1	1
Arctiodactyla	16	12	5
Number of species	55	81	26
Biomass (kg/ha)	30–50	25–40	5–10

Sources: *Taï*, Anonym (1979a) and Merz (1982); *Makokou*, Anonym (1987); *Lamto*, Bourliére et al. (1974).

APPENDIX A.12

Barro Colorado Island, Panama, tropical forest in South America.

TABLE A.12
Estimated dung production of nonvolant terrestrial mammals.

Species	Weight (kg)	Population Size	Biomass (kg/ha)	Dung Production (kg/ha/yr)
Tapirus bairdii	300.0	8	1.53	2.23
Odocoileus virginianus	40.0	10	0.25	0.68 (a)
Tayassu tajacu	23.0	140	2.05	0.52 (b)
Mazama americana	15.0	30	0.29	1.28
Felis pardalis	13.5*	12	0.10	0.46
Agouti paca	8.0	600	3.06	17.87 (c)
Alouatta palliata	5.5	1,300	4.55	54.20 (d)
Ateles geoffroyi	5.0	14	0.04	0.26
Dasypus novemcinctus	5.0*	800	2.55	16.93
Eira barbera	4.5*	25	0.07	0.48
Tamandua mexicana	4.5*	80	0.23	0.90 (e)
Choloepus hoffmanni	3.5	1,500	3.34	3.05 (f)
Coendou rothschildi	3.0*	150	0.29	2.32
Nasua narica	3.0	360	0.69	5.53
Potos flavus	3.0*	300	0.57	4.57
Bradypus infuscatus	2.8	8,000	14.27	13.02 (f)
Cebus capucinus	2.6	140	0.23	1.94
Dasyprocta punctata	2.0	1,500	1.91	14.64 (c)

TABLE A.12 (*cont.*)

Species	Weight (kg)	Population Size	Biomass (kg/ha)	Dung Production (kg/ha/yr)
Philander opossum	1.4	200	0.18	1.91
Sylvilagus brasiliensis	1.3*	100	0.08	0.87
Didelphis marsupialis	1.0	700	0.45	5.42
Aotus trivirgatus	0.8	40	0.02	0.26
Proechimys semispinosus	0.8	2,800	1.43	18.70
Saguinus geoffroyi	0.8	40	0.02	0.26
Caluromys derbianus	0.35*	300	0.07	1.24
Sciurus granatensis	0.25	2,700	0.43	8.65
Cyclopes didactylus	0.23	—	0.13	0.24 (e)
Heteromys desmarestianus	0.1	1,000	0.06	1.69
Oryzomys spp.	0.07	2,000	0.09	2.90
Marmosa robinsoni	0.06	400	0.02	0.68

Notes: Weights from Eisenberg & Thorington (1973), except those marked with an asterisk (*), which are averages computed from Nowak & Paradiso (1983). Estimates of population size from Glanz (1982), with the exception of *Heteromys*, which is from Eisenberg & Thorington (1973). Dung production given in kg dry weight/hectare/year. Unless otherwise indicated, dung production estimated from body weight (W in g), using equation $y = 0.85 \, W^{-0.37}$ (Blueweiss et al. 1978). Estimates converted to dry weight assuming 50% water content. (a) McCullough (1982) gives average daily fecal output of 300 g dry weight for white-tailed deer, *Odocoileus virginianus*. (b) Zervanos & Hadley (1973) calculated fecal water loss of 0.21% body weight/day (with fecal water content averaging 75%) for collared peccary, *Tayassu tajacu*, which corresponds to fecal output of 0.07% body weight/day (dry weight). (c) Assuming 50% digestive efficiency, Smythe et al. (1982) estimated that a 3 kg agouti, *Dasyprocta punctata*, and a 9 kg paca, *Agouti paca*, consume 128 and 290 g dry weight of food/day, respectively. These values correspond to defecation rates of 21 and 16 g dry weight/kg body weight per day, respectively. (d) Nagy & Milton (1979) calculated that howler monkey, *Alouatta palliata*, consumes an average of 53.5 g dry mass/kg of monkey per day, of which 61% is voided as feces. (e) Montgomery (1985) estimated an output of 80 g dry dung every two days for banded anteater, *Tamandua mexicana*, with average weight of 3.73 kg. Adult silky anteaters, *Cyclopes didactylus*, produced an average of 1.17 g dry dung per day. (f) Montgomery and Sunquist (1975) determined a defecation rate of 2.5 g dry feces/kg body weight per day for three-toed sloth, *Bradypus infuscatus*. Similar rates are assumed for the two-toed sloth, *Choloepus hoffmanni*.

APPENDIX A.13

Niamey, Niger, arid regions.

Cattle (zebu) are kept by pastoral populations.

APPENDIX A.14

Vanoise National Park, France, and Grisons National Park, Switzerland, montane regions.

Large herbivores (>200 kg). Biomass 33.0 kg/ha (data for 1979). Cattle (2,700) remain inside the park limits (52,839 ha) from May to September, occupying 62% of the total area available.

Medium-sized mammals (>20 kg). Biomass 37.9 kg/ha on average. Domestic mammals (32.6 kg/ha, data for 1979) remain inside the park limits from May to September. Wild mammals (5.3 kg/ha, data for 1986) remain in the park all year round, using 90% of its total area (10% is ice and permanent snow). The most important medium-sized mammals are sheep (18,000; 32.0 kg/ha); goats (540; 0.6 kg/ha); wild goats (*Capra ibex*, 700; 0.9 kg/ha); chamois (*Rupicapra rupicapra*, 4,700; 4.4 kg/ha).

Small mammals (<20 kg). Biomass is unknown. The main small mammals are marmots (*Marmota marmota*, several thousand) and hares (*Lepus timidus*, scarce).

Large herbivores (>200 kg). Domestic mammals are forbidden inside the park limits (17,000 ha). Cattle and sheep are abundant outside.

Medium-sized mammals (>20 kg). Biomass 12.4 kg/ha on average (data for 1985), consisting of red deer (*Cervus elaphus*, 2,130; 9.1 kg/ha); chamois (*Rupicapra rupicapra*, 1,230; 2.3 kg/ha); wild goats (*Capra ibex*, 238; 0.9 kg/ha); and roe deer (*Capreolus capreolus*, 57; 0.1 kg/ha).

Small mammals (<20 kg). Biomass unknown. Small mammals include foxes (*Vulpes vulpes*, 70–90 individuals); marmots (*Marmota marmota*, 800–1,000 individuals distributed at fifty sites); and hares (*Lepus timidus*, 500–600).

Appendix A.15

Badgingarra National Park, southwestern Australia
(30°24′S, 115°27′E).

Western gray kangaroo (40 kg), a herbivore, biomass 12–20 kg/ha, dung production 16–27 kg/ha/yr (G.W. Arnold, pers. comm.). Other species are present but have low biomass: rabbits, foxes, cats, native marsupials (small species), and reptiles.

Appendix B

APPENDIX B gives ecological information for the species recorded in the study localities described and discussed in the chapters in Part 2. See Appendix A for resource availabilities in these localities and the map on pp. 72–73 for their climates. The list of species for each locality is as complete as possible; but because the different studies have lasted for different lengths of time and have varied in sampling intensity, the localities should be compared with caution. Somewhat different types of information are available for different localities, hence the data cannot be presented in a uniform format.

The following abbreviations are used where applicable: LE = length in mm; FWt = fresh weight in mg; DWt = dry weight in mg; N/D = diel activity: N, nocturnal (or crepuscular), D, diurnal; LF = larval food: C, coprophagous, S, saprophagous (decaying plant material).

APPENDIX B.5

Wytham Wood, near Oxford, southern England,
north temperate region.

APPENDIX B.6

Southern France, south temperate region.

APPENDIX B.7

Santa Cruz Acatlan, Mexico, subtropical
North America.

APPENDIX B.8

Mkuzi Game Reserve, southern Africa,
subtropical savannas.

APPENDIX B.9

Part 1
Abokouamekro, Ivory Coast, West Africa,
tropical savannas.

Part 2
Six savanna localities in Ivory Coast,
West Africa, 1980–81.

APPENDIX B.10

*Dumoga-Bone National Park, North Sulawesi,
tropical forests in South-East Asia.*

APPENDIX B.11

*Part 1
Taï, Ivory Coast, tropical forests in Africa.*

*Part 2
Makokou, Gabon, tropical forests in Africa.*

*Part 3
Taï and Makokou.*

APPENDIX B.12

Barro Colorado Island, Panama, American tropical forests.

APPENDIX B.13

Niamey, Niger, arid regions.

APPENDIX B.14

*Vanoise National Park, French Alps,
montane dung beetles.*

APPENDIX B.15

*Badgingarra National Park, southwestern
Australia (30°24'S, 115°27'E).*

NOTE: Appendix numbers refer to corresponding chapters. Tables relating to
these appendixes begin on the next page.

TABLE B.5
Beetles caught in the pastures near Wytham Wood.

Species	LE	DWt	LF	N/D	Apr	May	Jun	Jul	Aug	Sep	Oct	Woodland Preference
								Seasonal Occurrence in the Pastures				
1. Geotrupes stercorarius (L.)	21.0		C	N	1		1		1			0.00
2. Geotrupes spiniger Marsh.	21.0		C	N					5	25	16	0.00
3. Aphodius sphacelatus (Panz.)	5.0		S?	D	235	40	16			1	58	0.06
4. Aphodius prodromus (Brahm)	5.5	4	S?	D	2,707	1,744	521	20		2	342	0.59
5. Aphodius constans Duft.	5.0		C	D	27	9						0.00
6. Aphodius paykulli Bedel	4.0		C	D	1							0.00
7. Aphodius luridus (Fabr.)	7.5		C	D	13	35	50	15				0.00
8. Aphodius granarius (L.)	4.0		S,C	D	7	25	61					0.00
9. Aphodius ater (DeGeer)	5.0	6	C	D	190	1,097	1,130	365	25	2		0.08
10. Aphodius fimetarius (L.)	7.0	9	C	D	104	151	172	72	75	57	13	0.02
11. Aphodius equestris (Panz.)	5.0		C	D	19	1,859	699	51	160	268	208	0.43
12. Aphodius pusillus (Herbst)	4.0	2	C	D	6	848	403	68	15	40		0.13
13. Aphodius borealis Gyll.	4.0	2	C	D		42	62	54	16	2	1	0.37
14. Aphodius fossor (L.)	10.5	40	C	D		79	348	293	144	4		0.01
15. Aphodius haemorrhoidalis (L.)	4.5	4	C	D		30	323	127	45	29	2	0.00
16. Aphodius rufipes (L.)	11.0	33	C	N		7	357	314	1,026	1,052	18	0.35
17. Aphodius rufescens Fabr.	6.5	7	C	N				236	431	434	21	0.17
18. Aphodius coenosus (Panz.)	4.0		C	D				1				0.00
19. Aphodius scybalarius (Fabr.)	6.5		C	D				(1)				0.00
20. Aphodius aestivalis Steph.	7.0	11	C	D					8	2		0.00
21. Aphodius contaminatus (Herbst)	6.0	6	S?	D					92	1,715	727	0.00
22. Aphodius ictericus (Laich.)	5.0	4	C	D						4		0.00
23. Aphodius porcus (Fabr.)	5.0		C(P)	D	1					2	2	0.00
24. Aphodius obliteratus Panz.	5.5		S?	D		20	7	34	2	4	632	0.02
25. Oxyomus silvestris (Scopoli)	3.0		?	?							1	0.00

Notes: Beetles were trapped from April to October in 1977 with large pitfalls baited with cattle pats (1.5 kg fresh wt). Trapping was conducted with five pitfalls in five pastures. Bait was renewed usually twice per month. To get an estimate of numbers of beetles immigrating to one dung pat during two weeks, divide figures by 60. Trapping was also conducted in the nearby woodland (Wytham Wood) in 1977; woodland preference index gives frequency of species in the woodland in comparison with pastures (ranges from 0 to 1). One species, *Aphodius foetidus*, was caught from the center of Oxford but not from the pastures.

TABLE B.6
Numbers of dung beetles trapped in an open Garrigue site (Viols-le-Fort, altitude 240 m, 20 km north of Montpellier).

Species	I	II	III	IV	LE	DWt	DP
1. *Sisyphus schaefferi* (L.)	49	46	96		8–10	29	S,M
2. *Scarabaeus laticollis* (L.)	21	13	23		15–25	172	C,S
3. *Scarabaeus typhon* Fischer			1		20–28	353	S
4. *Copris hispanus* (L.)	5	4			15–20	513	C,S
5. *Euoniticellus fulvus* (Goeze)		5			7–11	25	C,H
6. *Caccobius schreberi* (L.)	1	17	9		4–7	7	C,H,S
7. *Euonthophagus amyntas* (Ol.)		1			7–12	27	S
8. *Onthophagus emarginatus* Mulsant	28	7	2		4–7	7	R
9. *Onthophagus furcatus* (F.)	1	1	5		3–4	3	S
10. *Onthophagus verticicornis* (Laich.)	7	8			6–9	18	S,M
11. *Onthophagus joannae* Goljan	130	101	98	1	4–6	6	S
12. *Onthophagus coenobita* (Herbst)	97	29	17	3	6–10	21	M
13. *Onthophagus lemur* (F.)	1,685	497	8	1	5–8	13	S,M
14. *Onthophagus maki* (Ill.)	67	153	111		4–7	10	S
15. *Onthophagus vacca* (L.)	18	5	2		7–13	41	C
16. *Aphodius subterraneus* (L.)	1				5–7	7	C
17. *Aphodius haemorrhoidalis* (L.)	5	53			4–5	4	C,S
18. *Aphodius luridus* (F.)	31				6–9	12	S
19. *Aphodius paracoenosus* Balth. & Hrub.	13	3			4–5	2	S
20. *Aphodius biguttatus* Germar	10	6	8		2–3	1	S
21. *Aphodius fimetarius* (L.)	2	3			5–8	9	C,S,H
22. *Aphodius constans* Duft.	107	4	4	98	5–6	4	C
23. *Aphodius granarius* (L.)	3				3–5	4	C
24. *Geotrupes niger* Marsh.		2	10		12–23	143	M,C
Seasonal distribution of beetles (%)	61	26	10	3			
Shannon-Wiever diversity	1.7	2.5	2.7	0.4			
Equitability	0.4	0.6	0.7	0.2			

Notes: Two dung beetle traps of standard design (Lumaret 1979) were set up and baited with a mixture of sheep, horse, and human dung (ratio 2:2:1). Figures given are average numbers of individuals per trap. Traps were examined three times per month during one year. Roman numerals I to IV denote the 3-month periods March–May, June–August, September–November, and December–February (spring, summer, autumn, and winter in the Mediterranean climate). Dung-type preference (DP): C, cattle; H, horse; S, sheep; R, rabbit; M, human.

Table B.7
Dung beetle samples from Santa Cruz Acatlan.

Species	DWt	LE	N	N_{mean}	N_{max}	Seasonal Occurrence									N/D	Hab	LF
						Mar	Apr	May	Jun	Jul	Aug	Sep	Oct	Nov			
1. Aphodius lividus (Olivier)	0.005	5	360	2.14	8.36	6	10	21	117	72	50	44	27	12	N	P	C,S
2. Onthophagus mexicanus Bates	0.13	7	255	1.52	6.35		2	17	89	52	45	31	17	2	D	G	C
3. Onthophagus lecontei Harold	0.011	6	11	0.06	0.36				5	4	2				N	G	C
4. Phanaeus palliatus Sturm	0.33	16	135	0.80	3.36			10	47	38	28	9	3		D	G	C
5. Phanaeus quadridens Say	0.27	20	11	0.06	0.36				5	3	2	1			D	G	C
6. Phanaeus adonis Harold*	0.19	14	1	0.005	0.005				1						D	G	C
7. Dichotomius carolinus (Linnaeus)	2.08	26	18	0.11	0.57			2	5	4	3	2	2		N	P	C
8. Ceratotrupes fronticornis (Erichson)*	0.32	22	1	0.005	0.005					1					N	F	C

Notes: Monthly samples were collected with two traps baited with human dung, each trapping session consisting of 7 days and 6 nights (total of 72 trap-nights and 84 trap-days in 1975). Species believed not to be breeding locally are marked with an asterisk (*). N_{mean} is mean number of individuals collected per trap-day (average for whole year), while N_{max} is maximum abundance of the species (mean number of individuals per trap-day in June). Habitat (Hab) abbreviations: P, disturbed; G, grassland; F, forest.

TABLE B.8
Habitat associations of dung beetles in Mkuzi Game Reserve.

Species	DWt	Number Caught	N/D	G%	Average Number per Trap			
					Sand	Duplex	Loam	Clay
FG I								
1. *Anachalcos convexus* Boheman	668	650	N	49	13.8	22.2	15.8	7.7
2. *Kheper clericus* (Boheman)	1,135	4	D	50	0.0	0.0	0.0	0.3
3. *Kheper cupreus* (Castelnau)	995	1	N	100	0.1	0.0	0.0	0.0
4. *Kheper lamarcki* (Macleay)	1,216	136	D	57	8.8	2.0	0.0	0.5
5. *Kheper nigroaeneus* (Boheman)	1,135	1,189	D	55	31.7	27.6	46.6	8.8
6. *Pachylomera femoralis* Kirby	1,507	2,775	D	56	208.7	22.0	0.3	0.4
7. *Scarabaeus galenus* (Westwood)	1,000	3	D	67	0.3	0.0	0.0	0.0
8. *Scarabaeus goryi* Castelnau	1,071	103	N	40	8.3	0.2	0.0	0.1
9. *Scarabaeus zambesianus* Péringuey	500	532	N	58	43.0	1.3	0.0	0.0
FG II								
10. *Allogymnopleurus consocius* Péringuey	95	1,744	D	67	0.2	0.4	1.4	143.8
11. *Allogymnopleurus thalassinus* (Klug)	85	3,553	D	97	285.8	10.1	0.3	0.1
12. *Garreta nitens* (Olivier)	207	138	D	54	0.1	2.4	7.6	3.9
13. *Garreta unicolor* (Fahraeus)	264	35	D	69	0.0	0.0	4.4	0.0
14. *Gymnopleurus virens* Erichson	480	89	D	80	0.3	1.4	1.4	4.8
15. *Odontoloma* spp.	1	3	D	0	0.0	0.3	0.0	0.0
16. *Scarabaeus ebenus* Klug	300	1	N	100	0.0	0.1	0.0	0.0
17. *Sisyphus alveatus* Boucomont	28	1	D	100	0.0	0.0	0.1	0.0
18. *Sisyphus calcaratus* Klug	12	607	D	57	3.8	24.3	12.3	14.3
19. *Sisyphus fasciculatus* Boheman	41	14	D	14	0.0	0.0	0.0	1.2
20. *Sisyphus fortuitus* Péringuey	50	279	D	34	10.8	3.8	8.5	3.1
21. *Sisyphus gazanus* Arrow	11	28	D	4	0.1	0.3	1.0	1.3
22. *Sisyphus goryi* von Harold	10	864	D	73	0.4	11.8	63.1	17.7
23. *Sisyphus infuscatus* Klug	32	249	D	51	2.0	9.3	7.5	4.5
24. *Sisyphus mirabilis* Arrow	40	2,265	D	23	116.2	29.4	11.9	35.3
25. *Sisyphus rubrus* Paschalidis	27	74	D	81	0.1	0.8	6.5	1.0

TABLE B.8 (cont.)

Species	DWt	Number Caught	N/D	G%	Average Number per Trap			
					Sand	Duplex	Loam	Clay
26. *Sisyphus seminulum* Gerstaecker	4	2,325	D	3	36.8	45.5	57.0	73.4
27. *Sisyphus sordidus* Boheman	21	1,670	D	62	136.8	2.3	0.0	0.1
28. *Sisyphus spinipes* (Thunberg)	23	148	D	55	2.8	4.3	4.6	2.2
29. *Sisyphus* sp. y	8	12	D	0	0.3	0.0	0.1	0.7
FG III								
30. *Catharsius* near *pandion* Harold	300	12	N	83	1.0	0.0	0.0	0.0
31. *Catharsius pandion* Harold	300	1	N	0	0.1	0.0	0.0	0.0
32. *Catharsius philus* Kolbe	235	26	N	31	0.0	1.8	0.1	0.3
33. *Catharsius* sp.	350	11	N	18	0.8	0.1	0.0	0.0
34. *Catharsius tricornutus* (De Geer)	384	32	N	38	2.6	0.1	0.0	0.0
35. *Copris amyntor* Klug	112	328	N	52	1.0	12.6	4.1	11.0
36. *Copris elphenor* Klug	389	37	N	76	1.4	0.7	1.0	0.3
37. *Copris evanidus* Klug	20	39	N	69	0.0	1.6	1.1	0.9
38. *Copris fallaciosus* Gillet	417	1	N	0	0.1	0.0	0.0	0.0
39. *Copris mesacanthus* Harold	90	30	N	43	0.4	0.6	1.6	0.4
40. *Copris* near *macer* Péringuey	80	5	N	60	0.3	0.2	0.0	0.0
41. *Copris puncticollis* Boheman	80	4	N	100	0.3	0.0	0.0	0.0
42. *Heliocopris atropos* Boheman	1,800	3	N	100	0.3	0.0	0.0	0.0
43. *Heliocopris hamadryas* (Fabricius)	2,280	2	N	50	0.1	0.0	0.1	0.0
44. *Heliocopris neptunus* Boheman	1,770	13	N	31	0.0	0.3	1.1	0.0
45. *Metacatharsius pseudoopaca* Ferriera	80	176	N	61	14.3	0.4	0.0	0.0
46. *Metacatharsius troglodytes* (Boheman)	40	213	N	90	16.3	1.5	0.0	0.0
FG IV + VI								
47. *Cyptochirus ambiguus* (Kirby)	31	78	D	41	1.2	2.7	1.9	1.4
48. *Euoniticellus intermedius* (Reiche)	17	15	D	100	0.1	0.4	0.4	0.5
49. *Liatongus militaris* (Castelnau)	26	31	N	87	0.3	0.8	1.5	0.4
50. *Milichus apicalis* (Fahraeus)	14	407	N	37	1.6	10.0	29.4	2.8

Species			N/D					
51. *Onitis alexis* Klug	129	9	N	67	0.1	0.3	0.6	0.0
52. *Onitis deceptor* Péringuey	351	2	N	100	0.2	0.0	0.0	0.0
53. *Onitis fulgidus* Klug	204	349	N	28	5.7	6.3	21.5	2.8
54. *Onitis uncinatus* Klug	324	22	N	77	0.0	0.9	0.8	0.4
55. *Onitis picticollis* Boheman	80	5	N	100	0.0	0.1	0.5	0.0
56. *Onitis viridulus* Boheman	221	1	N	0	0.0	0.1	0.0	0.0
57. *Onthophagus aciculatus* Fahraeus	55	6,701	D	29	462.9	93.9	0.4	1.3
58. *Onthophagus acquepubens* d'Orbigny	12	30	D	0	0.5	0.3	0.5	1.3
59. *Onthophagus aeruginosus* Roth	11	858	D	22	14.8	30.3	13.9	17.3
60. *Onthophagus alcyonides* d'Orbigny	27	182	D	63	5.4	3.2	2.5	4.9
61. *Onthophagus aureiceps* d'Orbigny	42	3,639	D	49	279.3	23.7	0.1	0.2
62. *Onthophagus beiranus* Péringuey	15	667	N	48	0.1	7.3	2.3	46.7
63. *Onthophagus* near *carbonarius* Klug	25	3	N	67	0.1	0.0	0.3	0.0
64. *Onthophagus* near *carbonarius* Klug	25	348	N	62	0.1	0.4	42.6	0.1
65. *Onthophagus carbonarius* Klug	25	8,359	N	52	109.3	336.2	151.4	150.2
66. *Onthophagus corniculiger* d'Orbigny	25	1	N	100	0.0	0.0	0.1	0.0
67. *Onthophagus cribripennis* d'Orbigny	16	5	D	0	0.0	0.0	0.0	0.4
68. *Onthophagus depressus* Harold	15	1	N	100	0.1	0.0	0.0	0.0
69. *Onthophagus dives* Harold	39	1,772	D	19	125.0	22.5	0.1	0.1
70. *Onthophagus ebenus* Péringuey	33	14	N	50	0.1	0.3	0.8	0.3
71. *Onthophagus gazella* (F.)	52	102	N	72	1.6	3.1	3.0	1.8
72. *Onthophagus interstitialis* Fahraeus	22	2,824	N	64	2.3	55.8	238.8	18.0
73. *Onthophagus lacustris* Harold	35	39	N	0	3.3	0.0	0.0	0.0
74. *Onthophagus* near *leroyi* d'Orbigny	12	1,522	N	26	10.9	52.2	44.6	34.0
75. *Onthophagus obtusicornis* Fahraeus	21	50	D	66	0.9	3.2	0.0	0.1
76. *Onthophagus* near *pallidipennis* Fahraeus	10	82	D	61	4.1	2.5	0.3	0.1
77. *Onthophagus pallidipennis* Fahraeus	5	1,625	D	39	87.4	25.5	18.0	10.5
78. *Onthophagus plebejus* Klug	25	714	N	48	20.8	30.6	2.0	6.8
79. *Onthophagus quadrituber* d'Orbigny	104	30	D	43	0.0	0.3	0.0	2.3
80. *Onthophagus stigmosus* d'Orbigny	9	619	D	22	43.9	4.0	1.6	2.6
81. *Onthophagus tersidorsis* d'Orbigny	79	407	D	33	3.4	8.0	32.3	1.0
82. *Pedaria* sp.1	10	126	N	64	2.0	6.8	0.5	1.4
83. *Pedaria* sp.2	12	22	N	91	0.0	0.2	2.4	0.1

TABLE B.8 (cont.)

Species	DWt	Number Caught	N/D	G%	Sand	Duplex	Loam	Clay
						Average Number per Trap		
84. *Pedaria* sp.3	17	288	N	75	23.3	0.6	0.1	0.0
85. *Pedaria* sp.4	17	213	N	74	17.3	0.4	0.0	0.0
86. *Pedaria* sp.5	12	9	N	56	0.6	0.1	0.1	0.0
87. *Pedaria* sp.6	12	4	N	75	0.0	0.1	0.4	0.0
88. *Phalops flavocinctus* (Klug)	36	182	D	87	0.4	3.3	9.4	5.3
89. *Sarophorus costatus* (Fahraeus)	23	447	D	70	0.0	1.6	46.5	4.7
FG V + VI								
90. *Caccobius nigritulus* Klug	4	18	D	56	1.4	0.1	0.0	0.0
91. *Caccobius viridicollis* Fahraeus	3	324	D	69	21.4	4.8	0.3	0.6
92. *Caccobius* sp.1	3	13	D	39	0.0	0.0	1.3	0.3
93. *Caccobius* sp.2	3	17	N	29	1.2	0.3	0.0	0.0
94. *Caccobius* sp.3	4	1,306	D	38	106.3	2.4	0.0	0.1
95. *Caccobius* sp.4	3	49	D	2	4.1	0.0	0.0	0.0
96. *Caccobius* sp.5	1	71	N	70	5.5	0.4	0.0	0.0
97. *Caccobius* sp.6	4	2	N	100	0.2	0.0	0.0	0.0
98. *Caccobius* sp.7	4	3	N	0	0.0	0.0	0.1	0.2
99. *Drepanocerus fastiditus* Péringuey	5	4	D	75	0.2	0.0	0.3	0.0
100. *Drepanocerus freyi* Janssens	2	11	D	73	0.1	0.5	0.5	0.0
101. *Drepanocerus impressicollis* Boheman	10	134	D	8	0.0	0.4	0.4	0.5
102. *Drepanocerus kirbyi* (Kirby)	4	603	D	28	7.6	22.8	16.1	9.1
103. *Drepanocerus laticollis* Fahraeus	4	49	D	29	0.3	0.9	4.1	0.1
104. *Drepanocerus patrizii* (Boucomont)	8	10	D	70	0.0	0.0	1.3	0.0
105. *Euoniticellus zumpti* Janssens	4	1	D	100	0.0	0.1	0.0	0.0

No. Species	n	N/D	G%				
106. *Onthophagus apiciosus* d'Orbigny	17	N	25	0.1	0.2	0.4	0.2
107. *Onthophagus flavolimbatus* d'Orbigny	126	D	68	3.1	6.1	0.1	1.3
108. *Onthophagus lamelliger* Gerstaecker	33,111	N	32	186.8	1,180.4	354.6	1,155.6
109. *Onthophagus* near *sugillatus* Klug	424	D	25	34.7	0.6	0.0	0.1
110. *Onthophagus sugillatus* Klug	39,026	D	37	1,245.1	1,465.8	330.8	320.8
111. *Onthophagus* near *vinctus* Erichson	14	N	57	0.2	0.9	0.0	0.1
112. *Onthophagus vinctus* Erichson	132	N	55	7.6	1.5	0.0	1.9
113. *Onthophagus pugionatus* Fahraeus	32	N	22	0.4	0.9	0.5	1.0
114. *Onthophagus pullus* Roth	1,725	D	65	134.1	6.3	3.9	0.8
115. *Onthophagus signatus* Fahraeus	1,631	D	79	132.0	3.6	0.1	0.3
116. *Onthophagus stellio* Erichson	7,080	N	63	117.3	334.0	8.5	133.1
117. *Onthophagus verticallis* Fahraeus	27	N	70	1.4	0.7	0.3	0.0
118. *Tiniocellus spinipes* (Roth)	810	D	60	5.4	17.3	56.0	7.4
FG VII							
119. *Oniticellus formosus* Chevrolat	5	D	100	0.0	0.0	0.3	0.3
120. *Oniticellus planatus* Castelnau	3	D	67	0.0	0.1	0.3	0.0
Total	139,998		42	4,205	4,030	1,711	2,291

Source: Doube, Macqueen, and Davis (unpubl.).

Notes: Pitfall traps were baited with a mixture of pig, horse, and cattle dung for 24 hours on two occasions in December 1981. Trapping was conducted in bushveld and grassveld on four soil types. Overall, 88 traps were set and 120 species and 139,998 individuals were caught. G%, percentage of individuals caught from grassveld; FG, functional group of beetles (Chapter 8).

TABLE B.9(1)

List of Scarabaeidae dung beetles collected at Abokouamekro, 1980–81.

	D/N	FWt	Human Dung					
			I	II	III	IV	V	VI
1. *Scarabaeus goryi* (Laporte)	N	1,675	15	32	4	1		4
2. *Scarabaeus palemo* Olivier	D	540		120	25			
3. *Allogymnopleurus umbrinus* (Gerst.)	D	140	2	397	2	1	1	
4. *Garreta nitens* (Olivier)	D	310		2				
5. *Gymnopleurus coerulescens* (Ol.)	D	62	9	1,479	147	6	2	
6. *Gymnopleurus fulgidus* (Olivier)	D	62	17	92	2		5	13
7. *Gymnopleurus profanus* (Fabricius)	N	115	5	177	22	5		
8. *Gymnopleurus puncticollis* Gillet	D	130		1,618	39			
9. *Anachalcos aurescens* Bates	N	1,738			1			
10. *Anachalcos suturalis* Janssens	N	517			9			
11. *Neosisyphus armatus* (Gory)	D	49	6	16	10	6	2	11
12. *Neosisyphus baoule* Cambefort	D	80			1			
13. *Neosisyphus gladiator* Arrow	D	155			2			
14. *Neosisyphus paschalidisae* Camb.	D	104					1	
15. *Sisyphus biarmatus* Felsche	D	46		42	3	5	8	
16. *Sisyphus costatus* Thunberg	D	14		77	10	29	9	
17. *Sisyphus desaegeri* Haaf	D	68				6		
18. *Sisyphus goryi* Harold	D	17	308	1,835	918	545	220	411
19. *Sisyphus seminulum* Gerstaecker	D	8	175	1,123	2,325	1,926	1,650	707
20. *Catharsius crassicornis* Gillet	N	1,221	12	39	36	8	8	4
21. *Catharsius eteocles* Laporte	N	1,267		3	2	1	1	
22. *Catharsius sesostris* Waterhouse	N	566		2	2			
23. *Copris carmelita* Fabricius	N	464			1	2	1	
24. *Copris coriarius* Gillet	N	194		1				
25. *Copris interioris* Kolbe	N	365		30			4	1
26. *Copris orion* Klug	N	216						
27. *Copris renwarti* Nguyen & Camb.	N	853	1	7	1	4	1	
28. *Litocopris punctiventris* Wat.	N	136						
29. *Metacatharsius abortivus* (Fairm.)	N	15		312	20			
30. *Metacatharsius inermis* (Laporte)	N	130		5				
31. *Metacatharsius tchadensis* Balth.	N	40		36	3			
32. *Coptorhina subaenea* Janssens	D	30						
33. *Heliocopris antenor* (Olivier)	N	5,800		4				
34. *Pedaria coprinarum* Cambefort	N	35		18	1	5	2	2
35. *Pedaria dedei* Cambefort	N	35		1				
36. *Pedaria durandi* Paul., C. & M.	N	35						
37. *Pedaria humana* Cambefort	N	35		23	5	1	1	
38. *Pedaria renwarti* Cambefort	N	35		91	10	6	4	
39. *Onitis alexis* Klug	N	375						
40. *Onitis cupreus* Laporte	N	463		4		7	1	
41. *Onitis reichei* Van Lansberge	N	411						
42. *Cyptochirus distinctus* (Janssens)	D	145						

	Cattle Dung						C	V
	I	II	III	IV	V	VI	C	V
1. *Scarabaeus goryi* (Laporte)								
2. *Scarabaeus palemo* Olivier								
3. *Allogymnopleurus umbrinus* (Gerst.)								
4. *Garreta nitens* (Olivier)								
5. *Gymnopleurus coerulescens* (Ol.)		1						
6. *Gymnopleurus fulgidus* (Olivier)								
7. *Gymnopleurus profanus* (Fabricius)								
8. *Gymnopleurus puncticollis* Gillet								
9. *Anachalcos aurescens* Bates		1						
10. *Anachalcos suturalis* Janssens	1	4	25	19	11	1		
11. *Neosisyphus armatus* (Gory)								
12. *Neosisyphus baoule* Cambefort								
13. *Neosisyphus gladiator* Arrow								
14. *Neosisyphus paschalidisae* Camb.		5	10	30	56			
15. *Sisyphus biarmatus* Felsche								
16. *Sisyphus costatus* Thunberg								
17. *Sisyphus desaegeri* Haaf								
18. *Sisyphus goryi* Harold	178	699	137	32	89	499	10	
19. *Sisyphus seminulum* Gerstaecker	34	468	106	3	90	43	744	68
20. *Catharsius crassicornis* Gillet	25	18	4	12	6	42		
21. *Catharsius eteocles* Laporte								
22. *Catharsius sesostris* Waterhouse								
23. *Copris carmelita* Fabricius								
24. *Copris coriarius* Gillet								
25. *Copris interioris* Kolbe	6	216	58	48	103	2		
26. *Copris orion* Klug	28	17	22	13	24	11		
27. *Copris renwarti* Nguyen & Camb.	3	27	3	6	9			
28. *Litocopris punctiventris* Wat.	1	1	1	1	1	6		
29. *Metacatharsius abortivus* (Fairm.)								
30. *Metacatharsius inermis* (Laporte)		1						
31. *Metacatharsius tchadensis* Balth.		2						
32. *Coptorhina subaenea* Janssens								8
33. *Heliocopris antenor* (Olivier)		4	2					
34. *Pedaria coprinarum* Cambefort	51	99	119	138	164	168		
35. *Pedaria dedei* Cambefort	2	12	37	21	26	21		
36. *Pedaria durandi* Paul., C. & M.	4	6	4	13	8	4		
37. *Pedaria humana* Cambefort		3						
38. *Pedaria renwarti* Cambefort	6	65	2	2	2	2		
39. *Onitis alexis* Klug	6	6	6	4	2	4		
40. *Onitis cupreus* Laporte								
41. *Onitis reichei* Van Lansberge	12	3	10	7	9	3		
42. *Cyptochirus distinctus* (Janssens)		1	2	13	2	1		

TABLE B.9(1) (*cont.*)

	D/N	FWt	I	II	III	IV	V	VI
					Human Dung			
43. *Drepanocerus caelatus* (Gerst.)	D				5	2	1	
44. *Drepanocerus endroedyi* Endrödi	D	5	2		4	1	1	2
45. *Drepanocerus laticollis* Fahraeus	D	8	28	19	11	20	58	96
46. *Drepanocerus sulcicollis* (Laporte)	D	8						
47. *Euoniticellus intermedius* (Reiche)	D	38	6	2				6
48. *Euoniticellus parvus* (Kraatz)	D	5	17	11		10	6	38
49. *Oniticellus formosus* Chevrolat	D	73						
50. *Oniticellus planatus* Laporte	D	78						
51. *Tiniocellus spinipes* (Roth)	D	18	567	275	171	189	520	573
52. *Caccobius auberti* d'Orbigny	D	5	86	1,771	165	403	401	110
53. *Caccobius cavatus* d'Orbigny	N	5		102	8	16	6	1
54. *Caccobius ferrugineus* (Fahraeus)	N	8		37	2			
55. *Caccobius ivorensis* Cambefort	D	7	17	402	63	246	137	117
56. *Caccobius lamottei* Cambefort	D	8		1				
57. *Caccobius mirabilepunctatus* Camb.	D	12		23	2			
58. *Caccobius punctatissimus* Harold	N	13		17	5			
59. *Cambefortius lamizanai* (Cambefort)	N	50		23				
60. *Cleptocacc. balthasarianus* (Camb.)	D	4	71	4				
61. *Cleptocacc. convexifrons* (Raffr.)	D	4		114	13			
62. *Cleptocaccobius dorbignyi* Camb.	D	4		42				
63. *Cleptocacc. signaticollis* (d'Orb.)	D	4	11	766	73			
64. *Cleptocaccobius uniseries* (d'Orb.)	D	4	211	855	209	149	62	67
65. *Digitonthophagus gazella* (F.)	N	129	27	222	1			3
66. *Euonthophagus carbonarius* (Klug)	N	45	31	98	20	28	2	34
67. *Milichus apicalis* (Fahraeus)	N	31	1	8		1	1	
68. *Milichus serratus* d'Orbigny	N	32		15			1	
69. *Onthophagus altidorsis* d'Orbigny	N	24	1	5	30	11	11	6
70. *Onthophagus atridorsis* d'Orbigny	D	3		13	6	34	7	2
71. *Onthophagus baoule* Cambefort	D	8		54	10	5		
72. *Onthophagus basakata* Walt. & Camb.	D	4			5			1
73. *Onthophagus bidens* (Olivier)	D	35	18	681	47	16	54	36
74. *Onthophagus bidentifrons* d'Orbigny	D	16		2				2
75. *Onthophagus bimarginatus* d'Orbigny	N	104				1		
76. *Onthophagus bituberculatus* (Ol.)	N	61	2	2			9	10
77. *Onthophagus borassi* Cambefort	D	7		1	5	28	1	1
78. *Onthophagus callosipennis* Bouc.	N	21						
79. *Onthophagus cornifrons* Thomson	N	21	2	51	8			
80. *Onthophagus cribellum* d'Orbigny	D	18		1				
81. *Onthophagus cupreus* Harold	D	105		164	20			
82. *Onthophagus cyanochlorus* d'Orb.	D	12		54	5			
83. *Onthophagus fimetarius* (Roth)	N	27		9	4	70	2	
84. *Onthophagus flaviclava* d'Orbigny	D	14	86	214	107	30	6	77
85. *Onthophagus flexicornis* d'Orbigny	N	31				2		

	Cattle Dung							
	I	II	III	IV	V	VI	C	V
43. *Drepanocerus caelatus* (Gerst.)			1	2				
44. *Drepanocerus endroedyi* Endrödi	17	7	22	100	233	75		
45. *Drepanocerus laticollis* Fahraeus	610	142	53	592	736	911		
46. *Drepanocerus sulcicollis* (Laporte)	1			10	4	5		
47. *Euoniticellus intermedius* (Reiche)	244	414	422	86	89	864		
48. *Euoniticellus parvus* (Kraatz)	68	27	67	344	695	51		
49. *Oniticellus formosus* Chevrolat	20	2		13	3	3		
50. *Oniticellus planatus* Laporte	4	1	2	13	28	7		
51. *Tiniocellus spinipes* (Roth)	54	35	16	69	771	134		
52. *Caccobius auberti* d'Orbigny	7	149	101	208	1,392	25		
53. *Caccobius cavatus* d'Orbigny	2	8	16	22	2	8		
54. *Caccobius ferrugineus* (Fahraeus)								
55. *Caccobius ivorensis* Cambefort			3	11	25	3	9	
56. *Caccobius lamottei* Cambefort								
57. *Caccobius mirabilepunctatus* Camb.		1	1					
58. *Caccobius punctatissimus* Harold		2						
59. *Cambefortius lamizanai* (Cambefort)								
60. *Cleptocacc. balthasarianus* (Camb.)	14	1						
61. *Cleptocacc. convexifrons* (Raffr.)		36	7					
62. *Cleptocaccobius dorbignyi* Camb.								
63. *Cleptocacc. signaticollis* (d'Orb.)	4	29						
64. *Cleptocaccobius uniseries* (d'Orb.)	24	5	31	385	345	9		
65. *Digitonthophagus gazella* (F.)	319	239	43	51		12		
66. *Euonthophagus carbonarius* (Klug)	1	1	1	1		5		
67. *Milichus apicalis* (Fahraeus)	9	4	1	3	10	29		
68. *Milichus serratus* d'Orbigny	5	21	16	19	21	9		
69. *Onthophagus altidorsis* d'Orbigny							16	
70. *Onthophagus atridorsis* d'Orbigny							1	
71. *Onthophagus baoule* Cambefort								
72. *Onthophagus basakata* Walt. & Camb.								
73. *Onthophagus bidens* (Olivier)						2		
74. *Onthophagus bidentifrons* d'Orbigny								
75. *Onthophagus bimarginatus* d'Orbigny								
76. *Onthophagus bituberculatus* (Ol.)								
77. *Onthophagus borassi* Cambefort								
78. *Onthophagus callosipennis* Bouc.								1
79. *Onthophagus cornifrons* Thomson							1	1
80. *Onthophagus cribellum* d'Orbigny			3					
81. *Onthophagus cupreus* Harold						2		
82. *Onthophagus cyanochlorus* d'Orb.								
83. *Onthophagus fimetarius* (Roth)		3	2	5				
84. *Onthophagus flaviclava* d'Orbigny	16		1			1	15	1
85. *Onthophagus flexicornis* d'Orbigny								

TABLE B.9(1) (*cont.*)

	D/N	FWt	Human Dung					
			I	II	III	IV	V	VI
86. *Onthophagus ghanensis* Balthasar	N	22						
87. *Onthophagus imbellis* d'Orbigny	D	14						
88. *Onthophagus juvencus* Klug	N	25	3	4	3	42	3	3
89. *Onthophagus lamottellus* Cambefort	D	4				4		
90. *Onthophagus lamtoensis* Cambefort	N	44		60	62	59	2	
91. *Onthophagus latigibber* d'Orbigny	N	50				3		
92. *Onthophagus lioides* d'Orbigny	D	6	34	342	15			6
93. *Onthophagus longipilis* d'Orbigny	D	35		5	16			
94. *Onthophagus loudetiae* Cambefort	D	5		2				1
95. *Onthophagus marahouensis* Camb.	D	4	11	34				1
96. *Onthophagus marginifer* Frey	D	16	3	12	3	5		7
97. *Onthophagus mocquerysi* d'Orbigny	N	62		4	5			
98. *Onthophagus mucronatus* Thomson	D	11	104	695	338	366	212	87
99. *Onthophagus mucronifer* d'Orbigny	N	10						
100. *Onthophagus naevius* d'Orbigny	N	19		1	9	10	4	
101. *Onthophagus picatus* d'Orbigny	N	27		1				
102. *Onthophagus pullus* Roth	D	3		1				
103. *Onthophagus reticulatus* d'Orb.	N	44		22	4			
104. *Onthophagus rufonotatus* d'Orb.	D	48	34	781	294	650	1,260	146
105. *Onthophagus rufostillans* d'Orb.	N	77		4		22	1	
106. *Onthophagus sanguineus* d'Orbigny	D	3		1				
107. *Onthophagus sanguinolentus* d'Orb.	N	11		2				
108. *Onthophagus savanicola* Cambefort	D	38			1	1		
109. *Onthophagus semivirescens* d'Orb.	N	18		1	1	1		1
110. *Onthophagus sinuosus* d'Orbigny	D	7		20	3			
111. *Onthophagus subsulcatus* d'Orbigny	D	13		1				
112. *Onthophagus tersipennis* d'Orbigny	D	3	24	134	11		2	
113. *Onthophagus tripartitus* d'Orbigny	N	17		18		22	2	
114. *Onthophagus ulula* Balthasar	N	15		7	5			7
115. *Onthophagus variegatus* (Fabricius)	N	14	6	8	1			4
116. *Onthophagus vinctus* Erichson	N	19	34	56	24	37	8	29
117. *Onthophagus vuattouxi* Cambefort	D	31	2	3	12	9	3	2
118. *Onthophagus vultuosus* d'Orbigny	N	18	1		3			
119. *Phalops iphis* (Olivier)	D	124		1				
120. *Phalops vanellus* (Van Lansberge)	D	68		48				
121. *Proagoderus auratus* (Fabricius)	D	213		32	5		1	
122. *Proagoderus cambeforti* Palestrini	D	182				1		
123. *Strandius obliquus* (Olivier)	N	204		1				

Notes: The following data are given for each species: diel activity, fresh weight, number of individuals collected in human dung-baited traps in the six bimonthly periods (I: February–March, etc.); the same for beetles extracted from cattle dung pats; and the numbers of beetles caught in carrion-baited traps (C) and in vegetable material (V).

	Cattle Dung							
	I	II	III	IV	V	VI	C	V
86. *Onthophagus ghanensis* Balthasar								4
87. *Onthophagus imbellis* d'Orbigny								1
88. *Onthophagus juvencus* Klug	71	127	19	322	491	96		
89. *Onthophagus lamottellus* Cambefort								1
90. *Onthophagus lamtoensis* Cambefort			1				25	
91. *Onthophagus latigibber* d'Orbigny							41	
92. *Onthophagus lioides* d'Orbigny		5						
93. *Onthophagus longipilis* d'Orbigny								
94. *Onthophagus loudetiae* Cambefort								
95. *Onthophagus marahouensis* Camb.			1					
96. *Onthophagus marginifer* Frey		8	11	16	5			
97. *Onthophagus mocquerysi* d'Orbigny							1	
98. *Onthophagus mucronatus* Thomson	3		1				21	
99. *Onthophagus mucronifer* d'Orbigny	1							
100. *Onthophagus naevius* d'Orbigny							2	
101. *Onthophagus picatus* d'Orbigny								
102. *Onthophagus pullus* Roth								
103. *Onthophagus reticulatus* d'Orb.								
104. *Onthophagus rufonotatus* d'Orb.	7	5	17	11	57	8	185	1
105. *Onthophagus rufostillans* d'Orb.						2		
106. *Onthophagus sanguineus* d'Orbigny								
107. *Onthophagus sanguinolentus* d'Orb.								
108. *Onthophagus savanicola* Cambefort								
109. *Onthophagus semivirescens* d'Orb.								
110. *Onthophagus sinuosus* d'Orbigny								
111. *Onthophagus subsulcatus* d'Orbigny								
112. *Onthophagus tersipennis* d'Orbigny		9						
113. *Onthophagus tripartitus* d'Orbigny								
114. *Onthophagus ulula* Balthasar	12	5	2	1	1			
115. *Onthophagus variegatus* (Fabricius)	2	1	2	1		4		
116. *Onthophagus vinctus* Erichson		9	2	2	2	2		
117. *Onthophagus vuattouxi* Cambefort							26	
118. *Onthophagus vultuosus* d'Orbigny							8	
119. *Phalops iphis* (Olivier)								
120. *Phalops vanellus* (Van Lansberge)		5						
121. *Proagoderus auratus* (Fabricius)		1						
122. *Proagoderus cambeforti* Palestrini								
123. *Strandius obliquus* (Olivier)								

Table B.9(2)
Numbers of Scarabaeidae dung beetles collected in six savanna localities, 1980–81.

Species	Localities					
	1	2	3	4	5	6
1. *Kheper festivus* (Harold)					4	
2. *Kheper subaeneus* (Harold)					1	
3. *Scarabaeus goryi* (Laporte)			57		19	3
4. *Scarabaeus palemo* Olivier		17	145		40	
5. *Allogymnopleurus umbrinus* (Gerst.)	30	1	436		6	13
6. *Allogymnopleurus youngai* Endrödi					8	32
7. *Garreta azureus* (F.)					4	
8. *Garreta nitens* (Ol.)	460		2		593	30
9. *Gymnopleurus coerulescens* (Ol.)	428	2,676	1,644		168	122
10. *Gymnopleurus fulgidus* (Ol.)			129			
11. *Gymnopleurus profanus* (F.)	7	271	209		67	89
12. *Gymnopleurus puncticollis* Gillet			1,657		1,677	
13. *Anachalcos aurescens* Bates		4	1			29
14. *Anachalcos convexus* Boheman				7	30	61
15. *Anachalcos cupreus* (F.)					3	
16. *Anachalcos suturalis* Janssens		5	10		7	8
17. *Odontoloma relictum* Cambefort		1				3
18. *Neosisyphus armatus* (Gory)	16	114	112	6	20	139
19. *Neosisyphus baoule* Cambefort	3		1			
20. *Neosisyphus gladiator* (Arrow)	4		2	11	19	45
21. *Neosisyphus paschalidisae* Camb.	18		102	122	12	189
22. *Sisyphus biarmatus* Felsche		5,375	61	222	24	
23. *Sisyphus costatus* Thunberg	42	4,549	126	75	4	811
24. *Sisyphus desaegeri* Haaf		54				
25. *Sisyphus gazanus* Arrow		264	6	14		
26. *Sisyphus goryi* Harold	1,279	4,973	5,881	933	2,901	3,821
27. *Sisyphus tobi* Cambefort						145
28. *Sisyphus seminulum* Gerst.	2,800	16,679	8,718	1,032	418	2,822

#	Species						
29.	*Catharsius crassicornis* Gillet		18		209		
30.	*Catharsius eteocles* (Laporte)	20	50	29	9	2	
31.	*Catharsius fastidiosus* Thomson	1				31	
32.	*Catharsius pseudolycaon* Ferreira	4	4				
33.	*Catharsius sesostris* Waterh.	5	150		4	8	267
34.	*Copris angustus* Nguyen		11				
35.	*Copris carmelita* F.	14	15	3	4	7	2
36.	*Copris coriarius* Gillet	25	4		1	26	
37.	*Copris eburneus* Nguyen		6	9			
38.	*Copris gazellarum* Gillet	49					
39.	*Copris interioris* Kolbe	145		1	468	2	326
40.	*Copris megaceratoides* Waterh.	1	2				
41.	*Copris moffartsi* Gillet		3				
42.	*Copris orion* Klug		4	36	115	2	4
43.	*Copris orphanus* Guérin		1				
44.	*Copris renwarti* Ng.& Camb.		7				
45.	*Litocopris muticus* Boh.	5	5	8	42	5	
46.	*Litocopris punctiventris* Waterh.	1			11		31
47.	*Metacatharsius abortivus* (Fairm.)		13	189	332	1,418	2
48.	*Metacatharsius inermis* (Laporte)	1	18		6	2	
49.	*Metacatharsius tchadensis* Balth.		14		41		
50.	*Coptorhina subaenea* Janssens				8	36	
51.	*Delopleurus gilleti* Janssens	2				1	
52.	*Heliocopris antenor* (Ol.)	1	5		10		
53.	*Heliocopris colossus* Bates		3				
54.	*Heliocopris haroldi* Kolbe		2				
55.	*Heliocopris myrmidon* Kolbe	9	1				1
56.	*Pedaria coprinarum* Cambefort	1	22	2	768		78
57.	*Pedaria criberrima* Waterh.	2					2
58.	*Pedaria decorsei* Bouc.	10	7				
59.	*Pedaria dedei* Cambefort		1		120		
60.	*Pedaria durandi* P.,C. & M.		1	1	38		5
61.	*Pedaria fernandezi* Cambefort	14					
62.	*Pedaria humana* Cambefort	7	8		33		40

TABLE B.9(2) (cont.)

Species	Localities					
	1	2	3	4	5	6
63. *Pedaria ouangoensis* Cambefort						1
64. *Pedaria renwarti* Cambefort		2	190	2	87	
65. *Heteronitis pauliani* Janssens					2	
66. *Onitis aeneus* Van Lansberge						1
67. *Onitis affinis* Felsche						1
68. *Onitis alexis* Klug	1	2	28	6		51
69. *Onitis cupreus* Laporte	2	121	12	17	1	56
70. *Onitis lobi* Cambefort						1
71. *Onitis multidentatus* Gillet					1	
72. *Onitis nigeriensis* Ferreira					2	24
73. *Onitis occidentalis* Gillet					12	1
74. *Onitis reichei* Van Lansb.	28	2	44	3	4	31
75. *Onitis robustus* Boheman						1
76. *Onitis sibuensis* Janssens						7
77. *Onitis sphinx* (F.)					4	2
78. *Onitis violaceus* Van Lansb.						1
79. *Cyptochirus distinctus* (Janss.)	3	2	18	13	7	5
80. *Drepanocerus bechynei* Janssens						18
81. *Drepanocerus caelatus* (Gerst.)	2	57	3	12		4
82. *Drepanocerus endroedyi* Endrödi	52	177	465	265	9	122
83. *Drepanocerus laticollis* Fahr.	2		3,276	143	8	
84. *Drepanocerus marshalli* Bouc.	5			66	2	11
85. *Drepanocerus saegeri* Balth.					3	
86. *Drepanocerus sulcicollis* (Lap.)		2	19			19
87. *Euoniticellus fumigatus* (Bouc.)					3	8
88. *Euoniticellus intermedius* (Reiche)	239	19	2,132	451	1	621
89. *Euoniticellus kawanus* (Janssens)					1	
90. *Euoniticellus nasicornis* (Reiche)					8	

91. *Euoniticellus parvus* (Kraatz)	188	25	1,334	979	9	245
92. *Euoniticellus tibatensis* (Kolbe)	157				3	
93. *Liatongus fulvostriatus* d'Orb.						75
94. *Oniticellus formosus* Chevr.	3	1	41	6	1	4
95. *Oniticellus planatus* Lap.			55	1	1	679
96. *Tiniocellus spinipes* (Roth)	832	1,400	3,374	1,206	592	6
97. *Caccobius anthracites* d'Orb.					43	
98. *Caccobius auberti* d'Orb.	818	669	4,818	181	228	85
99. *Caccobius cavatus* d'Orb.	1		191	1	93	2
100. *Caccobius ferrugineus* (Fahr.)			38		24	26
101. *Caccobius inops* Péringuey				3	37	7
102. *Caccobius ivorensis* Camb.	20	2,655	1,032	80	101	1
103. *Caccobius lamottei* Camb.			1			
104. *Caccobius mirabilepunctatus* Camb.	4	16			107	
105. *Caccobius pentagonus* d'Orb.	12		27	16	1	
106. *Caccobius punctatissimus* Harold		3	24		225	39
107. *Cambefortius bidentiger* (d'Orb.)	1					
108. *Cambefortius lamizanai* (Camb.)	6	3	23			3
109. *Cleptocaccobius balthasarianus* (Cambefort)	4		90			131
110. *Cleptocaccobius convexifrons* (Raffray)	196		170		171	60
111. *Cleptocaccobius dorbignyi* Camb.			42	151	82	1,659
112. *Cleptocaccobius signaticollis* (d'Orb.)	52	362	883		122	108
113. *Cleptocaccobius uniseries* (d'Orb.)	287	3,011	2,352	115	651	1,071
114. *Diastellopalpus lamellicollis* (Qued.)	5					
115. *Digitonthophagus gazella* (F.)	1,172	53	917	7	135	177
116. *Euonthophagus carbonarius* (Klug)		6	222		2	179
117. *Hyalonthophagus nigroviolaceus* (d'Orb.)					14	5
118. *Hyalonthophagus pseudoalcyon* (d'Orb.)					42	
119. *Milichus apicalis* (Fahr.)	479	13	67		1	11
120. *Milichus serratus* d'Orb.	93		107	1	2	139
121. *Onthophagus altidorsis* d'Orb.	46	542	80	14	1	307
122. *Onthophagus antennalis* Frey						4
123. *Onthophagus atridorsis* d'Orb.	8	993	63	109	49	451
124. *Onthophagus baoule* Camb.		33	69		4	4

Table B.9(2) (cont.)

Species	Localities					
	1	*2*	*3*	*4*	*5*	*6*
125. *Onthophagus basakata* Walt. & Camb.						3
126. *Onthophagus bidens* (Ol.)	320	73	854	29	35	67
127. *Onthophagus bidentifrons* d'Orb.		4	3			
128. *Onthophagus bimarginatus* d'Orb.			1			
129. *Onthophagus bituberculatus* (Ol.)			23	1		
130. *Onthophagus borassi* Camb.	3	367	36	5		36
131. *Onthophagus callosipennis* Bouc.		2	1			6
132. *Onthophagus chrysoderus* d'Orb.						3
133. *Onthophagus civettae* Camb.						3
134. *Onthophagus clementianus* Camb.	2,066				72	353
135. *Onthophagus cornifrons* Thoms.	1	46	63		2	84
136. *Onthophagus cribellum* d'Orb.		2	4			1
137. *Onthophagus cupreovirens* d'Orb.						
138. *Onthophagus cupreus* Harold	9	6	186		61	1
139. *Onthophagus cyanochlorus* d'Orb.		42	59		12	201
140. *Onthophagus depressipennis* Camb.					6	1
141. *Onthophagus fimetarius* Roth		139	95	2	30	14
142. *Onthophagus fitiniensis* Camb.	24					14
143. *Onthophagus flaviclava* d'Orb.	431	466	554	10	63	619
144. *Onthophagus flexicornis* d'Orb.	6	70	2	1	2	35
145. *Onthophagus fumatus* d'Orb.	1					
146. *Onthophagus ghanensis* Balth.			4			2
147. *Onthophagus grandifrons* d'Orb.	28					
148. *Onthophagus imbellis* d'Orb.	1		1			
149. *Onthophagus iulicola* Camb.	3					11
150. *Onthophagus juvencus* Klug	63	39	1,184	71	6	1
151. *Onthophagus lamottellus* Camb.		67	5	2		
152. *Onthophagus lamtoensis* Camb.	16	512	209		4	19

Species						
153. *Onthophagus latigibber* d'Orb.	137			1		56
154. *Onthophagus lemagneni* d'Orb.		2	44			1
155. *Onthophagus lioides* d'Orb.	17		398		276	3
156. *Onthophagus lobi* Camb.	18				2	6
157. *Onthophagus longipilis* d'Orb.			21		1	57
158. *Onthophagus loudetiae* Camb.		15	3			8
159. *Onthophagus loveni* Gillet					10	
160. *Onthophagus loxodontae* Camb.					3	
161. *Onthophagus lutaticollis* d'Orb.	164				8	527
162. *Onthophagus mankonoensis* Balth.	1			1		
163. *Onthophagus marahouensis* Camb.			47		7	
164. *Onthophagus marginifer* Frey	1	70				
165. *Onthophagus mediofuscatus* d'Orb.					36	
166. *Onthophagus micros* d'Orb.					3	
167. *Onthophagus miles* d'Orb.					2	*
168. *Onthophagus mocquerysi* d'Orb.	5	9	10			4
169. *Onthophagus mucronatus* Thoms.	89	2,729	1,827	34	374	186
170. *Onthophagus mucronifer* d'Orb.		108	1		20	61
171. *Onthophagus naevius* d'Orb.		152	26	2		2
172. *Onthophagus picatus* d'Orb.	4	20	1	2		
173. *Onthophagus pullus* Roth		2	1		192	201
174. *Onthophagus reticulatus* d'Orb.	143	3	26		1	36
175. *Onthophagus rougonorum* Camb.						111
176. *Onthophagus rubricatus* d'Orb.	1					8
177. *Onthophagus rufonotatus* d'Orb.	912	2,288	3,456	397	669	50
178. *Onthophagus rufostillans* d'Orb.	1	258	29	2		29
179. *Onthophagus rugosipennis* Frey					39	7
180. *Onthophagus sanguineus* d'Orb.			1			
181. *Onthophagus sanguinolentus* d'Orb.		25	2			8
182. *Onthophagus savanicola* Camb.		4	1			
183. *Onthophagus sellatulus* d'Orb.						101
184. *Onthophagus semivirescens* d'Orb.	5	21	3			4
185. *Onthophagus sinuosus* d'Orb.		1	23		1	8
186. *Onthophagus subsulcatus* d'Orb.		7	1			1

TABLE B.9(2) *(cont.)*

Species	Localities 1	2	3	4	5	6
187. *Onthophagus subulifer* d'Orb.	7				15	
188. *Onthophagus tersipennis* d'Orb.		13			77	26
189. *Onthophagus triacanthus* Laporte			180		124	
190. *Onthophagus trinominatus* Goid.	60					9
191. *Onthophagus tripartitus* d'Orb.	9	187	42		33	125
192. *Onthophagus ulula* Balth.	93	4	39		18	860
193. *Onthophagus variegatus* (F.)		5	30		5	
194. *Onthophagus vinctus* Er.		160	204		5	
195. *Onthophagus vuattouxi* Camb.		115	42	2		
196. *Onthophagus vultuosus* d'Orb.		23	12		1	4
197. *Onthophagus* n. sp.						1
198. *Phalops batesi* Harold					2	1
199. *Phalops iphis* (Ol.)			1			11
200. *Phalops vanellus* Van Lansb.	106		53	2	91	2
201. *Proagoderus auratus* (F.)		69	37	35	231	17
202. *Proagoderus cambeforti* Palestr.			1	43		
203. *Proagoderus* n. sp.						
204. *Pseudosaproecius cylindroides* (d'Orb.)		2				
205. *Pseudosaproecius validicornis* (Qued.)		2				
206. *Strandius obliquus* (Ol.)			1	1	92	
207. *Strandius* sp.					56	
Number of species	85	91	123	63	132	139
Number of individuals	15,308	54,751	53,648	7,002	12,159	19,162

Notes: Beetles were trapped in the following localities: 1, Sipilou; 2, Lamto; 3, Abokouamekro; 4, La Marahoué; 5, Kakpin; and 6, Wango Fitini. Trapping periods as follows: Lamto, every month of the year; La Marahoué, October 1980; Kakpin, May 1981; Sipilou, four times once every three months, 1980–81; Abokouamekro and Wango Fitini, six times once every two months, in 1980–81.

TABLE B.10(1)

List of dung and carrion beetles collected in the Dumoga-Bone National Park in North Sulawesi in 1985 (families Scarabaeidae, Aphodiidae, Geotrupidae, Hybosoridae and Silphidae).

Species	LE	Food Selection	Elevational Distribution (meters)
Scarabaeidae			
1. *Copris calvus* Sharp	21	Co	200–1,150
2. *Copris macacus* Lansb.	11	Co	90–800
3. *Copris paramacacus* MS		Co	950–1,600
4. *Copris saundersi* Harold	27	Co	300–1,100
5. *Onthophagus aereomaculatus* Bouc.	5	Ge	10–650
6. *Onthophagus aper* Sharp	7	Ne	85–1,200
7. *Onthophagus aureopilosus* Bouc.	8	Co	85–1,120
8. *Onthophagus begoniophilus* MS		Ne	700–800
9. *Onthophagus bisscrutator* MS	6	Ne	85–1,200
10. *Onthophagus bongkudai* MS		Ge	230–1,780
11. *Onthophagus curvicarinatus* Bouc.	17	Co	350–1,150
12. *Onthophagus forsteni* Lansb.	8	Co	220–1,100
13. *Onthophagus fulvus* Sharp	5	Co	85–1,200
14. *Onthophagus fuscostriatus* Bouc.	6	Co	85–850
15. *Onthophagus hollowayi* MS	4	Ne	200–550
16. *Onthophagus holosericeus* Harold		Ge	200–1,000
17. *Onthophagus limbatus* Herbst.		Co	10–1,200
18. *Onthophagus magnipygus* Bouc.	6	Co	200–800
19. *Onthophagus mentaveiensis* Bouc.	7	Ge	200–1,340
20. *Onthophagus moajat* MS		Co	230–1,780
21. *Onthophagus rectecornutus* Lansb.		Co	10–250
22. *Onthophagus ribbei* Bouc.	10	Co	85–1,200
23. *Onthophagus rosenbergi* MS		Ge	600–1,150
24. *Onthophagus sangirensis* Bouc.	6	Co	85–1,200
25. *Onthophagus sarasini* MS		Ge	600–1,600
26. *Onthophagus scrutator* Harold		Ge	250–1,200
27. *Onthophagus sembeli* MS	5	Ge	200–600
28. *Onthophagus sinagai* MS		Co	200–1,150
29. *Onthophagus spiculatus* Bouc.	4	Co	200–700
30. *Onthophagus sulawesiensis* MS		Ne	800–1,150
31. *Onthophagus toraut* MS	3	Ge	85–1,150
32. *Onthophagus travestitus* MS	6	Co	220–1,150
33. *Onthophagus tumpah* MS	3	Co	200–1,600
34. *Onthophagus wallacei* Harold		Co	10–1,340
35. *Paragymnopleurus planus* (Sharp)	19	Co	200–1,150
36. *Cyobius knighti* MS	4		
37. *Haroldius cambeforti* MS			

TABLE B.10(1) (*cont.*)

Species	LE	Food Selection	Elevational Distribution (meters)
38. *Haroldius celebensis* MS			
39. *Panelus gonipa* MS			
Aphodiidae			
40. *Aphodius* sp. 3	4	Co	200
41. *Aphodius* sp. 4	3	Co	200–450
42. *Aphodius* sp. 5	6	Co	200–1,150
43. *Aphodius* sp. 6		Co	
Geotrupidae			
44. *Bolbochromus celebensis* Bouc.	9		
45. *Bolbochromus secundus* MS			
Hybosoridae			
46. *Phaeochrous emarginatus* Cast.	11	Ne	200–950
Silphidae			
47. *Nicrophorus distinctus* Grouv.	25	Ne	200–1,300
48. *Nicrophorus celebianus* MS		Ne	1,400–1,750
49. *Diamesus osculus*		Ne	
50. *Chrysosilpha renatae* (Portevin)		Ne	200–600

Notes: Several species are denoted by their manuscript names. These species will be published formally by Krikken & Huijbregts (in prep.). Some additional species are known from other localities in North Sulawesi. Food selection by adult beetles is coded as Co, primarily coprophagous species; Ne, primarily necrophagous species; and Ge, generalist.

TABLE B.10(2)

Species caught by British Museum team (on their Plot A) in three flight interception traps (FIT).

Species	Seasonal Occurrence (number of samples)											
	Feb (10	Mar 10	Apr 10	May 5	Jun 2	Jul 3	Aug 3	Sep 3	Oct 3	Nov 3)	x	CV
2. C. macacus	1										*	316
5. O. aereomaculatus	1		2	2				3	3	10	2	141
6. O. aper	101	64	210	158	350	66	73	163	203	483	187	72
7. O. aureopilosus	36	52	36	40	140	46	33	100	133	83	70	59
9. O. bisscrutator	36	47	167	60	440	123	96	50	50	210	127	96
12. O. forsteni			1	2	10	3		3		6	2	128
13. O. fulvus	14	8	26	42	135	23	30	40	133	56	50	90
14. O. fuscostriatus	75	55	86	68	555	136	143	446	383	583	253	84
15. O. hollowayi	2	4	2	2	15	6	6	3	3		1	115
18. O. magnipygus	4	3	9	2					6		5	75
19. O. mentaveiensis	87	89	107	34	95	100	120	123	40	166	96	40
20. O. moajat			1			3				3	2	134
21. O. rectecornutus		1						10	6	3	1	241
22. O. ribbei	7	8	9	6	80	6	10	36	16		17	132
24. O. sangirensis	166	133	303	308	1,330	710	590	643	540	683	540	64
27. O. sembeli	32	22	60	60	175	90	36	76	90	90	73	59
29. O. spiculatus	1		3	4	10	3		10	26	23	8	118
31. O. toraut	82	32	119	104	450	236	136	143	170	246	172	68
32. O. travestitus	2	29	56	12	35	66	23	26	56	20	32	63
33. O. tumpah	3					23	3		3		3	218
34. O. wallacei			1	2			3				*	183
35. P. planus	4	3	1		5	6	10	20	66	36	15	138

Table B.10(2) (cont.)

Species	Feb (10	Mar 10	Apr 10	May 5	Jun 2	Jul 3	Aug 3	Sep 3	Oct 3	Nov 3)	x	CV
							Seasonal Occurrence (number of samples)					
36. C. knighti	8	3	6	2	2	3			3		2	108
37. H. cambeforti			4								*	316
42. Aphodius sp.5								3		3	*	210
43. Aphodius sp. 6					5	6		3			1	169
44. B. celebensis	109	20	21	18	65	86	33	66	26	50	49	63

Notes: Results are given separately for February to November (June sample was collected 6–13 July, September sample 2–9 October). Number of samples is the number of weekly FIT samples. The figures given are averages per sample multiplied by 10. The last two columns give the mean and the coefficient of variation (SD/mean) across the 10 months.

* Average is less than 1.

TABLE B.11(1)

List of Scarabaeidae dung beetles collected in Taï forest, 1980–81.

Species	HU	EL	BU	MI	CA	VE	LF	FWt	D/N	Seasonal Occurrence					
										I	II	III	IV	V	VI
1. *Garreta diffinis* (Waterhouse)	2							350	D				2		
2. *Neosisyphus angulicollis* (Felsche)	69	6		1	2		1	50	D	9	12	10	23	23	2
3. *Neosisyphus tai* Cambefort		9						76	D	4	3			2	
4. *Sisyphus eburneus* Cambefort	142	6		2			9	10	D	16	58		50	14	21
5. *Sisyphus latus* Boucomont	10						20	15	D		7		21	2	
6. *Catharsius ninus* Gillet	1							255	N		1				
7. *Copris amabilis* Kolbe	7	2	1	4				101	N	7		4		3	
8. *Copris camerunus* Felsche		20						442	N		2		13	5	
9. *Copris phylax* Gillet		5						860	N	1	1		2	1	
10. *Copris tridens* Felsche		70						353	N	7	13	5		41	4
11. *Copris truncatus* Felsche		2	1				2	917	N	3				2	
12. *Pseudopedaria grossa* (Thomson)	5						2	230	N				5	2	
13. *Heliocopris dianae* Hope	16	9						3,110	N	2	3	3	11	5	1
14. *Paraphytus bechynei* (Balthasar)						1							1		
15. *Paraphytus sancyi* Paulian						24							24		
16. *Paraphytus tai* Cambefort						7							7		
17. *Alloniitis nasutus* (Felsche)		1						514	N	1					
18. *Lophodonitis carinatus* (Felsche)	2	1						370	N	1				2	
19. *Onitis nemoralis* Gillet		12						441	N	3	1			6	2
20. *Onitis subcrenatus* Kolbe		41	22					506	N	1	14	8		38	2
21. *Drepanocerus strigatus*		6	1					4	D		7				
22. *Drepanoplatynus gilleti* Boucomont		18	3					49	D	6	15				
23. *Euoniticellus tai* Cambefort		32						10	D	21	2	1		8	
24. *Liatongus sjostedti* (Felsche)		204	1	13				63	D		181	6	10	21	
25. *Oniticellus pseudoplanatus* Balthasar		346	61	16				72	D	35	104	46	97	101	40
26. *Tiniocellus panthera* (Boucomont)		11						70	D		6	1		4	
27. *Amietina eburnea* Cambefort					2		1	4	D				2	1	

TABLE B.11(1) (cont.)

| | | | | | | | | | | Seasonal Occurrence | | | | | |
Species	HU	EL	BU	MI	CA	VE	LF	FWt	D/N	I	II	III	IV	V	VI
28. Caccobius cribrarius Boucomont		83	1					10	D	26	19	4		26	9
29. Caccobius cyclotis Cambefort		12						9	N	11	1				
30. Caccobius elephantinus Balthasar		380						6	D	12	98	99	147	18	6
31. Caccobius tai Cambefort	1	43	10					5	N	19	25	4	5	1	
32. Diastellopalpus conradti d'Orbigny								277	N				1		
33. Diastellopalpus laevibasis d'Orbigny	3	20		1				447	N		3	10	9	1	
34. Diastellopalpus noctis (Thomson)	25	2		1				256	D	2	7	2	11	6	
35. Diastellopalpus tridens (Fabricius)	24	6						380	N	1	1	2	20	5	1
36. Milichus inaequalis Boucomont		1						25	N			1			
37. Milichus merzi Cambefort	3	33						57	N	33	1	2			
38. Mimonthophagus apicehirtus (d'Orb.)		6						110	N		1	1	3	1	
39. Onthophagus androgynus d'Orbigny				1				15	D		1				
40. Onthophagus atronitidus d'Orbigny	165	2		1				23	D	14	6	53	40	34	21
41. Onthophagus bartosi Balthasar					35			49	N		3	7	25		
42. Onthophagus cephalophi Cambefort				1				21	N		1				
43. Onthophagus curvifrons d'Orbigny	1	1						30	N		2				
44. Onthophagus densipilis d'Orbigny	87			1				29	D	6	3	29	42	7	1
45. Onthophagus denticulatus d'Orbigny	26			2			2	28	D				22	8	
46. Onthophagus denudatus d'Orbigny	49	1		6				23	N	3	6	4	41	2	
47. Onthophagus depilis d'Orbigny	1		2	23				69	N	2	23				
48. Onthophagus deplanatus Lansberge	7						1	71	N	1	3		4		1
49. Onthophagus feai d'Orbigny	2							28	D		1				
50. Onthophagus foulliouxi Cambefort	4				8			51	N			8		4	
51. Onthophagus fuscatus d'Orbigny	298	3			3			38	N	10	17	84	102	82	9
52. Onthophagus fuscidorsis d'Orbigny	97	1						27	N	3	4	23	47	20	1
53. Onthophagus gravoti d'Orbigny	14							6	D	1	1	2	10		
54. Onthophagus hilaridis Cambefort	5							9	D				5		
55. Onthophagus hilaris d'Orbigny	1	5		20			10	13	D	16	3		10	1	6

#	Species	HU	EL	BU	MI	CA	VE	LF	wt	diel	I	II	III	IV	V	VI
56	*Onthophagus ieti* Cambefort	1	4						19	N	1		1		3	
57	*Onthophagus infaustus* Cambefort							1	7	D		1			2	
58	*Onthophagus kindianus* Frey	10	1						67	N	1	1	1	4	4	2
59	*Onthophagus laeviceps* d'Orbigny	114			4			3	15	D	7	4	45	45	51	12
60	*Onthophagus laminosus* d'Orbigny	4							33	N	3			1	1	
61	*Onthophagus liberianus* Van Lansberge	67		13	2	63				N	4	18	34		20	2
62	*Onthophagus matae* Cambefort	1							20	N	1		1			
63	*Onthophagus orthocerus* Thomson	2			1				39	N	3	3			2	
64	*Onthophagus pleurogonus* d'Orbigny	4							25	N	1	1	1			
65	*Onthophagus rectorispauliani* Cambefor				8				11	D		2	2	4		
66	*Onthophagus rufopygus* Frey	123		5	16				80	N	6	15	13	44	48	2
67	*Onthophagus semiviridis* d'Orbigny	44	1						67	D	7	3	24	21	6	
68	*Onthophagus strictestriatus* d'Orbigny	32	2	1					58	N	3	4	4	22	1	1
69	*Onthophagus synceri* Cambefort	1		3	1				48	N	3	1			1	
70	*Onthophagus taiensis* Cambefort	3							12	D		3	3			
71	*Onthophagus tenuistriatus* d'Orbigny	26	1	1					62	N	2	4	8	14		
72	*Onthophagus vesanus* Balthasar		58	1					13	D	5	8	7	34	2	3
73	*Proagoderus ritsemai* Van Lansberge	7	1					1	159	D	1	7				
74	*Strandius tigrinus* (d'Orbigny)		17		2				171	N	1	2	3		14	
75	*Tomogonus crassus* d'Orbigny		11		3				35	N	1	3	7			
	Number of individuals	1,506	146	106	120	78	32	50			315	730	502	1,050	642	149
	Number of species	43	44	11	23	9	3	10								
	Number of species in a sample of 78	20	18	10	19	9	—	—								

Notes: Food selection given as the numbers of individuals caught with the following baits: HU, human dung; EL, elephant dung; BU, buffalo dung; MI, miscellaneous dung; CA, carrion; VE, vegetable stuffs; LF, specimens found sitting on leaves. Also included are numbers of specimens collected in six bimonthly periods (I = February–March, etc.), individual fresh weight, and diel activity.

TABLE B.11(2)

List of Scarabaeidae dung beetles collected in Makokou, 1981–83.

Species	HU	EL	MI	CA	VE	LI	LF
1. *Garreta ebenus* Janssens	3						
2. *Sisyphus arboreus* Walter	109						
3. *Neosisyphus angulicollis* (Felsche)	13						
4. *Neosisyphus basilewskyi* (Balthasar)	1						
5. *Catharsius gorilla* Thomson	74	5	1				
6. *Catharsius gorilloides* Felsche	10						
7. *Catharsius lycaon* Kolbe	27	1	1				
8. *Catharsius ninus* Gillet	1						
9. *Copris amabilis* Kolbe	2	2					
10. *Copris colmanti* Gillet						1	
11. *Copris jahi* Nguyen & Cambefort	2	2	7				
12. *Pseudopedaria grossa* (Thomson)	11						2
13. *Pseudopedaria villiersi* Walter	6	1					
14. *Heliocopris coronatus* Felsche	2						
15. *Heliocopris mutabilis* Kolbe		1	2				
16. *Paraphytus aphodioides* Boucomont					3		
17. *Paraphytus bechynei* (Balthasar)					3		
18. *Pedaria ovata* Boucomont	11						
19. *Onitis androcles* Janssens		2					
20. *Onitis artuosus* Gillet						1	
21. *Onitis subcrenatus* Kolbe		1					
22. *Lophodonitis carinatus* (Felsche)	8	2					
23. *Liatongus sjostedti* (Felsche)		3					
24. *Oniticellus pseudoplanatus* Balthasar	1	2	4				
25. *Tiniocellus panthera* (Boucomont)		1					
26. *Amietina larrochei* Cambefort				1			
27. *Caccobius elephantinus* Balthasar	1	4					
28. *Diastellopalpus anthonyi* Walter	1	1	1				
29. *Diastellopalpus conradti* d'Orbigny	11						
30. *Diastellopalpus gilleti* d'Orbigny	35	2	3				
31. *Diastellopalpus laevibasis* d'Orbigny		2					
32. *Diastellopalpus murrayi* (Harold)	8	1	2				
33. *Diastellopalpus sulciger* Kolbe	13	1	1				
34. *Heteroclitopus punctulatus* Boucomont						1	
35. *Mimonthophagus apicehirtus* (d'Orbigny)		18					
36. *Onthophagus ahenomicans* d'Orbigny	15						
37. *Onthophagus belinga* Walter	1						
38. *Onthophagus biplagiatus* Thomson	2	8	3				5
39. *Onthophagus densipilis* d'Orbigny	90	1	1				
40. *Onthophagus denudatus* d'Orbigny		3	2				
41. *Onthophagus depilis* d'Orbigny	7		10				
42. *Onthophagus erectinasus* d'Orbigny	3			2			

TABLE B.11(2) (*cont.*)

Species	HU	EL	MI	CA	VE	LI	LF
43. *Onthophagus fasciculiger* d'Orbigny	86						
44. *Onthophagus filicornis* Harold	7						
45. *Onthophagus fuscidorsis* d'Orbigny	358	10	1	1			4
46. *Onthophagus gabonensis* Walter				43			
47. *Onthophagus gibbidorsis* d'Orbigny				55			
48. *Onthophagus girardinae* Walter	1						6
49. *Onthophagus graniceps* d'Orbigny	7						
50. *Onthophagus grassei* Walter			1				1
51. *Onthophagus gravoti* d'Orbigny	4						
52. *Onthophagus justei* Walter	24		1				7
53. *Onthophagus laetus* d'Orbigny							1
54. *Onthophagus laeviceps* d'Orbigny	80						
55. *Onthophagus laminosus* d'Orbigny	15						
56. *Onthophagus liberianus* Lansberge	2		2				
57. *Onthophagus makokou* Walter	1						
58. *Onthophagus mpassa* Walter	1						
59. *Onthophagus orthocerus* Thomson	11	1	2				
60. *Onthophagus possoi* Walter	64						
61. *Onthophagus strictestriatus* d'Orbigny	1						
62. *Onthophagus umbratus* d'Orbigny	96	5	3	1			
63. *Proagoderus opulentus* d'Orbigny	14	2					
64. *Proagoderus semiiris* (Thomson)	65	8	4	2			
65. *Stiptopodius trituberculatus* Frey						1	
66. *Tomogonus crassus* d'Orbigny	50						
Number of individuals	1,355	90	52	105	6	4	26
Number of species	47	27	20	7	2	4	7
Number of species in a sample of 78	21.3	25.5		6.1			

Notes: This table lists the numbers of beetles caught with the following bait types: HU, human dung; EL, elephant dung; MI, miscellaneous dung; CA, carrion; VE, vegetable matter; LI, beetles collected at light; LF, beetles found sitting on leaves.

TABLE B.11(3)

Numbers of Aphodiinae collected in Taï (top) and in Makokou (bottom).

1. *Aphodius formosus* Paulian	21
2. *Aphodius cambeforti* Dellacasa	3
3. *Aphodius heleninae* Dellacasa	19
4. *Aphodius kumasianus* Endrödi	21
5. *Aphodius motoensis* Endrödi	7
6. *Aphodius egregius* Petrovitz	49
7. *Aphodius principalis* Harold	31
8. *Aphoidus renaudi* Clément	2
9. *Aphodius hartwigi* Petrovitz	1
10. *Aphodius anthrax* Gerstaecker	4
11. *Aphodius* n. sp.	1
12. *Aphodius aequatorialis* Petrovitz	2
13. *Aphodius maldesi* Bordat	43
14. *Aphodius schoutedeni* Boucomont	113
15. *Lorditomaeus cambeforti Bordat*	2
1. *Aphodius wittei* Paulian	1
2. *Aphodius plicatus* Endrödi	7
3. *Aphodius gilleti* Schmidt	27
4. *Aphodius* cf. *motoensis* Endrödi	1
5. *Aphodius principalis* Harold	13
6. *Aphodius szekessyi* Endrödi	1
7. *Aphodius astacus* Boucomont	8
8. *Aphodius motoi* Paulian	4
9. *Aphodius* n. sp.	11
10. *Aphodius* n. sp.	111
11. *Aphodius aequatorialis* Petrovitz	23
12. *Aphodius humator* Endrödi	14
13. *Aphodius multicarinatus* Endrödi	4
14. *Aphodius* sp. 1	1
15. *Aphodius* sp. 2	1
16. *Aphodius schoutedeni* Boucomont	2
17. *Aphodius baloghi* Endrödi	1

TABLE B.12
Dung beetles of Barro Colorado Island.

Species	DWt	LE	SDT	LDT	FIT	Total	N/D	Relocation Technique	Food Preference
1. Ateuchus aeneomicans (Harold)	8	5	4	4	209	222	D	T	D
2. Ateuchus calcaratus (Harold)	7	5	2	1	22	25	D	T	D
3. Ateuchus candezei (Harold)	16	6.5	19	18	94	132	N	T	D/C
4. Canthidium angusticeps Bates	15	7	0	0	5	5	D	?	?
5. Canthidium ardens Bates	4	4.5	3	1	432	500	D	C	Rodent dung
6. Canthidium aurifex Bates	5	4.5	11	0	3,312	3,620	D	C	Rodent dung
7. Canthidium centrale Boucomont	34	9	17	0	146	163	N	T	D/C
8. Canthidium elegantulum Balthasar	8	6	5	0	84	101	D	?	D?
9. Canthidium haroldi Preudhomme	23	8	22	0	1,846	2,118	D	C	Rodent dung
10. Canthidium hespenheidei H.& Y.	2	3.5	0	0	3	5	D	?	D?
11. Canthidium perceptibile H.& Y.	4	4.5	0	0	65	66	D	?	D?
12. Canthon aequinoctialis Harold	50	12	1,656	2,169	2,199	6,028	N	R	D/C
13. Canthon angustatus Harold	14	5.5	9	1	193	402	D	R	Howler dung
14. Canthon cyanellus sallei Harold	30	7.5	28	0	342	370	D	R	C/D
15. Canthon juvencus Harold	3	4	0	0	0	0	D	R	D
16. Canthon lamprimus Bates	6	5	53	16	628	756	D	R	D
17. Canthon moniliatus Bates	12	7	16	2	33	51	D	R	C/D
18. Canthon morsei Howden	6	5.5	0	0	57	57	D	R	D
19. Canthon mutabilis Lucas	?	6	0	0	0	0	D	R	D
20. Canthon septemmaculatus (Latreille)	42	10	81	21	132	251	D	R	D/C
21. Canthon subhyalinus Harold	9	5	0	0	9	60	D	R	Howler dung
22. Canthon viridis meridionalis M., H.& H.	6	5	0	0	0	0	D	R	D
23. Copris incertus Say	175	16	0	0	0	0	N	T	D

TABLE B.12 (cont.)

Species	DWt	LE	SDT	LDT	FIT	Total	N/D	Relocation Technique	Food Preference
24. *Copris lugubris* Boheman	186	16	0	1	0	1	N	T	D
25. *Coprophanaeus telamon corythus* (Harold)	525	23	0	0	0	0	C/N	T	C/D
26. *Deltochilum pseudoparile* Paulian	47	12	0	0	0	1	C/N	R	C/D
27. *Deltochilum valgum acropyge* Bates	70	12	0	0	1	1	N?	R?	?
28. *Dichotomius agenor* (Harold)	235	17	0	0	0	0	N	T	D
29. *Dichotomius femoratus* H. & Y.	103	14	0	0	1	2	N	T?	D
30. *Dichotomius satanas* (Harold)	267	20	20	79	21	120	N	T	D/C
31. *Eurysternus caribaeus* (Herbst)	105	16	2	4	12	19	A/C	T	D
32. *Eurysternus foedus* Guerin-Meneville	158	16	1	13	6	20	A?/C?	T	D
33. *Eurysternus plebejus* Harold	12	9	12	40	124	180	D	T	D
34. *Ontherus brevipennis* Harold	75	13	0	0	0	23	N?	?	Ant nests
35. *Ontherus sirius* H. & Y.	?	14	0	0	0	0	N?	?	?
36. *Onthocharis panamensis* Paulian	2	4	0	0	65	65	D	?	?
37. *Onthophagus acuminatus* Harold	8	6	67	127	453	656	D	T/K	D/C
38. *Onthophagus coscineus* Bates	2	3.5	38	65	796	913	D	T	D
39. *Onthophagus crinitus panamensis* Bates	32	9	5	23	81	110	D	T	D/C
40. *Onthophagus dicranius* Bates	6	5.5	0	0	323	323	D	?	D/F
41. *Onthophagus lebasi* Boucomont	8	5	4	17	19	42	D	?	?
42. *Onthophagus praecellens* Bates	17	6	29	12	134	175	D	T	D/C/F
43. *Onthophagus sharpi* Harold	12	7.5	0	2	164	170	D	T	D/C/F
44. *Onthophagus stockwelli* H. & Y.	24	8	11	15	115	141	D	T	D
45. *Oxysternon silenus* Castelnau	130	16	0	0	0	1	D	T	D
46. *Pedaridium bottimeri* H. & Y.	2	3.5	0	0	0	0	N?	?	Phoretic?

No. & Species	Dry weight	Length	SDT	LDT	FIT	Total	N?	?	Phoretic?
47. *Pedaridium brevisetosum* H.& Y.	2	3.5	0	0	0	0	N	K	D
48. *Pedaridium pilosum* (Robinson)	2	3.5	0	14	6	20	D	T/O	D
49. *Phanaeus howdeni* Arnaud	298	16	0	0	66	66	D	T/O	D/C
50. *Phanaeus pyrois* Bates	192	15	34	0	394	428	N	T/K	D/C
51. *Scatimus ovatus* Harold	7	6	48	87	33	168	D	T/O	D
52. *Sulcophanaeus cupricollis* (Nevinson)	192	19	0	0	21	27	N	?	D
53. *Uroxys gatunensis* H.& Y.	10	5	0	0	0	0	N	?	?
54. *Uroxys gorgon* Arrow	46	10	0	0	1	15	N	?	Sloth dung
55. *Uroxys macrocularis* H.& Y.	3	3.5	2	1	29	56	N	?	D?
56. *Uroxys metagorgon* H.& Y.	14	7	0	0	0	2	N	?	Sloth dung
57. *Uroxys microcularis* H.& Y.	3	3.5	0	0	0	0	N	?	D?
58. *Uroxys micros* Bates	3	4	58	101	594	789	N	T/K?	D
59. *Uroxys platypyga* H.& Y.	3	4	13	8	374	464	N	?	D?

Notes: Nomenclature follows Howden & Young (1981) and Howden & Gill (1987). Dry weights and lengths are averages for samples of 40 specimens (or less if fewer than 40 specimens were available). SDT, number of beetles collected in small dung traps (bait 2 cm^3, total sampling effort 118 trap-nights); LDT, number of beetles collected in large dung traps (bait 200 cm^3, total sampling effort 18 trap-nights); FIT, numbers of beetles collected with flight interception traps (total sampling effort 152 trap-nights); Total, numbers of beetles collected by all methods. Diel activity: A, auroral; C, crepuscular; D, diurnal; N, nocturnal. Food relocation techniques (see text for details): K, kleptoparasite; C, pellet roller; O, dung pusher; T, tunneler; R, roller. Food preference: C, carrion; D, dung; F, fruit. See Halffter et al. (1980) for details on *Eurysternus* behavior.

TABLE B.13
Ecological data for Scarabaeidae and Aphodiidae collected in Niamey (Niger), 1978–79.

Species	LE	FWt	Dung Type Preference	Seasonal Occurrence											
				Jan	Feb	Mar	Apr	May	Jun	Jul	Aug	Sep	Oct	Nov	Dec
1. *Metacatharsius ferrugineus* (Ol.)	10	110	Z			1			2						
2. *Metacatharsius inermis* (Cast.)	12	130	ZB						1						
3. *Pedaria nigra* Cast.	6	30	Z						1		1				
4. *Onitis alexis* Kl.	15	570	ZBE	11	4	5				5		3	9	5	3
5. *Onitis unguiculatus* (Ol.)	20	640	Z							2					
6. *Drepanocerus bechynei* Jans.	4	5	ZBE						1	1			8	2	1
7. *Euoniticellus intermedius* (Reiche)	8	52	ZBE	14	2					4	4	1	32	40	66
8. *Oniticellus formosus* Chevr.	10	82	ZBE	7	2	1			1				3	5	4
9. *Caccobius dorsalis* Har.	5	14	Z						1	6	1				
10. *Caccobius punctatissimus* Har.	5	13	ZE							1	1				
11. *Onthophagus bidens* (Ol.)	7	35	ZBEHA			1									
12. *Onthophagus bituberculatus* (Ol.)	9	34	Z	1		1	1	1			1	3	2	5	1
13. *Digitonthophagus gazella* (Fabr.)	10	130	ZBE			2			9	64	26	10	3	1	
14. *Onthophagus juvencus* Kl.	6	29	ZBEA											1	
15. *Onthophagus micros* d'Orb.	4	3	ZB				2		1			1	1		
16. *Onthophagus tersipennis* d'Orb.	4	3	ZBH						1			1			
17. *Onthophagus variegatus* (Fabr.)	5	14	ZBE	4	9	13	5		5	1			6	1	
18. *Onthophagus vinctus* Er.	6	19	ZBEH						3				2	1	
19. *Phalops vanellus* Lansb.	12	68	ZB								1		1		
20. *Proagoderus laticollis* Kl.	13		ZB										2		
21. *Megatelus dimidiatus* (Roth)	5	10	ZBE										2		
22. *Pleuraphodius utae* (End.)	3	2	Z							2					2

#	Species			Dung											
23.	*Pharaphodius desertus* (Kl.)	5	6	Z					3						1
24.	*Pharaphodius discolor* (Er.)	7	9	ZBE				5		41					
25.	*Pharaphodius nigeriensis* Paul.	7	12	ZBE				5							
26.	*Trichaphodius foveiventris* Paul.	4	2	Z				7	2	3					
27.	*Nialus bayeri* Endr.	6	2	Z	23	19	11	1	10	105	12	57	10	22	34
28.	*Nialus lividus* (Ol.) *sublividus* Balt.	4	3	Z	11	44	113	125	11	105	170	24	3	155	38
29.	*Nialus lividus* (Ol.) *paralividus* Balth.	5	11	Z	5		1	1			3	2	5	5	4
30.	*Nialus nigrita* (Fabr.)	4	3	ZBE	8	4	12	2	26		5	5	28	5	1
31.	*Calaphodius moestus* (Fabr.)	7	11	Z	1		2	4		3	1	1	33	1	2
32.	*Koshantschikovius splendens* (Balth.)	3	2	Z					14	4					
33.	*Orodalus parvulus* (Har.)	3	2	Z				5	6	295	41	4	6	5	
34.	*Mesontoplatys mbaoensis* (Petr.)	3	2	Z	1								5	8	2
35.	*Mesontoplatys rougoni* (Petr.)	3	2	ZBE				3	9	1	1		6	1	
36.	*Mesontoplatys simplicius* (Petr.)	3	2	ZB			19	1	1	1	3	5	8	1	
37.	*Lorditomaeus tenuis* Sch.	4	2	Z								1			
38.	*Psammodius desertorum* (Fair.)	4	2	Z					1						
39.	*Rhyssemus africanus* Petr.	4	3	ZBE			3		3			5			
40.	*Rhyssemus ritsemae* Cl.	3	3	ZBE			3	3		6	6	12	1		

Notes: Five dung pats were removed once per month 1, 3, 5, and 10 days after their deposition (total of 240 zebu droppings). Preference for type of dung is based on information from the entire Niger. Dung-type preferences: A, antelope; B, buffalo; E, elephant; H, human; Z, zebu.

TABLE B.14
Numbers of dung beetles of Vanoise National Park.

Species	Month	Site 1	2	3	4	5	6	7
1. *Geotrupes stercorarius* (L.)								
LE: 16–25; DWt: 3,803	Jun	9		1				
c, h, ho; co	Jul	2						
N; p	Aug	1		1				
	Sep	18	2	6				
2. *Anoplotrupes stercorosus* (Scriba)								
LE: 12–19; DWt: 773	Jun	18	20					
c, h, ho, s; co	Jul	3	31		3		7	
D; p, f	Aug	1	19		10		2	
	Sep	2	3		6		9	
3. *Onthophagus baraudi* Nicolas								
LE: 5–7; DWt: 75.5	Jun	191	575					
c, s, ho, m; co	Jul	86	243				21	
D; p	Aug	5	10				1	
	Sep		1					
4. *Onthophagus fracticornis* (Preyssl.)								
LE: 7–10; DWt: 100	Jun	143	168	5			117	
c, h, ho, m, s; co	Jul	5	249	4			3	
D; p	Aug	19	43	67				
	Sep	272	1,074	84	1			
5. *Aphodius aestivalis* Stephens								
LE: 6–9; DWt: 117	Jun							
c, ho; co	Jul							
p	Aug	1						
	Sep	1						
6. *Heptaulacus villosus* (Gyll.)								
LE: 3.5–4.5; DWt: 8	Jun							
c	Jul							
	Aug	1	1					
	Sep							
7. *Heptaulucus carinatus* (Germer)								
LE: 3.5–5; DWt: 13	Jun							
c, s, h, m; co	Jul	8	22					
p	Aug	359	207		4	9	2	
	Sep					1		

Table B.14 (*cont.*)

Species	Month	Site 1	2	3	4	5	6	7
8. *Aphodius putridus* (Fourcroy)								
LE: 2.5–3; DWt: 5.5	Jun	23						
s, c, ho; co	Jul	107						
	Aug	11	1					
	Sep	1						
9. *Aphodius pusillus* (Herbst)								
LE: 3–4.5; DWt: 13	Jun	490	88					
s, c; co	Jul	489	9					
D; p	Aug							
	Sep							
10. *Aphodius rufus* (Moll)								
LE: 5–7; DWt: 45.5	Jun							
c, ho; co	Jul	7						
p	Aug	142	2				1	
	Sep	6	1		1			
11. *Aphodius ater* (DeGeer)								
LE: 3–6; DWt: 21	Jun		45		1			
c, ho, s; co	Jul						1	
p	Aug							
	Sep							
12. *Aphodius corvinus* Er.								
LE: 3–4; DWt: 13	Jun	47	7	1				
s; co	Jul	13	5					
p	Aug	32	6					
	Sep	157	9					
13. *Aphodius erraticus* (L.)								
LE: 6–9; DWt: 101	Jun	656	628					
c, s; co	Jul	416	382	80			3	
p	Aug	81	38	6		1	1	
	Sep	30	44	1				
14. *Aphodius amblyodon* K. Daniel								
LE: 3.5–5; DWt: 13	Jun			252	2			
c; de	Jul			47	78			
D; p	Aug				1			
	Sep			2	1			

TABLE B.14 (*cont.*)

Species	Month	Site						
		1	*2*	*3*	*4*	*5*	*6*	*7*
15. *Aphodius depressus* (Kugel.)								
LE: 6–9; DWt: 65.5	Jun	3,670	112	1				
c, s, h, m; co	Jul	45	34	58	214	85	65	
D; p, f	Aug	43	1	1	16		1	
	Sep	16					1	
16. *Aphodius rufipes* (L.)								
LE: 11–13; DWt: 198	Jun	10						
c, ho; co	Jul	58	64	316	12	8		
p	Aug	376	9	153	49	2	2	
	Sep	20	12	19				
17. *Aphodius tenellus* Say								
LE: 4–5; DWt: 16.5	Jun	1		4				
c, s; co	Jul			2				
p	Aug	1		1		7		
	Sep	15	28	564	146	88		2
18. *Aphodius fimetarius* (L.)								
LE: 5–8; DWt: 99	Jun	48	7	28	1	1		
c, ho, s; co, de	Jul	5	1	1	1	5	6	3
p	Aug	71	1	55	13	4	4	
	Sep	145	147	531	8	10	20	
19. *Aphodius haemorrhoidalis* (L.)								
LE: 4–5; DWt: 40	Jun	183	26					
c, s, ho; co	Jul	272	137	20	9	279	6	1
p	Aug	166	34	11	102	98	1	
	Sep	17	4					
20. *Aphodius alpinus* (Scop.)								
LE: 5–7; DWt: 34	Jun			1				
c, m, s, ho; co	Jul		7	257	324	173	158	44
p	Aug	102	79	397	511	777	641	104
	Sep	6	32	47	26	21	68	20
21. *Aphodius satyrus* Reitter								
LE: 5–6.5; DWt: 45.5	Jun	1	2					
s, c, m; co	Jul		10		92	1,055	440	1
p, f	Aug	3	27	1	279	1,581	206	6
	Sep	1	15	1	8	82	192	4

TABLE B.14 (*cont.*)

Species	Month	Site						
		1	*2*	*3*	*4*	*5*	*6*	*7*
22. *Aphodius germandi* Nicolas & Riboulet								
LE: 6.5–8.5; DWt: 60	Jun	5	42	2	1	1		
c, m, s, ho; co	Jul	1	2			16	969	128
D; p	Aug		1		1	2	79	10
	Sep				1	1	13	3
23. *Aphodius abdominalis* Bonelli (= *A. mixtus* A. & G.B. Villa)								
LE: 5–6.5; DWt: 37	Jun					1		
c, s; de	Jul	1		5	17	73	139	115
D; p	Aug					2	6	39
	Sep					1		
24. *Aphodius obscurus* (Fabr.)								
LE: 6–8; DWt: 59	Jun	562	586	56	2			
c, m, s, ho; co	Jul	97	381	670	814	1,545	12,174	3,313
D; p	Aug	201	959	91	813	1,497	4,298	1,591
	Sep	13	193	74	15	39	1,202	733

Notes: Study sites: 1, Bessans (altitude 1,750 m); 2, Bonneval (1960 m); 3, La Ramasse (2,000 m); 4, Grand Plan (2,230 m); 5, Pont du Montet (2,410 m); 6, Les Roches (2,440 m); 7, Balcon du Montet (2,760 m). Dung-type preferences (adult beetles): c, cattle; h, human; ho, horse; m, marmot; s, sheep. Larval food habits: co, coprophagous; de, detritivorous. Habitat selection: p, pasture; f, forest.

TABLE B.15
Dung beetles of Badgingarra National Park.

Species	LE	N	Seasonality	N/D	Hab
1. *Onthophagus ferox* Harold	16	178	··*****	N	PHF
2. *Onthophagus rupicapra* Waterhouse	7	415	·******	?	H
3. *Onthophagus evanidus* Harold	4	20	·******	D	HF
4. *Euoniticellus intermedius* (Reiche)	9	1	*······	D	P
5. *Aphodius* sp. 81-168	3	6,923	··*****	?	HF
6. *Aphodius* sp. 81-169	3	344	···***·	?	HF
7. *Aphodius* sp. 86-0009	5	1,027	···***·	?	H
8. *Aphodius* sp. 86–0010	4	9	···*···	?	H
9. *Aphodius pseudolividus* Balthasar	5	1	·*·····	D	P
10. *Aphodius frenchi* Blackburn	4	7	···****	?	P
11. *Aphodius granarius* (L.)	7	9	···****	?	PH
12. *Proctophanes (?) sculptus* Hope	5	6	··**···	?	PH

Notes: Vegetation in park is scrub heath mosaic on lateritic sandplain. Sampling was carried out with five dung-baited pitfall traps 10 m apart in a line. Trap positions were permanent and traps were set for one week every month, November 1982–October 1984. In each trap 7 ml of fresh human dung was used as bait, placed on wooden skewers over a waxed paper cup 70 x 80 mm, sunk so that the rim was level with the soil surface. Ethylene glycol was placed in bottom of each trap as an odorless preservative. Trap data from Ridsdill-Smith & Hall (1984b). Habitat preference data are from a comparison of trapping results in pasture (P), heath (H) and open jarrah forest (F); none of dung beetle species trapped at Badgingarra showed a preference for closed karri forest (Ridsdill-Smith et al. 1983; Matthews 1972). All species preferred human dung over carrion baits (Ridsdill-Smith et al. 1983), but were trapped at both human- and cattle-dung baits. Species code numbers are Australian National Insect Collection voucher numbers. N, total number of individuals collected. Dots (·) and asterisks (*) under "Seasonality" represent absence or presence of the species in samples from March through September.

References

Absy, M. L. 1985. Palynology of Amazonia: The history of the forests as revealed by the palynological record. In G. T. Prance and T. E. Lovejoy, eds., pp. 72–82. *Key Environments: Amazonia*. Pergamon Press, New York.

Allsop, P. G. 1984. Checklist of the Hybosorinae (Coleoptera: Scarabaeidae). *Coleopt. Bull.* 38:105–17.

Anderson, J. M., and M. J. Coe. 1974. Decomposition of elephant dung in an arid, tropical environment. *Oecologia (Berl.)* 14:111–25.

Anderson, J. R., and S. Loomis. 1978. Exotic dung beetles in pasture and range land ecosystems. *Calif. Agric.* 32:31–32.

Anderson, R. S. 1982. Resource partitioning in the carrion beetle (Coleoptera: Silphidae) fauna of southern Ontario: Ecological and evolutionary considerations. *Can. J. Zool.* 60:1314–25.

Anderson, R. M., D. Gordon, M. J. Crawley, and M. P. Hassell. 1982. Variability in the abundance of animal and plant species. *Nature* 296:245–48.

Andrewartha, H. G., and L. C. Birch. 1954. *The Distribution and Abundance of Animals*. Univ. of Chicago Press, Chicago.

Anduaga, S., G. Halffter, and C. Huerta. 1987. Adaptaciones ecologicas de la reproduccion en Copris (Coleoptera: Scarabaeidae: Scarabaeinae). *Boll. Mus. Reg. Sci. Nat. Torino* 5:45–65.

Anonym. 1974. *Soil Map of the World*. FAO/UNESCO.

Anonym. 1979a. État actuel des parcs nationaux de la Comoé et de Taï ainsi que de la réserve d'Azagny et propositions visant à leur conservation et à leur développement aux fins de promotion du tourisme. Vol. 2, P.N. de la Comoé. Vol. 3, P.N. de Taï. FGU-Kronberg, Kronberg.

Anonym. 1979b. *Le climat de la Côte d'Ivoire*. ASECNA, Services météorologiques. Abidjan.

Anonym. 1987. *Makokou, Gabon: A Research Station in Tropical Forest Ecology*. Overview and Publications (1962–1986). IRET/ECOTROP (CNRS)/UNESCO, Paris.

Arrow, G. J. 1903. On the laparostict lamellicorn Coleoptera of Grenada and St. Vincent (West Indies). *Trans. Entomol. Soc. London* 1903:509–20.

———. 1926. Mimicry in Coleoptera. *Proc. Entomol. Soc. London* 1:18–19.

———. 1931. Coleoptera Lamellicornia. Part 3 (Coprinae). In *The Fauna of British India, Including Ceylon and Burma*. Taylor and Francis, London.

———. 1932. New species of lamellicorn beetles (subfam. Coprinae) from South America. *Stylops* 1:223–26.

Ashworth, A. C. 1977. A Late Wisconsin Coleopterous assemblage from southern Ontario and its environmental significance. *Can. J. Earth Sci.* 14:625–34.

Atkinson, W. D., and B. Shorrocks. 1981. Competition on a divided and ephemeral resource: A simulation model. *J. Anim. Ecol.* 50:461–71.

———. 1984. Aggregation of larval Diptera over discrete and ephemeral breeding sites: The implications for coexistence. *Amer. Nat.* 124:336–51.

Avila, J. M., and F. Pascual. 1981. Contribucion al conocimiento de los Escarabeidos coprophagos de Sierra Nevada: Muestro preliminar (Coleoptera: Scarabaeoidea). *Trab. Monogr. Dep. Zool. Univ. Granada (N.S.)* 4:93–105.

———. 1986. Contribucion al estudio de los escarabeidos coprofagos de Sierra Nevada. II. Relaciones con la vertiente, naturaleza del suelo y el grado de dureza, humedad y vegetacion del sustrato (Coleoptera, Scarabaeoidea). *Bolm. Soc. Port. Entomol.* 3:1–14.

———. 1987. Contribucion al estudio de los escarabeidos (Col. Scarabaeidae) coprofagos de Sierra Nevada. I. Introduccion e inventario de especies. *Boletin Asoc. Esp. Entomol.* 11:81–86.

Axelrod, D. I. 1967. Quaternary extinctions of large mammals. *Univ. Calif. Publ. in Geol. Sci.* 74:1–42.

———. 1979. Age and origin of Sonoran desert vegetation. *Occ. Pap. Calif. Acad. Sci.* 132:1–74.

Baker, C. W. 1968. Larval taxonomy of the Trogidae in North America with notes on biologies and life histories (Coleoptera: Scarabaeidae). *U.S. Nat. Mus. Bull.* 279:1–79.

Balseinte, R. 1955. La pluviosité en Savoie. *Rev. Géogr. Alp.* 43:299–355.

Balthasar, V. 1963. *Monographie der Scarabaeidae und Aphodiidae der Palaearktischen und Orientalischen Region (Coleoptera: Lamellicornia)*, vols. 1–3. Verlag Tschechosl. Akad. Wissenschaft, Prague.

———. 1967. The scientific results of the Hungarian soil zoological expedition to the Brazzaville-Congo. 22. Scarabaeinae und Coprinae (Coleoptera) (131. Beitrag zur Kenntnis der Scarabaeoidea). *Opusc. Zool. Budapest* 7:47–73.

———. 1970. Neue *Onthophagus*-Arter von Neu-Guinea und der benachbarter Inseln. *Acta Entomol. Mus. Nat. Pragae,* 38:61–408.

Baraud, J. 1977. Coléoptères Scarabaeoidea. Faune de l'Europe occidentale: Belgique, France, Grande Bretagne, Italie, Péninsule ibérique. Supplément à la Nouvelle Revue d'Entomologie. Vol. 7, fasc. 1. *Publ. Nouv. Rev. Entomol.* 4. Toulouse.

———. 1985. Coléoptères Scarabaeoidea. Faune du Nord de l'Afrique, du Maroc au Sinaï. *Encycl. Entomol.* 46. Editions Lechevalier, Paris.

Barkhouse, J., and T. J. Ridsdill-Smith. 1986. Effect of soil moisture on brood ball production by *Onthophagus binodis* Thunberg and *Euoniticellus intermedius* (Reiche) (Coleoptera: Scarabaeinae). *J. Entomol. Soc. Aust.* 25:75–78.

Barnes, R.F.W. 1983. Effects of elephant browsing on woodlands in a Tanzanian National Park: Measurements, models and management. *J. Appl. Ecol.* 20:521–40.

Bartholomew, G. A., and B. Heinrich. 1978. Endothermy in African dung beetles during flight, ball making and ball rolling. *J. Experim. Biol.* 73:65–83.

Bates, H. W. 1886. Pectinicornia and Lamellicornia. In *Biologia Centrali-Americana*, Insecta, Coleoptera. Vol. 2, part 2. London.

Beaver, R. A. 1979. Host specificity of temperate and tropical animals. *Nature* 281:139–41.

Begum, J., and J. R. Oppenheimer. 1981. Bangladesh dung beetles (Scarabaeidae and Trogidae): Seasonality, habitat, food and partial distribution. *Bangladesh J. Zool.* 9:9–15.

Bell, R.H.V. 1982. The effect of soil nutrient availability on community structure in

African ecosystems. In B. J. Huntley and B. H. Walker, eds., *Ecology of Tropical Savannas*, pp. 193–216. Springer-Verlag, Berlin.

Bellés, X., and M. E. Favila. 1984. Protection chimique du nid chez *Canthon cyanellus cyanellus* LeConte (Col. Scarabaeidae). *Bull. Soc. Entomol. France* 88:602–607.

Bender, E. A., T. J. Case, and M. E. Gilpin. 1984. Perturbation experiments in community ecology: Theory and practice. *Ecology* 65:1–13.

Benitez, J. C., and M. Martinez. 1982. Analisis del processo de degeneracion testicular en *Canthon cyanellus* LeConte (Coleoptera: Scarabaeinae). *Fol. Entomol. Mex.* 54:55–56.

Bernon, G. 1980. A trap for monitoring Coleoptera and phoretic mites associated with dung. *Coleopt. Bull.* 34:389–91.

———. 1981. "Species Abundance and Diversity of the Coleoptera Component of a South African Cow Dung Community, and Associated Insect Predators." Ph.D. diss., Univ. of Bowling Green, Ohio.

Besuchet, C. 1983. Coléoptères des Alpes suisses atteignant ou dépassant l'altitude de 3000 m. *Bull. Rom. Entomol.* 1:167–76.

Bigalke, R. C. 1972. The contemporary mammal fauna of Africa. In A. Keast, F. C. Erk, and B. Glass, eds., *Evolution, Mammals and Southern Continents*, pp. 141–94. State Univ. of New York Press, Albany.

Binaghi, G., G. Dellacasa, and R. Poggi. 1969. Nuovi caratteri diagnostici per la determinazione degli *Onthophagus* del gruppo *ovatus* (L.) e geonemia controllata delle specie italiane del gruppo (Coleoptera, Scarabaeidae). *Mem. Soc. Entomol. Ital.* 48:29–46.

Birks, H.J.B. 1986. Late-Quaternary biotic changes in terrestrial and lacustrine environments, with particular reference to north-west Europe. In B. E. Berglund, ed., *Handbook of Holocene Palaeoecology and Palaeohydrology*, pp. 39–56. Wiley, New York.

Blank, R. H., H. Black, and M. H. Olsen. 1983. Preliminary investigations of dung removal and flight biology of the Mexican dung beetle *Copris incertus* in Northland (Coleoptera: Scarabaeidae). *N.Z. Entomol.* 7:360–64.

Blueweiss, L., H. Fox, V. Kudzma, D. Nakashima, R. Peters, and S. Sams. 1978. Relationships between body size and some life history parameters. *Oecologia (Berl.)* 37:257–72.

Blume, R. R. 1970. Insects associated with bovine droppings in Kerr and Bexar counties, Texas. *J. Econ. Entomol.* 63:1023–24.

———. 1972. Additional insects associated with bovine droppings in Kerr and Bexar counties, Texas. *J. Econ. Entomol.* 65:621.

———. 1985. A checklist, distributional record, and annotated bibliography of the insects associated with bovine droppings on pastures in America North of Mexico. *Suppl. to the Southwest. Entomol.* 9:1–55.

Blume, R. R., and A. Aga. 1975. *Onthophagus gazella*: Mass rearing and laboratory biology. *Env. Entomol.* 4:735–36.

———. 1976. *Phanaeus difformis* Leconte (Coleoptera: Scarabaeidae): Clarification of published descriptions, notes on biology, and distribution in Texas. *Coleopt. Bull.* 30:199–205.

Blume, R. R., and A. Aga. 1978. *Onthophagus gazella*: Progress of experimental releases in south Texas. *Fol. Entomol. Mex.* 39–40:190–91.

Blume, R. R., A. Aga, D. D. Oehler, and R. L. Younger. 1974. *Onthophagus gazella*: A non-target arthropod for the evaluation of bovine feces containing methoprene. *Env. Entomol.* 3:947–49.

Blume, R. R., S. E. Kunz, B. F. Hogan, and J. J. Matter. 1970. Biological and ecological investigations of horn flies in central Texas: Influence of other insects in cattle manure. *J. Econ. Entomol.* 63:1121–23.

Blume, R. R., J. J. Matter, and J. L. Eschle. 1973. *Onthophagus gazella*: Effect on survival of horn flies in the laboratory. *Env. Entomol.* 2:811–13.

Booth, A. H. 1954. The Dahomey gap and the mammalian fauna of the West African forests. *Rev. Zool. Bot. Afr.* 50:305–14.

Bordat, P. 1983. Contribution la connaissance des Aphodiidae de la Côte d'Ivoire. *Rev. Fr. Entomol. (N.S.)* 5:64–70.

Borgia, G. 1980. Size and density-related changes in male behaviour in the fly *Scatophaga stercoraria*. *Behaviour* 75:185–206.

———. 1981. Mate selection in the fly *Scatophaga stercoraria*: Female choice in a male-controlled system. *Anim. Beh.* 29:71–80.

Bornemissza, G. F. 1960. Could dung eating insects improve our pasture? *J. Aust. Inst. Agric. Sci.* 75:257–60.

———. 1968. Studies on the histerid beetle *Pachylister chinensis* in Fiji, and its possible value in control of buffalo fly in Australia. *Austr. J. Zool.* 16:673–88.

———. 1969. A new type of brood care observed in the dung beetle *Oniticellus cinctus* (Scarabaeidae). *Pedobiologia* 9:223–25.

———. 1970a. An effect of dung beetle activity on plant yield. *Pedobiologia* 10:1–7.

———. 1970b. Insectary studies on the control of dung breeding flies by the activity of the dung beetle, *Onthophagus gazella* F. (Coleoptera: Scarabaeinae). *J. Aust. Entomol. Soc.* 9:31–41.

———. 1976. The Australian dung beetle project, 1965–1975. *Austr. Meat Res. Comm. Rev.* 30:1–30.

———. 1979. The Australian Dung Beetle Research Unit in Pretoria. *S. Afr. J. Sci.* 75:257–60.

Bortesi, O., and M. Zunino. 1974. Les résultats de l'expédition entomologique tchécoslovaque-iranienne à l'Iran en 1970. No. 10: Les *Onthophagus* du sous-genre *Euonthophagus* Balth. (Coleoptera, Scarabaeoidea). *Acta Entomol. Mus. Nat. Pragae* 6:105–107.

Boucomont, A. 1932. Synopsis des *Onthophagus* d'Amérique du Sud. *Ann. Soc. Entomol. Fr.* 101:293–332.

Bourlière, F., ed. 1983. *Tropical Savannas*. Ecosystems of the World, vol. 13. Elsevier, Amsterdam.

Bourlière, F., and M. Hadley. 1983. Present-day savannas: An overview. In F. Bourlière, ed., *Tropical Savannas*. Ecosystems of the World, vol. 13, pp. 1–17. Elsevier, Amsterdam.

Bourlière, F., E. Minner, and R. Vuattoux. 1974. Les grands mammifères de la région de Lamto. *Bull. Liais. Cherch. Lamto, Numero Special* 4:93–104.

Bowers, M. A., and J. H. Brown. 1982. Body size and coexistence in desert rodents: Chance or community structure? *Ecology* 63:391–400.

Braack, L.E.O. 1984. "An Ecological Investigation of the Insects Associated with Exposed Carcasses in Northern Kruger National Park; A Study of Populations and Communities." Ph.D. diss., Univ. of Natal, Pietermaritzburg, South Africa.

Brain, C. K. 1981. Hominid evolution and climatic change. *S. Afr. J. Sci.* 77:104–105.

Breymeyer, A., and B. Zacharieva-Stoilova. 1975. Scarabaeidae in two mountain pastures in Poland and Bulgaria. *Bull. L'Acad. Pol. Sc. Sér. Cl. II.* 23:173–80.

Breytenbach, W., and G. J. Breytenbach. 1986. Seasonal patterns in dung-feeding Scarabaeidae in the southern Cape. *J. S. Afr. Entomol. Soc.* 49:359–66.

Britton, E. B. 1970. Coleoptera. In *The Insects of Australia.* Melbourne Univ. Press, Australia.

Brown, K. S., Jr. 1982. Paleoecology and regional patterns of evolution in neotropical forest butterflies. In G. T. Prance, ed., *Biological Diversification in the Tropics*, pp. 255–308. Columbia Univ. Press, New York.

Brown, L. 1980. Aggression and mating success in males of the forked fungus beetle *Bolitotherus cornutus* (Coleoptera: Tenebrionidae). *Proc. Entomol. Soc. Wash.* 82:430–34.

Brown, W. J. 1940. Notes on the American distribution of some species of Coleoptera common to the European and North American continents. *Can. Entomol.* 72:65–75.

Brussaard, L. 1983. Reproductive behaviour and development of the dung beetle *Typhaeus typhoeus* (Coleoptera, Geotrupidae). *Tijdschr. Entomol.* 126:203–31.

———. 1985a. "A Pedobiological Study of the Dung Beetle *Typhaeus typhoeus*." Ph.D. diss., Univ. of Wageningen, The Netherlands.

———. 1985b. Back-filling of burrows by the scarab beetles *Lethrus apterus* and *Typhaeus typhoeus* (Coleoptera, Geotrupidae). *Pedobiologia* 28:327–32.

———. 1987. Kleptocopry of *Aphodius coenosus* (Coleoptera, Aphodiidae) in nests of *Typhaeus typhoeus* (Coleoptera, Geotrupidae) and its effect on soil morphology. *Biol. Fertil. Soils* 3:117–19.

Burger, B. V., Z. Munro, M. Roeth, H.S.C. Spies, V. Truter, G. D. Tribe, and R. M. Crewe. 1983. Composition of the heterogeneous sex pheromone secretion of the dung beetle *Kheper lamarcki*. *Z. Naturforsch. Sect. C Biosci.* 38:848–55.

Bush, A. O., and J. C. Holmes. 1986. Intestinal helminths of lesser scaup ducks: An interactive community. *Can. J. Zool.* 64:142–52.

Caballe, G. 1978. Essai sur la géographie forestière du Gabon. *Adansonia, Sér. 2,* 17:425–40.

Caillol, H. 1908. Catalogue des Coléoptères, part 2. *Ann. Soc. Sci. Nat. Provence, Marseille*, pp. 359–456.

———. 1954. Catalogue des Coléoptères de Provence, part 5 (Additions and corrections), pp. 507–36. Museum National d'Histoire Naturelle, Paris.

Cambefort, Y. 1980. Données préliminaires sur l'écologie des Scarabaeinae coprophages de Lamto (Insecta, Coleoptera, Scarabaeoidea). *Ann. Univ. Abidjan, Sér. E (Ecologie)* 13:61–79.

———. 1981. La nidification du genre *Cyptochirus* (Coleoptera, Scarabaeidae). *C. R. Acad. Sci. Paris III* 292:379–81.

———. 1982a. Nidification behavior of Old World *Oniticellini* (Coleoptera: Scarabaeidae). In G. Halffter and W. D. Edmonds, eds., *The Nesting Behavior of Dung*

Beetles (Scarabaeinae): An Ecological and Evolutionary Approach, pp. 141–45. Publ. 10, Inst. Ecol. Mexico.

———. 1982b. Les coléoptères Scarabaeidae S. Str. de Lamto (Côte-d'Ivoire): Structure des peuplements et rôle dans l'écosysteme. *Ann. Entomol. Soc. Fr.* 18:433–59.

———. 1982c. *Pedaria* nouveaux ou peu connus de Côte-d'Ivoire (Coleoptera, Scarabaeidae). *Rev. Fr. Entomol. (N.S.)* 4:57–62.

———. 1984. Étude écologique des Coléoptères Scarabaeidae de Côte d'Ivoire. *Trav. Cherch. Lamto* 3:1–294.

———. 1985. Les Coléoptères Scarabaeidae du Parc national de Taï (Côte d'Ivoire). *Rev. Fr. Entomol. (N.S.)* 7:337–42.

———. 1986. Rôle des coléoptères Scarabaeidae dans l'enfouissement des excréments en savane guinéenne de Côte d'Ivoire. *Acta Oecol., Oecol. Gen.* 7:17–25.

———. 1987a. Aulonocnemidae. In *Faune de Madagascar*, vol. 69. Paris.

———. 1987b. Le scarabée dans l'Egypte ancienne. Origine et signification du symbole. *Rev. Hist. Religions* 204:3–46.

Cambefort, Y., and J.-P. Lumaret. 1984. Nidification et larves des *Oniticellini* afrotropicaux (Col. Scarabaeidae). *Bull. Soc. Entomol. Fr.* 88:542–69.

Cambefort, Y., and P. Walter. 1985. Description du nid et de la larve de *Paraphytus aphodioides* Boucomont et notes sur l'origine de la coprophagie et l'évolution des Coléoptères Scarabaeidae s. str. *Ann. Soc. Entomol. Fr. (N.S.)* 21:351–56.

Campbell, M. M. 1976. Periodicity of 4 Diptera and 1 Coleoptera on fresh cow dung in southeast Queensland, Australia. *J. Nat. Hist.* 10:601–606.

Campbell, T. G. 1938. Recent investigations on the buffalo fly (*Lyperosia exigua* De Meijere) and its parasites in North Australia. *J. Counc. Sci. Industr. Res.* 11:77–83.

Carne, P. B. 1950. The morphology of the immature stages of *Aphodius howitti* Hope (Coleoptera, Scarabaeidae, Aphodiinae). *Proc. Linn. Soc. New South Wales* 75:158–66.

———. 1965. A revision of the genus *Elephastomus* Macleay (Coleoptera: Geotrupidae). *J. Entomol. Qld.* 4:3–13.

Carpaneto, G. M. 1974. Note sulla distribuzione geografica ed ecologica dei Coleotteri Scarabaeoidea Laparosticti nell'Italia appenninica (I contributo). *Boll. Assoc. Romana Entomol.* 29:32–54.

———. 1981. Distribution patterns and zoogeographical analysis of dung beetles in Greece (Coleoptera, Scarabaeoidea). *Biologia Gallo-Hellenica* 10:229–44.

———. 1986. I Coleotteri Scarabeoidei delle zoocenosi coprofaghe nel Parco nazionale del Circeo. *Atti Conv. Asp. Faun. Probl. Zool. PN Circ., Sab.* 1984:37–75

Carpaneto, G. M., and E. Piattella. 1986. Studio ecologico su una comunita di Coleotteri Scarabeoidei coprofagi nei Monti Cimini. *Boll. Assoc. Romana Entomol.* 40:31–58.

Case, T. J. 1983. Sympatry and size similarity in *Cnemidophorus*. In R. B. Huey, E. R. Pianka, and T. W. Schoener, eds., *Lizard Ecology*, pp. 297–325. Harvard Univ. Press, Cambridge, Mass.

Case, T. J., M. E. Gilpin, and J. M. Diamond. 1979. Overexploitation, interference competition, and excess density compensation in insular faunas. *Amer. Nat.* 113:843–54.

Caufield, C. 1984. *In the Rainforest*. Univ. of Chicago Press, Chicago.

Caussanel, C. 1984. Comportement maternel et repos ovarien chez *Labidura riparia* Pallas (Derm. Labiduridae). *Bull. Soc. Entomol. Fr.* 88:522–31.

Cervenka, V. J. 1986. "A Survey of Insects Associated with Bovine Dung in Minnesota." M.Sc. diss., Univ. of Minnesota, Minneapolis.

César, J. 1977. L'estimation de la charge optimale des pâturages guinéens. *Publ. C.R.Z. Bouaké, Côte d'Ivoire.*

Chapman, T. A. 1869. *Aphodius porcus*, a cuckoo parasite on *Geotrupes stercorarius. Ent. Mon. Mag.* 5:273–76.

Charles Dominique, P., and M. Hladik. 1971. Le lépilémur du sud de Madagascar: Écologie, alimentation et vie sociale. *Terre et Vie* 1971:3–36.

Charpentier, R. 1968. Élevage aseptique d'un Coléoptère coprophage: *Aphodius constans* Duft. (Col., Scarabaeidae). *Ann. Epiphyt.* 19:533–38.

Chesson, P. L. 1985. Coexistence of competitors in spatially and temporally varying environments: A look at the combined effects of different sorts of variability. *Theor. Pop. Biol.* 28:263–87.

Chesson, P. L., and T. J. Case. 1986. Overview. Nonequilibrium community theories: Chance, variability, history, and coexistence. In J. Diamond and T. J. Case, eds., *Community Ecology*, pp. 229–39. Harper & Row, New York.

Chesson, P. L., and N. Huntly. 1988. Community consequences of life-history traits in a variable environment. *Ann. Zool. Fenn.* 25:5–16.

Chesson, P. L., and W. W. Murdoch. 1986. Aggregation of risk: Relationships among host-parasitoid models. *Amer. Nat.* 127:696–715.

Chesson, P. L., and R. R. Warner. 1981. Environmental variability promotes coexistence in lottery competitive systems. *Amer. Nat.* 117:923–43.

Christensen, C. M., and R. C. Dobson. 1976. Biological and ecological studies on *Aphodius distinctus* (Coleoptera: Scarabaeidae). *Amer. Midl. Nat.* 95:242–49.

Coe, M. J., D. H. Cumming, and J. Phillipson. 1976. Biomass and production of large African herbivores in relation to rainfall and primary production. *Oecologia (Berl.)* 22:341–54.

Cole, M. 1986. *The Savannas*. Academic Press, London.

Collinet, J., B. Monteny, and B. Pouyaud. 1984. Le milieu physique. In: *Recherche et aménagement en milieu tropical humide: Le Projet Taï de Côte d'Ivoire.* UNESCO, Notes techniques MAB 15:35–58.

Comins, H. N., and M. P. Hassell. 1987. The dynamics of predation and competition in patchy environments. *Theor. Pop. Biol.* 31:393–421.

Connell, J. H. 1983. On the prevalence and relative importance of interspecific competition: Evidence from field experiments. *Amer. Nat.* 122:661–96.

Connor, F., and E. D. McCoy. 1979. The statistics and biology of the species-area relationship. *Amer. Nat.* 113:791–833.

Coope, G. R. 1962. A Pleistocene coleopterous fauna with arctic affinities from Fladbury, Worcestershire. *Q. J. Geol. Soc. London* 118:103–23.

———. 1970. Interpretations of Quaternary insect fossils. *Ann. Rev. Entomol.* 15:97–120.

———. 1973. Tibetan species of dung beetle from late Pleistocene deposits in England. *Nature* 245:335–36.

———. 1975. Mid-Weichselian climatic changes in western Europe, re-interpreted from Coleopteran assemblages. In R. P. Suggate, and M. M. Creswell, eds., *Quaternary Studies*, pp. 101–108. The Royal Soc. of New Zealand, Wellington.

———. 1977. Quaternary Coleoptera as aids in the interpretation of environmental history. In F. W. Shotton, ed., *British Quaternary Studies*, pp. 56–68. Clarendon Press, Oxford.

———. 1978. Constancy of insect species versus inconstancy of Quaternary environments. In L. A. Mound, and N. Waloff, eds., *Diversity of Insect Faunas*, pp. 176–87. Blackwell, Oxford.

Coope, G. R., and R. B. Angus. 1975. An ecological study of a temperate interlude in the middle of the last glaciation, based on fossil Coleoptera from Isleworth, Middlesex. *J. Anim. Ecol.* 44:365–91.

Cornaby, B. W. 1974. Carrion reduction by animals in contrasting tropical habitats. *Biotropica* 6:51–63.

Costa, M. 1964. Descriptions of the hitherto unknown mites associated with *Copris hispanus* (L.) (Coleoptera: Scarabaeidae) in Israel. *J. Linn. Soc. (Zool.)* 45:25–45.

———. 1969. The association between mesostigmatic mites and coprid beetles. *Acarologia* 11:411–28.

Coutinho, L. M. 1982. Ecological effect of fire in Brazilian Cerrado. In B. J. Huntley, and B. H. Walker, eds., *Ecology of Tropical Savannas*, pp. 273–91. Springer-Verlag, Berlin.

Crawley, M. J. 1987. What makes a community invasible? In A. J. Gray, M. J. Crawley, and P. J. Edwards, eds., *Colonization, Succession and Stability*, pp. 429–53. Blackwell, Oxford.

Crowson, R. A. 1981. *The Biology of the Coleoptera*. Academic Press, London.

Cumming, D.H.M. 1982. The influence of large herbivores on savanna structure in Africa. In B. J. Huntley and B. H. Walker, eds., *Ecology of Tropical Savannas*, pp. 217–45. Springer-Verlag, Berlin.

Curtsinger, J. W. 1986. Stay times in *Scatophaga* and the theory of evolutionary stable strategies. *Amer. Nat.* 128:130–36.

Danks, H. V. 1987. *Insect Dormancy: An Ecological Perspective*. Biological Survey of Canada, no. 1. Ottawa.

D'Arcy, W. G. 1977. Endangered landscapes in Panama and Central America: The threat to plant species. In G. T. Prance and T. S. Eliot, eds., *Extinction Is Forever*, pp. 89–104. New York Botanical Garden, New York.

Darlington, P. J., Jr. 1957. *Zoogeography: The Geographical Distribution of Animals*. Wiley, New York.

Davis, A.L.V. 1977. "The Endocoprid Dung Beetles of Southern Africa (Coleoptera: Scarabaeidae)." M.Sc. thesis, Rhodes Univ., Grahamstown, South Africa.

———. 1987. Geographical distribution of dung beetles (Coleoptera: Scarabaeidae) and their seasonal activity in south-western Cape Province. *J. S. Afr. Entomol. Soc.* 50:275–85.

———. 1989. Nesting of Afrotropical *Oniticellus* (Coleoptera: Scarabaeidae) and its evolutionary trend from soil to dung. *Ecol. Entomol.* 14:11–21.

Davis, A.L.V., B. M. Doube, and P. McLennan. 1988. Habitat associations and seasonal abundance of coprophilous Coleoptera (Staphylinidae, Hydrophilidae, Histeridae) in the Hluhluwe Region of South Africa. *Bull. Entomol. Res.* 78:425–34.

Davis, L. V. 1966. Feeding habits and seasonal distribution of scarab beetles in the North Carolina piedmont. *J. Elisha Mitchell Sci. Soc.* 82:212–20.

Delcourt, H. R. 1987. The impact of prehistoric agriculture and land occupation on natural vegetation. *TREE* 2:39–44.

Delcourt, P. A., and H. R. Delcourt. 1987. Long-term forest dynamics of the temperate zone. *Ecol. Stud.* 63. Springer-Verlag, New York.

Dellacasa, G. 1983. Sistematica e nomenclatura degli Aphodiini italiani (Coleoptera Scarabaeidae: Aphodiinae). Monografie 1, *Mus. Reg. di Scienze Nat. Edit.*, Turin.

———. 1986. A world-wide revision of *Aphodius* sharing a large scutellum (Coleoptera Scarabaeidae: Aphodiinae). *Frust. Entomol.* 7–8:173–282.

Dellacasa, M. 1988. Contribution to a world-wide catalogue to Aegialiidae, Aphodiidae, Aulonocnemidae, Termitotrogidae (Coleoptera Scarabaeoidea). *Mem. Soc. Entomol. Ital.* 66: 3–456; 67:3–231.

Deloya, C., G. Ruiz-Lizárraga, and M. A. Morón. 1987. Análisis de la entomofauna necrófila en la región de Jojutla, Morelos, México. *Fol. Entomol. Mex.* 73:154–67.

Denno, R. F., and W. R. Cothran. 1976. Competitive interactions and ecological strategies in sarcophagid and calliphorid flies inhabiting rabbit carrion. *Ann. Entomol. Soc. Amer.* 69:103–13.

Desière, M. 1970. Essai d'écologie quantitative sur une population de *Geotrupes stercorosus* (Scriba) (Coléoptère Lamellicorne). *Bull. Inst. Roy. Soc. Nat. Belg.* 46:1–14.

Dethier, M. 1985. Coléoptères des pelouses alpines au Parc national suisse. *Bull. Soc. Entomol. Suisse* 58:47–67.

Devineau, J. L., L. Lecordier, and R. Vuattoux. 1984. Evolution de la diversité spécifique du peuplement ligneux dans une succession préforestière de colonisation d'une savane protégée des feux (Lamto, Côte d'Ivoire). *Candollea*, 39:103–34.

Diamond, J. M. 1972. *Avifauna of the Eastern Highlands of New Guinea*. Nuttall Ornithological Club, Cambridge, Mass.

———. 1973. Distributional ecology of New Guinea birds. *Science* 179:759–69.

———. 1978. Niche shifts and the rediscovery of interspecific competition. *Amer. Scient.* 66:322–31.

Diamond, J. M., and A. G. Marshall. 1977. Niche shifts in New Hebridean birds. *Emu* 77:61–72.

Dickinson, C. H., and G.J.F. Pugh, eds. 1974. *Biology of Plant Litter Decomposition*. Academic Press, New York.

Donovan, C. H. 1979. "Indigenous Dung Beetles Near Armidale Northern N.S.W.: Seasonal and Diurnal Activity Patterns over Four Years." Ph.D. diss., Univ. of New England, Armidale, Australia.

Dorst, J., and P. Dandelot. 1970. *A Field Guide to the Larger Mammals of Africa*. Collins, London.

Doube, B. M. 1983. The habitat preference of some bovine dung beetles (Coleoptera: Scarabaeidae) in Hluhluwe Game Reserve, South Africa. *Bull. Entomol. Res.* 73:357–71.

———. 1986. Biological control of the buffalo fly in Australia: The potential of the southern African dung fauna. *Misc. Publ. Entomol. Soc. Amer.* 61:16–34.

———. 1987. Spatial and temporal organisation in communities associated with dung pads and carcasses. In J.H.R. Gee and P. S. Giller, eds., *Organisation of Communities: Past and Present*, pp. 253–80. Blackwells, London.

Doube, B. M., and F. Moola. 1987. Effects of intraspecific larval competition on the development of the African buffalo fly *Haematobia thirouxi potans*. *Entomol. Exper. Appl.* 43:145–51.

———. 1988. The effect of the activity of the African dung beetle *Catharsius tricornutus* De Geer (Coleoptera: Scarabaeidae) upon the survival and size of the African buffalo fly *Haematobia thirouxi potans* (Bezzi (Diptera: Muscidae) in bovine dung. *Bull. Entomol. Res.* 78:63–73.

Doube, B. M., P. S. Giller, and F. Moola. 1988. Dung burial strategies in some South African coprine and onitine dung beetles (Scarabaeidae, Scarabaeinae). *Ecol. Entomol.* 13:251–61.

Doube, B. M., and K. A. Huxham. 1987. Laboratory assessment of predation on immature stages of *Haematobia thirouxi potans* (Bezzi) (Diptera: Muscidae) by some beetles from the southern African dung fauna. *J. S. Afr. Entomol. Soc.* 50:475–80.

Doube, B. M., A. Macqueen, and H.A.C. Fay. 1988. Effects of dung fauna on survival and size of buffalo flies (*Haematobia* spp.) breeding in the field in South Africa and Australia. *J. Appl. Ecol.* 25:523–36.

Doube, B. M., A. Macqueen, and K. A. Huxham. 1986. Aspects of the predatory activity of *Macrocheles peregrinus* (Acarina: Macrochelidae) on two species of *Haematobia* fly (Diptera: Muscidae). *Misc. Publ. Entomol. Soc. Amer.* 61:32–41.

Downes, W. 1928. On the occurrence of *Aphodius pardalis* Lec. as a pest of lawns in British Columbia. *Ann. Rep. Entomol. Soc. Ontario* 58:59–61.

Downing, J. A. 1986. Spatial heterogeneity: Evolved behaviour or mathematical artefact? *Nature* 323:255–57.

Dubost, G. 1984. Comparison of the diets of frugivorous forest ruminants of Gabon. *J. Mammal.* 65:298–316.

Dunkle, S. W., and J. J. Belwood. 1982. Bat predation on Odonata. *Odonatologica* 11:225–29.

Durden, L. A. 1986. Rats and ectoparasites on Project Wallace, *Antenna* 10:29–30.

During, C., and W. C. Weeda. 1973. Some effects of the cattle dung on soil properties, pasture production and nutrient uptakes. I. Dung as a source of phosphorus. *N.Z. J. Agric. Res.* 16:423–30.

East, R. 1984. Rainfall, soil nutrient status and biomass of large African savanna mammals. *Afr. J. Ecol.* 22:245–70.

Edmonds, W. D. 1972. Comparative skeletal morphology, systematics and evolution of the Phanaeine dung beetles (Coleoptera: Scarabaeidae). *Univ. Kansas Sci. Bull.* 49:731–874.

———. 1974. Internal anatomy of *Coprophanaeus lancifer* (L.) (Coleoptera: Scarabaeidae). *Int. J. Insect Morphol. Embryol.* 3:257–72.

Edmonds, W. D., and G. Halffter. 1972. A taxonomic and biological study of the immature stages of some New World Scarabaeinae (Coleoptera: Scarabaeidae). *An. Esc. Nac. Cienc. Biol. Mex.* 19:85–122.

Edwards, P. B. 1984. Field ecology and reproductive behaviour of the dung beetle *Kheper nigroaeneus* in Southern Africa. *17th Int. Congr. Entomol., Hamburg, Abstr. Vol.*, p. 338.

————. 1986a. Phenology and field biology of the dung beetle *Onitis caffer* Boheman (Coleoptera: Scarabaeidae) in southern Africa. *Bull. Entomol. Res.* 76:433–46.

————. 1986b. Development and larval diapause in the southern African dung beetle *Onitis caffer* Boheman (Coleoptera: Scarabaeidae). *Bull. Entomol. Res.* 76:109–17.

————. 1988a. Field ecology of a brood-caring dung beetle *Kheper nigroaeneus*— habitat predictability and life history strategy. *Oecologia (Berl.)* 75:527–34.

————. 1988b. Contribution of the female parent to survival of laboratory-reared offspring of the dung beetle *Kheper nigroaeneus* (Boheman) (Coleoptera: Scarabaeidae). *J. Aust. Entomol. Soc.* 27:223–37.

Edwards, P. B., and H. H. Aschenborn. 1987. Patterns of nesting and dung burial in *Onitis* dung beetles: Implications for pasture productivity and fly control. *J. Appl. Ecol.* 24:837–52.

————. 1988. Male reproductive behaviour of the African ball-rolling dung beetle *Kheper nigroaeneus* (Scarabaeidae). *Coleopt. Bull.* 42:17–27.

————. 1989. Maternal care of a single offspring in the dung beetle *Kheper nigroaeneus*: The consequences of extreme parental investment. *J. Nat. Hist.* 23:17–27.

Eisenberg, J. F. 1980. The density and biomass of tropical mammals. In M. Soule, and B. Wilcox, eds., *Conservation Biology: An Evolutionary-Ecological Perspective*, pp. 35–55. Sinauer, Sunderland, Mass.

Eisenberg, J. F., and R. W. Thorington, Jr. 1973. A preliminary analysis of a neotropical mammal fauna. *Biotropica* 5:150–61.

Elton, C. 1949. Population interspersion: An essay on animal community patterns. *J. Ecol.* 37:1–23.

Emberger, L. 1954. Une classification biogéographique des climats. *Trav. Lab. Bot. Geol. Univ. Montpellier, Ser. Bot.* 7:3–43.

Endrody-Younga, S. 1982. An annotated checklist of dung-associated beetles of the Savannah Ecosystem Project study area Nylsvlei. *S. Afr. Natl. Programmes Report*, no. 59.

Engen, S. 1978. *Stochastic Abundance Models*. Chapman & Hall, London.

Fabre, J.-H. 1897. *Souvenirs entomologiques*. 5th series. Delagrave, Paris.

————. 1910. *Souvenirs entomologiques*. 10th series. Delagrave, Paris.

————. 1918. *The Sacred Beetle and Others*. Translated by A. T. de Mattos. London.

Fay, H.A.C. 1986. Fauna induced mortality in *Haematobia thirouxi potans* (Bezzi) (Diptera: Muscidae) in buffalo dung in relation to soil and vegetation type. *Misc. Publ. Entomol. Soc. Amer.* 61:142–49.

Fay, H.A.C., and B. M. Doube. 1983. The effect of some coprophagous and predatory beetles on the survival of immature stages of the African buffalo fly, *Haematobia thirouxi potans*, in bovine dung. *Z. Ang. Entomol.* 95:460–66.

————. 1987. Aspects of the adult population dynamics of *Haematobia thirouxi potans* (Bezzi) (Diptera: Muscidae) in southern Africa. *Bull. Entomol. Res.* 77:135–44.

Feeley, J. M. 1980. Did iron age man have a role in the history of Zululand wilderness landscapes? *S. Afr. J. Sci.* 76:150–52.

Ferrar, P. 1975. Disintegration of dung pads in north Queensland before the introduction of exotic dung beetles. *Aust. J. Exper. Agric. An. Husb.* 15:325–29.

Ferreira, M. C. 1972. Os Escarabideos de Africa (Sul do Saara). I. *Revta Entomol. Moçamb.* 11:5–1088.

Figg, D. E., R. D. Hall, and G. D. Thomas. 1983. Insect parasites associated with Diptera developing in bovine dung pats on Central Missouri pastures. *Env. Entomol.* 12:961–66.

Fincher, G. T. 1972. Notes on the biology of *Phanaeus vindex* (Coleoptera: Scarabaeidae). *J. Georgia Entomol. Soc.* 7:128–33.

―――. 1973a. Nidification and reproduction of *Phanaeus* spp. in three textural classes of soil (Coleoptera: Scarabaeidae). *Coleopt. Bull.* 27:33–37.

―――. 1973b. Dung beetles as biological control agents for gastrointestinal parasites of livestock. *J. Parasitol.* 59:396–99.

―――. 1975a. Dung beetles of Blackbeard Island (Coleoptera: Scarabaeidae). *Coleopt. Bull.* 29:319–20.

―――. 1975b. Effect of dung beetle activity on the number of nematode parasites acquired by grazing cattle. *J. Parasitol.* 61:759–66.

―――. 1981. The potential value of dung beetles in pasture ecosystems. *J. Georgia Entomol. Soc.* 16:316–33.

―――. 1986. Importation, colonization, and release of dung-burying scarabs. *Misc. Publ. Entomol. Soc. Amer.* 62:69–76.

Fincher, G. T., and R. E. Woodruff. 1975. A European dung beetle, *Onthophagus taurus* Schreber, new to the U.S. (Coleoptera: Scarabaeidae). *Coleopt. Bull.* 29:349–50.

―――. 1979. Dung beetles of Cumberland Island, Georgia (Coleoptera: Scarabaeidae). *Coleopt. Bull.* 33:69–70.

Fincher, G. T., R. R. Blume, J. S. Hunter, and K. R. Beerwinckle. 1986. Seasonal distribution and diel flight activity of dung-feeding scarabs in open and wooded pasture in east-central Texas. *Southwest. Entomol., Suppl.* 10:1–35.

Fincher, G. T., R. Davis, and T. B. Stewart. 1971. Flight activity of coprophagous beetles on a swine pasture. *Ann. Entomol. Soc. Amer.* 64:855–60.

Fincher, G. T., T. B. Stewart, and R. Davis. 1969. Beetle intermediate hosts for swine spirurids in southern Georgia. *J. Parasitol.* 55:355–58.

―――. 1970. Attraction of coprophagous beetles to feces of various animals. *J. Parasitol.* 56:378–83.

Fincher, G. T., T. B. Stewart, and J. S. Hunter. 1983. The 1981 distribution of *Onthophagus gazella* Fabricius from releases in Texas and *Onthophagus taurus* Schreber from an unknown release in Florida (Coleoptera: Scarabaeidae). *Coleopt. Bull.* 37:159–63.

Forsyth, A. B., and R. J. Robertson. 1975. K-reproductive strategy and larval behavior of the pitcher plant sarcophagid fly, *Blaesoxipha fletcher*. *Can. J. Zool.* 53:174–79.

Foster, R. B., and N.V.L. Brokaw. 1982. Structure and history of the vegetation of Barro Colorado Island. In E. G. Leigh, Jr., A. S. Rand, and D. M. Windsor, eds., *The Ecology of a Tropical Forest*, pp. 67–81. Smithsonian Institution Press, Washington, D.C.

Fournier, A. 1982. "Cycle saisonnier de la biomasse et démographie des feuilles de quelques graminées dans les savanes Guinéennes de Ouango Fitini (Côte d'Ivoire)." 3d cycle diss., Univ. of Montpellier, France.

Frenguelli, G. 1938. Nidi fossili di Scarabeidi e Vespidi. *Boll. Soc. Geol. Ital.* 57:77–96.

Frey, G. 1961. Onthophagini (Coleoptera Lamellicornia). In *Exploration du Parc na-*

tional de la Garamba, Mission H. De Saeger, Institut des Parcs nationaux du Congo Belge, Fascicule 21:69–98.

Frison, M. 1967. *Caractéristiques physico-chimiques du lisier*. Journées d'information sur le lisier, Lyon.

Fugiyama, I. 1968. A Miocene fossil of tropical dung beetle from Noto, Japan. *Bull. Nat. Sci. Mus. Tokyo* 11:201–11.

Galante, E. 1979. Los Scarabaeoidea de las heces de vacuno de la provincia de Salamanca (Col.). II. Familia Scarabaeidae. *Bol. Assoc. Esp. Entomol.* 3:129–52.

————. 1983. Primera contribucion al conocimiento de los escarabeidos (Col., Scarabaeoidea) del Pireneo Altoaragonés. *Bol. Assoc. Esp. Entomol.* 7:19–29.

Gams, H. 1935. Pflanzenleben des Glocknergebietes. Kurze Eläuterungen der Vegetationskarte. *Z. Dtsch.-Österr. Alpenver.* 66:157–76.

Gardner, J.C.M. 1935. Immature stages of Indian Coleoptera (16) (Scarabaeoidea). *Indian Forest Records (N.S.). Entomology* 1:1–33.

Gause, G. F. 1934. *The Struggle for Existence*. Williams and Wilkins, Baltimore, Md.

Gensac, P. 1978. Observations thermométriques de 1973 à 1976 dans le parc national de la Vanoise. Conséquences biologiques. *Trav. Scient. Parc Nat. Vanoise* 9:9–24.

Gill, B. D. 1986. "Foraging Behaviour of Tropical Forest Scarabaeinae in Panama." Ph.D. diss., Carleton Univ., Ottawa, Canada.

Giller, P. S., and B. M. Doube. 1989. Experimental analysis of inter- and intraspecific competition in dung beetle communities. *J. Anim. Ecol.* 58:129–44.

Gillon, Y. 1983. The fire problem in tropical savannas. In F. Bourlière, ed., *Tropical Savannas*, pp. 617–42. Ecosystems of the World, vol. 13. Elsevier, Amsterdam.

Gilpin, M. E. 1990. Extinction of finite metapopulations in correlated environments. In B. Shorrocks and I. Swingland, eds., *Patchy Environments*, pp. 177–86. Oxford Univ. Press, Oxford.

Glanz, W. E. 1982. The terrestrial mammal fauna of Barro Colorado Island: Censuses and long-term changes. In E. G. Leigh, Jr., A. S. Rand, and D. M. Windsor, eds., *The Ecology of a Tropical Forest*, pp. 455–68. Smithsonian Institution Press, Washington, D.C.

Goggio, D. E. 1926. Studii sulla vita dell' *Ateuchus semipunctatus* Fabr. *Arch. Zool. Ital. (N.S.)* 11:1–44.

Goidanich, A., and C. E. Malan. 1962. Sulla fonte di alimentazione e sulla microflora aerobica del nido pedotrofico e dell'apparato digerente delle larve di Scarabei coprofagi (Coleoptera Scarabaeidae). *Atti Accad. Sc. Torino* 96:575–628.

————. 1964. Sulla nidificazione pedotrofica di alcune specie di *Onthophagus* europei e sulla microflora aerobica dell'apparato digerente della larva di *Onthophagus taurus* Schreber (Coleoptera Scarabaeidae). *Ann. Fac. Sc. Agr. Univ. Torino* 2:213–378.

González-Quintero, L. 1980. Paleoecología de un sector costero de Guerrero, México (3000 años). III. Coloquio sobre paleobotánica y palinología. *SEP-INAH, Colección Científica. Dept. de Prehistoria*, pp. 133–57.

Gordon, R. D. 1983. Studies on the genus *Aphodius* of the USA and Canada (Coleoptera: Scarabaeidae). 7. Food and habitat distribution, key to eastern species. *Proc. Entomol. Soc. Wash.* 85:633–52.

Gordon, R. D., and O. L. Cartwright. 1974. Survey of food preferences of some No. American Canthonini (Coleoptera: Scarabaeidae). *Entomol. News.* 85:181–85.

Greathead, D. J., and J. Monty. 1982. Biological control of stable flies (*Stomoxys* spp.): Results from Mauritius in relation to fly control in dispersed breeding sites. *Biocontrol News Inf.* 3:105–109.

Grebenscikov, I. 1985. Bemerkungen zu Verbreitung und Systematik der aus der Mongolischen Volksrepublik bekannten laparosticten Scarabaeoidea (Coleoptera). Ergebnisse der Mongolisch-Deutschen Biologischen Expeditionen seit 1962, no. 139. *Mitt. Zool. Mus. Berl.* 61:105–36.

Green, R. F. 1986. Does aggregation prevent competitive exclusion? A response to Atkinson and Shorrocks. *Amer. Nat.* 128:301–304.

Greenslade, P.J.M. 1983. Adversity selection and the habitat templet. *Amer. Nat.* 22:352–56.

Guillaumet, J. L., and E. Adjanohoun. 1971. La végétation de la Côte d'Ivoire. In *Le milieu naturel en Côte d'Ivoire. Mémoires ORSTOM* 50:157–263.

Gutierrez, J., A. Macqueen, and L. O. Brun. 1988. Essais d'introduction de quatre espèces de bousiers Scarabaeinae en Nouvelle Calédonie et au Vanuatu. *Acta. Oecol., Oecol. Appl.* 9:39–53.

Haaf, E. 1955. Über die Gattung *Sisyphus* Latr. (Col. Scarab.). *Entomol. Arb. Mus. Frey* 6:341–81.

Haffer, J. 1969. Speciation in Amazonian forest birds. *Science* 165:131–37.

———. 1974. Avian speciation in tropical South America. *Nuttal Ornith. Club*, no. 14, Cambridge, Mass.

———. 1979. Quaternary biogeography of tropical lowland South America. In W. E. Duellman, ed., *The South American Herpetofauna: Its Origin, Evolution and Dispersal.* Univ. of Kansas Mus. Nat. Hist. Monogr. 7, pp. 107–39.

Halffter, G. 1961. Monografia de las especies norteamericanas del genero *Canthon* Hoffsg. (Coleopt., Scarab.). *Ciencia (Mex.)* 20:225–320.

———. 1962. Explicación preliminar de la distribución geográfica de los Scarabaeidae mexicanos. *Act. Zool. Mex.* 5:1–17.

———. 1964. La entomofauna americana, ideas acerca de su origen y distribución. *Fol. Entomol. Mex.* 6:1–108.

———. 1972. Eléments anciens de l'entomofaune Néotropicale: Ses implications biogéographiques. In *Biogéographie et Liaisons Intercontinentales au Cours du Mésozoïque.* 17th Congr. Int. Zool., Monte-Carlo 1:1–40.

———. 1974. Eléments anciens de l'entomofaune Neotropicale: Ses implications biogéographiques. *Quaest. Entomol.* 10:223–62.

———. 1976. Distribución de los insectos en la zona de Transición Mexicana. Relaciones con la entomofauna de Norteamérica. *Fol. Entomol. Mex.* 35:1–64.

———. 1977. Evolution of nidification in the Scarabaeinae (Coleoptera, Scarabaeidae). *Quaest. Entomol.* 13:231–53.

———. 1978. Un nuevo patrón de dispersión en la Zona de Transición Mexicana: El Mesoamericano de montaña. *Fol. Entomol. Mex.* 39–40:219–26.

———. 1982. Evolved relations between reproductive and subsocial behaviors in Coleoptera. In M. D. Breed, C. D. Michener, and H. E. Evans, eds., *The Biology of Social Insects*, pp. 164–70. Westview Press, Boulder, Colo.

———. 1987. Biogeography of the montane entomofauna of Mexico and Central America. *Ann. Rev. Entomol.* 32:95–114.

Halffter, G., and W. D. Edmonds. 1981. Evolucion de la nidificacion y de la cooperacion bisexual en Scarabaeinae (Ins.: Col.). *An. Esc. Nac. Cienc. Biol. Mex.* 25:117–44.

———. 1982. *The Nesting Behavior of Dung Beetles (Scarabaeinae): An Ecological and Evolutive Approach*. Instituto de Ecología, México, D.F.

Halffter, G., and V. Halffter. 1989. Behavioral evolution of the non-rolling roller beetles (Coleoptern: Scarabaeidae: Scarabaeinae). *Aeta Zool. Mex. (N.S.)* 32:1–53.

Halffter, G., and I. López. 1972. Nidificación y comportamiento sexual en *Phanaeus* (Coleop., Scarabaeidae). *Fol. Entomol. Mex.* 23–24:105–108.

———. 1977. Development of the ovary and mating behavior in *Phanaeus*. *Ann. Entomol. Soc. Amer.* 70:203–13.

Halffter, G., and A. Martinez. 1977. Revision monografica de los Canthonina americanos, IV parte. Clave para generos y subgeneros. *Fol. Entomol. Mex.* 38:29–107.

Halffter, G., and E. G. Matthews. 1966. The natural history of dung beetles of the subfamily Scarabaeinae (Coleoptera: Scarabaeidae). *Fol. Entomol. Mex.* 12–14:1–312.

———. 1971. The natural history of dung beetles: A supplement on associated biota. *Rev. Lat.-Amer. Microbiol.* 13:147–64.

Halffter, G., S. Anduaga, and C. Huerta. 1982. Nidificacion en *Nicrophorus* (Coleoptera, Silphidae). *Fol. Entomol. Mex.* 54:50–52.

———. 1984. Nidification des *Nicrophorus* (Col. Silphidae). *Bull. Soc. Entomol. Fr.* 88:648–66.

Halffter, G., V. Halffter, and I. Lopez. 1974. *Phanaeus* behavior: Food transportation and bisexual cooperation. *Env. Entomol.* 3:341–45.

Halffter, G., V. Halffter, and C. Huerta. 1980. Mating and nesting behavior of *Eurysternus* (Coleoptera: Scarabaeinae). *Quaest. Entomol.* 16:599–620.

———. 1984. Comportement sexuel et nidification chez *Canthon cyanellus cyanellus* LeConte (Col. Scarabaeidae). *Bull. Soc. Entomol. Fr.* 88:585–94.

Halffter, V., Y. Lopez-Guerrero, and G. Halffter. 1985. Nesting and ovarian development in *Geotrupes cavicollis* Bates (Coleoptera: Scarabaeidae). *Acta Zool. Mex. (N.S.)* 7:1–28.

Hall, M. 1979a. The Umfolozi, Hluhluwe and Corridor Reserves during the Iron Age. *Lammergeyer* 27:28–40.

———. 1979b. The influence of prehistoric man on the vegetation of the Umfolozi-Corridor-Hluhluwe Complex. In *Workshop on the Vegetation Dynamics of the Hluhluwe-Corridor-Umfolozi Complex*, Hluhluwe, Aug. 10–12, 1979. Blackwell, Oxford.

Hamilton, W. D. 1978. Evolution and diversity under bark. In L. A. Mound, and N. Waloff, eds., *Diversity of Insect Faunas*, pp. 154–75. Blackwell, Oxford.

Hammer, O. 1941. Biological and ecological investigations of flies associated with pasturing cattle and their excrement. *Vidensk. Medd. Dansk Naturhist. For.* 105:141–393.

Hammond, P. M. 1976. Kleptoparasitic behaviour of *Onthophagus suturalis* Peringuey (Coleoptera: Scarabaeidae) and other dung-beetles. *Coleopt. Bull.* 30:245–49.

Handschin, E. 1963. Die Coleopteren des schweizerischen Nationalparks und seiner Umgebung. *Erglb. Wiss. Unters. Schweitz. Nat. Park* 8:1–302.

Hanski, I. 1976. Assimilation by *Lucilia illustris* (Diptera) larvae in constant and changing temperatures. *Oikos* 27:288–99.

———. 1979. "The Community of Coprophagous Beetles." D.Phil. thesis, Univ. of Oxford, England.

———. 1980a. The community of coprophagous beetles (Coleoptera, Scarabaeidae and Hydrophilidae) in northern Europe. *Ann. Entomol. Fenn.* 46:57–74.

———. 1980b. Spatial patterns and movements in coprophagous beetles. *Oikos* 34:293–310.

———. 1980c. Patterns of beetle succession in droppings. *Ann. Zool. Fenn.* 17:17–25.

———. 1980d. Migration to and from cow droppings by coprophagous beetles. *Ann. Zool. Fenn.* 17:11–16.

———. 1980e. Spatial variation in the timing of the seasonal occurrence in coprophagous beetles. *Oikos* 34:311–21.

———. 1980f. Movement patterns in dung beetles and in the dung fly. *Anim. Behav.* 28:953–64.

———. 1981. Coexistence of competitors in patchy environment with and without predation. *Oikos* 37:306–12.

———. 1982. Dynamics of regional distribution: The core and satellite species hypothesis. *Oikos* 38:210–21.

———. 1983. Distributional ecology and abundance of dung and carrion-feeding beetles (Scarabaeidae) in tropical rain forests in Sarawak, Borneo. *Acta Zool. Fenn.* 167:1–45.

———. 1986. Individual behaviour, population dynamics and community structure of *Aphodius* (Scarabaeidae) in Europe. *Acta Oecol., Oecol. Gen.* 7:171–87.

———. 1987a. Nutritional ecology of dung- and carrion-feeding insects. In F. Slansky, Jr., and J. G. Rodríguez, eds., *Nutritional Ecology of Insects, Mites, and Spiders*, pp. 837–84. Wiley, New York.

———. 1987b. Colonisation of ephemeral habitats. In A. J. Gray, M. J. Crawley, and P. J. Edwards, eds., *Colonisation, Succession and Stability*, pp. 155–85. Blackwell, Oxford.

———. 1987c. Carrion fly community dynamics: Patchiness, seasonality and coexistence. *Ecol. Entomol.* 12:257–66.

———. 1987d. Cross-correlation in population dynamics and the slope of spatial variance-mean regressions. *Oikos* 50:148–51.

———. 1988. Four kinds of extra long diapause in insects: A review of theory and observations. *Ann. Zool. Fenn.* 25:37–53.

———. 1989a. Dung Beetles. In H. Lieth and J. A. Wagner, eds., *Ecosystems of the World, 14b, Tropical Forests*, pp. 489–511. Elsevier, Amsterdam.

———. 1989b. Metapopulation dynamics: Does it help to have more of the same? *TREE* 4:113–14.

———. 1989c. Fungivory: Fungi, Insects and Ecology. In N. Wilding, N. M. Collins, P. M. Hammond, and J. F. Webber, eds., *Insect-Fungus Interactions*, pp. 25–68. Academic Press, London.

———. 1990. Dung and carrion insects. In B. Shorrocks and I. Swingland, eds., *Living in a Patchy Environment*, pp. 127–45. Oxford Univ. Press, Oxford.

————. 1991. Insectivorous mammals. In M. J. Crawley, ed., *Ecology of Natural Enemies*. Blackwell, Oxford, in press.

Hanski, I., and P. M. Hammond. 1986. Assemblages of carrion and dung Staphylinidae in tropical rain forests in Sarawak, Borneo. *Ann. Entomol. Fenn.* 52:1–19.

Hanski, I., and H. Koskela. 1977. Niche relations amongst dung-inhabiting beetles. *Oecologia (Berl.)* 28:203–31.

————. 1979. Resource partitioning in six guilds of dung-inhabiting beetles (Coleoptera). *Ann. Entomol. Fenn.* 45:1–12.

Hanski, I., and S. Kuusela, 1983. Dung beetle communities in the Åland archipelago. *Acta Entomol. Fenn.* 42:36–42.

Hanski, I., and J. Niemelä. 1990. Elevational distributions of dung and carrion beetles in northern Sulawesi. In W. J. Knight and J. D. Holloway, eds., *Insects and the Rain Forests of South East Asia (Wallacea)*, pp. 145–52. The Royal Entomol. Soc., London.

Hardin, G. 1960. The competitive exclusion principle. *Science* 131:1292–97.

Harper, J. E., and J. Webster. 1964. An experimental analysis of the coprophilous fungus succession. *Trans. Br. Mycol. Soc.* 47:511–30.

Harrington, H. J. 1962. Paleogeographic development of South America. *Bull. Amer. Assoc. Pet. Geol.* 46:1773–1814.

Harris, L. R., and R. R. Blume. 1986. Beneficial insects inhabiting bovine droppings in the United States. In R. S. Patterson and D. A. Rutz, eds., *Biological Control of Muscoid Flies. Misc. Publ. Entomol. Soc. Amer.* 62:10–15.

Harvey, P. H., R. K. Colwell, J. W. Silvertown, and R. M. May. 1983. Null models in ecology. *Ann. Rev. Ecol. Syst.* 14:189–212.

Hassell, M. P. 1978. *The Dynamics of Arthropod Predator-Prey Systems*. Princeton Univ. Press, Princeton, N.J.

Hassell, M. P., and R. M. May. 1973. Stability in insect host-parasite models. *J. Anim. Ecol.* 42:693–726.

————. 1985. From individual behaviour to population dynamics. In R. M. Sibly and R. H. Smith, eds., *Behavioural Ecology*, pp. 3–32. Blackwell, Oxford.

Hata, K., and W. D. Edmonds. 1983. Structure and function of the mandibles of adult dung beetles (Coleoptera: Scarabaeidae). *Int. J. Insect Morphol. Embryol.* 12:1–12.

Heaney, L. R. 1986. Biogeography of mammals in SE Asia: Estimates of rates of colonization, extinction and speciation. *Biol. J. Linn. Soc.* 28:127–65.

Heinrich, B., and G. A. Bartholomew. 1979a. Roles of endothermy and size in inter- and intraspecific competition for elephant dung in an African dung beetle, *Scarabaeus laevistriatus. Physiol. Zool.* 52:484–96.

————. 1979b. The ecology of the African dung beetle. *Sci. Amer.* 235:118–26.

Hendrichs, J. 1975. A note on the occurrence of *Sphaeridium scarabaeoides* L. near Mexico City (Col., Hydrophilidae). *Coleopt. Bull.* 29:171.

Henry, K. R. 1983. Introduction of dung beetles into South Australia, 1970–1983. *South Austr. Dep. of Agric. Technical Report*.

Herschkovitz, P. 1972. The recent mammals of the Neotropical region: A zoogeographic and ecological review. In A. Keast, F. C. Erk, and B. Glass, eds., *Evolution, Mammals and Southern Continents*, pp. 311–431. State Univ. of New York Press, Albany.

Heymons, R., and H. von Lengerken. 1929. Biologische Untersuchungen an Coprophagen Lamellicornien. I. Nahrungserwerb und Fortpflanzungsbiologie der Gattung *Scarabaeus* L. *Morph. Ökol. Tiere* 14:531–613.

Hoffstetter, R. 1972. Relationships, origins and history of the ceboid monkeys and caviomorph rodents: A modern reinterpretation. *Evol. Biol.* 6:323–47.

———. 1974. Phylogeny and geographical deployment of the Primates. *J. Human Evol.* 3:327–50.

———. 1976. Histoire des mammifères et dérive des continents. *La Recherche* 7, 64:124–38.

Holdridge, L. R. 1967. *Life Zone Ecology.* Tropical Science Center, San Jose, Costa Rica.

Holdridge, L. R., W. C. Grenke, W. H. Hatheway, T. Liang, and J. Tosi, Jr. 1971. *Forest Environments in Tropical Life Zones: A Pilot Study.* Pergamon Press, Oxford.

Holm, E., and J. F. Kirsten. 1979. Pre-adaptation and speed mimicry among Namib Desert scarabaeids with orange elytra. *J. Arid Env.* 2:263–71.

Holm, E., and M.M.H. Wallace. 1987. The influence of superphosphate on the establishment of introduced dung beetles in south-eastern Australia. *J. Aust. Inst. Agric. Sci.* 53:202–204.

Holter, P. 1975. Energy budget of a natural population of *Aphodius rufipes* larvae (Scarabaeidae). *Oikos* 26:177–86.

———. 1979. Abundance and reproductive strategy of the dung beetle *Aphodius rufipes* (L.) (Scarabaeidae). *Ecol. Entomol.* 4:317–26.

———. 1982. Resource utilization and local coexistence in a guild of scarabaeid dung beetles (*Aphodius* spp.). *Oikos* 9:213–27.

Honda, H. 1927. Interesting instincts of *Gymnopleurus sinuatus* Ol. *Proc. Imp. Acad. Tokyo* 3:684–86.

Horion, A. 1950. Discontinuierliche Ost-West Verbreitung mitteleuropäischer Käfer. *Proc. 7th Int. Congr. Entomol., Stockholm*, p. 408.

Houston, W.W.K. 1986. Exocrine glands in the forelegs of dung beetles in the genus *Onitis* F. (Coleoptera: Scarabaeidae). *J. Aust. Entomol. Soc.* 25:161–69.

Houston, W.W.K., and P. McIntyre. 1985. The daily onset of flight in the crepuscular dung beetle *Onitis alexis* (Coleoptera: Scarabaeidae). *Entomol. Exp. Appl.* 39:223–32.

Howden, H. F. 1952. A new name for *Geotrupes (Peltotrupes) chalybaeus* LeConte with a description of the larva and its biology. *Coleopt. Bull.* 6:41–48.

———. 1955a. The biology and taxonomy of the North American beetles of the subfamily Geotrupinae with revisions of the genera *Bolbocerosoma, Eucanthus, Geotrupes* and *Peltotrupes* (Scarabaeidae). *Proc. U.S. Nat. Mus.* 104:151–319.

———. 1955b. Cases of interspecific "parasitism" in Scarabaeidae (Coleoptera). *J. Tennessee Acad. Sci.* 30:64–66.

———. 1965. A second New World species of *Sisyphus* Latreille (Coleoptera: Scarabaeidae). *Can. Entomol.* 97:842–44.

———. 1971. Five unusual genera of New World Scarabaeidae (Coleoptera). *Can. Entomol.* 103:1463–71.

———. 1973. Revision of the New World genus *Cryptocanthon* Balthasar (Coleoptera: Scarabaeidae). *Can. J. Zool.* 51:39–48.

————. 1976. New species in the genera *Bdelyropsis*, *Cryptocanthon* and *Drepano-cerus* (Coleoptera: Scarabaeidae). *Proc. Entomol. Soc. Wash.* 78:95–103.

————. 1981. Zoogeography of some Australian Coleoptera as exemplified by the Scarabaeoidea. In A. Keast, ed., *Ecological Biogeography of Australia*, pp. 1009–35. Dr. W. Junk Publishers, The Hague.

————. 1985. Expansion and contraction cycles, endemism and area: The taxon cycle brought full circle. In G. E. Ball, ed., *Taxonomy, Phylogeny and Zoogeography of Beetles and Ants*, pp. 473–87. Dr. W. Junk Publishers, Dordrecht.

Howden, H. F., and O. L. Cartwright. 1963. Scarab beetles of the genus *Onthophagus* Latreille North of Mexico (Coleoptera: Scarabaeidae). *Proc. U.S. Nat. Mus.* 3467:1–135.

Howden, H. F., and B. D. Gill. 1987. New species and new records of Panamanian and Costa Rican Scarabaeinae (Coleoptera: Scarabaeidae). *Coleopt. Bull.* 41:201–24.

Howden, H. F., and V. G. Nealis. 1975. Effects of clearing in a tropical rain forest on the composition of the coprophagous scarab beetle fauna (Coleoptera). *Biotropica* 7:77–83.

————. 1978. Observations on height of perching in some tropical dung beetles (Scarabaeidae). *Biotropica* 10:43–46.

Howden, H. F., and S. B. Peck. 1987. Adult habits, larval morphology, and phylogenetic placement of *Taurocerastes patagonicus* Philippi (Scarabaeidae: Geotrupinae). *Can. J. Zool.* 65:329–32.

Howden, H. F., and C. H. Scholtz. 1986. Changes in a Texas dung beetle community between 1975 and 1985 (Coleoptera: Scarabaeidae, Scarabaeinae). *Coleopt. Bull.* 40:313–16.

————. 1987. A revision of the African genus *Odontoloma* Boheman (Coleoptera: Scarabaeidae: Scarabaeinae). *J. Entomol. Soc. S. Afr.* 50:155–92.

Howden, H. F., and O. P. Young. 1981. Panamanian Scarabaeinae: Taxonomy, distribution, and habits (Coleoptera, Scarabaeidae). *Contr. Amer. Entomol. Inst.* 18:1–204.

Huerta, C., S. Anduaga, and G. Halffter. 1981. Relaciones entre nidificacion y ovario en Copris (Coleoptera Scarabaeidae Scarabaeinae). *Fol. Entomol. Mex.* 47:139–70.

Hughes, R. D. 1975. Assessment of the burial of cattle dung by Australian dung beetles. *J. Aust. Entomol. Soc.* 14:128–34.

Hughes, R. D., and J. Walker. 1970. The role of food in the population dynamics of the Australian bushfly. In A. Watson, ed., *Animal Populations in Relation to Their Food Resources*, pp. 255–69. Blackwell, Oxford.

Hughes, R. D., P. M. Greenham, M. Tyndale-Biscoe, and J. M. Walker. 1972. A synopsis of observations on the biology of the Australian bush fly (*Musca vetustissima* Walker). *J. Austr. Entomol. Soc.* 11:311–31.

Hughes, R. D., M. Tyndale-Biscoe, and J. Walker. 1978. Effects of introduced dung beetles (Coleoptera: Scarabaeinae) on the breeding and abundance of the Australian bush fly *Musca vetustissima* (Diptera: Muscidae). *Bull. Entomol. Res.* 68:361–72.

Huijbregts, J. 1984. *Bdelyrus geijskesi*, a new scarab (Coleoptera: Scarabaeidae) from Suriname associated with Bromeliaceae. *Zool. Mededel.* 59:61–67.

Huntley, B. J., and B. H. Walker, eds. 1982. *Ecology of Tropical Savannas*. Springer-Verlag, Berlin.

Ives, A. R. 1988a. Covariance, coexistence and the population dynamics of two competitors using a patchy resource. *J. Theor. Biol.* 133:108–15.

———. 1988b. Aggregation and the coexistence of competitors. *Ann. Zool. Fenn.* 25:75–88.

———. 1990. Aggregation and coexistence in a carrion-fly community. *Ecol. Monogr*, in press.

Ives, A. R., and R. M. May. 1985. Competition within and between species in a patchy environment: Relations between microscopic and macroscopic models. *J. Theor. Biol.* 115:65–92.

Janssens, A. 1949. Contribution à l'étude des Coléoptères Lamellicornes. XIII. Table synoptique et essai de classification pratique des Coléoptères Scarabaeidae. *Bull. Inst. Roy. Sci. Nat. Belg.* 25, 15:1–30.

———. 1953. *Oniticellini* (Coleoptera Lamellicornia). In *Exploration du Parc national de l'Upemba*, Mission G. F. DeWitte. Institut des Parcs nationaux du Congo Belge. Fascicule 11.

———. 1954. Contribution à l'étude des Coléoptères Lamellicornes. XVII. Description d'une nouvelle et curieuse espéce d'*Onthophagus* du Parc national de l'Upemba et considérations sur l'évolution et la convergence de certains Coléoptères Scarabaeinae. *Vol. Jubil. V. Van Straelen, Bruxelles* 2:971–76.

———. 1960. *Faune de Belgique. Insectes Coléoptères Lamellicornes*. Patrimoine Institut Royal Sci. Nat. Belgique édit., Brussels.

Janzen, D. H. 1983a. Seasonal change in abundance of large nocturnal dung beetles (Scarabaeidae) in a Costa Rican deciduous forest and adjacent horse pasture. *Oikos* 41:274–83.

——— 1983b. Insects at carrion and dung. In D. H. Janzen, ed., *Costa Rican Natural History*, pp. 640–42. Univ. of Chicago Press, Chicago.

———. 1983c. *Eulissus chalybaeus* (Abejon Culebra, Green Rove Beetle). In D. H. Janzen, ed., *Costa Rican Natural History*, pp. 721–22. Univ. of Chicago Press, Chicago.

———. 1986. Guanacaste National Park—A tropical Christmas catalog. *Biotropica* 18:272.

Jeannel, R. 1942. *La genèse des faunes terrestres*. Presses Univ. de France, Paris.

Jerath, M. L., and P. O. Ritcher. 1959. Biology of Aphodiinae with special reference to Oregon. *Pan-Pac. Entomol.* 35:169–75.

Jessop, L. 1985. An identification guide to Eurysternine dung beetles (Coleoptera, Scarabaeidae). *J. Nat. Hist.* 19:1087–1111.

Johnson, C. 1962. The Scarabaeoid Coleoptera fauna of Lancashire and Cheshire and its apparent changes over the last 100 years. *The Entomol.* 95:153–65.

Johnson, C. G. 1969. *Migration and Dispersal of Insects by Flight*. Methuen, London.

Johnson, K. A., and W. G. Whitford. 1975. Foraging ecology and relative importance of subterranean termites in Chihuahuan desert ecosystems. *Env. Entomol.* 4:66–70.

Kareiva, P., and G. Odell. 1987. Swarms of predators exhibit "preytaxis" if individual predators use area-restricted search. *Amer. Nat.* 130:233–70.

Keast, A. 1972. Continental drift and the biota of the mammals on southern continents.

In A. Keast, F. C. Erk, and B. Glass, eds., *Evolution, Mammals and Southern Continents*, pp. 23–87. State Univ. of New York Press, Albany.

Kessler, H., and E. U. Balsbaugh. 1972. Succession of adult Coleoptera in bovine manure in East Central South-Dakota. *Ann. Entomol. Soc. Amer.* 65:1333–36.

Kingston, T. J. 1977. "Natural Manuring by Elephants in Tsavo National Park, Kenya." D.Phil. thesis, Univ. of Oxford, England.

Kingston, T. J., and M. Coe. 1977. The biology of a giant dung-beetle (*Heliocopris dilloni*) (Coleoptera, Scarabaeidae). *J. Zool. Lond.* 181:243–63.

Kirk, A. A. 1983. The biology of *Bubas bison* (L.) (Coleoptera: Scarabaeidae) in southern France and its potential for recycling dung in Australia. *Bull. Entomol. Res.* 73:129–36.

Kirk, A. A., and J. E. Feehan. 1984. A method for increased production of eggs of *Copris hispanus* L. and *C. lunaris* L. (Coleoptera, Scarabaeidae). *J. Aust. Entomol. Soc.* 23:293–94.

Kirk, A. A., and J.-P. Lumaret. 1991. The importation of Mediterranean-adapted dung beetles (Coleoptera, Scarabaeidae) from the northern hemisphere to other parts of the world. In R. H. Groves and F. Di Castri, eds., *Biogeography of Mediterranean Invasions*, pp. 409–20. Cambridge Univ. Press, Cambridge, Eng.

Kirk, A. A., and T. J. Ridsdill-Smith. 1986. Dung beetle distribution patterns in the Iberian peninsula. *Entomophaga* 31:183–90.

Klemperer, H. G. 1978. The repair of larval cells and other larval activities in *Geotrupes spiniger* Marsham and other species (Coleoptera, Scarabaeidae). *Ecol. Entomol.* 3:119–31.

———. 1979. An analysis of the nesting behaviour of *Geotrupes spiniger* Marsham (Coleoptera, Scarabaeidae). *Ecol. Entomol.* 4:133–50.

———. 1981. Nest construction and larval behaviour of *Bubas bison* (L.) and *Bubas bubalus* (Ol.) (Coleoptera, Scarabaeidae). *Ecol. Entomol.* 6:23–33.

———. 1982a. Normal and atypical nesting behaviour of *Copris lunaris* (L.): Comparison with related species (Coleoptera, Scarabaeidae). *Ecol. Entomol.* 7:69–83.

———. 1982b. Parental behaviour in *Copris lunaris* (Coleoptera, Scarabaeidae): Care and defence of brood balls and nest. *Ecol. Entomol.* 7:155–67.

———. 1982c. Nest construction and larval behaviour of *Onitis belial* and *Onitis ion* (Coleoptera, Scarabaeidae). *Ecol. Entomol.* 7:291–97.

———. 1983a. The evolution of parental behaviour in Scarabaeinae (Coleoptera, Scarabaeidae): An experimental approach. *Ecol. Entomol.* 8:49–59.

———. 1983b. Brood ball construction by the non-brooding Coprini *Sulcophanaeus carnifex* and *Dichotomius torulosus* (Coleoptera, Scarabaeidae). *Ecol. Entomol.* 8:61–68.

———. 1983c. Subsocial behaviour in *Oniticellus cinctus* (Coleoptera, Scarabaeidae): Effect of the brood on parental care and oviposition. *Physiol. Entomol.* 8:393–402.

———. 1984. Nest construction, fighting, and larval behaviour in a geotrupine dung beetle, *Ceratophyus hoffmannseggi* (Coleoptera: Scarabaeidae). *J. Zool. Soc. Lond.* 204:119–27.

Klemperer, H. G., and R. Boulton. 1976. Brood burrow construction and brood care by *Heliocopris japetus* (Klug) and *H. hamadryas* (Fabricius) (Coleoptera, Scarabaeidae). *Ecol. Entomol.* 1:19–29.

Klemperer, H. G., and J.-P. Lumaret. 1985. Life cycle and behaviour of the flightless beetles *Thorectes sericeus* Jekel, *T. albarracinus* Wagner, and *T. laevigatus cobosi* Baraud (Col. Geotrupidae). *Ann. Soc. Entomol. Fr. (N.S.)* 21:425–31.

Kneidel, K. A. 1985. Patchiness, aggregation, and the coexistence of competitors for ephemeral resources. *Ecol. Entomol.* 10:441–48.

Knight, W. J. 1988. *Project Wallace Report*. Roy. Entomol. Soc., London.

Kohlmann, B. 1979. Some notes on the biology of *Euphoria inda* (Linné) (Coleoptera: Scarabaeidae). *Pan-Pac. Entomol.* 55:279–83.

———. 1984. Biosistemática de las especies norteamericanas del género *Ateuchus* (Coleoptera: Scarabaeidae). *Fol. Entomol. Mex.* 60:3–81.

Kohlmann, B., and G. Halffter. 1988. Cladistic and biogeographical analysis of *Ateuchus* (Coleoptera: Scarabaeidae) of Mexico and the United States. *Fol. Entomol. Mex.* 74:109–30.

Kohlmann, B., and S. Sánchez-Colón. 1984. Structure of a Scarabaeinae community: A numerical-behavioural study (Coleoptera: Scarabaeinae). *Acta Zool. Mex.* 2:1–27.

Koskela, H. 1972. Habitat selection of dung-inhabiting staphylinids (Coleoptera) in relation to age of the dung. *Ann. Zool. Feun.* 9:156–71.

———. 1979. Patterns of diel flight activity in dung inhabiting beetles: An ecological analysis. *Oikos* 33:419–39.

Koskela, H., and I. Hanski. 1977. Structure and succession in a beetle community inhabiting cow dung. *Ann. Zool. Fenn.* 14:204–23.

Kotler, B. P., and J. S. Brown. 1988. Environmental heterogeneity and the coexistence of desert rodents. *Ann. Rev. Ecol. Syst.* 19:281–308.

Krebs, J. R., and N. B. Davies. 1981. *An Introduction to Behavioural Ecology*. Blackwell, Oxford.

Krikken, J. 1971. The characters of *Cyobius wallacei* Sharp, a little-known onthophagine scarab from the Malay Archipelago (Coleoptera: Scarabaeidae). *Entomol. Berichten* 31:22–28.

———. 1977. Some new and otherwise noteworthy species of *Onthophagus* Latreille from the Indo-Australian archipelago (Coleoptera: Scarabaeidae). *Zool. Mededel.* 52:169–184.

———. 1978. Interessante *Aphodius*-soorten (Coleoptera: Scarabaeidae) uit mest van Nederlands grofwild. *Zool. Bijdr.* 23:137–47.

———. 1983. An interesting case of camouflage in African dung-beetles of the genus *Drepanocerus* (Coleoptera: Scarabaeidae). *Entomol. Ber. Amsterdam* 43:90–92.

———. 1984. A new key to the suprageneric taxa in the beetle family Cetoniidae, with annotated lists of the known genera. *Zool. Verhand.* 210:1–75.

Kronblad, W. 1971. *Aphodius* spp. in algal waste (Coleoptera: Scarabaeidae): Some interesting finds from the region of Vetland in the highlands of Småland, Sweden. *Entomol. Tidskr.* 92:281–82.

Kuijten, P. J. 1978. Revision of the Indo-Australian species of the genus *Phaeochrous* Castenau, 1840 (Coleoptera: Scarabaeidae, Hybosorinae), with notes on the African species. *Zool. Verhand.* 165:1–40.

———. 1981. Revision of the genus *Phaeochroops* Candze (Coleoptera: Scarabaeidae, Hybosorinae). *Zool. Verhand.* 183:1–76.

Lacey, C. J., J. Walker, and I. R. Noble. 1982. Fire in Australian tropical savannas. In B. J. Huntley, and B. H. Walker, eds., *Ecology of Tropical Savannas*, pp. 246–72. Springer-Verlag, Berlin.

La Greca, M. 1964. La categorie corologiche degli elementi faunistici italiani. *Mem. Soc. Entomol. Ital.* 43:147–65.

Lamotte, M. 1978. La savane préforestière de Lamto, Côte d'Ivoire. In M. Lamotte and F. Bourlière, eds., *Problèmes d'écologie: Structure et fonctionnement des écosystèmes terrestres*, pp. 231–311. Masson, Paris.

Lancaster, J. L., R. R. Blume, and J. S. Simco. 1976. Laboratory evaluation of *Onthophagus gazella* F. against *Musca autumnalis* De Geer. *Southw. Entomol.* 1:111–13.

Lande, R. 1987. Extinction thresholds in demographic models of territorial populations. *Amer. Nat.* 130:624–35.

Landin, B.-O. 1960. The Lamellicorn beetles of the Azores (Coleoptera) with some reflexions on the classification of certain Aphodiini. *Bol. Mus. Munic. Funchal* 13:49–84.

———. 1961. Ecological studies on dung beetles (Col. Scarabaeidae). *Opusc. Entomol. Suppl.* 19:1–228.

———. 1968. The diel flight activity of dung-beetles (Coleoptera, Scarabaeidae). *Opusc. Entomol. Suppl.* 32:1–172.

Landin, J. 1967. On the relationship between the microclimate in cow droppings and some species of *Sphaeridium* (Col. Hydrophilidae). *Opusc. Entomol.* 32:207–12.

Laurence, B. R. 1954. The larval inhabitants of cow pats. *J. Anim. Ecol.* 23:234–60.

Lecordier, C. 1974. Le climat de la région de Lamto. *Bull. Liais. Cherch. Lamto, Numéro Special* 1:45–103.

Lee, J. M., and Y.-S. Peng. 1982. Influence of manure availability and nesting density on the progeny of *Onthophagus gazella*. *Env. Entomol.* 11:38–41.

Legner, E. F. 1986. The requirement for reassessment of interactions among dung beetles, symbovine flies, and natural enemies. In R. S. Patterson and D. A. Rutz, eds., *Biological Control of Muscoid Flies. Misc. Publ. Entomol. Soc. Amer.* 61:120–31.

Leigh, E. G., Jr., A. S. Rand, and D. M. Windsor, eds. 1982. *The Ecology of a Tropical Forest*. Smithsonian Institution Press, Washington, D.C.

Levins, R. 1969. Some demographic and genetic consequences of environmental heterogeneity for biological control. *Bull. Entomol. Soc. Amer.* 15:237–40.

———. 1970. Extinction. In M. Gerstenhaber, ed., *Some Mathematical Problems in Biology*, pp. 77–107. American Mathematical Society, Providence, R.I.

Levot, G. W., K. A. Brown, and E. Shipp. 1979. Larval growth characteristics of some calliphorid and sarcophagid Diptera. *Bull. Entomol. Res.* 69:469–76.

Lindquist, A. W. 1933. Amounts of dung buried and soil excavated by certain Coprini (Scarabaeidae). *J. Kansas Entomol. Soc.* 6:109–25.

Lindroth, C. H. 1948. Interglacial insect remains from Sweden. *Årsbok Sver. Geol. Undersökn. (C)* 42:1–29.

Livingstone, D. A. 1975. Late quaternary climatic changes in Africa. *Ann. Rev. Ecol. Syst.* 6:249–80.

Lloyd, M., and J. White. 1980. On reconciling patchy microspatial distributions with competition models. *Amer. Nat.* 115:29–44.

Luederwaldt, H. 1911. Os insectos necrofagos Paulistas. *Rev. Mus. Paulista* 8:414–33.

———. 1914. Biologia de varias especies de *Pinotus* de S. Paulo. *Rev. Mus. Paulista* 9:365–66.

———. 1931. O genero *Ontherus* (Coleopt.) (Lamell.-Coprid.-Pinot.) com una chave para a determinação dos pinotides americanos. *Rev. Mus. Paulista* 17:363–422.

Lumaret, J.-P. 1975. Étude des conditions de ponte et de développement larvaire d'*Aphodius* (*Agrilinus*) *constans* Dft. (Coléoptères Scarabaeidae) dans la nature et au laboratoire. *Vie et Milieu C,* 25:267–82.

———. 1978. "Biogéographie et écologie des Scarabéides coprophages du sud de la France." Sc.D. diss., Univ. Sci. Techn. Languedoc, Montpellier, France.

———. 1978–79a. Biogéographie et écologie des Scarabéides coprophages du sud de la France. I. Méthodologie et modèles de répartition. *Vie et Milieu C,* 28–29:1–34.

———. 1978–79b. Biogéographie et écologie des Scarabéides coprophages du sud de la France. II. Analyse synécologique des répartitions. *Vie et Milieu C,* 28–29:179–201.

———. 1979. Un piège attractif pour la capture des insectes coprophages et nécrophages. *L'Entomologiste* 35:57–60.

———. 1980a. Analyse des communautés de Scarabéides coprophages dans le maquis corse et étude de leur rôle dans l'utilisation des excréments. *Ecol. Medit.* 5:50–58.

———. 1980b. *Les Bousiers.* Balland éd., Paris.

———. 1983a. Structure des peuplements de coprophages Scarabaeidae en région méditerranéenne française: Relations entre les conditions écologiques et quelques paramètres biologiques des espèces. *Bull. Soc. Entomol. Fr.* 88:481–95.

———. 1983b. La nidification des *Trox* (Col. Scarabaeoidea Trogidae). *Bull. Soc. Entomol. Fr.* 88:594–96.

———. 1986. Toxicité de certains helminthicides vis-à-vis des insectes coprophages et conséquences sur la disparition des excréments de la surface du sol. *Acta Oec., Oecol. Appl.* 7:313–24.

———. 1987. Use of excrement by dung beetles in drought affected areas. In *Proceedings 5th International Conference on Mediterranean-Climate Ecosystems*, Medecos V: Time-scales of Water Stress Response of Mediterranean Biota. Montpellier, France, July 15–21, 1987.

Lumaret, J.-P., and M. Bertrand. 1985. L'effet de reposoir sur les Arthropodes édaphiques, conséquence d'une accumulation excessive d'excréments dans les zônes pâturées. *Bull. Ecol.* 16:55–62.

Lumaret, J.-P., and Y. Cambefort. 1985. Description de la larve de *Campsiura trivittata* (Moser) (Coleoptera, Cetoniidae). *Nouv. Rev. Entomol. (N.S.)* 2:319–23.

Lumaret, J.-P., and A. A. Kirk. 1987. Ecology of dung beetles in the French Mediterranean region (Coleoptera, Scarabaeidae). *Acta Zool. Mex. (N.S.)* 24:1–60.

Lumaret, J.-P., and N. Stiernet. 1984. Contribution à l'étude de la faune des Alpes suisses. Description de la larve d'*Aphodius (Agolius) abdominalis* Bonelli, 1812 (Coleoptera, Aphodiidae). *Bull. Soc. Entomol. Suisse* 57:335–40.

———. 1990. Inventaire et distribution des Scarabéides coprophages dans le massif de la Vanoise. *Trav. Sci. Parc Nat. Vanoise* 17:193–228.

MacArthur, R. H. 1972. *Geographical Ecology*. Harper & Row, New York.

MacArthur, R. H., and E. O. Wilson. 1967. *The Theory of Island Biogeography*. Princeton Univ. Press, Princeton, N.J..

McCullough, D. R. 1982. White-tailed deer pellet-group weights. *J. Wildl. Manage*. 46:829–32.

McDougall, I., and F. H. Chamalaun. 1969. Isotopic dating and geomagnetic polarity studies on volcanic rocks from Mauritius, Indian Ocean. *Geol. Soc. Amer*. 80:1419–42.

McIntosh, R. P. 1985. *The Background of Ecology: Concept and Theory*. Cambridge Univ. Press, Cambridge, England.

MacNaughton, S. J., and N. J. Georgiadis. 1986. Ecology of African grazing and browsing mammals. *Ann. Rev. Ecol. Syst*. 17:39–65.

Macqueen, A., and B. P. Beirne. 1974. Insects and mites associated with fresh cattle dung in the southern interior of British Columbia. *J. Entomol. Soc. Brit. Col*. 71:5–9.

———. 1975a. *Haematobia irritans* (Diptera: Muscidae) from cattle dung in South Central British Columbia, Canada. *Can. Entomol*. 107:1255–64.

———. 1975b. Dung burial activity and fly control potential of *Onthophagus nuchicornis* (Coleoptera: Scarabaeinae) in British Columbia (Canada). *Can. Entomol*. 107:1215–20.

———. 1975c. Influence of other insects on production of horn fly *Haematobia irritans* (Diptera: Muscidae), from cattle dung in South-Central British Columbia. *Can. Entomol*. 107:1255–64.

Macqueen, A., M.M.H. Wallace, and B. M. Doube. 1986. Seasonal changes in favourability of cattle dung in central Queensland for three species of dung inhabiting insect. *J. Aust. Entomol. Soc*. 25:23–29.

McQuillan, P. B., and J. E. Ireson. 1983. Some aspects of the invertebrate fauna and its role in Tasmanian (Australia) pastures. In K. E. Lee, ed., *Proc. 3rd Australasian Conf. on Grassland Invertebrate Ecology*, pp. 101–106. Adelaide, Nov. 30–Dec. 4, 1981. Govt. Print. Div., Plympton, S.A., Australia.

Main, H. 1917. On rearing beetles of the genus *Geotrupes*. *Proc. So. London Entomol. Soc., 1916* 17:18–22.

Mäkelä, I. 1983. "Pellon ja Metsän Rapakärpäslajisto (Diptera, Sphaeroceridae) sekä Lennon Kausirytmiikka." M.Sc. thesis, Univ. of Jyväskylä, Finland.

Maley, J., and D. A. Livingstone. 1983. Extension d'un élément montagnard dans le sud du Ghana (Afrique de l'Ouest) au Pléistocène supérieur et l'Holocène inférieur: Premières données polliniques. *C.R. Acad. Sci. Paris III* 296:761–66.

Mani, M. S. 1968. *Ecology and Biogeography of High Altitude Insects*. Dr. W. Junk Publishers, The Hague.

Mani, M. S., and S. Singh. S. 1961. Entomological Survey of Himalaya. Part 26: A contribution to our knowledge of the geography of the high altitude insects of the nival zones from the north-west Himalaya, part 2. *J. Bombay Nat. Hist. Soc*. 58:724–48.

Marsch, E. 1982. Experimentelle Analyse des Verhaltens von *Scarabaeus sacer* L. beim Nahrungserwerb. *Bonn Zool. Monogr*. 17:1–79.

Marshall, L. G., R. F. Butler, R. E. Drake, G. H. Curtis, and R. H. Tedford. 1979. Calibration of the Great American interchange: A radioisotope chronology for Late

Tertiary interchange of faunas between the Americas. *Science (N.Y.)* 204:272–79.

Martin, P. S., and R. G. Klein, eds. 1984. *Quaternary Extinctions: A Prehistoric Revolution.* Univ. of Arizona Press, Tucson.

Martínez, A. 1952. Scarabaeidae nuevos o poco conocidos III. *Mis. Est. Pat. Reg. Argentina* 81–82:53–118.

———. 1959. Catálogo de los Scarabaeidae Argentinos (Coleoptera). *Rev. Mus. Cienc. Nat. Bernardino Rivadavia* 5:1–126.

———. 1988. Notas sobre *Eurysternus* Dalman (Coleoptera, Scarabaeidae). *Entomol. Basil.* 12:279–304.

Martínez, A., G. Halffter, and F. S. Pereira. 1964. Notes on the genus *Canthidium* Erichson and allied genera. Part 1. (Col. Scarabaeidae). *Studia Entomol.* 7:161–78.

Martínez, I., and C. Caussanel. 1984. Modification de la pars intercerebralis, des corpora allata, des gonades et comportement reproducteur chez *Canthon cyanellus*. *C.R. Acad. Sci. Paris, III* 14:597–602.

Martin-Piera, F. 1982. "Los Scarabaeinae (Col., Scarabaeoidea) de la Peninsula Iberica e Islas Baleares." Ph.D. diss., Univ. of Complutense, Madrid, Spain.

———. 1983. Composicion sistematica y origen biogeografico de la fauna iberica de Onthophagini (Coleoptera, Scarabaeoidea). *Boll. Mus. Reg. Sci. Nat. Torino* 1:165–200.

———. 1986. The palearctic species of the subgenus *Parentius* Zunino, 1979 (Coleoptera, Scarabaeoidea, Onthophagini). *Boll. Mus. Reg. Sci. Nat. Torino* 4:77–122.

Masters, W. M. 1980. Insect disturbance stridulation: Characterization of airborne and vibrational components of the sound. *J. Comp. Physiol.* 135:259–68.

Masumoto, K. 1973. Observation of the nidification of *Synapsis davidi* Fairmaire. *Entomol. Rev. Japan* 25:60–62.

Matthews, E. G. 1962. A revision of the genus *Copris* Müller of the Western Hemisphere (Coleoptera, Scarabaeidae). *Entomol. Amer.* 41:1–139.

———. 1963. Observations on the ball-rolling behavior of *Canthon pilularius* (L.) (Coleoptera, Scarabaeidae). *Psyche* 70:75–93.

———. 1965. The taxonomy, geographical distribution, and feeding habits of the Canthonines of Puerto Rico (Coleoptera: Scarabaeidae). *Trans. Amer. Entomol. Soc.* 91:431–65.

———. 1966. A taxonomic and zoogeographic survey of the Scarabaeinae of the Antilles (Coleoptera: Scarabaeidae). *Mem. Amer. Entomol. Soc.* 21:1–134.

———. 1972. A revision of the scarabaeine dung beetles of Australia. I. Tribe Onthophagini. *Aust. J. Zool. Suppl. Ser.* 9:1–330.

———. 1974. A revision of the scarabaeine dung beetles of Australia. II. Tribe Scarabaeini. *Aust. J. Zool. Suppl. Ser.* 24:1–211.

———. 1976. A revision of the scarabaeine dung beetles of Australia. III. Tribe Coprini. *Aust. J. Zool. Suppl. Ser.* 38:1–52.

Matthews, E. G., and G. Halffter. 1968. New data on American *Copris* with discussion of a fossil species (Coleopt. Scarab.). *Ciencia Mex.* 26:147–62.

Matthews, E. G., and Z. Stebnicka. 1986. A review of *Demarziella* Balthasar, with a transfer from Aphodiinae to Scarabaeinae (Coleoptera: Scarabaeidae). *Aust. J. Zool.* 34:449–61.

Matthiessen, J. N. 1982. The role of seasonal changes in cattle dung in the population dynamics of the bush fly in south-western Australia. *Proc. 3rd Aust. Conf. Grassl. Invert. Ecol., Adelaide,* pp. 221–26.

———. 1983. The seasonal distribution and characteristics of bush fly *Musca vetustissima* Walker populations in south-western Australia. *Aust. J. Ecol.* 8:383–94.

Matthiessen, J. N., and L. Hayles. 1983. Seasonal changes in characteristics of cattle dung as a resource for an insect in southwestern Australia. *Aust. J. Ecol.* 8:9–16.

Matthiessen, J. N., G. P. Hall, and V. H. Chewings. 1986. Seasonal abundance of *Musca vetustissima* Walker and other cattle dung fauna in central Australia. *J. Aust. Entomol. Soc.* 25:141–47.

May, R. M. 1973. *Complexity and Stability in Model Ecosystems.* Princeton Univ. Press, Princeton, N.J..

———. 1977. Thresholds and breakpoints in ecosystems with a multiplicity of stable states. *Nature* 269:471–77.

———. 1978. Host-parasitoid systems in patchy environments: A phenomenological model. *J. Anim. Ecol.* 47:833–43.

Mayr, E., and J. M. Diamond. 1976. Birds on islands in the sky: Origin of the montane avifauna of northern Melanesia. *Proc. Natl. Acad. Sci. USA* 73:1765–69.

Merritt, R. W. 1974. "The Species Diversity and Abundance of Insects Inhabiting Cattle Droppings." Ph.D. diss., Univ. of California, Berkeley.

Merritt, R. W., and J. R. Anderson. 1977. The effects of different pasture and rangeland ecosystems on the animal dynamics of insects in cattle droppings. *Hilgardia* 45:31–71.

Merz, G. 1982. "Zur Oekologie des Waldelefanten (Loxodonta africana cyclotis Matschie, 1900)." Ph.D. diss., Heidelberg Univ., FRG.

Mesa, M. M. 1985. "Contribució al coneixement dels Escarabeids de Catalunya. Estudi especial dels gèneres Aphodius Ill. i Onthophagus Latr." Ph.D. diss., Univ. of Barcelona, Spain.

Miller, A. 1954. Dung beetles (Coleoptera, Scarabaeidae) and other insects in relation to human feces in a hookworm area of southern Georgia. *Amer. J. Trop. Med. Hygiene* 3:372–78.

———. 1961. The mouthparts and digestive tract of adult dung beetles (Coleoptera: Scarabaeidae), with reference to the ingestion of helminth eggs. *J. Parasitol.* 47:735–44.

Miller, S. E. 1983. Late Quaternary insects of Rancho La Brea and McKittrick, California. *Quat. Res.* 20:90–104.

Miller, S. E., R. D. Gordon, and H. F. Howden. 1981. Re-evaluation of Pleistocene scarab beetles from Rancho La Brea, California (Coleoptera: Scarabaeidae). *Proc. Entomol. Soc. Wash.* 83:625–30.

Mittal, I. C. 1981a. Distributional patterns in Scarabaeidae (Coleoptera): A study on northwest India. *Entomol. Blätter* 77:75–85.

———. 1981b. Scarabaeids of Haryana and surrounding areas. *Bull. Entomol.* 22:35–40.

———. 1986. Dung beetles attracted to human faeces. *Entomol. Blätter* 82:55–64.

Mohr, C. O. 1943. Cattle droppings as ecological units. *Ecol. Monogr.* 13:275–309.

Monnier, Y. 1968. Les effets du feu de brousse sur une savane préforestière de Côte d'Ivoire. *Études Éburnéennes* 9:1–260.

Monod, T. 1957. Les grandes divisions chorologiques de l'Afrique. *Publ. Cons. Sci. Afr. S. Sahara* 24:1–147.

Monteith, G. B., and R. I. Storey. 1981. The biology of *Cephalodesmius*, a genus of dung beetles which synthesizes "dung" from plant material (Coleoptera: Scarabaeidae: Scarabaeinae). *Mem. Queensland Mus.* 20:253–77.

Montgomery, G. G. 1985. Impact of vermiliguas (*Cyclopes*, Tamandua: Xenarthra = Edentata) on arboreal ant populations. In G. G. Montgomery, ed., *The Evolution and Ecology of Armadillos, Sloths, and Vermilinguas*, pp. 351–63. Smithsonian Institution Press, Washington, D.C.

Montgomery, G. G., and M. E. Sunquist. 1975. Impact of sloths on neotropical forest energy flow and nutrient cycling. In F. B. Golley and E. Medina, eds., *Tropical Ecological Systems: Trends in Terrestrial and Aquatic Research*, pp. 69–98. Springer-Verlag, New York.

Moon, R. D., E. C. Loomis, and J. R. Anderson. 1980. Influence of two species of dung beetles on larvae of face fly. *Env. Entomol.* 9:607–12.

Morgan, A. 1972. The fossil occurrence of *Helophorus arcticus* Brown (Coleoptera, Hydrophilidae) in Pleistocene deposits of the Scarborough Bluffs, Ontario. *Can. J. Zool.* 50:555–58.

Morley, R. J., and J. R. Flenley. 1987. Late Cenozoic vegetational and environmental changes in the Malay Archipelago. In T. C. Whitmore, ed., *Biogeographical Evolution of the Malay Archipelago*, pp. 50–59. Clarendon Press, Oxford.

Morón, M. A. 1979. Fauna de coleópteros lamelicornios de la estación de biología tropical, "Los Tuxtlas," Veracruz, UNAM. México. *An. Inst. Biol. Univ. Nal. Autón. México* 50:375–454.

———. 1987. The necrophagous Scarabaeinae beetles (Coleoptera: Scarabaeidae) from a coffee plantation in Chiapas, Mexico: Habits and phenology. *Coleopt. Bull.* 41:225–32.

Morón, M. A., and J. A. López-Méndez. 1985. Análisis de la entomofauna necrófila de un cafetal en el Soconusco, Chiapas, México. *Fol. Entomol. Mex.* 63:47–59.

Morón, M. A., and R. A. Terrón. 1984. Distribución altitudinal y estacional de los insectos necrófilos en la Sierra Norte de Hidalgo, México. *Acta Zool. Mex. (N.S.)* 3:1–47.

Morón, M. A., and S. Zaragoza. 1976. Coleópteros Melolonthidae y Scarabaeidae de Villa de Allende, Estado de México. *An. Inst. Biol. Univ. Nal. Autón. México* 47:83–118.

Morón, M. A., J. F. Camal, and O. Canul. 1986. Análisis de la entomofauna necrófila del área norte de la reserva de la biosfera "Sian Ka'an," Quintana Roo, México. *Fol. Entomol. Mex.* 69:83–98.

Morón, M. A., F. J. Villalobos, and C. Deloya. C. 1985. Fauna de coleópteros lamelicornios de Boca del Chajul, Chiapas, México. *Fol. Entomol. Mex.* 66:57–118.

Morton, J. K. 1972. Phytogeography of the West African mountains. In D. H. Valentine, ed., *Taxonomy, Phytogeography and Evolution*, pp. 221–36. Academic Press, New York.

Mostert, L. E., and C. H. Scholtz. 1986. Systematics of the subtribe Scarabaeina (Coleoptera: Scarabaeidae). *Entomol. Mem. Dep. Agric. Wat. Supply, Reb. S. Afr.* 65:1–25.

Murdoch, W. W., and A. Oaten. 1975. Predation and population stability. *Adv. Ecol. Res.* 9:1–131.

Musser, G. G. 1987. The mammals of Sulawesi. In T. C. Whitmore, ed., *Biogeographical Evolution of the Malay Archipelago*, pp. 71–93. Oxford, Clarendon Press.

Myers, N. 1984. *The Primary Source: Tropical Forests and Our Future.* Norton, New York.

Nagel, P. 1986. Die Methode der Arealsystemanalyse als Beitrag zur Rekonstruktion der Landschaftsgenese im tropischen Africa. *Geomethodica* 11:145–76.

Nagy, K. A., and K. Milton. 1979. Energy metabolism and food consumption by wild howler monkeys (*Alouatta palliata*). *Ecology* 60:475–80.

Nealis, V. G. 1977. Habitat associations and community analysis of south Texas dung beetles (Coleoptera: Scarabaeinae). *Can. J. Zool.* 55:138–47.

Neuhaus, W. 1983. Die Ausbreitung von Pheromonen im Wind. *Zool. Jahrb. (Allg. Zool. Physiol. Tiere)* 87:443–45.

Nibaruta, G., M. Desière, and R. Debaere. 1980. Étude comparée de la composition chimique des excréments de quelques grands mammifères herbivores africains. *Acta Zool. Pathol. Antverp.* 75:59–70.

Nikolajev, G. V. 1966. *Lethrus jacobsoni* Sem. et Medv. (Coleoptera, Scarabaeidae), pest of vine-shoots in south Kazakhstan. *Entomol. Obozr.* 45:814–18.

Nikolajev, G. V., and Z. Puntsagdulam. 1984. *Lamellicorns (Coleoptera, Scarabaeoidea) of the Mongolian People's Republic*, 9:90–294. Nacekomie Mongolii, Nayka edit., Leningrad.

Nowak, R. M., and J. L. Paradiso. 1983. *Walker's Mammals of the World.* 4th ed. Johns Hopkins Univ. Press, Baltimore, Md.

Nummelin, M., and I. Hanski. 1989. Dung beetles of the Kibale Forest, Uganda: Comparison between virgin and managed forests. *J. Trop. Ecol.* 5:349–52.

Oberholzer, J. J. 1958. A description of the third stage larva of *Onitis caffer* Bohem. (Copridae: Col.) with notes on its biology. *S. Afr. J. Agric. Sci.* 1:415–22.

Olechowicz, E. 1974. Analysis of a sheep pasture ecosystem in the Pieniny mountains (the Carpathians). X. Sheep dung and the fauna colonizing it. *Ekol. Polska* 22:589–616.

Oppenheimer, J. R. 1977. Ecology of dung beetles (Scarabaeidae: Coprinae) in two villages of West Bengal. *Rec. Zool. Surv. India* 72:389–98.

Oppenheimer, J. R., and J. Begum. 1978. Ecology of some dung beetles (Scarabaeidae and Aphodiidae) in Dacca district. *Bangladesh J. Zool.* 6:23–29.

Ord, J. K., G. P. Patil, and C. Taillie, eds., 1980. *Statistical Distributions in Ecological Work.* Statistical Ecology Series 4, ICPH, Maryland.

Osberg, D. C. 1988. "The Influence of Soil Type on the Potential Distribution of Two Species of Scarabaeine Dung Beetle." M.Sc. diss., Univ. of Witwatersrand, Johannesburg, South Africa.

Otronen, M. 1984a. Male contests for territories and females in the fly *Dryomyza anilis. Anim. Beh.* 32:891–98.

———. 1984b. The effect of difference in body size on the male territorial system in the fly *Dryomyza anilis. Anim. Beh.* 32:882–90.

———. 1988. Intra- and intersexual interactions at breeding burrows in the horned beetle, *Coprophanaeus ensifer. Anim. Beh.* 36:741–48.

Otronen, M., and I. Hanski. 1983. Movement patterns in *Sphaeridium*: Differences between species, sexes, and feeding and breeding individuals. *J. Anim. Ecol.* 52:663–80.

Owen-Smith, R. N. 1987. Pleistocene extinctions: The pivotal role of megaherbivores. *Paleobiology* 13:351.

————. 1988. *Megaherbivores*. Cambridge Univ. Press, Cambridge, England.

Paarmann, W., and N. E. Stork. 1987. Seasonality of ground beetles (Coleoptera: Carabidae) in the rain forests of N. Sulawesi (Indonesia). *Insect Sci. Applic.* 8:483–87.

Palestrini, C. 1985. Problemi filogenetici e biogeografici del popolamento australiano di Onthophagini (Coleoptera, Scarabaeidae). *Atti XIV Congr. Naz. Ital. Entomol.*, pp. 249–53.

Palestrini, C., and M. Zunino. 1987. The biological meaning of sounds produced by nesting and subsocial Lamellicorn beetles. In *Ethological Perspectives in Social and Presocial Arthropods*. *Publ. Ist. Entomol. Univ. Pavia* 36:81–85.

Palestrini, C., R. Piazza, and M. Zunino. 1988. Segnali sonori in tre specie di Geotrupini (Coleoptera, Scarabaeoidea, Geotrupini). *Boll. Soc. Ital. Genova* 119:139–51.

Palestrini, C., A. Simonis, and M. Zunino. 1987. Modelli di distribuzione dell'entomofauna della Zona di Transizione Cinese, analisi di esempli e ipotesi sulle sue origini. *Biogeographia* 11:195–209.

Palmer, W. A., and D. E. Bay. 1984. A computer simulation model for describing the relative abundance of the horn fly, *Haematobia irritans irritans* L., under various ecological and pest management regimes. *Protec. Ecol.* 7:27–35.

Palmer, W. A., D. E. Bay, and P.J.H. Sharpe. 1981. Influence of temperature on the development and survival of the immature stages of the horn fly, *Haematobia irritans irritans* L. *Protec. Ecol.* 3:299–309.

Papp, L. 1976. Ecological and zoogeographical data on flies developing in excrement droppings (Diptera). *Acta Zool. Acad. Sci. Hung.* 22:119–38.

Parker, G. A. 1970. The reproductive behaviour and the nature of sexual selection in *Scatophaga stercoraria* L. (Diptera: Scatophagidae). II. The fertilization rate and the spatial and temporal relationships of each sex around the site of mating and oviposition. *J. Anim. Ecol.* 39:205–28.

———— 1974. The reproductive behaviour and the nature of sexual selection in *Scatophaga stercoraria* L. (Diptera: Scatophagidae). IX. Spatial distribution of fertilization rates and evolution of male search strategy within the reproductive area. *Evolution* 28:93–108.

————. 1978. Searching for mates. In J. R. Krebs, and N. B. Davies, eds., *Behavioural Ecology*, pp. 214–44. Blackwell, Oxford.

Parker, G. A., and R. A. Stuart. 1976. Animal behaviour as a strategy optimizer: Evolution of resource assessment strategies and optimal emigration thresholds. *Amer. Nat.* 110:1055–76.

Paschalidis, K. M. 1974. "The Genus *Sisyphus* Latr. (Coleoptera, Scarabaeidae) in Southern Africa." M.Sc. diss., Rhodes Univ., Grahamstown, South Africa.

Patil, G. P., E. C. Pielou, and W. E. Waters. 1971. Spatial patterns and statistical distributions. *Statistical Ecology*, vol. 1. Penn. State Univ. Press, Pennsylvania.

Patterson, B., and R. Pascual. 1968. Evolution of mammals on southern continents. V. The fossil mammal fauna of South America. *Q. Rev. Biol.* 43:409–51.

────. 1972. The fossil mammal fauna of South America. In A. Keast, F. C. Erk, and B. Glass, eds., *Evolution, Mammals and Southern Continents*, pp. 247–309. State Univ. of New York Press, Albany.

Patterson, R. S., and D. A. Rutz, eds., 1986. Biological control of muscoid flies. *Misc. Publ. Entomol. Soc. Amer.* 61:1–174.

Paulian, R. 1943. *Les Coléoptères: Formes, moeurs, rôles*. Payot, Paris.

────. 1945. *Coléoptères Scarabéides de l'Indochine (Faune de l'empire français, 3)*. Larose, Paris.

────. 1960. Onthophagini. In R. Paulian and E. Lebis, eds., *Insectes Coléoptères Scarabaeidae, Scarabaeini, Onthophagini et Helictopleurina*. Faune de Madagascar, 11. Tananarive, Madagascar.

────. 1961. *La zoogéographie de Madagascar et des îles voisines*. Faune de Madagascar, 13. Tananarive, Madagascar.

────. 1984. I. Systématique. In R. Paulian and D. Pluot-Sigwalt, Les Canthonines de Nouvelles Calédonie (Coleoptera, Scarabaeidae). Etude systématique et biogéographique. *Bull. Mus. Natl. Hist. Nat., Paris, 4e sér., sect. A.* 6:1091–1133.

────. 1985. Les Coléoptères Scarabaeidae Canthonines de Nouvelle-Guinee. *Ann. Soc. Entomol. Fr. (N.S.)* 21:219–38.

────. 1986. Un nouveau genre de Canthonine de Nouvelle Calédonie (Coleoptera, Scarabaeidae). *Rev. Fr. Entomol. (N. S.)* 8:5–8.

────. 1987a. *Onthobium* nouveaux ou peu connus de Nouvelle Calédonie (Col. Scarab.). *Bull. Soc. Entomol. Fr.* 91:229–36.

────. 1987b. Canthonina néo-calédoniens nouveaux ou peu connus (Coleoptera, Scarabaeidae). *Nouv. Rev. Entomol. (N.S.)* 4:247–51.

Paulian, R., and J. Baraud. 1982. *Lucanoidea et Scarabaeoidea. Faune des Coléoptères de France*. Encycl. Entomol. 43. Editions Lechevalier, Paris.

Paulian, R., J. P. Lumaret, and G. B. Monteith. 1984. La larve du genre *Cephalodesmius* Westwood (Col. Scarabaeidae). *Bull. Soc. Entomol. Fr.* 88:635–48.

Peck, S. B., and A. E. Davies. 1980. Collecting small beetles with large-area "window" traps. *Coleopt. Bull.* 34:237–39.

Peck, S. B., and A. Forsyth. 1982. Composition, structure, and competitive behaviour in a guild of Ecuadorian rain forest dung beetles (Coleoptera, Scarabaeidae). *Can. J. Zool.* 60:1624–34.

Peck, S. B., and H. F. Howden. 1984. Response of a dung beetle guild to different sizes of dung bait in a Panamanian rainforest. *Biotropica* 16:235–38.

Pellew, R.A.P. 1983. The impact of elephant, giraffe and fire upon the *Acacia tortilis* woodlands of the Serengeti. *Afr. J. Ecol.* 21:41–74.

Pennycuick, C. J. 1982. *Animal Flight*. Studies in Biology, no. 33. Edward Arnold, London.

Pereira, F. S., and G. Halffter. 1961. Nuevos datos sobre Lamellicornia Mexicanos con algunas observaciones sobre saprofagia. *Rev. Brasil. Entomol.* 10:53–66.

Pereira, F. S., and A. Martínez. 1956. Os gêneros de Canthonini Americanos. *Rev. Brasil. Entomol.* 6:91–192.

Pereira, F. S., M. A. Vulcano, and A. Martínez. 1960. O gênero *Bdelyrus* Harold, 1869. *Actas y Trabajos del Primer Congr. Sudamericano de Zool.* 3:155–64.

Perez-Inigo, C. 1971. Diptera and Coleoptera pseudoparasites of the human intestine. *Graellsia* 27:161–76.

Peters, R. H. 1983. *The Ecological Implications of Body Size*. Cambridge Univ. Press, Cambridge, England.

Petrovitz, R. 1959. Scarabaeidae (Col.). Contribution à l'étude de la faune d'Afghanistan 12. *Kungl. Fysiogr. Sällsk. i Lund Förhandl.* 29:103–11.

———. 1961. Scarabaeidae (Col.) II. Contribution à l'étude de la faune d'Afghanistan, 59. *Kungl. Fysiogr. Sällsk. i Lund Förhandl.* 31:31–45.

———. 1965. Österreichische entomologische Expeditionen nach Persien und Afghanistan. Beiträge zur Coleopterologie. Teil II. Lamellicornia. *Ann. Naturhistor. Mus. Wien* 68:671–94.

———. 1967. Österreichische entomologische Expeditionen nach Persien und Afghanistan. Beiträge zur Coleopterologie. Teil XI. Lamellicornia. *Ann. Naturhistor. Mus. Wien* 70:479–90.

Peyre de Fabrègues, B. 1980. *Végétation*. In *Atlas du Niger*. Editions Jeune Afrique, Paris.

Pianka, E. R. 1974. *Evolutionary Ecology*. Harper & Row, New York.

———. 1975. Niche relations of desert lizards. In M. L. Cody and J. M. Diamond, eds., *Ecology and Evolution of Communities*, pp. 292–314. Harvard Univ. Press, Cambridge, Mass.

Pielou, E. C. 1975. *Ecological Diversity*. Wiley, New York.

———. 1977. *Mathematical Ecology*. Wiley, New York.

Pierotti, H. 1977. Contributo alla conoscenza degli *Aphodius* della Calabria e del Pollino (Coleoptera Aphodiidae). *Boll. Soc. Entomol. Ital., Genova* 109:173–98.

Pires, J. M., and G. T. Prance. 1977. The Amazon forest: A natural heritage to be preserved. In G. T. Prance and T. S. Eliot, eds., *Extinction Is Forever*, pp. 158–94. New York Botanical Garden, New York.

———. 1985. The vegetation types of the Brazilian Amazon. In G. T. Prance and T. E. Lovejoy, eds., *Key Environments: Amazonia*, pp. 109–45. Pergamon Press, New York.

Pirone, D. 1974. "Ecology of Necrophilous and Carpophilous Coleoptera in a Southern New York Woodland." Ph. D. diss., Fordham Univ., New York.

Pittino, R. 1980. Aphodiidae interessanti della regione sardo-corsa (Coleoptera, Scarabaeoidea). *Boll. Soc. Entomol. Ital., Genova* 112:127–34.

Pluot, D. 1979. Évolution régressive des ovarioles chez les Coléoptères Scarabaeinae. *Ann. Soc. Entomol. Fr. (N.S.)* 15:575–88.

Pluot-Sigwalt, D. 1982. Diversité et dimorphisme sexuel de glandes tégumentaires abdominales chez les Coléoptères Scarabaeidae. *C.R. Acad. Sci. Paris III* 294:945–48.

———. 1984. Les glandes tégumentaires des Coléoptères Scarabaeidae: Répartition des glandes sternales et pygidiales dans la famille. *Bull. Soc. Entomol. Fr.* 88:597–602.

———. 1986. Les glandes tégumentaires des Coléoptères Scarabaeidae: Structure et diversité des canalicules. *Ann. Soc. Entomol. Fr. (N.S.)* 22:163–82.

———. 1988. Le système des glandes tégumentaires des Scarabaeidae rouleurs, particulièrement chez deux espèces de *Canthon* (Coleoptera). *Fol. Entomol. Mex.* 74:79–108.

Poorbaugh, J. H., J. R. Anderson, and J. F. Burger. 1968. The insect inhabitants of undisturbed cattle droppings in northern California. *Calif. Vector Views* 15:17–35.

Popovici-Baznosanu, A. 1932. Beiträge zur Kenntnis des Rebschneiders *Lethrus apterus* Laxm. *Zool. Anz.* 100:3–13.

Porter, D. A. 1939. Some new intermediate hosts of the swine stomach worms, *Ascarops strongylina* and *Physocephalus sexalatus. Proc. Helm. Soc. Wash.* 6:79–80.

Prance, G. T. 1973. Phytogeographic support for the theory of Pleistocene forest refuges in the Amazon Basin, based on evidence from distributional patterns in Caryocaraceae, Chrysobalanaceae, Dichapetalaceae and Lecythidaceae. *Acta Amazonica* 3:5–28.

————. 1985. The changing forests. In G. T. Prance and T. E. Lovejoy, eds., *Key Environments: Amazonia*, pp. 146–65. Pergamon Press, New York.

Prasse, J. 1958. Die Kämpfe der Pillenwälzer *Sisyphus schaefferi* L. und *Gymnopleurus geoffroyi* Fuessl. (Col. Scarab.). *Wiss. Zeitschr. Martin-Luther-Univ. Halle-Witt.* 7:89–92.

Price, P. W. 1980. *Evolutionary Biology of Parasites*. Princton Univ. Press, Princeton, N.J.

Pulliam, H. R. 1988. Sources, sinks, and population regulation. *Amer. Nat.* 132:652–61.

Pyhälahti, A. 1934. "Lehmänlannassa esiintyvät kovakuoriaiset." M.Sc. diss., Univ. of Turku, Finland.

Quinn, J. F., and A. Hastings. 1987. Extinction in subdivided habitats. *Cons. Biol.* 1:198–208.

Rainio, M. 1966. Abundance and phenology of some coprophagous beetles in different kinds of dung. *Ann. Zool. Fenn.* 3:88–98.

Ratcliffe, B. C. 1980. New species of Coprini (Coleoptera: Scarabaeidae: Scarabaeinae) taken from the pelage of Three Toed Sloths (*Bradypus tridactylus* L.) (Edentata: Bradypodidae) in central Amazonia with a brief commentary on scarab-sloth relationships. *Coleopt. Bull.* 34:337–50.

————. 1983. *Trox hamatus* Robinson (Trogidae) using a *Canthon* (Scarabaeinae) brood ball and a new record of North American *Trox* (Coleoptera: Scarabaeidae). *Trans. Nebraska Acad. Sci.* 11:53–55.

Raven, P. H., and D. I. Axelrod. 1975. History of the flora and fauna of Latin America. *Amer. Sci.* 63:420–29.

Read, J. D. 1976. "A Study of the Dung and Carrion Beetles (Coleoptera: Scarabaeidae) of Papua New Guinea, and the Influence of Elevation on Species Composition." B.Sc. thesis, Carleton Univ., Ottawa, Canada.

Reinig, W. F. 1931. Entomologische Ergebnisse der deutsch-russischen Alai-Pamir-Expedition 1928 (II): Coleoptera II. Tenebrionidae. *Mitt. Münch. Entomol. Ges.* 21:865–912.

Rembialkowska, E. 1982. Energy balance of the developmental period of *Geotrupes stercorosus* Scriba (Scarabaeidae, Coleoptera). *Ekol. Polska* 30:393–427.

Reyes-Castillo, P., and G. Halffter. 1984. La structure sociale chez les Passalidae (Col.). *Bull. Soc. Entomol. Fr.* 88:619–35.

Ricou, G. E. 1984. "Recyclage des Fèces et Faune Associée dans les Écosystèmes: Pâturages d'Altitude et Garrigues." Ph.D. diss., Univ. of Rennes, France.

Ridsdill-Smith, T. J. 1981. Some effects of three species of dung beetle (Coleoptera:

Scarabaeidae) in south-western Australia on the survival of the bush fly *Musca vetustissima* Walker (Diptera: Muscidae) in dung pads. *Bull. Entomol. Res.* 71:425–33.

———. 1986. The effect of seasonal changes in cattle dung on egg production by two species of dung beetles (Coleoptera: Scarabaeidae) in south-western Australia. *Bull. Entomol. Res.* 76:63–68.

———. 1988. The Scarabaeine dung beetles. In R. L. Specht, ed., *Mediterranean-type Ecosystems: A Data Source Book*, pp. 216–18. Kluwer, Dordrecht.

———. 1990. Competition in dung breeding insects. In W. J. Bailey and T. J. Ridsdill-Smith, eds., *Reproductive Behaviour in Insects—Individuals and Populations*. Chapman & Hall, London.

Ridsdill-Smith, T. J., and G. P. Hall. 1984a. Beetles and mites attracted to fresh cattle dung in south-western Australian pastures. *CSIRO Division of Entomology Report No.* 34:1–29.

———. 1984b. Seasonal patterns of adult dung beetle activity in south-western Australia. In B. Dell, ed., MEDECOS IV. *Proc. 4th Int. Conf. Med. Ecosystems*, Perth, pp. 139–40.

Ridsdill-Smith, T. J., and A. A. Kirk. 1985. Selecting dung beetles (Scarabaeidae) from Spain for bush fly control in south western Australia. *Entomophaga* 30:217–23.

Ridsdill-Smith, T. J., and J. N. Matthiessen. 1984. Field assessments of the impact of night-flying dung beetles (Coleoptera: Scarabaeidae) on the bush fly, *Musca vetustissima* Walker (Diptera: Muscidae), in south-western Australia. *Bull. Entomol. Res.* 74:191–94.

———. 1988. Bush fly, *Musca vetustissima* Walker (Diptera: Muscidae) control in relation to seasonal abundance of scarabaeine dung beetles (Coleoptera: Scarabaeidae) in south-western Australia. *Bull. Entomol. Res.* 78:633–39.

Ridsdill-Smith, T. J., G. P. Hall, and G. F. Craig. 1982. Effect of population density on reproduction and dung dispersal by the dung beetle *Onthophagus binodis* in the laboratory. *Entomol. Exper. Appl.* 32:80–85.

Ridsdill-Smith, T. J., G. P. Hall, and T. A. Weir. 1989. A field guide to the dung beetles (Scarabaeidae: Scarabaeinae and Aphodiinae) in pastures in south-western Australia. *J. Roy. Soc. WA.* 71:49–58.

Ridsdill-Smith, T. J., L. Hayles, and M. L. Palmer. 1986. Competition between the bush fly and dung beetles in dung of differing characteristics. *Entomol. Exp. Appl.* 41:83–90.

Ridsdill-Smith, T. J., T. A. Weir, and S. B. Peck. 1983. Dung beetles (Scarabaeidae: Scarabaeinae and Aphodiinae) active in forest habitats in south-western Australia during winter. *J. Aust. Entomol. Soc.* 22:307–309.

Ritcher, P. O. 1966. White grubs and their allies. A study of North American Scarabaeoid larvae. *Oregon State Monogr. Stud. Entomol.* 4:1–219.

Ritcher, P. O., and H. E. Morrison. 1955. *Aphodius pardalis* Lec. A new turf pest. *J. Econ. Entomol.* 48:476.

Ritchie, J. C., and C. V. Haynes. 1987. Holocene vegetation zonation in the eastern Sahara. *Nature* 330:645–47.

Robinson, M. 1948. Remarks on a few Scarabaeidae (Coleoptera). *Entomol. News* 59:175–77.

Robinson, M. H., and B. Robinson. 1970. Prey caught by a sample population of the spider *Argiope argentata* (Araneae: Araneidae) in Panama: A year's census data. *Zool. J. Linn. Soc.* 49:345–57.

Rojewski, C. 1983. Observations on the nesting behaviour of *Aphodius erraticus* (L.) (Coleoptera, Scarabaeidae). *Bull. Entomol. Pol.* 53:271–79.

Root, R. B. 1967. The niche exploitation pattern of the blue-grey gnat catcher. *Ecol. Monogr.* 37:317–50.

Rosenzweig, M. L., and Z. Abramsky. 1986. Centrifugal community organization. *Oikos* 46:339–54.

Roth, J. P., G. T. Fincher, and J. W. Summerlin. 1983. Competition and predation as mortality factors of the horn fly, *Haematobia irritans* (L.) (Diptera: Muscidae) in a Central Texas pasture habitat. *Env. Entomol.* 12:106–109.

Roughgarden, J. 1979. *Theory of Population Genetics and Evolutionary Ecology: An Introduction.* Macmillan, New York.

Roughgarden, J., S. D. Gaines, and S. W. Pacala. 1987. Supply side ecology: The role of physical transport processes. In J.H.R. Gee and P. S. Giller, eds., *Organization of Communities. Past and Present*, pp. 491–518. Blackwell, Oxford.

Rougon, C., and D. Rougon. 1980. Contribution à la biologie des Coléoptères coprophages en région sahélienne. Étude du développement d'*Onthophagus gazella* F. (Coleoptera: Scarabaeidae). *Rev. Ecol. Biol. Sol.* 17:379–92.

Rougon, D. 1987. "Coléoptères Coprophiles en zone sahélienne: Étude biocénotique, comportement nidificateur, intervention dans le recyclage de la matière organique du sol." Sc.D. thesis, Univ. of Orléans, France.

Rougon, D., and C. Rougon. 1980. Le cleptoparasitisme en zone sahélienne: Phénomène adaptatif d'insectes Coléoptères coprophages Scarabaeidae aux climats arides et semi-arides. *C.R. Acad. Sci. Paris D* 291:417–19.

——. 1982a. Le comportement nidificateur d'*Onitis alexis* Klug en région sahélienne (Col. Scarabaeidae Onitini). *Bull. Soc. Entomol. Fr.* 87:15–19.

——. 1982b. Le comportement nidificateur des Coléoptères Scarabaeinae Oniticellini en zone sahélienne. *Bull. Soc. Entomol. Fr.* 87:272–79.

——. 1983. Nidification des Scarabaeidae et cleptoparasitisme des Aphodiidae en zone sahélienne (Niger). Leur rôle dans la fertilisation des sols sableux. *Bull. Soc. Entomol. Fr.* 88:496–513.

Rummel, J., and J. Roughgarden. 1985. A theory of faunal buildup for competition communities. *Evolution* 39:1009–33.

Rzedowski, J. 1978. *Vegetación de México.* Editorial Limusa, México, D.F.

Saint-Vil, J. 1977. Les climats du Gabon. *Ann. Univ. Natl. Gabon* 1:101–25.

Sale, P. F. 1977. Maintenance of high diversity in coral reef fish communities. *Amer. Nat.* 111:337–59.

——. 1979. Recruitment, loss, and coexistence in a guild of territorial coral reef fishes. *Oecologia (Berl.)* 42:159–77.

Sands, P., and R. D. Hughes. 1976. A simulation model of seasonal changes in the value of cattle dung as a food resource for an insect. *Agric. Meteor.* 17:161–83.

Sato, H., and M. Imamori. 1986a. Nidification of an African ball-rolling scarab, *Scarabaeus platynotus* Bates (Coleoptera, Scarabaeidae). *Kontyu* 54:203–207.

——. 1986b. Production of two brood pears from one dung ball in an African ball-roller, *Scarabaeus aegyptiorum* (Coleoptera, Scarabaeidae). *Kontyu* 54:381–85.

Sato, H., and M. Imamori. 1987. Nesting behaviour of a subsocial African ball-roller *Kheper platynotus* (Coleoptera: Scarabaeidae). *Ecol. Entomol.* 12:415–25.

———. 1988. Further observations on the nesting behaviour of a subsocial ball-rolling scarab, *Kheper aegyptiorum. Kontyu* 56:873–78.

Schmidt-Nielsen, K. 1984. *Scaling: Why Is Animal Size So Important?* Cambridge Univ. Press, Cambridge, England.

Schnell, R. 1976. *Flore et végétation de l'Afrique tropicale.* Vol. 1. Gauthier-Villars, Paris.

Schoener, T. W. 1974. Resource partitioning in ecological communities. *Science* 185:27–39.

———. 1983. Field experiments on interspecific competition. *Amer. Nat.* 122:240–85.

———. 1984. Size differences among sympatric, bird-eating hawks: A worldwide survey. In D. R. Strong, D. Simberloff, L. G. Abele, and A. B. Thistle, eds., *Ecological Communities: Conceptual Issues and the Evidence*, pp. 254–81. Princeton Univ. Press, Princeton, N.J.

Schoenly, K. 1983. Arthropods associated with bovine and equine dung in an ungrazed Chihuahuan desert ecosystem. *Ann. Entom. Soc. Amer.* 76:790–96.

Scholtz, C., and H. F. Howden. 1987a. A revision of the African Canthonina (Coleoptera: Scarabaeidae: Scarabaeinae). *J. Entomol. Soc. S. Afr.* 50:75–119.

———. 1987b. A revision of the Southern African genus *Epirinus* Reiche (Coleoptera: Scarabaeidae: Scarabaeinae). *J. Entomol. Soc. S. Afr.* 50:121–54.

Scholtz, C. H., and W. M. de Villiers. 1983. *Dung Beetles.* Sigma Press, Pretoria, South Africa.

Scudo, F. M., and J. R. Ziegler. 1978. *The Golden Age of Theoretical Ecology: 1923–1940.* Lecture Notes in Biomath. 22. Springer-Verlag, Berlin.

Servant, M., and S. Servant-Vildary. 1980. L'environnement quaternaire du bassin du Tchad. In M.A.J. Williams, and H. Faure, eds., *The Sahara and the Nile*, pp. 133–62. Maisonneuve et Larose, Paris.

Seymour, J. 1980. Dung beetles get a little help from their friends. *Ecos, CSIRO Env. Res.* 26:20–25.

Shelly, T. E. 1985. Ecological comparisons of robber fly species (Diptera: Asilidae) coexisting in a Neotropical forest. *Oecologia (Berl.)* 67:57–70.

Shorrocks, B., and J. Rosewell. 1986. Guild size in drosophilids: A simulation model. *J. Anim. Ecol.* 55:527–42.

———. 1987. Spatial patchiness and community structure: Coexistence and guild size of drosophilids on ephemeral resources. In J.H.R. Gee and P. S. Giller, eds., *Organization of Communities: Past and Present*, pp. 29–51. Blackwell, Oxford.

———. 1988. Aggregation does prevent competitive exclusion: A response to Green. *Amer. Nat.* 131:765–71.

Shorrocks, B., J. Rosewell, K. Edwards, and W. Atkinson. 1984. Interspecific competition is not a major organizing force in many insect communities. *Nature* 310:310–12.

Shubeck, P. P. 1976. An alternative to pitfall traps in carrion beetle studies (Coleoptera). *Entomol. News* 87:5–6.

Simberloff, D. 1979. Rarefaction as a distribution-free method of expressing and estimating diversity. In J. F. Gressle, G. P. Patil, W. Smith, and C. Taillie, eds., *Ecological Diversity in Theory and Practice*, pp. 159–76. Statistical Ecology Series 6, ICPH, Maryland.

———. 1983. Sizes of coexisting species. In D. J. Futuyma and M. Slatkin, eds., *Coevolution*, pp. 404–30. Sinauer, Sunderland, Mass.

Simberloff, D., and W. Boecklen. 1981. Santa Rosalia reconsidered: Size ratios and competition. *Evolution* 35:1206–28.

Simonis, A. 1985. Un nuovo genere e tre nuove specie di *Drepanocerina* (Coleoptera, Scarabaeidae: Oniticellini). *Rev. Suisse Zool.* 92:93–104.

Simonis, A., and Y. Cambefort. 1984. Nouvelles observations sur *Drepanoplatynus gilleti* Boucomont (Coleoptera, Scarabaeoidea, Scarabaeidae). *Ann. Soc. Entomol. Fr. (N.S.)* 20:105–10.

Sinclair, A.R.E. 1983. The adaptations of African ungulates and their effects on community functions. In F. Bourlière, ed., *Tropical Savannas: Ecosystems of the World*, vol. 13, pp. 401–25. Elsevier, Amsterdam.

Singh, G., and E. A. Geissler. 1985. Late cenozoic history of vegetation, fire, lake levels and fire at Lake George, New South Whales, Australia. *Phil. Trans. Roy. Soc. Lond., Ser. B* 311:379–447.

Singh, G., A. P. Kershaw, and R. Clark. 1981. Quaternary vegetation and fire history in Australia. In A. M. Gill, R. H. Groves, and I. R. Noble, eds., *Fire and the Australian Biota*, pp. 23–54. Canberra, Australian Academy of Science.

Skidmore, P. 1985. *The Insects of the Cowdung Community*. Field Studies Council, Richmond Publ. Co., Surrey, England.

Smythe, N., W. E. Glanz, and E. G. Leigh, Jr. 1982. Population regulation in some terrestrial frugivores. In E. G. Leigh, Jr., A. S. Rand, and D. M. Windsor, eds., *The Ecology of a Tropical Forest*, pp. 227–38. Smithsonian Institution Press, Washington, D.C.

Southwood, T.R.E. 1976. *Ecological Methods*. Chapman & Hall, London.

———. 1977. Habitat, the templet for ecological strategies? *J. Anim. Ecol.* 46:337–65.

———. 1987. Tactics, strategies and templets. *Oikos* 52:3–18.

Southwood, T.R.E., V. C. Moran, and C.E.J. Kennedy. 1982. The richness, abundance and biomass of the arthropod communities on trees. *J. Anim. Ecol.* 51:635–49.

Springett, B. P. 1968. Aspects of the relationship between burying beetles, *Necrophorus* spp. and the mite, *Poecilochirus necrophori* Vitz. *J. Anim. Ecol.* 37:417–24.

Ståhls, G., M.E.B. Ribeiro, and I. Hanski. 1989. Fungivorous *Pegomya* flies: Spatial and temporal variation in a guild of competitors. *Ann. Zool. Fenn.* 26:103–12.

Stebnicka, Z. 1975. Neue Scarabaeidenarten (Coleoptera) aus Asien. *Bull. Acad. Pol. Sci. Sr. Sci. Biol. II* 23:185–89.

———. 1976. *Klucze do Oznaczania owadów Polski. XIX. Coleoptera.* 28a. Scarabaeidae. Polskie Towarzystwo Entomologiczne, no. 89, Panstwowe Wydawnictwo Naukowe, Warsaw.

———. 1985. A new genus and species of Aulonocnemidae from India with notes on comparative morphology (Coleoptera: Scarabaeidae). *Rev. Suisse Zool.* 92:649–58.

Stebnicka, Z. 1986. Revision of the Aphodiinae of the Nepal-Himalayas (Coleoptera: Scarabaeidae). *Stuttgarter Beitr. Naturk., Ser. A* 397:1–51.

Stephens, D. W., and J. R. Krebs. 1986. *Foraging Theory*. Princeton Univ. Press, Princeton, N.J..

Stevenson, B. G., and D. L. Dindal. 1985. Growth and development of *Aphodius* beetles (Scarabaeidae) in laboratory microcosms of cow dung. *Coleopt. Bull.* 39:215–20.

Stewart, T. B. 1967. Food preferences of coprophagous beetles with special reference to *Phanaeus* spp. *J. Georgia Entomol. Soc.* 2:69–77.

Stewart, T. B., and K. M. Kent. 1963. Beetles serving as intermediate hosts of swine nematodes in southern Georgia. *J. Parasit.* 49:158–59.

Stickler, P. D. 1979. "The ecology of the Scarabaeinae (Coleoptera) in the high veld of South Africa." Ph.D. diss., Rhodes Univ., Grahamstown, South Africa.

Storey, R. I. 1977. Six new species of *Onthophagus* Latreille (Coleoptera: Scarabaeidae) from Australia. *J. Aust. Entomol. Soc.* 16:313–20.

———. 1984. A new species of *Aptenocanthon* Matthews from north Queensland (Coleoptera: Scarabaeidae: Scarabaeinae). *Mem. Qd. Mus.* 21:387–90.

———. 1986. A new flightless species of *Aulacopris* White from north Queensland (Coleoptera: Scarabaeidae: Scarabaeinae). *Mem. Qd. Mus.* 22:197–203.

Storey, R. I., and T. A. Weir. 1988. New localities and biological notes for the genus *Onthophagus* Latreille (Coleoptera: Scarabaeidae) in Australia. *Aust. Entomol. Mag.* 15:17–24.

———. 1989. New species of *Onthophagus* Latreille (Coleoptera: Scarabaeidae) from Australia. *Invert. Taxon*, in press.

Stork, N. E. 1987. Guild structure of arthropods from Bornean rain forest trees. *Ecol. Entomol.* 12:69–80.

Strong, D. R., J. H. Lawton, and T.R.E. Southwood. 1984. *Insects on Plants*. Blackwell, Oxford.

Strong, D. R., D. Simberloff, L. G. Abele, and A. B. Thistle, eds., 1984. *Ecological Communities: Conceptual Issues and the Evidence*. Princeton Univ. Press, Princeton, N.J.

Stumpf, I.V.K. 1986a. Study of the scarab fauna in Mandirituba, Parana, Brazil. *Acta Biol. Par.* 15:125–53.

———. 1986b. Scarabs from Mandirituba, Parana, Brazil. *Acta Biol. Par.* 15:179–216.

Stumpf, I.V.K., E. Luz, and V. R. Tonin. 1986a. Biology of *Ateuchus apicatus* Harold, 1867. *Acta Biol. Par.* 15:63–85.

———. 1986b. Biology of *Ateuchus mutilatus* Harold, 1867. *Acta Biol. Par.* 15:155–77.

Sudhaus, W. 1981. Successions of nematodes in cow droppings. *Pedobiologia* 21:271–97.

Swan, D. C. 1934. A scarab beetle (*Aphodius tasmaniae* Hope) destructive to pastures in the southeast of South Australia. *J. Dept. Agr. South Austr.* 37:1149–56.

Tauber, M. J., C. A. Tauber, and S. Masaki. 1986. *Seasonal Adaptations of Insects*. Oxford Univ. Press, Oxford.

Taylor, L. R. 1961. Aggregation, variance and the mean. *Nature* 189:732–35.

————. 1986. Synoptic dynamics, migration and the Rothamsted Insect Survey. *J. Anim. Ecol.* 55:1–38.

Taylor, L. R., I. P. Woiwod, and J. N. Perry. 1978. The density-dependence of spatial behaviour and the rarity of randomness. *J. Anim. Ecol.* 47:383–406.

Terborgh, J. 1971. Distribution on environmental gradients: Theory and a preliminary interpretation of distributional patterns in the avifauna of the Cordillera Vilcabamba, Peru. *Ecology* 52:23–40.

————. 1977. Bird species diversity on an Andean elevational gradient. *Ecology* 58:1007–19.

————. 1985. The role of ecotones in the distribution of Andean birds. *Ecology* 66:1237–46.

Thérond, J. 1975. Catalogue des Coléoptères de la Camargue et du Gard. Part 1. *Mém. Soc. Sci. Nat. Nimes* 10:1–410.

————. 1980. Supplément au catalogue des Coléoptères de la Camargue et du Gard. Addenda et Corrigenda. II. *Bull. Soc. Sci. Nat. Nimes* 56:1–18.

Thomas, W. P. 1960. Notes on a preliminary investigation into the habits and life cycle of *Copris incertus* Say (Coleoptera: Scarabaeidae) in New Zealand. *N.Z. J. Sci.* 3:8–14.

Thòrarinsson, K. 1986. Population density and movement: A critique of delta-models. *Oikos* 46:70–81.

Toledo, V. M. 1976. "Los Cambios Climáticos del Pleistoceno y sus Efectos Sobre la Vegetación Natural Cálida y Húmeda de México." M.Sc. diss., Fac. Ciencias, UNAM, México.

Tribe, G. D. 1975. Pheromone release by dung beetles (Coleoptera: Scarabaeidae). *S. Afr. J. Sci.* 71:277–78.

————. 1976. "The Ecology and Ethology of Ball-rolling Dung Beetles (Coleoptera: Scarabaeidae)." M.Sc. thesis, Univ. of Natal, Pietermaritzburg, South Africa.

Trollope, W.S.W. 1982. Ecological effects of fire in South African savannas. In B. J. Huntley, and B. H. Walker, eds., *Ecology of Tropical Savannas*, pp. 292–306. Springer-Verlag, Berlin.

Tyndale-Biscoe, H. 1971. *Life of Marsupials*. Edward Arnold, Australia.

Tyndale-Biscoe, M. 1978. Physiological age-grading in females of the dung beetle *Euoniticellus intermedius* (Reiche) (Coleoptera: Scarabaeidae). *Bull. Entomol. Res.* 68:207–17.

————. 1983. Effects of ovarian condition on nesting behaviour in a brood-caring dung beetle *Copris diversus* Waterhouse (Coleoptera: Scarabaeidae). *Bull. Entomol. Res.* 73:45–53.

————. 1984. Adaptive significance of brood care of *Copris diversus* Waterhouse (Coleoptera: Scarabaeidae). *Bull. Entomol. Res.* 74:453–61.

————. 1985. "An Ecological Study of Two Dung Beetles (Coleoptera: Scarabaeidae) with Contrasting Phenologies." Ph.D. diss., James Cook Univ., Townesville, Australia.

————. 1988. The phenology of *Onitis alexis* (Coleoptera: Scarabaeidae) in the Araluen Valley: Survival in a marginal environment. *Aust. J. Ecol.* 13:431–43.

Tyndale-Biscoe, M., and R. D. Hughes. 1969. Changes in the female reproductive system as age indicators in the bushfly *Musca vetustissima* Wlk. *Bull. Entomol. Res.* 59:129–41.

Tyndale-Biscoe, M., M.M.H. Wallace, and J. M. Walker. 1981. An ecological study of an Australian dung beetle, *Onthophagus granulatus* Boheman (Coleoptera: Scarabaeidae) using physiological age grading techniques. *Bull. Entomol. Res.* 71:137–52.

UNESCO. 1979. *Tropical grazing land ecosystems*. Natural Resources Research 16, UNESCO/UNEP/FAO.

Vadon, J. 1947. Les Epilissiens de Madagascar (Coleoptera, Scarabaeidae, Canthoniini). II. Biologie. *Bull. Acad. Malgache, 1944–1945*, 26:173–74.

Valiela, I. 1969. An experimental study of the mortality factors of larval *Musca autumnalis* De Geer. *Ecol. Monogr.* 39:199–220.

———. 1974. Composition, food webs and population limitation in dung arthropod communities during invasion and succession. *Amer. Midl. Nat.* 92:370–85.

Van der Hammen, T. 1983. The palaeoecology and palaeogeography of savannas. In F. Bourlière, ed., *Tropical Savannas: Ecosystems of the World*, vol. 13, pp. 19–35. Elsevier, Amsterdam.

Vanzolini, P. E., and E. E. Williams. 1970. South American anoles: Geographic differentiation and evolution of the *Anolis chrysolepis* species group (Sauria, Iguanidae). *Arq. Zool. São Paulo* 19:1–298.

Vinson, J. 1946. On *Nesosisyphus*, a new genus of coprine beetles from Mauritius. *Proc. Roy. Entomol. Soc. London* 15:89–96.

———. 1951. Le cas des *Sisyphes* mauriciens (Insectes Coléoptères). *Proc. Roy. Soc. Arts Sci. Mauritius* 1:105–22.

Von Koenigswald, G.H.R. 1967. An upper Eocene mammal of the family Anthracotheriidae from the island of Timor. *Proc. Kon. Ned. Akad. Wetensch. (B)* 70:528–33.

Vuattoux, R. 1970. Observations sur l'évolution des strates arborée et arbustive dans la savane de Lamto (Côte d'Ivoire). *Ann. Univ. Abidjan, E.* 3:285–315.

———. 1976. Contribution à l'étude de l'évolution des strates arborée et arbustive dans la savane de Lamto (Côte d'Ivoire). *Ann. Univ. Abidjan, C.* 12:35–63.

Waage, J. K., and R. C. Best. 1985. Arthropod associates of sloths. In G. G. Montgomery, ed., *The Evolution and Ecology of Armadillos, Sloths, and Vermilinguas*, pp. 319–22. Smithsonian Institution Press, Washington, D.C.

Wallace, M.M.H., and E. Holm. 1983. Establishment of dispersal of the introduced predatory mite, *Macrocheles peregrinus* Krantz, in Australia. *J. Austr. Entomol. Soc.* 22:345–48.

Wallace, M.M.H., and M. Tyndale-Biscoe. 1983. Attempts to measure the influence of dung beetles (Coleoptera: Scarabaeidae) on the field mortality of the bush fly *Musca vetustissima* Walker (Diptera: Muscidae) in south-eastern Australia. *Bull. Entomol. Res.* 73:33–44.

Walter, H., and H. Lieth. 1964. *Klimadiagram-Weltatlas*, Part 2. Gustav Fischer, Jena.

Walter, P. 1977. Répartition des Scarabaeidae coprophages dans les diverses formations végétales du plateau Bateke, Zaïre. *Geo-Trop-Eco* 1:259–75.

———. 1978. "Recherches écologiques et biologiques sur les Scarabéides Coprophages d'une savane du Zaïre." Sc. D. thesis, Univ. of Montpellier, France.

————. 1980. Comportement de recherche et d'exploitation d'une masse stercorale chez quelques coprophages afro-tropicaux (Col. Scarabaeidae). *Ann. Soc. Entomol. Fr. (N.S.)* 16:307–23.

————. 1983. La part de la nécrophagie dans le régime alimentaire des Scarabéides coprophages afro-tropicaux. *Bull. Soc. Zool. Fr.* 108:397–402.

————. 1984a. Contribution à la connaissance des Scarabéides coprophages du Gabon (Col.). 2. Présence de populations dans la canopée de la forêt gabonaise. *Bull. Soc. Entomol. Fr.* 88:514–21.

————. 1984b. Contribution à la connaissance des Scarabéides coprophages du Gabon (Col.). 3. Données préliminaires sur la faune forestière de Bifoun. *Ann. Univ. Nat. Gabon* 5:15–20.

————. 1987. Contribution à la connaissance des Scarabéides coprophages du Gabon (Col.). 5. Nidification et morphologie larvaire de *Tiniocellus panthera* (Boucomont). *Ann. Soc. Entomol. Fr. (N.S.)* 23:309–14.

Waterhouse, D. F. 1974. The biological control of dung. *Sci. Amer.* 230:100–109.

Watson, H. K., and A. W. McDonald. 1983. Vegetation changes in the Hluhluwe-Umfolozi Game Reserve Complex from 1937 to 1975. *Bothalia* 14:265–69.

White, E. 1960. The natural history of some species of *Aphodius* (Col., Scarabaeidae) in the northern Pennines. *Entomol. Monthly Mag.* 66:25–30.

White, F. 1965. The savanna woodlands of the Zambezian and Sudanian domains. *Webbia* 19:651–81.

————. 1976. The vegetation map of Africa. The history of a completed project. *Boissiera* 24b:659–66.

————. 1983. *The Vegetation of Africa*. UNESCO, Paris.

Whitmore, T. C. 1975. *Tropical Rain Forests of the Far East*. Clarendon Press, Oxford.

————, ed., 1987. *Biogeographical Evolution of the Malay Archipelago*. Clarendon Press, Oxford.

Wille, A. 1973. Observations on the behavior of a tropical rain forest dung beetle, *Megathoposoma candezei* (Harold), Coleoptera: Scarabaeidae. *Rev. Biol. Trop.* 21:41–57.

Wille, A., E. Orozco, G. Fuentes, and E. Solís. 1974. Additional observations on the behavior of a tropical forest dung beetle, *Megathoposoma candezei* (Coleoptera: Scarabaeidae). *Rev. Biol. Trop.* 22:129–33.

Williams, C. B. 1964. *Patterns in the Balance of Nature*. Academic Press, London.

Williams, G. A. 1979. Scarabaeidae (Coleoptera) from the Harrington District of coastal northern New South Wales, with special reference to a littoral rainforest habitat. *Aust. Entomol. Mag.* 5:103–108.

Williams, G. A., and T. Williams. 1982. A survey of the Aphodiinae, Hybosorinae and Scarabaeinae (Coleoptera: Scarabaeidae) from small wet forests of coastal New South Wales. Part 1: Nowra to Newcastle. *Aust. Entomol. Mag.* 9:42–48.

————. 1983a. A survey of the Aphodiinae, Hybosorinae and Scarabaeinae (Coleoptera: Scarabaeidae) from small wet forests of coastal New South Wales. Part 2: Barrington Tops to the Comboyne Plateau. *Vict. Natur.* 100:25–30.

————. 1983b. A survey of the Aphodiinae, Hybosorinae and Scarabaeinae (Coleoptera: Scarabaeidae) from small wet forests of coastal New South Wales. Part 3: Buladelah to Taree. *Vict. Natur.* 100:98–105.

Williams, G. A., and T. Williams. 1983c. A survey of the Aphodiinae, Hybosorinae and Scarabaeinae (Coleoptera: Scarabaeidae) from small wet forests of coastal New South Wales. Part 4: Lansdowne State Forest. *Vict. Natur.* 100:146–54.

———. 1984. A survey of the Aphodiinae, Hybosorinae and Scarabaeinae (Coleoptera: Scarabaeidae) from small wet forests of coastal New South Wales. Part 5: Littoral rainforests from Myall Lakes to Crowdy Bay National Park. *Vict. Natur.* 101:127–35.

Wilson, D. S. 1975. The adequacy of body size as a niche difference. *Amer. Nat.* 109:769–84.

———. 1980. *The Natural Selection of Populations and Communities.* The Benjamin/ Cummings Publ. Co., Menlo Park, Calif.

———. 1982. Genetic polymorphism for carrier preference in a phoretic mite. *Ann. Entomol. Soc. Amer.* 75:293–96.

Wilson, D. S., W. G. Knollenberg, and J. Fudge. 1984. Species packing and temperature dependent competition amongst burying beetles (Silphidae: *Nicrophorus*). *Ecol. Entomol.* 9:205–16.

Wilson, J. W. 1932. Coleoptera and Diptera collected from a New Jersey sheep pasture. *J. N.J. Entomol. Soc.* 40:77–93.

Wolcott, G. N. 1922. Insect parasite introduction in Porto Rico. *J. Dept. Agric. Porto Rico* 6:5–20.

Wolda, H. 1978a. Fluctuations in abundance of tropical insects. *Amer. Nat.* 112:1017–45.

———. 1978b. Seasonal fluctuations in rainfall, food and abundance of tropical insects. *J. Anim. Ecol.* 47:369–81.

———. 1982. Seasonality of Homoptera on Barro Colorado Island. In E. G. Leigh, Jr., A. S. Rand, and D. M. Windsor, eds., *The Ecology of a Tropical Forest*, pp. 319–30. Smithsonian Institution Press, Washington, D.C.

———. 1983. Spatial and temporal variation in abundance in tropical animals. In S. L. Sutton, T. C. Whitmore, and A. C. Chadwick, eds., *Tropical Rain Forest: Ecology and Management*, pp. 93–105. Blackwell, Oxford.

———. 1988. Insect seasonality. *Ann. Rev. Ecol. Syst.* 19:1–18.

Wolda, H., and M. Estribi. 1985. Seasonal distribution of the large sloth beetle *Uroxys gorgon* Arrow (Scarabaeidae: Scarabaeinae) in light traps in Panama. In G. G. Montgomery, ed., *The Evolution and Ecology of Armadillos, Sloths, and Vermilinguas*, pp. 319–22. Smithsonian Institution Press, Washington, D.C.

Woodruff, R. E. 1973. *The Scarab Beetles of Florida.* Arthropods of Florida and Neighboring Land Areas. Vol. 8. Gainesville, Florida.

Wright, E. J., and P. Mueller. 1989. Laboratory studies of host finding, acceptance and suitability of the dung-breeding fly, *Haematobia thirouxi potans* (Muscidae: Diptera) by *Aleochara* sp. (Coleoptera: Staphylinidae). *Entomophaga* 34:61–71.

Wright, E. J., P. Mueller, and J. D. Kerr. 1989. Agents for biological control of novel hosts: Assessing an aleocharine parasitoid of dung-breeding flies. *J. Appl. Ecol.* 26:453–61.

Wulff, E. V. 1937. Essay on dividing the world into phytogeographical regions according to the numerical distribution of species. *Bull. Appl. Bot. Genet. Plant Breed. (Leningrad)* 1:315–68.

Yasuda, H. 1984. Seasonal changes in the number and species of Scarabaeid dung beetles in the Middle Part of Japan. *Jap. J. Appl. Entomol. Zool.* 28:217–22.

———. 1986. Fecundity of two dung beetle species, *Onthophagus lenzi* Harold and *Liatongus phanaeoides* Westwood (Coleoptera: Scarabaeidae). *Appl. Entomol. Zool.* 21:177–79.

———. 1987. Reproductive properties of two sympatric dung beetles, *Aphodius haroldianus* and *A. elegans* (Coleoptera: Scarabaeidae). *Res. Popul. Ecol.* 29:179–87.

Yoshida, N., and H. Katakura. 1985. Life cycles of *Aphodius* dung beetles (Scarabaeidae, Coleoptera) in Sapporo, northern Japan. *Env. Sci. Hokkaido* 8:209–29.

Young, O. P. 1978. "Resource Partitioning in a Neotropical Necrophagous Scarab Guild." Ph.D. diss., Univ. of Maryland, College Park.

———. 1980a. Predation by tiger beetles (Coleoptera: Cicindelidae) on dung beetles (Coleoptera: Scarabaeidae) in Panama. *Coleopt. Bull.* 34:63–65.

———. 1980b. Bone burial by a Neotropical beetle (Coleoptera: Scarabaeidae). *Coleopt. Bull.* 34:253–55.

———. 1981. The attraction of neotropical Scarabaeinae (Coleoptera: Scarabaeidae) to reptile and amphibian fecal material. *Coleopt. Bull.* 35:345–48.

———. 1983. The distribution and ecology of *Coilodes castanea* (Coleoptera: Scarabaeidae: Hybosorinae). *Coleopt. Bull.* 37:247–53.

———. 1984. Perching of neotropical dung beetles on leaf surfaces: An example of behavioral thermoregulation? *Biotropica* 16:324–27.

Young, R. M. 1969. Ecosystem economy: *Onthophagus* using a *Canthon* brood ball. *Coleopt. Bull.* 23:24–25.

Zervanos, S. M., and N. F. Hadley. 1973. Adaptational biology and energy relationships of the collared peccary (*Tayassu tajacu*). *Ecology* 54:759–74.

Zunino, M. 1981a. Note su alcuni *Onthophagus americani* e descrizione di nuove specie (Coleoptera, Scarabaeidae). *Boll. Mus. Zool. Univ. Torino* 6:75–86.

———. 1981b. Onthophagini. In W. Wittmer and W. Büttiker, eds., *Fauna of Saudi Arabia*, pp. 408–16. Basel.

———. 1984a. Sistematica generica dei Geotrupinae (Coleoptera Scarabaeoidea: Geotrupidae), filogenesi della sottofamiglia e considerazioni biogeografiche. *Boll. Mus. Reg. Sci. Nat. Torino* 2:9–162.

———. 1984b. Essai préliminaire sur l'évolution des armures génitales des Scarabaeinae, par rapport à la taxonomie du groupe et l'évolution du comportement de nidification (Col. Scarabaeidae). *Bull. Soc. Entomol. Fr.* 88:531–42.

———. 1984c. Analisi sistematica e zoogeografica della sottofamiglia Taurocerastinae Germain (Coleoptera, Scarabaeoidea: Geotrupidae). *Boll. Mus. Reg. Sci. Nat. Torino* 2:445–464.

———. 1985. Las relaciones taxonomicas de los Phanaeina (Coleoptera, Scarabaeinae) y sus implicaciones biogeograficas. *Fol. Entomol. Mex.* 64:101–15.

Zunino, M., and G. Halffter. 1981. Descrizione di *Onthophagus micropterus* n.sp. (Coleoptera, Scarabaeidae), note sulla sua distribuzione geografica e sulla riduzione alare nel genere. *Boll. Mus. Zool. Univ. Torino* 8:95–110.

Zunino, M., and C. Palestrini. 1986. El comportamiento telefagico de *Trypocopris pyrenaeus* (Charp.) adulto. *Graellsia* 52:205–16.

Index of the Genera in Scarabaeidae

For each genus is given: the author's name, the year of publication, the tribe (CA, Canthonini; CO, Coprini; DI, Dichotomiini; EC, Eucraniini; ER, Eurysternini; GY, Gymnopleurini; OC, Oniticellini; OP, Onthophagini; OT, Onitini; PH, Phanaeini; SC, Scarabaeini; SI, Sisyphini), the number of described species, and the page numbers where the genus is mentioned in this book. This information gives a visual impression of our ecological knowledge of the genera.

Total number of genera : 234
Total number of species: 4,940

Index

30; depth in soil, 84, 319; number of, 316; primitive, 29, 39; rate of construction, 323; types, 46–47

nesting: alternative behavior, 325; *Aphodius*, 39; behavioral advantages, 27; behavior, 10, 36, 42–43, 238–39; cost of, 40; experimental, 59; patterns, 44, 46; sequence, 45

New Caledonia, 62, 297

New Guinea, 62, 65, 182, 192, 194–95

New Zealand, 62, 367

Nialus, 236–37, 240, 415

niche: complementary, 87; differences, 85, 88; overdispersed, 346; overlap, 87

Nicrophorus, 7–8, 10, 15, 27, 180, 185, 189, 191, 195, 402

Nitidulidae, 123

nitrogen, 32

nocturnal. *See* activity

North America, Ch. 5, 297, 303, 343, 354, 357–58; subtropical, Ch. 7, 335, 373, 378, 382

old taxa, 53–54, 57

Oligocene, 51, 57

Oniticellini, 46–48, 55, 60, 161, 164, 166, 175–76, 203, 206, 210, 213, 289, 332, 336, 339

Onitini, 46–48, 55, 58–59, 65, 112, 164–65, 203, 336, 339

Onthophagini, 46–47, 53, 55, 60–62, 112, 161, 164–66, 176, 203, 206, 210, 213, 256–58, 289, 294, 336

Orodalus, 237

Orphnidae, 23

ovary, 39–40, 42–43

oviposition, 45; density-dependent, 83, 312, 315

Oxyomus, 107, 246

Oxytelinae, 233

pair bond, 29, 42

Palawan, 182

Panama, 57, 59, 309, 332, 341, 379

Pangaea, 52

parasite, 8, 19, 121; of dung beetles, 137

parasitic dung beetle. *See* kleptoparasite

parental: care, 36, 43, 328; care inhibits reproduction, 43; investment, 151

Passalidae, 36, 203

patchy habitat, 6–7, 283

Pectinicornia, 23

perching on vegetation, 41, 72, 184, 186, 205–206, 219–22, 355

Phaeochroops, 191, 196–97

Phaeochrous, 184–85, 187, 196–97, 402

Phanaeini, 46, 55, 58–59, 65, 126, 213, 219, 336

Pharaphodius, 415

phenology, 247; dung beetle types, 169

pheromone, 42–43, 46, 90, 290

Philippines, 183

Philonthus, 191

phoresy, 215, 225, 259

Phoridae, 123

Platydracus, 161, 191

Pleistocene. *See* extinction

Pleuraphodius, 414

Pleurosticti, 26

polyphagy, 7

predation, 314, 358; by dung beetles, 32; on dung beetles, 72, 370; frequency-dependent, 11

predator, 120–21, 220, 223–24, 250, 274, 359; of dung beetles, 137, 160–61, 165; generalist, 11

predatory: beetles, 17, 20, 356–59, 366; flies, 17

Proctophanes, 420

proteins, 26, 28, 32

Psammodius, 415

Psychodidae, 121

Pteromalidae, 137

Puerto Rico, 297

r-species, 82, 151–52, 302, 367

rain forest, 26–27, Ch. 10–12

rare species, 91, 94, 181, 187, 295, 335, 342

ratel, 49, 370

recognition: mate, 44; sex, 42; species, 42, 189, 291

relict, 45, 53, 60, 62, 64

reproductive strategy, 272

resource: in excess, 202; multidimensional partitioning, 345; partitioning, 14, 72, 141, 186, 330; patch heterogeneity, 290; selection in tropical forests, 333; spatial partitioning, 341; temporal partitioning, 336; utilization, 28, 227–28; roller, 36, 43, 161–63, 207, 227; competition in, 114, 169, 171; of dung pellets, 227; large, 139; population biology of, 370; small, 139; spatial distribution, 105